The Bright Superior Planets —1991 to 2000

■ Oppositions of Mars

Jan 7, 1993 in Gemini
Feb 11, 1995 in Leo

Mar 17, 1997 in Virgo
Apr 24, 1999 in Virgo

■ Oppositions of Jupiter

Jan 28, 1991 in Cancer
Feb 28, 1992 in Leo
Mar 30, 1993 in Virgo
Apr 30, 1994 in Libra
Jun 1, 1995 in Scorpius

Jul 4, 1996 in Sagittarius
Aug 9, 1997 in Capricornus
Sep 15, 1998 in Aquarius
Oct 23, 1999 in Pisces
Nov 27, 2000 in Taurus

■ Oppositions of Saturn

Jul 26, 1991 in Capricornus
Aug 7, 1992 in Capricornus
Aug 19, 1993 in Aquarius
Sep 1, 1994 in Aquarius
Sep 14, 1995 in Aquarius

Sep 26, 1996 in Pisces
Oct 9, 1997 in Pisces
Oct 23, 1998 in Pisces
Nov 6, 1999 in Aries
Nov 19, 2000 in Taurus

Since Mars, Jupiter, and Saturn are farther from the Sun than the Earth is, they can be seen all night long at certain times. When a planet is located directly opposite to the Sun in the sky, it is said to be at *opposition*. When this occurs, it rises at sunset, is highest in the sky at midnight, and sets at sunrise. The tables above list opposition dates for the three superior planets visible to the unaided eye and indicate the constellations (star patterns) in which the planets are located when at opposition.

INTRODUCTION TO THE SOLAR SYSTEM

INTRODUCTION TO

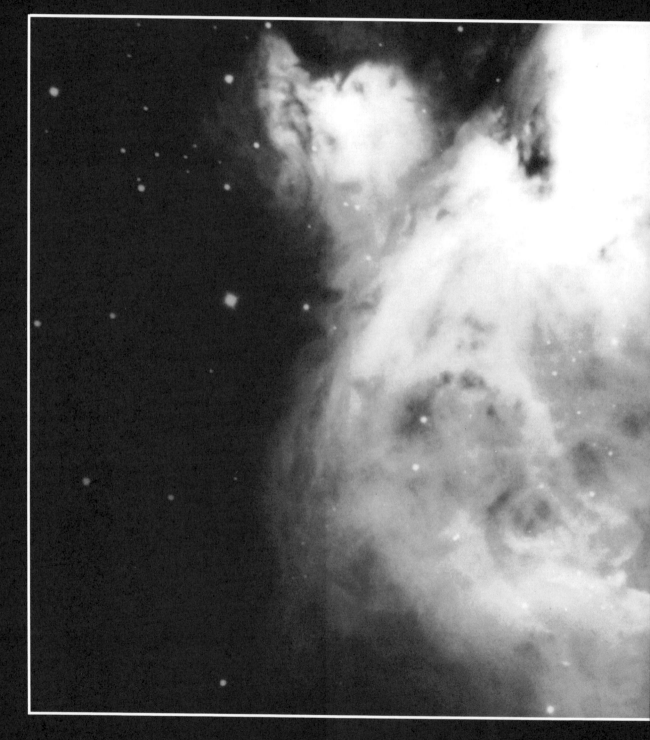

THE SOLAR SYSTEM

Jeffrey K. Wagner

Firelands College

SAUNDERS COLLEGE PUBLISHING

Philadelphia   Fort Worth   Chicago   San Francisco   Montreal   Toronto   London   Sydney   Tokyo

Copyright © 1991 by Saunders College Publishing, a
division of Holt, Rinehart and Winston, Inc.

Text Typeface: Optima
Compositor: Progressive Typographers
Acquisitions Editor: John J. Vondeling
Developmental Editor: Jennifer Bortel
Managing Editor: Carol Field
Project Editor: Margaret Mary Anderson
Copy Editor: Diane Lamsback
Manager of Art and Design: Carol Bleistine
Art and Design Coordinator: Doris Bruey
Text Designer: Tracy Baldwin
Cover Designer: Doris Bruey
Text Artwork: J & R Technical Services
Layout Artist: Anne O'Donnell
Director of EDP: Tim Frelick
Production Manager: Bob Butler
Marketing Manager: Marjorie Waldron

Cover Credits:
Front Cover
Top: Moon during partial eclipse of 8/16/70 (photograph
by George East, courtesy of Dennis Milon).
Middle: Earth, photographed by Apollo 17 (NASA).
Bottom: Mars, photographed by Viking (NASA/JPL).
Background: Uranus, photographed by Voyager 2 (NASA/
JPL).
Back Cover
Top: Neptune, photographed by Voyager 2 (NASA/JPL).
Middle: Moon, photographed by Apollo 17 (NASA).
Bottom: Saturn, photographed by Voyager 2 (NASA/JPL).

Printed in the United States of America

INTRODUCTION TO THE SOLAR SYSTEM

0-03-051899-7

Library of Congress Catalog Card Number: 90-052773

0123 016 987654321

Preface

Recent decades have seen rapid advances in all branches of astronomy. Although it is difficult to single out one facet of astronomical research that has seen the most progress, a case can be made for solar system studies, which have received a dramatic impetus during the space age. Space probes launched by the United States, the Soviet Union, and other nations have returned so many thousands of spectacular images to Earth that we now almost take them for granted. Although over two decades have passed since the first manned lunar landing, scientists are still making discoveries from the lunar samples returned to Earth by Project Apollo.

Astronomy courses are becoming increasingly popular among college students. Although many take introductory courses in general astronomy, other students are taking courses that deal with the solar system, either instead of taking, or after completing, a general astronomy course. A course about the solar system must be interdisciplinary, combining topics traditionally covered in astronomy courses with those once exclusively the domain of geology. Indeed, solar system courses are taught by geologists as frequently as by astronomers.

When I first began teaching a course about the solar system in 1983, I was unable to find a textbook that adequately covered both solar system astronomy and solar system geology without slighting important topics. As a result, I have written this text, which evolved from the course notes developed for my solar system course. It is my intention to convey the excitement of solar system studies to astronomers, geologists, and those with no science background at all. This book is designed for a one-semester or one-quarter solar system course for college undergraduates, even if they have had no previous training in either astronomy or geology.

After an introductory chapter, Chapter 2 surveys the solar system and describes its place in the galaxy and universe. Chapter 3 describes the historical development of our understanding of the solar system's dynamics, as well as the fundamentals of celestial mechanics. Chapters 4 and 5 tell how we study and explore the solar system, covering the topics of light, telescopes, and space travel.

Chapter 6 begins the study of specific objects with the Sun. The majority of the book deals with the planets and their satellites, and Chapter 7 introduces them. Chapters 8, 9, 10, and 11 treat topics common to all terrestrial planets, specifically rocks and minerals, interiors, surface features, and atmospheres. The next five chapters cover specific objects in an order corresponding to their distance from the Sun — Mercury, Venus,

Earth, Moon, and Mars. Chapter 17 describes features common to gas giant planets, whereas Chapter 18 describes rings. Chapters 19 through 25 cover each planet from Jupiter to Pluto, with separate chapters for the satellites of Jupiter and the satellites of Saturn.

The next three chapters cover the solar system's minor members—comets, asteroids, and meteorites. Chapter 29 discusses the origin of the solar system. The final chapter describes the search for other planetary systems and discusses the possibility that life exists elsewhere other than Earth.

Of course, it is not necessary to cover all of the chapters and sections, or to cover them in the order presented in the book. Although reference to previous material is necessary, I have attempted to minimize such references so that the order of coverage by an instructor can be as flexible as possible.

■ Features

I have included a number of features that I hope will be of help to students. Each chapter begins with a list of objectives and closes with a summary and list of vocabulary words. Each chapter also includes a set of review questions and a short list of additional readings.

I have used metric units throughout the book, but because most students are more familiar with English units of measurement, I have also given most dimensions in feet and miles as well. I have given the dates of birth and death for deceased scientists, not because they are necessarily worth remembering but so that students can better place them into historical context. I have included a number of equations that I consider important, but the level of mathematics is kept to basic algebra. Individual instructors may choose to supplement the mathematics in the book or, conversely, to eliminate it entirely.

■ Supplements

Introduction to the Solar System is accompanied by an Instructor's Manual, which includes advanced references and answers to end-of-chapter problems, and a separate set of 100 Overhead Transparency Acetates.

■ Acknowledgments

I wish to acknowledge the help of the following reviewers who read and commented on the entire manuscript: Bruce W. Hapke, University of Pittsburgh; Bryon D. Anderson, Kent State University; Barbara Anthony-Twarog, University of Kansas; Gordon Baird, University of Mississippi; László Baksay; William P. Bidelman, Case Western Reserve University; Donald J. Bord, University of Michigan; Larry Helfer, University of Rochester; Kenneth Janes, Boston University; V. Gordon Lind, Utah State University; David Martins, Georgia Institute of Technology; Bernard McNamara, New Mexico State University; George D. Nickas, Ithaca College; James N. Pierce, Mankato State University; Scott Roby, SUNY-Oswego; Roger L. Scott, Ball State University; and William Skinner, University of Tennessee. In addition, my wife, Barbara J. Wagner, and my mother, Betty Jane H. Wagner, read early drafts of the text and provided helpful comments from nonscientists' perspectives. The combined efforts of these reviewers made this book much, much better than it would have been otherwise.

I would also like to acknowledge the help given me by Jeff Holtmeier of Harcourt Brace Jovanovich, who guided the book through the review process and gave me encouragement when I needed it. John Vondeling, Margaret Mary Anderson, and Jennifer Bortel of Saunders also provided invaluable assistance. Although they are credited in the captions, I would also like to give a "thank you" to everyone who provided figures or gave permission to reproduce figures from other sources. Will Currie of Firelands College library provided much help in tracking down copies of books and journal articles.

There is nothing like writing a book to show one how *little* one knows about a topic! It is my sincere hope that this book will be helpful to many instructors and students. I will be most glad to receive comments from anyone who reads this book, and I promise to respond personally to all correspondence. Please write to Jeffrey K. Wagner, Firelands College, 901 Rye Beach Road, Huron, Ohio 44839.

Contents

1

Introduction to Solar System Science

CHAPTER OBJECTIVES

After studying this chapter, you should be able to

1. Explain how the science of planetary geology has derived from astronomy and geology.
2. Summarize the basic history of astronomy and geology.
3. Understand the nature of scientific inquiry and research.
4. Explain the value of studying the solar system.
5. Know the difference between qualitative and quantitative descriptions of solar system objects.

The space age has seen humanity's first efforts at firsthand exploration of the solar system. Here, Edwin Aldrin steps down toward the surface of the Moon during the Apollo 11 mission on July 20, 1969. (NASA Photograph.)

Astronomy, the scientific study of the universe, is generally considered to be the oldest of the sciences. Six thousand years ago, residents of the Near East gazed into the night sky and recorded their observations on clay tablets, some of which still survive today. In the 20th century, astronomers use multimillion dollar telescopes to peer into the deepest reaches of space, analyze data with advanced computers, and publish numerous articles in scientific journals to report their findings. Although today's researchers may appear to have little in common with their ancient predecessors, astronomers throughout the ages have shared the excitement of studying the vast universe, which is filled with numerous objects that can be seen in the sky but that generally are too distant to be examined firsthand. The 20th century, however, has added an especially exciting dimension to the study of the universe, because Earth's neighborhood, the solar system, is now being explored in ways undreamt of even a century ago. Humans are now sending instruments — and even people — into space to study its wonders firsthand!

The exploration of the solar system, however, has not been the exclusive domain of astronomers. Another science with a long heritage is also playing an important role: **geology**, the study of the Earth. Geology began when primitive humans started using materials from the Earth for various purposes and began to wonder about those materials as well as the Earth itself. The modern geologist has learned as much about the Earth as the astronomer has learned about the universe. When firsthand exploration of the Moon, Mars, Venus, and other somewhat Earth-like bodies began during the 1960s, geologists helped decipher their features and history, using the same techniques that they had perfected while studying the Earth. As a new science, **planetary geology**, or **planetology**, was born, its practitioners began looking for similarities among the various worlds that they were studying, as well as explanations for their many differences. (Other

1

sciences, such as meteorology, the study of atmospheres, are also part of solar system science, as we will see.)

■ The Development of Astronomy and Geology

Although the synthesis of astronomy and geology into planetary geology is a recent development, both sciences have very long histories. It is therefore not surprising that both of them have changed considerably over the years. It is worthwhile to examine briefly the history of astronomy and geology, not only to gain insight into the present states of these disciplines but also to see why changes have occurred in them. Many of the changes in astronomy and geology have been brought about by inventions of new types of scientific instrumentation, but others have resulted from changes in scientific philosophy and scientists' basic goals of what knowledge they seek to gain.

The earliest concern of astronomy was constructing models to explain the organization and structure of the universe. As we discuss in Chapter 3, the earliest such models placed the Earth in the center of the universe, and it was not until the Renaissance that evidence to the contrary finally altered science's perception of how the universe was organized. The desire to determine this organization had practical importance, because the basis for the motion of objects in the sky needed to be understood in order for predictions to be made about their future locations.

When the astronomical telescope came into widespread use during the 1600s, the emphasis of astronomy quickly shifted to describing the multitude of wonders now visible for the first time and to preparing catalogs of these celestial objects. Although the structure of the solar system was understood by this time, science's view of the rest of the universe remained uncertain. It was not until the late 1800s and the first half of the present century that advances led to the first real understanding of the overall structure of the universe.

As time passed, positional and descriptive astronomy were joined by attempts to understand what the various celestial objects were like and what physical laws governed their behavior. Astrophysics, the combination of astronomy and physics that seeks to do this, began in the 1800s and has flourished ever since. Astrophysics, like planetary geology, is an example of how sharp boundaries between scientific disciplines have disappeared in recent decades.

The development of geology over the centuries has been spectacular. Much of this progress was spurred by its practicality, as geological knowledge enabled humans to find and recover mineral wealth from within the Earth and understand both the beneficial and harmful processes occurring in the world around them.

The first major breakthrough in geology was made by ancient Greek naturalists, who discovered that changes occur in the Earth. This was a surprise to the majority of people, who believed the Earth to be unchanging. The evidence, however, was definite: Fossils of seashells and other ocean organisms could be found in rocks in mountains that were far away from the oceans.

Although the concept of change within the Earth was accepted, geologists disagreed on the rate at which this change occurred. Some believed that change, such as the formation of mountains or deep valleys, was the result of rapid, catastrophic processes. Others believed that change was constant and gradual, producing noticeable results only over incredibly long time spans. The latter view, clearly stated by Charles Lyell in the 1830s and now called the **Principle of Uniformity**, eventually prevailed and became a cornerstone of geological thought.

Meanwhile, geologists in Europe realized that rocks normally form sequentially in layers, and this understanding gave them the ability to reconstruct the geological history of our planet and, by looking at fossils in the rocks, the history of terrestrial life. This work was further enhanced in the early 20th century, when techniques using radioactive elements were developed for determining the actual ages of rocks.

The 1960s and 1970s witnessed two developments that further revolutionized geology. The first was the realization that the Earth's crust is not a single, unbroken layer but is divided into sections, or plates. The discovery that these plates undergo gradual horizontal movement explained

109

the development of the Earth's major surface features, such as mountain chains, and why some areas are particularly prone to earthquakes and volcanoes. The second development, mentioned previously, was the first application of geological research techniques to worlds other than the Earth.

Recent decades have been marked by a trend affecting all sciences, the acquisition of knowledge at such a rapid rate that today, no one can know everything about astronomy, geology, or any other discipline. For this reason, specialization is the order of the day, and many researchers spend much of their careers working in one small branch of a discipline.

Examining the Solar System

One of the first astronomical observations was undoubtedly of the various motions visible in the sky. Objects were seen to rise and set daily, and different stars were visible during different seasons of the year. Especially noticeable were the motions of certain celestial objects with respect to one another and the starry background. The Sun and Moon were obvious examples, but five bright, star-like objects also exhibited such motion. The latter objects are now called **planets**, a term derived from the Greek word meaning "wanderer."

Today we know that the Sun, Moon, Earth, and other planets, as well as other objects, are part of a celestial family known as the **solar system**. The solar system consists of the Sun (its central star) and all of the objects in orbit around it. (Some of the objects that orbit the Sun are orbited in turn by smaller bodies.) The term solar comes from the Latin word sol, which means "Sun." The motion of solar system objects in the sky presented a complex challenge to early observers, and, as already mentioned, unraveling the mechanics of the solar system was not completed until the 17th century. After that time, more distant objects began to attract the attention of astronomers, and interest in the solar system declined somewhat as startling discoveries were made about distant stars, galaxies, and the universe as a whole.

Since the 1950s, study of the solar system has increased at a rapid rate, and many breathtaking discoveries have been made. The main reason for the revival of interest in the solar system has been space exploration. Just as the brightness and conspicuous motions of solar system objects made them the first targets for stargazers, the proximity of our solar system neighbors made them the first targets for space probes and space travelers. Other than the Sun, whose inner workings had begun to be deciphered by astrophysicists early in the 20th century, studies of solar system objects had been limited to observing and mapping surface features. Now firsthand observations are revealing the secrets of these objects, including their compositions, what kinds of physical processes are at work in and on them, and their origins. In fact, the traditional techniques of astronomy quickly became inadequate once firsthand observation of other solar system bodies became possible in the space age. Geologists, who by that time had unlocked many of the Earth's secrets, applied the techniques that had worked so well in studying our planet to other objects, with equally enlightening results.

This book reflects the interdisciplinary nature of solar system studies. Some sections deal with what are considered "astronomical" concepts, whereas others discuss topics that once were the exclusive domain of geology. It is only by combining these two sciences into planetary geology that we can fully begin to study our solar system.

The Nature of Science

Any science textbook will obviously contain a large quantity of factual information. In addition to understanding this information, it is important for the reader to gain some appreciation of how it was obtained.

Science may be defined as the process of identifying and investigating the natural phenomena occurring around us and of attempting to understand and explain them. An important attribute of science is that it is flexible and changing. Its explanations of natural phenomena are not to be considered rigid and final but simply a "progress report" of what is known at any given time.

This is the major difference between science and religion. As Edward Rosen, an historian of astronomy, has written,

"Astronomy, like other natural sciences, does not deal with ultimate truths. These it leaves to theologians who are convinced that such final verities are accessible to themselves. For science, on the other hand, the goal is rather ever closer approximation to the truth. The history of science is full of examples of discarded ideas. These may have seemed suitable for a time, only to be replaced by more acceptable refinement."*

People often refer to the irreconcilability of science and religion without understanding this inherent difference in their philosophies.

One characteristic of science is that its practitioners generally proceed in an orderly fashion, making sure of one result before pushing on to explore new territory. In that way, scientific knowledge increases, with new discoveries built upon the foundation provided by earlier work. Sir Isaac Newton noted this by remarking that "If I have seen further than others, it is by standing upon the shoulders of giants." The step-by-step approach of science is called the **scientific method.** In short, it is a basic exercise in problem solving: The scientist's first step is to ask a question or formulate a problem to be solved. Good scientific researchers do not proceed haphazardly. The astronomer doesn't just peer through the telescope to see what can be seen; rather, plans are made to study certain objects in order to gain specific information. The question may be simple ("What is the composition of the Moon?") or complicated ("What process is responsible for the extreme eccentricity and inclination of Pluto's orbit?"), but it represents the scientist's starting point.

The next step may seem obvious, but it is very important: The researcher must examine what is already known about the problem at hand. This serves two purposes: It provides information upon which to build and helps avoid duplication and wasted time. Occasionally, in the

*Quotation from page 52 of Rosen, Edward. *Copernicus and the Scientific Revolution.* Robert E. Krieger Publishing Company, Malabar, Florida, 1984.

history of science, important breakthroughs have been announced, only to have the embarrassed researcher learn that someone had earlier obtained the same result!

At this point, the scientist may have enough information to propose one or more possible solutions to the problem. A solution proposed in this way is called a **hypothesis.** Even if a hypothesis is developed, a reputable scientist will never rush to announce it without first testing its validity. This is one of the key differences between true science and pseudoscience, which promulgates preposterous claims with no evidence for proof.

How do scientists gather the information necessary to test hypotheses and devise them in the first place? The methods vary, but generally they can be grouped into one of three categories. Some sciences are experimental, and knowledge is gained by simulating natural processes in the laboratory. Chemistry is a good example of an experimental science. However, most astronomical and many geological processes are impossible to duplicate in the laboratory, simply because of the high temperatures, huge dimensions, and long time periods involved. Astronomy and geology are observational sciences, in which natural processes are watched closely. Of course, there are some phenomena that cannot even be observed directly, such as energy production at the center of the Sun or the generation of the Earth's magnetic field in its core. Scientists gain information about these types of phenomena primarily by theoretical calculations using the laws of physics. It comes as a surprise to many people that some astronomers never use a telescope but spend their research time engaged in making incredibly complicated mathematical calculations! Obviously, in many cases, experiment, observation, and calculation complement one another to a large degree.

No matter which of the three procedures is used, some sort of experiment, observation, or theoretical calculation must be devised and conducted. This is a crucial part of the research process and must be planned and accomplished carefully. There are more astronomers seeking observing time on the world's large telescopes than time available. Therefore, an observing program must be well thought out, so that it can take full advantage of the time available. Similarly, considerable planning must be applied to spacecraft missions, in which a planet may be under study for only a few short days while the spacecraft is in its vicinity.

After an observation is made or a long and difficult calculation is completed, the last step in the scientific method is interpretation of the results. In many cases, the original hypothesis is found to be consistent with experiment or observation, but important information can be obtained even if it was proved incorrect, necessitating a change of ideas. At this point, the researcher(s) will submit a paper describing the findings to a scientific journal for publication, so that other scientists will become aware of the results. The number of scientific papers published annually in astronomy, geology, and related sciences is truly staggering and is continually increasing!

Several terms are used to describe solutions to scientific problems. In addition to a hypothesis, which is generally applied to the solution of a specific problem, there are **theories**, which are similar but are applied to solutions with more general applications. Examples are the theory of relativity in physics and the theory of evolution in biology, which are syntheses of many specific hypotheses. The term **law** is often applied to solutions of universal application, especially when they can be stated mathematically, as in Newton's laws of gravitation and motion.

As mentioned previously, a hypothesis, theory, or law is never considered ultimate truth that is proved correct for all time, because additional evidence might be found that requires its modification or replacement. Another complication in scientific research is that sometimes more than one theory explaining a phenomenon might be supported by the available evidence. How do scientists choose among competing theories? William of Ockham (or Occam, ca. 1280–1349) considered this problem, and his solution, usually called **Ockham's** (or **Occam's**) **Razor**, states that in such an instance, the preferred theory will be the one that makes the fewest and simplest assumptions. In other words, a theory requiring the acceptance of numerous assumptions is less useful than a simpler theory. A theory is also likely to be accepted if it accurately predicts future phenomena. Even with these guidelines, however, the best solution to a scientific problem is not always easy to select. The history of science, as well as contemporary planetary science, is filled with examples of lively debates between scientists who hold two or more opposing views, and bitter, long-lived feuds between scientists have occurred in some cases!

Even if a research project fails to verify a particular hypothesis or theory, or if totally unexpected results are obtained, the researcher has not wasted time and effort. The results may be used to formulate another working hypothesis, which further study might verify. The history of science contains examples of researchers who published results that they themselves did not fully understand but that were later interpreted by someone else who used them as a stepping stone to a new idea.

■ The Value of Astronomy

Two or three centuries ago, many astronomical discoveries were made by people whom we today would call amateurs. Many of these people were independently wealthy and possessed telescopes that were the finest of their day. In the 20th century, the situation is much different. Although there are many amateur astronomers, some of whom make significant discoveries, no individual can launch a space probe or afford a state-of-the-art telescope. Modern science in our country is largely supported by public money, some of which is used to build and operate large astronomical observatories and fund NASA space missions. As taxpayers, we might wonder what we are getting for our money.

Admittedly, astronomy does not have as many practical applications as geology and other sciences such as physics or chemistry; however, some applications do come to mind. For example, astronomy provides the basis for timekeeping and for determination of the calendar. Navigation has traditionally been accomplished using celestial objects as points of reference. The study of our Sun is very important because changes in its output could have serious effects on our planet's weather and climate. There is a possibility that other celestial objects could affect Earth as well. The theory that dinosaur extinction was caused by the impact of a meteorite is an example. The study of the atmospheres of other planets has become very important because it provides information that enables us to understand better the workings of the Earth's atmosphere.

Space exploration has benefited mankind immensely. Satellites have greatly facilitated long-distance communications, and weather satellites have saved many lives by giving advance warning of severe weather. The technology of computers and miniaturization developed for use in spacecraft has numerous terrestrial applications, including consumer electronics, robotics, and medicine. In the future, the rewards of space travel may be even greater. Space colonies may provide a new habitat for humans, and mining of the Moon or other celestial objects may yield new sources of scarce resources.

Although some might argue that we do not "need to know" what other planets are like, what powers the distant stars, or what formed the universe in the first place, seeking knowledge as an end in itself seems to be a requirement of human nature. The astronomer Herbert Friedman has noted that the greatest use of astronomy is the "exercise of the mind." This is certainly true, because in discussing astronomy, we often deal with huge numbers, large distances, and other quantities and concepts that are far outside the normal human experience. Astronomy certainly shows us how small and insignificant Earth and humanity are in comparison to the vastness of the universe! In the United States, the total amount of public money spent on science and space exploration is actually a very small percentage of the federal budget.

■ Introduction to Quantitative Measurement

When discussing solar system objects, we often use qualitative language such as "Jupiter is the largest planet" or "The Sun emits huge amounts of energy." However, scientists normally find quantitative terminology to be more useful, as in "Jupiter is 142,796 kilometers in diameter" or "The Sun emits 3.9×10^{33} ergs of energy per second." Accordingly, this book contains much quantitative information.

Certain physical quantities are important when discussing the properties of solar system objects. These include diameter, mass, density, temperature, and many others. Appendix 1 describes these quantities and the units used in measuring them. Scientists today use **metric system** units, and we use them in this book. However, because many readers will be more familiar with traditional English units of measurement, many quantities are listed using these units as well. Appendix 1 also describes the system of **scientific notation** used to express very large or small numbers, such as 3.9×10^{33}, which expresses the Sun's energy output in ergs.

■ Chapter Summary

Astronomy, the scientific study of the universe, and geology, the study of the Earth, are two physical sciences with long histories. Recent decades have seen the development of planetary geology, a modern synthesis of astronomy and geology. Although early astronomers paid considerable attention to solar system objects, more distant objects had received more study until the space age, when direct exploration of the solar system began.

Science attempts to understand natural phenomena by proceeding in a step-by-step fashion known as the scientific method. This has led to an orderly growth of knowledge over many centuries. Although astronomy may lack some of the tangible applications of geology and other sciences, it compensates by teaching us about the vast, exciting universe.

■ Chapter Vocabulary

astronomy
astrophysics
geology
hypothesis
law

metric system
Ockham's (or
    Occam's) Razor
planet
planetary geology

planetology
Principle of
    Uniformity
science

scientific method
scientific notation
solar system
theory

■ Review Questions

1. How have astronomy and geology changed since ancient times?
2. Why has geology become linked with astronomy?
3. Why has the study of the solar system advanced in recent years?
4. What is science, and in what ways does it differ from religion?
5. Summarize the steps involved in the scientific method.
6. What three procedures are used to test scientific hypotheses and gather new information?
7. What is the value of astronomy? What is the value of geology?

■ For Further Reading

Astronomy. AstroMedia, Milwaukee. (A monthly magazine.)

Berry, Arthur. *A Short History of Astronomy*. Dover, New York, 1961. (Reprint of 1898 edition.)

Earth Science. American Geological Institute, Alexandria, VA. (A quarterly magazine.)

Faul, Henry and Carol Faul. *It Began with a Stone: A History of Geology from the Stone Age to the Age of Plate Tectonics*. Wiley, New York, 1983.

Geology Today. Blackwell Scientific Publications, Oxford, England. (A bimonthly magazine.)

Harwit, Martin. *Cosmic Discovery: The Search, Scope and Heritage of Astronomy*. MIT Press, Cambridge, MA, 1984.

Mercury. Astronomical Society of the Pacific, San Francisco. (A bimonthly magazine.)

Pannekoek, Anton. *A History of Astronomy*. Dover, New York, 1989. (Reprint of 1961 edition.)

Sky and Telescope. Sky Publishing, Cambridge, MA. (A monthly magazine.)

2

An Overview of the Solar System and Its Place in the Universe

CHAPTER OBJECTIVES

After studying this chapter, you should be able to

1. Understand the basic organization of the solar system and how this organization manifests itself as we view solar system objects in the sky.
2. Understand the overall dimensions of the solar system and how distances within it are determined.
3. Describe the types of objects present in the solar system and summarize their properties.
4. Summarize the properties of the Milky Way and describe the solar system's place within it.
5. Understand the features of the universe beyond the Milky Way.
6. Explain why we believe that the Big Bang theory explains the ultimate origin of the universe.

To the average person, the vastness of the universe can scarcely be imagined. Photographs of the Earth in space, such as this Apollo 8 view with the Moon in the foreground, help us appreciate the relative insignificance of our home world. (NASA Photograph.)

Although our solar system is exceedingly large compared to the dimensions of even the largest objects that people encounter in everyday life, it is only a small part of the universe as a whole. The solar system contains a wide variety of interesting objects with which we will become well acquainted in this book. In this chapter, we examine the organization and dimensions of the solar system and briefly inventory its components. Then we describe how the solar system fits into the larger scheme of things.

■ The Solar System's Organization and Dimensions

As we look out at the celestial objects in the night sky, it is not immediately obvious how we might unravel the relationships among them. The fact that it required many centuries for astronomers to determine the basic organization of the solar system underscores this difficulty. Chapter 3 describes in detail how the solar system's organization was determined, but we summarize some of the results now.

Although celestial objects do not all lie at the same distance from the Earth, it is often convenient to picture the night sky as a sphere surrounding the Earth. Like our planet, this **celestial sphere** has poles and an equator, defined by projecting the Earth's poles and equator onto the sky (Figure 2.1). Careful observations of solar system objects reveal that the Sun, Moon, and planets all travel along approximately the same path in the sky, a great circle that, like the celestial equator, encircles the sky. The reason for this is that the Sun and most of the objects that orbit it lie in approximately a single plane called the **ecliptic plane**. As viewed in the sky from Earth, the ecliptic plane is inclined to the celestial equator with an angle of 23.5° between them. The constella-

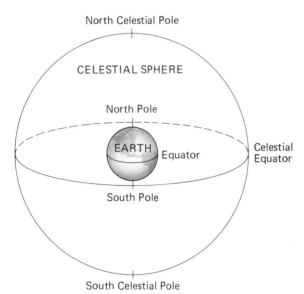

■ **Figure 2.1** The celestial sphere. Note that the locations of the celestial equator and celestial poles correspond to the locations of the Earth's equator and poles.

tions (groups of stars forming patterns in the sky named for animals, objects, and mythological people) through which the ecliptic plane passes are called **zodiacal constellations.** The word zodiac comes from Greek words meaning "circle of animals," because the 12 constellations that comprise the zodiac all originally represented living creatures.

As an additional example of the solar system's orderliness, nearly all objects orbit in the same direction (counterclockwise as seen from "above" the ecliptic plane, which, by definition, is the side in the direction of the north celestial pole), and most rotate on their axes in the same direction. (This direction is called **prograde,** whereas the opposite direction is called **retrograde.**) Although there are objects whose orbits are tilted with respect to the ecliptic plane or that orbit or rotate in the retrograde direction, the regimentation present in the solar system is an important clue in trying to understand its origin and history.

Just how large is the solar system? Distance determination is often difficult in astronomy, and many techniques are used. In the solar system, distance has traditionally been measured by using **triangulation** (observing an object from two different points on the Earth) (Figure 2.2). The use

of distant stars as a fixed reference frame allows the angle to be determined and distance calculated. In recent decades, this technique has been supplanted by the use of radar reflected from solar system objects, with the delay time used to calculate distance. The majority of the objects in the solar system with orbits in the ecliptic plane are located within 40 astronomical units of the Sun. (The **astronomical unit**, the basic distance unit used in solar system studies, is defined as the average distance between the Earth and the Sun and is equal to about 149.6 million kilometers or 92.9 million miles.) This distance is marked by the orbit of Pluto, the most distant planet yet known. Although many minor solar system objects have orbits that take them far above or below the ecliptic plane, the general distribution of the planets forms a disk about 80 astronomical units across (Figure 2.3).

No solar system members have yet been detected beyond Pluto, but there is considerable indirect evidence that a huge cloud of comets orbits the Sun at a distance of about 50,000 astronomical units, nearly one-fifth of the way to the nearest star. This **Oort cloud**, named for the astronomer who deduced its existence, is not confined to the ecliptic plane, but it is believed to be spherical in shape. Because there is no evidence that any objects associated with the Sun are located beyond the Oort cloud, it can be considered to mark the outer boundary of our solar system. Some of the Sun's neighboring stars may

■ **Figure 2.2** Triangulation. Observation of a celestial object, C, from two points on the Earth's surface, A and B, allows the distance to the object to be determined, because distance AB and angle ACB are known. In practice, points A and B can be two different places on the Earth's surface, or a single point moved over time by the Earth's rotation.

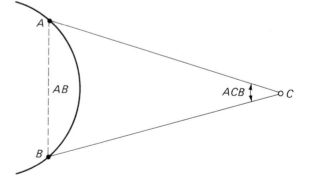

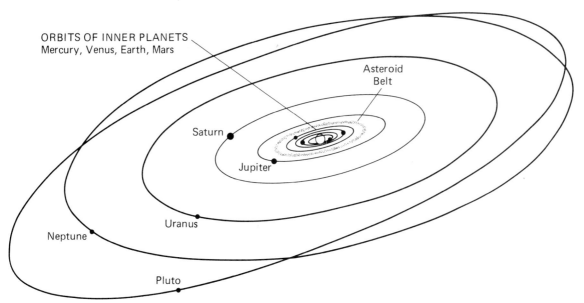

ORBITS OF INNER PLANETS
Mercury, Venus, Earth, Mars

Asteroid
Belt

Saturn

Jupiter

Uranus

Neptune

Pluto

■ **Figure 2.3**   Simplified view of the solar system. The Sun (*center*) and the orbits of the planets can be seen in this perspective view from just above the ecliptic plane. (Orbits not to scale.) (From Levin, Harold L. *Contemporary Physical Geology,* 3rd edition. Saunders College Publishing, Philadelphia, 1990.)

have Oort clouds of their own, located not too far beyond the one associated with our solar system.

■ Components of the Solar System

We devote much of this book to discussions of six types of objects found in the solar system, but it is useful to preview them now so that we have some understanding of their interrelationships (Figure 2.4). We define these six types of objects based upon their masses, diameters, compositions, and locations within the solar system.

At the center of the solar system is our **star**, the Sun. It is a huge, spherical object that is composed mostly of the gases hydrogen and helium and contains over 99% of the total mass in the solar system. Like other stars, it generates light and heat by nuclear fusion, which occurs deep in its interior. Comparison with other stars shows us that the Sun is a star of average mass, diameter, and luminosity.

The **planets** are relatively large, massive objects that orbit the Sun. Nine planets are known currently, but others may lie undiscovered in the far reaches of the solar system. Although each planet is a unique object, we may define three planetary types. The first is terrestrial, or earth-like, and it is characterized by a relatively small size, solid, rocky composition with metallic core, and possibly a surrounding atmosphere (Mercury, Venus, Earth, and Mars are examples). The second type of planet, the gas giant, is an enormous body composed primarily of hydrogen and helium gas (Jupiter, Saturn, Uranus, and Neptune are examples). The final planet, Pluto, is an icy planet composed of frozen gases and some rock.

Only five planets (Mercury, Venus, Mars, Jupiter, and Saturn) are bright enough to be seen easily without a telescope. When seen in the sky, these planets usually appear about as bright as the brightest stars, but planets can be distinguished from stars because planets, unlike stars, don't twinkle. The twinkling of stars is caused by rapid variation in position and brightness due to turbulence in the atmosphere as starlight passes through it, because the path of light from the point-like star constantly varies due to slight, constant atmospheric effects. Planets are so much closer to Earth that their light does not come from a single point, like stars, but from a measurable disk. Therefore, small shifts of their image, much

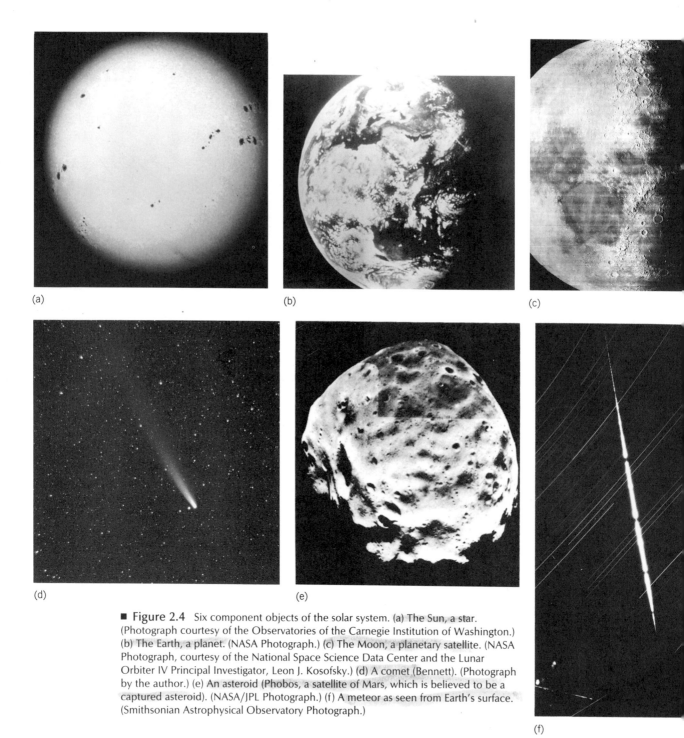

(a)

(b)

(c)

(d)

(e)

(f)

■ **Figure 2.4** Six component objects of the solar system. (a) The Sun, a star.
(Photograph courtesy of the Observatories of the Carnegie Institution of Washington.)
(b) The Earth, a planet. (NASA Photograph.) (c) The Moon, a planetary satellite. (NASA
Photograph, courtesy of the National Space Science Data Center and the Lunar
Orbiter IV Principal Investigator, Leon J. Kosofsky.) (d) A comet (Bennett). (Photograph
by the author.) (e) An asteroid (Phobos, a satellite of Mars, which is believed to be a
captured asteroid). (NASA/JPL Photograph.) (f) A meteor as seen from Earth's surface.
(Smithsonian Astrophysical Observatory Photograph.)

less than their disk diameter, will have little effect on their apparent position, and twinkling will not be seen unless they are very close to the horizon.

Planetary satellites are objects that orbit planets. Two planets have no satellites, whereas several others have over a dozen each. The total number of solar system satellites currently known is 60, many of which have recently been discovered by spacecraft exploring the outer planets. Some satellites are rocky objects that are similar to the terrestrial planets, whereas many satellites of the outer planets are either entirely icy or mixtures of rock and ice. Other than the fact that one group orbits planets instead of the Sun, there are no physical differences between satellites and planets.

Unlike the three preceding classes of objects, which contain relatively few total objects, each of the final groups contains myriad members whose total numbers are uncountable, for all practical purposes. In addition, members of these groups are relatively low in mass and small in diameter compared to the solar system's more substantial objects. **Comets** are small, icy objects whose orbits around the Sun are often unusual, because they generally have greatly elongated shapes and do not lie in the ecliptic plane. As discussed earlier, multitudes of comets reside in the Oort cloud at the outer edge of the solar system. One or two dozen comets pass through the inner solar system annually and are seen from Earth.

Asteroids are relatively small bodies, made of rock and metals, that orbit the Sun. They are relatively scarce in the inner solar system, and the majority of them orbit the Sun between the orbits of the planets Mars and Jupiter in an area called the asteroid belt. Asteroids appear to be ''leftovers'' from the time when the planets were forming, and they are probably fragments that failed to combine into a single, planet-sized body. Although some are larger, the vast majority of asteroids are only a few kilometers or tens of kilometers in diameter.

The final group of objects consists of the **meteoroids**, which are pieces of debris from comets or asteroids whose orbits will cause them to enter the Earth's atmosphere. As the fragments enter, brief flashes of light called **meteors** mark their fiery passage. Pieces of larger meteoroids may survive this passage and land on Earth's surface. These fragments, called **meteorites**, are valuable because they are easily obtained samples of extraterrestrial material. (It is perhaps confusing that three names are applied to what is essentially the same object, but there are occasions when the distinction is important.)

In addition to these objects, the solar system also contains interplanetary gas and dust, as well as energetic particles streaming outward from the Sun as the solar wind. Magnetic fields, generated inside the Sun and most planets, influence extended areas of space around many planets and satellites.

■ The Milky Way Galaxy

What lies beyond the outer edge of the solar system? Although this book generally is limited to discussions of solar system objects, it is important to understand how our solar system fits into the larger scheme of things. In fact, as we will learn in Chapter 29, knowledge of the universe as a whole is applicable to understanding how our solar system originated.

As we look out into the night sky, we see stars — 2500 to 3000 on a clear, moonless night far from city lights. The Sun, these stars, and approximately 200 billion others compose a larger system of celestial objects that we call the **Milky Way Galaxy.** Just as the various objects in the solar system revolve around the Sun, the Sun and the other stars of the Milky Way orbit around the galaxy's center.

A **galaxy** is a huge aggregate of stars bound together by their mutual gravitational attraction. Our galaxy was named for its misty, milky appearance as a faint band of light circling the sky, in similar fashion to the ecliptic plane. The milky band visible to the naked eye is the light from myriad stars that appear too faint to be individually visible without a telescope (Figure 2.5).

The structure and dimensions of our galaxy were realized early in the 20th century. The visual appearance of the Milky Way is itself a clue. It covers only part of the sky in a distinct band, because the overall structure of the galaxy is a flat

■ **Figure 2.5** Photograph of the Milky Way as it appears in the sky. This photograph was taken with a wide-field camera, which records most of the sky. Note that the Milky Way stretches from upper left to lower right, and its appearance is a result of its disk-shaped structure. (The three lines running through the picture are produced by the camera.) (Yerkes Observatory Photograph.)

disk. When we look out into space in the plane of the disk, where the majority of stars are located, the misty band of light is seen. If we look in a direction away from the plane, we look through a thinner layer of stars, which is insufficient to appear as a milky area.

Although the fact that the Milky Way is disk-shaped was first suggested several centuries ago, the size of the galaxy and the location of the Sun within it were not ascertained until the early 20th century. Prior to that, many astronomers had assumed that the Sun was at or near the center of the Milky Way. In the 1920s, **Harlow Shapley** (1885–1972) examined the distribution of globular clusters in the sky. These clusters, each containing at least hundreds of thousands of stars in a closely packed, spherical distribution, were considered by Shapley to form a uniform halo around the center of the Milky Way. Shapley determined that the Sun was not in the center of this distribution but offset to one side, and, therefore, not in the galaxy's center. Although Shapley's dimensions were incorrect because he was unaware of

the dimming of distant starlight caused by gas and dust between the stars and other observational complications, later researchers refined his measurements and determined our galaxy's dimensions (Figures 2.6 and 2.7 are edge and top views of the Milky Way, and Figures 2.8 and 2.9 are photographs of galaxies that resemble our own and appear in similar orientations.)

The disk of the Milky Way is about 80,000 to 100,000 light years across, but only about 2000 light years thick. (The **light year**, the fundamental distance unit for galactic astronomy, is defined as the distance that light travels in one year, about 9.5 trillion kilometers or 6 trillion miles.) The Sun is located about 25,000 to 30,000 light years away from the center of the galaxy and about 30 light years above the plane of the galaxy. The distance from the Sun to the center of the galaxy can be appreciated by noting that a jet airliner traveling 600 miles per hour would require 34 billion years to travel that distance! Is there a single central object around which the stars of the galaxy orbit? This question is difficult to answer because dust between us and the galaxy's center obscures our view. We know that the core of the galaxy contains a large concentration of stars, and there is evidence that the actual center of the galaxy may be a massive object known as a black hole, so named because its gravity is so powerful that it even prevents light from escaping. Stars in the

■ **Figure 2.6** Diagram showing the edge-on appearance of the Milky Way Galaxy. Note the Sun's location and the globular clusters above and below the disk. (From Abell, George, David Morrison, and Sidney Wolff. *Exploration of the Universe,* 5th edition. Saunders College Publishing, Philadelphia, 1987.)

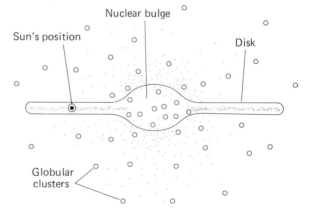

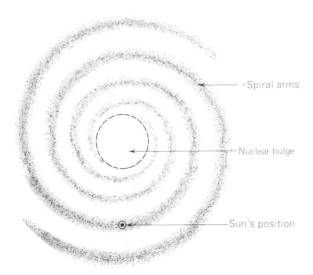

Spiral arms

Nuclear bulge

Sun's position

■ **Figure 2.7**   Diagram showing the face-on appearance of the Milky Way Galaxy. Note the Sun's location and the spiral structure. (From Abell, George, David Morrison, and Sidney Wolff. *Exploration of the Universe*, 5th edition. Saunders College Publishing, Philadelphia, 1987.)

■ **Figure 2.8**   The galaxy designated NGC 4565, which is located in the constellation Coma Berenices. NGC 4565 is similar to the Milky Way, and our galaxy would probably look like this if viewed edge-on. (Palomar Observatory Photograph.)

■ **Figure 2.9**   The galaxy designated M51, which is located in the constellation Canes Venatici. M51 is similar to the Milky Way, and our galaxy would probably look like this if viewed from far above its disk. The smaller galaxy at the bottom of the figure is an irregular companion galaxy of M51. (Palomar Observatory Photograph.)

center of the galaxy are much closer together than those in the vicinity of the Sun, and they form a large bulge that extends above and below the plane of the disk. In the vicinity of the Sun, stars are typically separated by several light years, and the closest star to the Sun is 4.3 light years away. An appreciation of this distance can be gained by noting that if both stars were scaled down to the size of peas, they would be located 145 kilometers (90 miles) apart!

The Sun, like all other stars in the Milky Way, orbits the galaxy's center. It takes approximately 200 to 250 million years for the Sun to complete its orbit. It and the other stars in the disk are not distributed uniformly but in a pinwheel-shaped fashion, forming what are called spiral arms. For this reason, our galaxy is called a **spiral galaxy.** The spiral arms contain open star clusters, groups of hundreds or thousands of stars that formed together and are held together by gravity. (Globular clusters are not found in the spiral arms.) Between the stars and clusters in the spiral arms is tenuous gas and dust, called the interstellar medium. In some areas, concentrations of the interstellar medium, called nebulae, exist. Many nebulae emit light due to excitation by nearby stars, and it is believed that stars and solar systems form from these nebulae. The spiral structure of the Milky Way was determined by tracing radio radiation emitted by the interstellar medium and the nebulae in the spiral arms.

How many of the Milky Way's other stars have planets? At this time, we simply do not know. Individual planets are too small and too faint to be seen directly around other stars. As we shall see in Chapter 30, considerable indirect evidence leads us to believe that other stars may have planets. In fact, many astronomers believe that such planets are probably very common. Although we cannot see planets in orbit around other stars, we can see stars orbiting stars. Alan H. Batten has estimated that 53% of the stars in our galaxy are binary (double) stars. There are even systems with three, four, five, or even six stars gravitationally bound together. Some researchers believe that when a star forms, either a multiple star system develops or a star and a planetary system form. Future research should reveal whether this is correct.

■ Other Galaxies

During the 18th and 19th centuries, as astronomers cataloged the numerous star clusters and nebulae in our galaxy, some nebulae were seen to have a spiral structure, identical to what we now know to be the shape of the Milky Way. Early in the 20th century, one of astronomy's main problems was ascertaining whether these "spiral nebulae" were part of the Milky Way or separate "island universes," as they were then called. By the 1920s, telescopes had improved enough that individual stars could be seen and studied in these "nebulae," proving that they were more than just clouds of gas and dust. In addition, these objects were found to be moving too fast to be part of the Milky Way. Consequently, the spiral nebulae were determined to be external galaxies, distant cousins of our own Milky Way.

A great deal has been learned about galaxies. In addition to spirals like our own, there are two other types of external galaxies. Elliptical galaxies have a spherical or oblong shape, and irregular galaxies have essentially no apparent form or structure. Counts of visible galaxies and estimates of the overall size of the universe have placed the total number of galaxies in the universe at about 100 billion. Although there is considerable variation in the sizes and number of stars in galaxies, the average number of stars in each of these galaxies is probably about the same as in our own. When described in these terms, the universe's immensity and complexity can easily be appreciated.

Although the universe is filled with galaxies, their distribution is not random. Most galaxies are members of galaxy clusters, which contain from a few dozen to thousands of members. The Milky Way and about two dozen of its nearby neighbors compose a small cluster called the **Local Group.** The relatively nearby Virgo cluster is about 50 million light years distant and contains approximately 2500 galaxies. As we look out at even more distant galaxies and clusters, we are also looking back in time, because we are seeing the objects as they were long ago, when their light left them. This means that we cannot see the entire universe as it is now; instead, we are seeing different objects in it at different times in their pasts.

The Expanding Universe and the Big Bang

In the 1920s, when the large-scale structure of the universe was starting to be understood, astronomers began investigating whether other galaxies are approaching or receding from the Milky Way. Measurement of velocities of galaxies in the direction along the line of sight, a quantity called **radial velocity**, is accomplished by examining the spectra of these galaxies. As we will learn in Chapter 4, if a light source is moving in the radial direction, the wavelengths of its spectral features will be shifted because of the Doppler effect. An object receding will experience a shift to longer (redder) wavelengths of light (a red shift), whereas an approaching object will experience a shift to shorter (bluer) wavelengths of light (a blue shift).

At first, researchers expected that other galaxies would be randomly approaching and receding. With the exception of other galaxies in the Local Group, however, all external galaxies were found to be moving away from the Milky Way. Even more surprisingly, the distant galaxies had larger red shifts than nearby ones. Near the end of the 1920s, the American astronomer **Edwin Hubble** (1889–1953) formulated a relationship now called **Hubble's law:** The velocity of recession of a galaxy is directly proportional to its distance from Earth.

What does this result tell us about the nature of the universe as a whole? The modern explanation is that the entire universe is expanding, with the expansion carrying the various clusters of galaxies farther away from each other. An observer in a distant galaxy would see the same expansion that we see; there is nothing unique about the location of the Milky Way or our view of the universe from it (Figure 2.10).

What is the cause of this expansion? A major area of research in modern astronomy is the attempt to understand the origin and history of the entire universe. The theory that is generally accepted today is called the **Big Bang** theory, because it proposes that the universe began when a small concentration of incredibly hot, dense material exploded in a fireball, from which the basic components of our universe were derived.

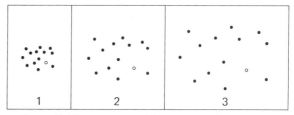

■ **Figure 2.10**   Expansion of the universe. The sequence from 1 to 3 shows how clusters of galaxies (represented by *solid dots*) get farther apart as time passes. (The *open dot* represents the Local Group, the galaxy cluster containing the Milky Way.)

Shortly after the initial Big Bang explosion, atoms, the basic building blocks of matter, began to form. Conditions were such that only the elements hydrogen, helium, and lithium formed at this time, because the universe soon cooled sufficiently that heavier elements could not be produced. The expanding matter was not uniformly distributed but separated into large clumps that eventually became clusters of galaxies. Individual galaxies then formed as further clumping occurred within this maelstrom of expanding material. Finally, individual stars formed within each galaxy. These early stars contained only hydrogen and helium, but nuclear reactions occurring within them produced the heavier elements that are now also present in the universe. (We discuss the elements and their origin in more detail in Chapter 8.)

Estimates of the present rate of expansion of the universe indicate that the Big Bang occurred sometime between 15 and 20 billion years ago. Studies of the ages of stars in globular clusters within our galaxy indicate that it formed between 12 and 15 billion years ago. As we will see in Chapter 6, the Sun is believed to be only about 5 billion years old, indicating that it did not form when the Milky Way originated. The Sun and solar system formed from the interstellar medium, which at that time included material produced within older stars and spewed out into space at the time of their demise. This "stellar recycling" allowed elements other than the primordial hydrogen and helium to be present in the material from which the solar system formed.

Although the Big Bang occurred so long ago that the interval of time since then can scarcely be comprehended, the event set the pattern for the universal expansion that continues to the present. One of the major unsolved questions of astronomy is whether the expansion will continue forever or gradually slow, stop, and reverse to universal contraction. The answer requires knowledge of the average density of material throughout the entire universe, a quantity that, understandably, has not yet been determined with sufficient accuracy. If there is enough matter in the universe as a whole, it would exert sufficient gravitational force to someday halt the expansion, but current results indicate that there may not be enough material in the universe to do so.

■ Chapter Summary

The solar system may be described as a distribution of celestial bodies that lie generally in one plane (the ecliptic plane) and extend outward about 40 astronomical units from the Sun. Much farther away, a spherical distribution of comets, the Oort cloud, marks the outer boundary of the solar system. The solar system contains a star (the Sun), nine planets, 60 planetary satellites, and countless comets, asteroids, and meteoroids. With the exception of planetary satellites, all objects in the solar system orbit the Sun.

The solar system is part of a larger entity, the Milky Way galaxy, which contains about 200 billion stars. The Sun orbits the center of the Milky Way in much the same way as the Earth orbits the Sun. The universe contains over 100 billion galaxies, which are grouped into clusters that are receding from one another because of the expansion of the universe. This expansion is a result of the fact that the universe was created about 15 to 20 billion years ago by an event called the Big Bang. The Big Bang created the major components of stars, hydrogen and helium, whereas heavier elements have since been created by the nuclear reactions inside stars.

■ Chapter Vocabulary

asteroid	galaxy	meteoroid	radial velocity
astronomical unit	Hubble, Edwin	Milky Way	retrograde
Big Bang	Hubble's law	Oort cloud	Shapley, Harlow
celestial sphere	light year	parallax	spiral galaxy
comet	Local Group	planet	star
constellation	meteor	planetary satellite	triangulation
ecliptic plane	meteorite	prograde	zodiac

■ Review Questions

1. What is the ecliptic plane, and how does it manifest itself in the sky?
2. How may one define the outer boundary of the solar system?
3. Describe the Oort cloud of comets.
4. List and describe the six types of objects in the solar system.
5. Explain how to tell the difference between a star and a planet in the sky.
6. Sketch and describe the Milky Way.
7. Indicate the Sun's approximate location in the Milky Way, and explain how it was determined.
8. In addition to those found in our solar system, what types of objects are found in the galaxy?
9. What is the arrangement and motion of the universe's galaxies?
10. Explain how the Big Bang theory seeks to explain the basic features of the universe.

■ For Further Reading

Barrow, John D. and Joseph Silk. *The Left Hand of Creation*. Basic Books, New York, 1983.

Beatty, J. Kelly and Andrew Chaikin, editors. *The New Solar System*. Sky Publishing, Cambridge, MA, 1990.

Berendzen, Richard et al. *Man Discovers the Galaxies*. Columbia University Press, New York, 1984.

Bok, Bart J. and Priscilla F. Bok. *The Milky Way*. Harvard University Press, Cambridge, MA, 1981.

Hodge, Paul W. *Galaxies*. Harvard University Press, Cambridge, MA, 1986.

Kivelson, Margaret G., editor. *The Solar System—Observations and Interpretations*. Prentice-Hall, Englewood Cliffs, NJ, 1986.

Reeves, Hubert. *Atoms of Silence*. MIT Press, Cambridge, MA, 1984.

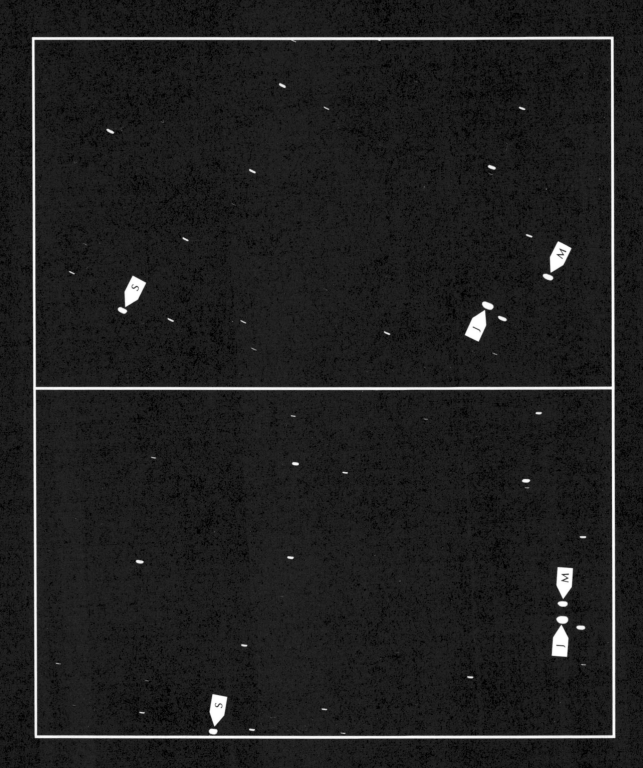

Celestial Mechanics

The planets Saturn, Jupiter, and Mars in the constellation Leo. **Top,** April 18, 1980. **Bottom,** May 3, 1980. Note the changes in their locations relative to each other and to the background stars. Celestial mechanics gives us the ability to understand the planets' motions and predict their locations. (Photographs by the author.)

One of the main concerns of early astronomers was determining the organization of the solar system and what motions occur within it. Two major, competing models were proposed in ancient times, and it was not until the 17th century that one of them, the heliocentric, gained universal acceptance. The study of orbital motion and the laws describing it is called **celestial mechanics.**

■ Early Theories of Orbital Motion

Long before astronomers possessed the equipment and techniques to learn about the composition and other physical properties of the solar system's objects, they were concerned primarily with studying their motions through the sky and being able to predict accurately their future positions. Our best records of this early search for understanding take us back to the ancient Greece of the pre-Christian era. Although the Greeks were not the first astronomers, much of the important early work of which records remain today was done by them.

As the Greeks sought the ability to explain the motion of the Sun, Moon, and planets through the sky and predict their future positions, a major question arose: Were these changes in the sky due to the objects' movement around a fixed, immobile Earth, or was the Earth moving through space with another object, the Sun, fixed in the center of the solar system? Although the answer is well known today, we must bear in mind that the Greeks were relying on naked-eye observations and were not aided by modern instruments!

Many of the ancient Greeks were proponents of the **geocentric,** or Earth-centered, point of view. The most notable of these was **Aristotle** (384–322 B.C.). He believed that if the Earth or-

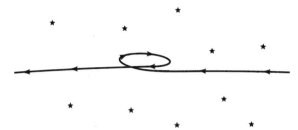

■ **Figure 3.1** Retrograde motion of a planet through the sky. The planet's apparent motion among the stars (*solid line*) is from west to east, except for the retrograde loop where the movement reverses for a time.

bited the Sun, our changing point of view would cause the stars to shift positions, a phenomenon that was not observed. (Today, this shift, called parallax, can be detected, but only with telescopes.) The geocentric concept of motion did have a major problem: As planets gradually move among the background stars, their normal west-to-east motion occasionally stops and reverses. After a period of this reverse motion, called retrograde motion, the planet once again changes direction and proceeds as before, creating a feature called a **retrograde loop** (Figure 3.1). This was hard to explain if a planet was orbiting the Earth in a circular orbit. (Because the circle is a perfect geometrical form, circular orbits were assumed.) To solve the problem, it was postulated that planets themselves did not orbit the Earth directly. A secondary orbit, called an **epicycle**, traveled around the Earth in a path called a **deferent**, and the planet moved in the epicycle. At the proper times, as the planet moved around in the epicycle, the planet would appear to be moving backward (Figure 3.2). Although proponents of the geocentric concept agreed that the Earth did not move through space, they disagreed about whether it rotated. Although some believed the Earth to be totally motionless, many Greeks believed that the Earth rotated, since the alternative, that the entire celestial sphere rotated around the Earth daily, seemed unreasonable.

A few Greeks, most notably **Aristarchus** (310–230 B.C.), proposed a **heliocentric** (Sun-centered) universe. Although this was a significant development, they were unable to convince members of the geocentric school because they

could not explain the absence of parallax. About 150 A.D., **Claudius Ptolemy** compiled a summary of all Greek astronomy. His work was destined to last until the Renaissance as the "final word" on all things astronomical. Unfortunately, Ptolemy advocated the geocentric concept, and the support it later received relegated the heliocentric system to the scientific scrap heap for centuries (Figure 3.3).

Following the time of Ptolemy, European astronomy was virtually dormant for many centuries. (Fortunately, Arabic astronomers preserved earlier learning and made discoveries of their own during this time.) When science began anew during the Renaissance, one of astronomy's first big accomplishments was determining, at

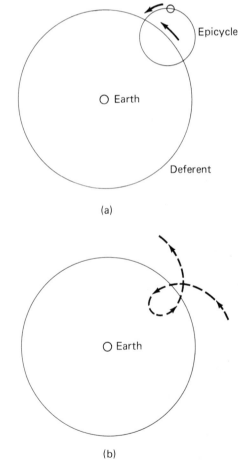

■ **Figure 3.2** Geocentric orbital theory. (a) The planet moves in an epicycle, which in turn follows a deferent around the Earth. (b) The path of the planet due to the combined motion.

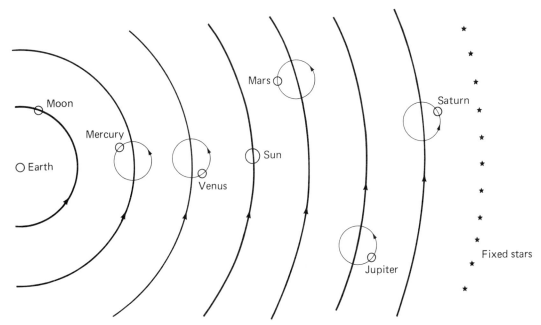

■ **Figure 3.3**   The geocentric system. This system, devised by the ancient Greeks, was accepted with few questions from ancient times until the 1500s.

last, the organization of the solar system. As a result of the work of several important astronomers, we now understand the laws that explain planetary motion.

■ Copernicus

Nicholas Copernicus (1473–1543) was a Polish cleric and scholar who made a careful study of Ptolemy's writings and decided that the geocentric concept was in error. In the early decades of the 1500s, he learned that certain ancient Greeks had proposed a moving Earth, and he soon became the first modern proponent of the heliocentric concept. In the year of his death, he published a book summarizing his work and opinions. Although Copernicus did not live to witness it, his influential book soon became the talk of the scientific community, and from its publication onward, there was never a time when the heliocentric concept was "forgotten," as it had been for so many centuries.

Copernicus understood that the heliocentric system adequately explained the backward mo-

tion of the planets in the sky. In order to account for the retrograde loops resulting from this motion, the geocentric concept had become inordinately complex. Copernicus explained that planets would move at different velocities depending on their distance from the Sun, and that retrograde motion was simply caused by the Earth either passing an outer planet or being passed by an inner one (Figure 3.4). Although refinements of Copernicus' basic ideas were made by later astronomers, he is still regarded as the father of the modern heliocentric concept.

■ Tycho

Tycho Brahe (1546–1601) was the last major astronomer to make naked-eye observations prior to the invention of the telescope. Before telescopes existed, astronomical observations consisted primarily of making accurate measurements of the positions of stars and planets. Tycho constructed precise instruments for making positional measurements, and his observational skill was unmatched. Tycho's observations of the Sun

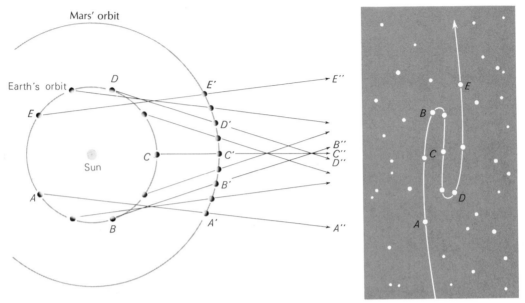

■ **Figure 3.4** True cause of retrograde motion. The faster-moving inner Earth at *A* through *E* passes the slower-moving outer Mars at *A'* through *E'*. The lines connecting the planets point toward Mars' observed sky locations, shown at right. (From Abell, George, David Morrison, and Sidney Wolff. *Exploration of the Universe*, 5th edition. Saunders College Publishing, Philadelphia, 1987.)

and planets covered a period of over 20 years and eventually were instrumental in solving the problem of the solar system's organization. However, this was not done by Tycho, who failed to accept the heliocentric concept because he could not measure any stellar parallax. Instead, Tycho developed a theory of the solar system that might best be described as a strange compromise: He believed that the Sun and Moon orbited the Earth, but that the planets in turn orbited the Sun! This hypothesis, unlike Tycho's data, had almost no effect on astronomical research.

In 1572, Tycho made an observation of special significance: He saw a bright new star, or supernova, at a place in the sky where no star had previously been observed. After more than a year of naked-eye visibility, the star faded away. The ancient Greeks had thought that the heavens were the center of perfection and that perfection could not change. This philosophy, along with the geocentric concept, was one of the long-lasting legacies of Greek thought. Tycho's discovery of the new star, however, was dramatic proof of

the changeability of the heavens, which had long been denied.

■ Kepler

Johannes Kepler (1571–1630) became a supporter of the Copernican system while attending college. He began his career as a teacher of mathematics and astronomy, but later obtained a position as Tycho's assistant. Following Tycho's death, Kepler came into possession of his numerous observations. The fact that Kepler had very accurate positions of the planets, covering a period of many years, allowed him to check the various planetary theories. If a theory did not predict locations matching Tycho's observations, Kepler knew it was incomplete. Such was the case even with the Copernican theory, which assumed circular orbits for the Earth and other planets. Kepler tried to use Copernicus' heliocentric model of the solar system to calculate Mars' position in the sky on the dates when its

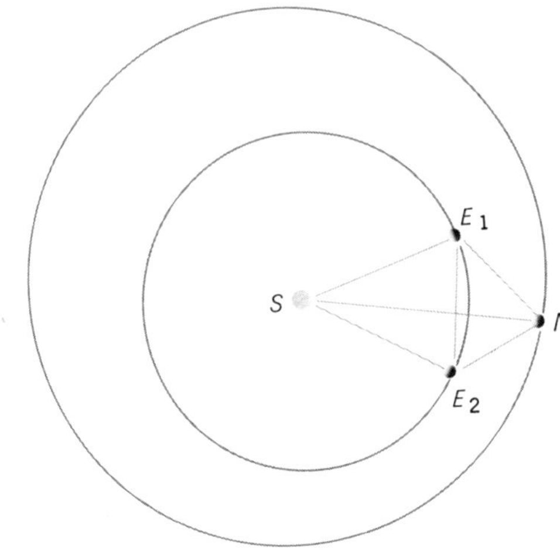

■ **Figure 3.5** Kepler's method of determining the form of Mars' orbit. Mars was observed at two dates 687 days apart. Because this is Mars' sidereal period, Mars was at the same point in its orbit (*M*) each time, but Earth was at different points (E_1 and E_2). Knowledge of the two observed Sun–Earth–Mars angles allowed the calculation of Mars' distance from the Sun. Repeating this process numerous times around the orbit allowed its elliptical form to be discovered. (From Abell, George, David Morrison, and Sidney Wolff. *Exploration of the Universe,* 5th edition. Saunders College Publishing, Philadelphia, 1987.)

precise location had been measured by Tycho. Unfortunately, Kepler was not able to reconcile the theory with the observations exactly, so he realized that there was a flaw somewhere in the Copernican theory. He soon realized that the error was the assumption of circular orbits. Kepler began investigating various noncircular orbits, such as oval and egg-shaped ones, but eventually determined that the true shape of planetary orbits is elliptical.

How did Kepler discover that planetary orbits are elliptical? Using Tycho's Martian positional data, Kepler made a careful plot of Mars' orbit in relation to Earth's. He did this by examining pairs of positional data separated by 687 days, which is the orbital period of Mars. (Copernicus had determined how to calculate true orbital periods of planets around the Sun from the apparent, or synodic, periods of the planets between their successive alignments with the Earth and the Sun.) Although Mars returns to the same point in its orbit after a 687-day interval, the Earth's location at these two times is different (Figure 3.5). Because Tycho's observational data allowed him to calculate the Sun–Earth–Mars angle at each time, Kepler was able to determine geometrically the distance between the Sun and Mars, in astronomical units. By repeating this analysis for the entire range of Mars' orbit, he could plot the Martian orbit and determine that its shape was elliptical.

■ The Ellipse

The circle is a very simple figure consisting of a set of points of equal distance (the radius) from a central point. Although a circle is easily drawn with a compass, the more complicated **ellipse** is more difficult to draw. It is based not on one central point but on two **focus points,** which lie along the line passing from one end of the ellipse to the other along its greatest dimension. This line is called the major axis; exactly half of it is called the **semi-major axis.** The line at right angles to it, spanning the shortest distance across the ellipse, is called the minor axis (Figure 3.6). Unlike circles, ellipses occur in a variety of shapes, from nearly circular to extremely elongated. The shape or degree of elongation of an ellipse is given by a quantity called **eccentricity** (Figure 3.7). The eccentricity, e, of an ellipse is given by the equation

$$e = \frac{f}{a},$$

where a is the ellipse's semi-major axis and f is the distance from the center to each focus point. An

■ **Figure 3.6** The ellipse. Points *F* mark the focus points, the *vertical line* marks the minor axis, and the *horizontal line* marks the major axis. The solid portion of the latter is the semi-major axis.

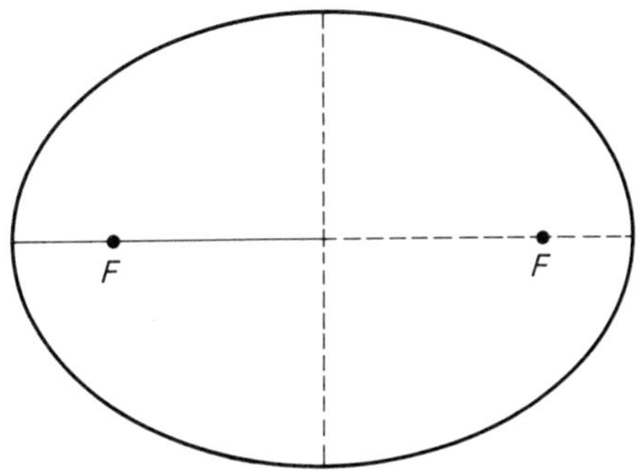

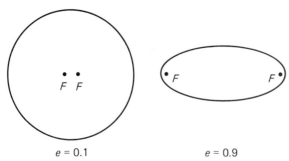

$e = 0.1$ $e = 0.9$

■ **Figure 3.7** Eccentricity. The ellipse on the left has an eccentricity of 0.1, and the one on the right, 0.9. (Their semi-major axes are identical.)

an ellipse can have any value between zero and one. A conic section with an eccentricity of exactly one is called a **parabola**, which resembles an ellipse but is open at one end. A more general type of open-ended conic section, the **hyperbola**, has an eccentricity greater than one. Solar system objects do not have parabolic or hyperbolic orbits, because such open-ended paths would take them out of the solar system forever. Some comets, however, appear to have such orbits, but it is believed that they are actually solar system members with orbital eccentricities close to, but not equal to, one.

ellipse may be drawn by placing two tacks in cardboard (these are the focus points) and placing a loose loop of string around them. Stretching the string tight with a pen or pencil and looping around the focus points produce an ellipse (Figure 3.8). Increasing the separation of the tacks increases the eccentricity, whereas using a larger or smaller loop of string changes the length of the major axis.

    Circles and ellipses are members of a family of curves collectively called **conic sections**, because they can be derived by slicing a plane through a cone (Figure 3.9). A circle has an eccentricity of exactly zero, whereas the eccentricity of

■ Kepler's Laws

Kepler expressed his discoveries as three great laws of planetary motion. He based his laws on observational evidence and did not understand completely what physical processes caused them to be true. For example, he did not understand the force responsible for holding the planets in their orbits. Later, Newton developed the laws of motion and gravity, which allowed Kepler's laws to be derived mathematically and understood physically. Another point not realized until Newton's time is that these laws are valid not only for

■ **Figure 3.8** How to draw an ellipse. The two tacks become the focus points, and the length of the loop of string determines the semi-major axis. The greater the separation between the tacks, the more eccentric the ellipse. (Photograph by the author.)

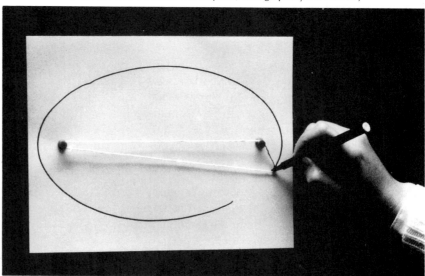

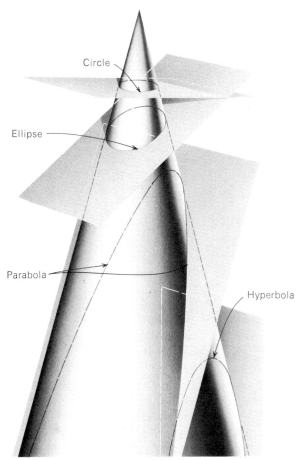

■ **Figure 3.9** How the four conic sections are formed by "slicing" a plane through a cone. Orbits can occur as any of these forms. (From Abell, George, David Morrison, and Sidney Wolff. *Exploration of the Universe,* 5th edition. Saunders College Publishing, Philadelphia, 1987.)

planets orbiting the Sun but for any objects in orbital motion, anywhere in the universe.

Kepler's first law states that planets orbit the Sun in elliptical orbits, with the Sun located at one focus point. Because the Sun does not lie in the center of the ellipse, the distance of a planet from the Sun varies. The point of closest approach is called **perihelion,** and the most distant point is called **aphelion.** Even though orbital distance is variable, the average distance of a planet from the Sun can easily be obtained and is equal to the semi-major axis of the orbit. Most planetary orbits have very low eccentricity and are nearly circular. (This is why it took so long for their elliptical nature to be discovered.) Other objects, especially comets, may have more eccentric orbits.

Kepler's second law deals with the rates of orbital motion. It says that a planet travels fastest when at perihelion and slowest when at aphelion. Kepler originally stated this graphically, showing an ellipse with lines connecting the planet and the Sun (Figure 3.10). Kepler calculated that the areas "swept out" by the line connecting the planet to the Sun would be equal at perihelion (left side of Figure 3.10) and at aphelion (right side of Figure 3.10). The planet would require the same amount of time to cover the indicated distances at both perihelion and aphelion, because its orbital velocity was not constant. For this reason, the second law is often called the law of equal areas.

Kepler's third law (also called the harmonic law) relates a planet's semi-major axis (a) and its orbital period (P). Kepler determined that the relationship is given by

$$P^2 = a^3,$$

if a is expressed in astronomical units and P in years. For example, Mars orbits the Sun at an average distance of 1.524 astronomical units in 1.881 years. Because both the cube of 1.524 and the square of 1.881 are about 3.54, the law is verified. (Both values are not exactly equal. This is because Kepler failed to realize that his form of the law was incomplete, because he attributed the small differences to observational errors. As we will see, Newton later determined the complete form of the third law.) A listing of the Kepler's third law quantities for all of the planets is given in Table 3.1.

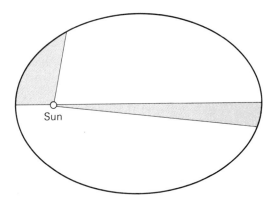

■ **Figure 3.10** The Law of Equal Areas. A graphical representation of Kepler's second law. See text for a full explanation.

■ Table 3.1   Figures Verifying Kepler's Third Law

Planet	Semi-Major Axis (AU)*	Cube of Semi-Major Axis	Period (yr)	Square of Period
Mercury	0.387	0.058	0.241	0.058
Venus	0.723	0.378	0.615	0.378
Earth	1.000	1.000	1.000	1.000
Mars	1.524	3.540	1.881	3.538
Jupiter	5.203	140.852	11.862	140.707
Saturn	9.555	872.352	29.458	867.774
Uranus	19.218	7097.81	84.012	7058.02
Neptune	30.110	27298.1	164.80	27152.7
Pluto	39.439	61344.8	247.69	61348.9

* AU = astronomical units.

■ Galileo

Galileo Galilei was an Italian mathematician, physicist, and astronomer who lived from 1564 to 1642. He was a contemporary of Kepler, and they corresponded, although they never met. By 1600, Galileo had accepted the Copernican explanation of solar system motion. Soon afterward, he began a revolution that would not only promote acceptance of the heliocentric theory but change the way in which astronomers study the universe.

About 1609, Galileo heard that Dutch spectacle makers had developed the first telescopes by combining two lenses to magnify distant objects. He quickly deduced how to build his own telescopes and began looking at objects in the night sky with them. Until that time, it was assumed that celestial objects were perfect bodies, without blemish of any sort. Galileo soon learned otherwise: He saw black markings, now called sunspots, on the Sun. He saw craters, mountains, and other irregularities on the surface of the Moon. He also saw that the Milky Way was actually made of numerous individual stars.

Two of Galileo's discoveries provided important evidence for proponents of the heliocentric theory. As Galileo observed the planet Venus, which is always close to the Sun in the sky, he noted that it exhibited phase changes similar to those of the Moon. According to the geocentric system, Venus and the Sun both orbited the Earth, with Venus closer to the Earth than the Sun.

As Figure 3.11 illustrates, a complete range of phases would be impossible in that case, but would be expected if both Venus and Earth orbit the Sun. Venus' observed phases prove that it is sometimes located on the opposite side of the Sun from the Earth. Galileo's most startling discovery was that the planet Jupiter was accompanied by four small companions whose positions changed daily. Galileo realized that these bodies were orbiting Jupiter. This showed clearly that the Earth was not the center of all orbital motion, as the geocentric theory proposed, but that an orbiting object could in turn have objects orbiting it. Opponents of Copernicus had argued that the Moon could not orbit a moving Earth because it would be "left behind." Jupiter and its satellites clearly dispelled that notion.

Galileo publicized his findings and invited others to look through his telescopes and see for themselves what was visible. In 1632, he published a book discussing the geocentric and heliocentric hypotheses, clearly supporting the latter. Unfortunately for Galileo, the ideas of Ptolemy had been accepted for so long that by his time they were actually considered religious dogma. The Catholic Church in Italy, believing the geocentric concept more in line with Biblical teaching, vigorously persecuted anyone with a dissenting view, and Galileo was no exception. Fortunately, he was only placed under house arrest after being forced to recant his views. (A contemporary, Giordano Bruno, had been executed for holding similar views.) History has, of course,

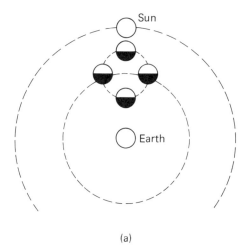

(a)

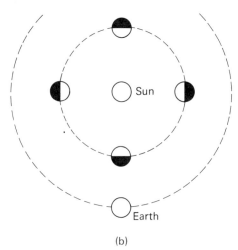

(b)

■ **Figure 3.11** Venus' phases. (a) The geocentric system would preclude a fully illuminated Venus. (b) The heliocentric system allows a complete range of phases. In both cases, Venus is never far from the Sun in the sky.

vindicated Galileo, and he was officially exonerated by the Catholic Church in 1982.

■ Newton and Orbital Motion

Sir Isaac Newton, who lived from 1643 to 1727, was one of the greatest scientists of all time. He invented the branch of mathematics we now call differential calculus (much to the chagrin of gen-

erations of math students!) and applied it to the solution of physical problems. As mentioned earlier, he developed the physics necessary to prove that Kepler's laws were correct.

Newton set forth three laws of motion. The first states that an object at rest tends to remain at rest and an object in motion tends to remain in motion, unless acted upon by an external force. We often refer to the property of an object that resists such changes as inertia. The second law states that if a force acts on an object, it will produce a change in velocity, or acceleration. This law is expressed by the equation

$$F = ma,$$

where F is force, m is mass, and a is acceleration. (It is important to note that a change in direction of motion is an acceleration, even if speed remains constant.) The third law states that if a force is applied, an equal but opposite force results. This is often stated as "for every action, there is an equal and opposite reaction" and expressed by the equation

$$F_{12} = -F_{21},$$

where F_{12} is the force exerted by object 1 on object 2 and F_{21} is the force exerted by object 2 on object 1. An example is the recoil produced when a rifle fires. The explosion of the gunpowder forces the bullet out of the muzzle, but it also "kicks" the gun back against the shooter. Similarly, a child jumping from a wagon pushes the wagon away in the opposite direction.

Newton realized that planets are constantly being accelerated because their velocities are always changing. He wondered what force was responsible for this acceleration, which was directed toward the Sun and held the planets in their orbits. (He coined the word centripetal, meaning "center seeking," to describe such a force.) Newton finally devised his theory of gravitation, which describes an attractive force between any objects with mass. Although the physical cause of this attraction is not fully understood even today, Newton accurately described its behavior. Two objects of mass, M_1 and M_2, separated by a distance R, will attract each other with a force F, given by the equation

$$F = \frac{GM_1M_2}{R^2},$$

where G is the gravitational constant defining the unit gravitational force for objects of unit mass at unit separation. (Unlike some other physical forces, gravity is always attractive.) Although it is one of the fundamental forces of nature, gravity is comparatively weak and is important only when very massive objects are involved. For example, two people have a gravitational attraction for each other, but it is so insignificant that neither are aware of it, due to the much greater gravitational pull of the Earth. Gravity makes up for this weakness by being able to act over very great distances, and it is gravity that holds together large systems such as the solar system and the Milky Way.

According to legend, Newton began thinking about gravitation while sitting under an apple tree. He saw an apple fall to the ground and wondered why the Moon, visible at that moment in the sky, did not also fall to Earth. Newton subsequently described gravity as the force that holds the Moon in place, and he developed the physics of orbital motion, which explain why the Moon keeps going. Newton realized that the Earth's gravity provides the centripetal acceleration that holds the Moon in orbit, whereas the Sun's gravity does the same for the planets. (To demonstrate centripetal force, twirl around your head an object tied to a string. You must hold on to the string to keep the object in "orbit," and if you release the string, the object will fly off in a straight line. The tension that you feel in the string as you twirl the object is the centripetal force in this case. For orbiting bodies, gravity replaces the string!) From the mathematical expressions of Newton's laws of motion and gravitation, Kepler's three laws can be derived.

Newton examined the energetics of orbiting bodies. An orbiting object's **kinetic energy** (*KE*), or energy of motion, depends upon its mass, m, and velocity, v, and is given by the equation

$$KE = \frac{mv^2}{2}$$

An object's gravitational **potential energy** (*PE*), or energy of position, depends upon its mass and distance from the Sun and is given by the equation

$$PE = -\frac{GM_1M_2}{R},$$

where G, M_1, M_2, and R refer to the same quantities as in the law of gravity. (Note that kinetic energy always has a positive value, whereas potential energy of orbiting bodies always has a negative value. In our discussion comparing energies, however, we will consider only absolute values — the magnitude of a quantity independent of its sign.) Because an object's velocity and distance from the Sun vary over the course of its orbit, neither kinetic energy nor potential energy are constant. However, an orbiting object's total energy, equal to the sum of kinetic and potential energy, is constant. Newton discovered that the total energy of an orbiting body determines what type of orbit it will follow.

Newton derived an extended version of Kepler's first law. Although Kepler had specified elliptical orbits, Newton determined that orbits following any conic section form were possible, depending upon the total energy of the orbiting body. If the object's kinetic energy is less than its potential energy, the total energy is negative and the object moves in an elliptical orbit. (The circle is a special case, having the lowest possible total energy for a given semi-major axis.) If the object's kinetic and potential energies are equal, its total energy is zero and it moves in a parabolic orbit. (True parabolic orbits are probably rare, because it is unlikely that the two energy values would be exactly equal.) If the object's kinetic energy is greater than its potential energy, the total energy is positive and it moves in a hyperbolic orbit.

Energy considerations also allow the derivation and understanding of Kepler's second law. When an object is at perihelion, its potential energy is smallest. Because total energy is constant, kinetic energy must be greatest at perihelion, requiring maximum velocity. Conversely, at aphelion, potential energy is greatest, so kinetic energy (and velocity) must be smallest.

Newton refined Kepler's third law by realizing that the mass of the Sun (M_S) and the mass of the planet (M_P) must be taken into account, making the complete form of the law

$$(M_S + M_P)P^2 = a^3,$$

where M_S and M_P are given in solar mass units (with $M_S = 1$), P in years, and a in astronomical units. In the solar system, the mass of even the largest planet, Jupiter, is so insignificant compared with that of the Sun that $(M_S + M_P)$ is virtually equal to 1. (The accurate value here is 1.001.) For this reason, Kepler had been unaware that masses needed to be taken into account in his third law. (If M, P, and a in Kepler's third law are expressed in standard quantities such as kilograms, seconds, and meters, respectively, a more general form of the law must be used:

$$(M_S + M_P)P^2 = \left(\frac{4\pi^2}{G}\right)a^3,$$

where G is the gravitational constant.)

Kepler's third law has become one of the most important laws in astronomy because it allows us to determine the masses of many different objects when orbital motion is involved. Although orbital period and semi-major axis can often be measured by observation, astronomers cannot "weigh" celestial objects directly. Because Kepler's third law contains the expression $(M_S + M_P)$, or the more general $(M_1 + M_2)$, individual masses cannot be determined. In the case of a small satellite orbiting a planet, however, the mass of the satellite is negligible, and the planet's mass can be calculated. In the case of a binary star or a pair of galaxies, the two masses usually will be more nearly equal, so measurement of the relative motions of the objects about their center of gravity is used to determine the ratio of their masses once the sum of the masses has been obtained.

In describing orbital motion around the Earth, Newton suggested the following experiment: Suppose a cannon were mounted atop an extremely tall mountain. A projectile fired from the cannon would go forward from its mouth, then arc downward and fall to Earth, pulled by gravity. If the projectile were fired with greater velocity, it would go farther before landing. Newton realized that given sufficient velocity, the projectile would begin arcing downward, with a curve exactly matching the curve of the Earth's surface (Figure 3.12). Therefore, it would continue "falling" around the Earth but would never get closer to the ground; it would be in orbit. In other words, the Moon in the sky and the apple falling from the tree are more alike than Newton originally thought! (Earth satellites are not fired from cannons but, of course, obtain their initial velocities from rocket engines. We might wonder

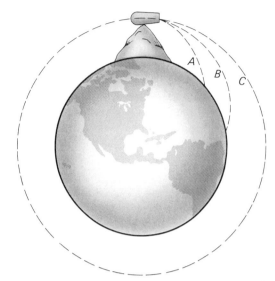

■ **Figure 3.12**   Newton's cannon. Normally, a projectile fired from a mountain-top cannon will fall to Earth (*A* and *B*), but with sufficient velocity, would continue "falling" in an orbit (*C*).

how the planets acquired their velocities and, thus, kinetic energies. As we see in Chapter 29, this occurred during the formation of the solar system.)

Using his laws of motion, Newton was able to calculate the velocity (V_O) necessary to stay in orbit. For the special case of a circular orbit, **orbital velocity** is given by the equation

$$V_O = \sqrt{\frac{GM}{R}},$$

where G is the gravitational constant, M is the mass of the object being orbited, and R is the distance of the orbit from the center of the body orbited. Orbital velocity for an Earth satellite 160 kilometers (100 miles) above the surface is about 28,500 kilometers per hour (17,700 miles per hour), and manned spacecraft, which usually orbit at this altitude, take about 90 minutes to complete an orbit. A higher satellite will travel slower and take longer to complete an orbit. Satellites that are 35,700 kilometers (22,200 miles) above the Earth travel 11,100 kilometers per hour (6870 miles per hour) and orbit once every 24 hours. Because they appear motionless from the Earth's surface, which rotates at the same rate, they are called **geosynchronous satellites**.

If the velocity of an object is sufficiently great, it will be able to escape entirely from a more massive body rather than move in orbit around it. This **escape velocity**, V_E (sometimes called parabolic velocity because it produces just enough kinetic energy for a parabolic orbit), is given by the equation

$$V_E = \sqrt{\frac{2GM}{R}},$$

where G is the gravitational constant, M is the mass of the object to be escaped, and R is the distance from the center of the object to be escaped (R is the object's radius if one is calculating escape velocity from its surface). Earth's escape velocity is 40,250 kilometers per hour (25,000 miles per hour), and rockets launched from the Earth toward other planets must attain this velocity.

■ Orbital Elements

Newton's work solved forever the problem of the overall dynamics of the solar system and drove the final nail into the coffin of the geocentric concept. Newton himself could have calculated trajectories for spaceflight to the Moon and planets, although he probably would have considered it no more than a mathematical exercise of no useful value. Many refinements to the field of celestial navigation have since been made, but all are based on Newton's work.

As mentioned in the last chapter, the total number of objects in the solar system is very large. Most of them orbit the Sun, and each has a different orbit. How can we describe these orbits? How do we keep track of all of the objects in the solar system and predict their locations? The quantities that describe an orbit and allow us to locate a solar system object at any given time are called its **orbital elements**. A set of seven orbital elements is normally used. Five describe the spatial orientation of the orbit (Figure 3.13), and two deal with time.

The first element, **semi-major axis** (a), describes the size of the orbit. Recall that this quantity is the average distance of an orbiting body from the object around which it orbits, over the course of the entire orbit. This distance is always that between the centers of the objects, not their surfaces. It is usually given in astronomical units.

The second element, **eccentricity** (e), describes the shape of the orbit. Values lie between zero and one, with most solar system objects other than comets having very small eccentricities. Eccentricity is a dimensionless quantity.

As noted earlier, most solar system objects lie in or near the ecliptic plane, which is defined by the orbit of the Earth around the Sun. All other objects have orbits that are tilted at least a few degrees from the ecliptic. The angle of this tilt is called the **inclination** (i), which is measured in degrees from 0° to 180°. (For a satellite orbiting a planet, inclination is often measured not with respect to ecliptic, but to the planet's equatorial plane, in which most satellite orbits lie.) If inclination is greater than 90°, the object is in a retro-

■ Figure 3.13 Orbital elements. See text for a full explanation.

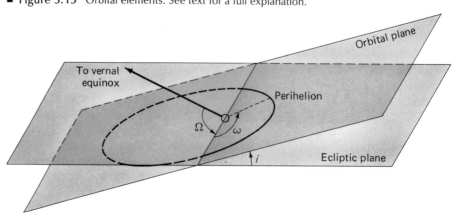

grade (backward) orbit and moves clockwise as seen from north of the ecliptic plane.

If an orbit is tilted with respect to the ecliptic plane, obviously the orbiting object will spend some time below the ecliptic and some time above. (These times are generally not equal, nor are the linear distances traveled the same.) The two points at which an orbit crosses the ecliptic are called **nodes,** and the line connecting them, which represents the intersection between the plane of the orbit and the plane of the ecliptic, is called the **line of nodes.** The next orbital element, **longitude of the ascending node** (Ω, the upper-case Greek letter Omega), is the orientation of the node where the planet crosses the ecliptic plane from south to north. The starting point for measuring this angle is the **vernal equinox,** one of two points in space where the ecliptic plane intersects the plane of the Earth's equator. (This is the point in the sky where the Sun crosses the celestial equator from south to north on the first day of spring. Recall that the celestial equator is the imaginary line in the sky formed by projecting the Earth's equator into space.) Longitude of ascending node is measured in degrees.

The fifth spatial element, **argument of perihelion** (ω, the lower-case Greek letter omega), measures the angle between the ascending node and the point of perihelion. It is measured in the direction the planet travels in its orbit. (Note that although the starting point of measurement is the ascending node, perihelion may occur in the part of the orbit that is south of the ecliptic plane.) For planets, the element called **longitude of perihelion** (π, the lower-case Greek letter pi), is used more often. This is the angle between the vernal equinox and the point of perihelion. It is equal to the longitude of the ascending node plus the argument of perihelion.

The first temporal orbital element is period (P), the time required for an object to complete one orbit. This is not an independent quantity, but it can be derived from the semi-major axis using Kepler's third law. Measuring period directly presents some difficulty, because the base for making observations (Earth), as well as the planet or other object, are both in motion. This must be taken into account, as orbital period is the time to return to the starting point in space along the orbit, not the time between alignments with the Earth. In practice, measurements of periods in the solar system use the distant stars as a nonmoving reference frame, and the period is also called **sidereal period** (from the Latin word sidus, meaning "star" or "constellation").

The orbital elements considered so far might be compared to a description of an automobile race course. The size and shape of the course, as well as the time required for a driver to complete a lap, can be accurately described. Suppose we wish to determine the location of a driver's car at any given moment. The course description and the lap time are not enough information. We also need to know the driver's location at any one given time in order to calculate the location at other times. Orbits are similar and require a final element, **time of perihelion passage** (T), which is usually expressed as a date. From this date, when the object was at perihelion, past and future perihelion passages can easily be calculated using the period. In addition, more complex calculations can be made to determine where the object will be in its orbit and where it will appear in Earth's sky at any time.

One of the more important practical problems of celestial mechanics is the determination of the orbital elements of newly discovered objects, such as asteroids and comets. The mathematical techniques for orbit determination based on three positions in the sky measured at reasonably separate intervals were developed in the 18th century, with the German mathematician **Karl F. Gauss** (1777–1855) perfecting the technique just after 1800. Today, computers greatly speed the process by carrying out such computations very quickly, once three positions of a newly found object have been recorded.

If only the Sun and the orbiting object were present, an object's orbital elements would not vary with time. Unfortunately, solar system motions are complicated by the fact that gravitational interactions of other bodies influence planetary movements and change orbital elements. These interactions, called **perturbations,** require that all of the other planets must be taken into account when calculating the position of any planet. Some theoretical calculations indicate that certain planets have elements that may vary cyclically. For example, the eccentricity of Mars' orbit is believed to vary from 0.004 to 0.141 over a

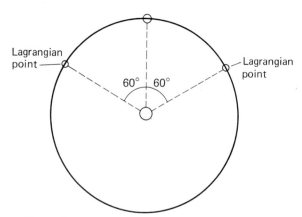

■ **Figure 3.14** Lagrangian points. If an object (*top*) is orbiting another object (*center*), the Lagrangian points lie 60 degrees ahead of and behind the orbiting object.

period of 10^5 to 10^6 years (the current value is 0.093) as a result of Jupiter's influence.

When an object is in an orbit, normally no other object can share that orbit. However, **J.L. Lagrange** (1736–1813) calculated that there are points in an orbit where another body may be located. There are two such relatively stable **Lagrangian points**, as they are called, 60° ahead of and behind an object in its orbit (Figure 3.14). There are several examples of solar system objects located at Lagrangian points. Swarms of asteroids lie at both Lagrangian points in the orbit of Jupiter, and two satellites of Saturn have Lagrangian satellites associated with them. The satellite Dione has a Lagrangian satellite orbiting 60° ahead of it, whereas Tethys has Lagrangian satellites at both positions. Although objects orbit at these Lagrangian positions, they do not remain fixed and may oscillate slightly from the exact 60° position.

■ Bode's Law

In the 18th century, a curious numerical relationship was deduced that seemed to indicate that the values of the semi-major axes of the planets are not random. First developed by **Johann Titius** (1729–1796) in 1766 and later popularized by **Johann Bode** (1747–1826), this relationship is now called **Bode's law**, even though Bode did not develop it and it is certainly not a "law" in the strict sense of the word.

Bode's law is based upon the sequence of numbers 0, 3, 6, 12, 24, 48, 96, 192, 384, and 768. (Note that each number after 3 is obtained by doubling the preceding one.) If four is added to each number and the result is divided by ten, the sequence becomes 0.4, 0.7, 1.0, 1.6, 2.8, 5.2, 10.0, 19.6, 38.8, and 77.2. This is remarkably close to the semi-major axes (in astronomical units) of the orbits of the planets known in the 1760s: Mercury, 0.39; Venus, 0.72; Earth, 1.0; Mars, 1.52; Jupiter, 5.2; and Saturn, 9.5. (Table 3.2 shows these values in tabular form, and Figure 3.15 illustrates the actual spacing of the planets.) This relationship suggested that something was "missing" between Mars and Jupiter, and after the discovery of Uranus at 19.2 astronomical units in 1781 strengthened the belief of astronomers that the law had some validity, a search for the missing object began. The object, the first asteroid, subsequently named Ceres, was found accidentally by someone not involved with the search, although the searchers eventually found other asteroids in similar orbits. Neptune and Pluto "break" Bode's law, because their semi-major axes are 30.1 and 39.4 astronomical units, respectively. (Although Bode's law is not completely valid, it can serve as an aid for remembering the semi-major axes of the inner planets.)

One today might wonder why such importance was placed on this relationship. The reason is that numerology was taken very seriously by many scientists several centuries ago. (For in-

■ **Table 3.2** Bode's Law

0 to 3, Then Doubled	Add 4	Divide By 10	Actual Semi-Major Axis (AU)*	Object
0	4	0.4	0.39	Mercury
3	7	0.7	0.72	Venus
6	10	1.0	1.0	Earth
12	16	1.6	1.52	Mars
24	28	2.8	2.77	Ceres
48	52	5.2	5.20	Jupiter
96	100	10.0	9.56	Saturn
192	196	19.6	19.22	Uranus
384	388	38.8	30.11	Neptune
768	772	77.2	39.44	Pluto

* AU = astronomical units.

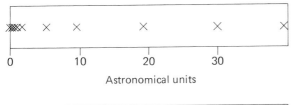

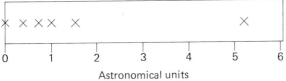

■ **Figure 3.15** Planetary spacing. The *top* figure indicates the location of the Sun and all planets, and the *bottom* figure shows the Sun and the planets out to and including Jupiter.

stance, Kepler predicted that Mars had two satellites, because the Earth had one and Jupiter had four!) Although no one has satisfactorily explained why this spacing of planetary orbits occurs, some theoretical models of solar system formation predict such spacings. It will certainly be interesting to discover whether similar relationships hold in planetary systems around other stars, if they exist.

■ The Planets in the Sky

Because the planets and other solar system objects are all orbiting the Sun and the Earth is also moving, the apparent motions of these objects in the sky are more complicated than might be expected. A good example is something as basic as orbital period. As mentioned earlier, the sidereal period of an object is the time required for it to complete one orbit, using the fixed stars as a reference. Because the Earth and planets are all moving, the time between a planet's close approach to a star in the sky and its return to that point after completing a circuit around the sky is not a sidereal period, because the Earth has moved and changed our line of sight to the star.

One quantity that is often used to describe planetary motion in the sky is the **synodic period**, the time required between successive alignments of the Sun, Earth, and planet in question (Figure 3.16). Suppose the Sun, Earth, and Jupiter were in a straight line, with Earth in the middle. In that case, called **opposition**, Jupiter would reach its

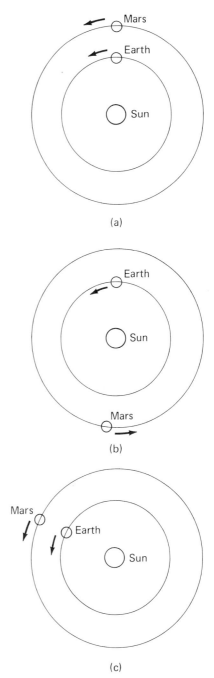

■ **Figure 3.16** Synodic period. In (a), Mars is at opposition as seen from the Earth. In (b), the Earth has completed an orbit one year later, but Mars has not yet completed an orbit. In (c), slightly over two years after (a), Mars is again at opposition as seen from the Earth.

highest point in the sky at midnight. Although Jupiter requires 12 years to orbit the Sun, the three objects would be aligned in opposition again in about 13 months. During this time, Earth would have completed one orbit plus one-twelfth, whereas Jupiter would have completed just over one-twelfth of its orbit. Therefore, we say that Jupiter's synodic period is 13 months.

Planets between the Earth and Sun are called **inferior planets**, whereas those beyond Earth are called **superior planets**. The relationships between sidereal period (*P*) and synodic period (*S*) were derived by Copernicus and are given by

$$\frac{1}{P} = 1 + \frac{1}{S}$$

for inferior planets, and by

$$\frac{1}{P} = 1 - \frac{1}{S}$$

for superior planets, where *P* and *S* are both measured in years.

Many other terms are used to describe various planetary locations in the sky (Figure 3.17). When a superior planet is diametrically opposed to the Earth, on the other side of the Sun, it is said to be in **conjunction**. When a superior planet is 90° from the Sun in the sky, it is said to be at **quadrature**. A planet at eastern quadrature will

be at its highest point in the sky at sunset; one at western quadrature, at sunrise. Inferior planets can be in conjunction either between Earth and Sun (inferior conjunction) or beyond the Sun (superior conjunction). Opposition cannot occur, but when an inferior planet is at its farthest angle from the Sun in the sky, it is said to be at **greatest elongation**. Mercury and Venus can be seen best at this time, when they rise earliest before sunrise or set latest after sunset, when they are at western and eastern elongation, respectively. (A planet's **elongation** is defined as the angle between it and the Sun as viewed from the Earth.)

Several phenomena occur because of various alignments among solar system objects (Figure 3.18). The first of these are **phases**, the changes in the apparent shape of an object due to the changing positions of it, the Sun, and the Earth. Lunar phases are visible to the naked eye, but a telescope is required to see planetary phases. Phases are most apparent in inferior planets, but Mars and Jupiter also exhibit less noticeable phase changes. Lunar phases are discussed in detail in Chapter 15.

Eclipses occur when one solar system body passes into the shadow of another. A familiar example occurs when the Moon's shadow touches the Earth, producing a solar eclipse at the point of shadow contact. A lunar eclipse occurs when the

■ **Figure 3.17** Special planetary configurations for superior planets (*left*) and inferior planets (*right*). (From Abell, George, David Morrison, and Sidney Wolff. *Exploration of the Universe*, 5th edition. Saunders College Publishing, Philadelphia, 1987.)

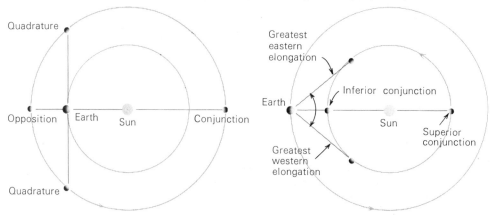

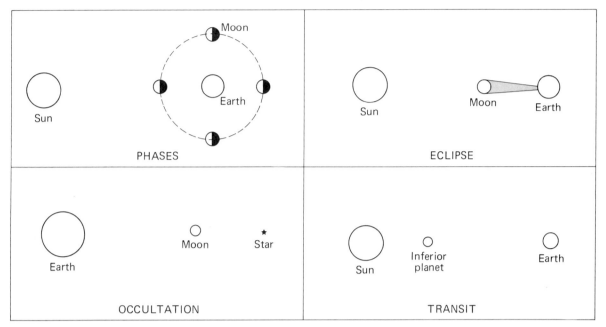

■ **Figure 3.18** Mutual phenomena. Phases involve changing views of an illuminated object. An eclipse occurs when an object's shadow touches another object. An occultation occurs when an object blocks the view of a star or planet. A transit occurs when an inferior planet passes between Earth and Sun.

Moon travels into the shadow of the Earth. These eclipses are rather infrequent because the Moon's orbit around Earth is inclined several degrees to the ecliptic. Eclipses involving planets and their satellites are visible telescopically. The small shadows of Jupiter's satellites can occasionally be seen on the planet's disk, and the satellites can be seen to disappear from view when their orbits carry them into Jupiter's shadow. In 1675, Olaus Roemer (1644–1710) demonstrated that the speed of light was finite by noting that the observed times of Jovian eclipse events varied from times predicted by celestial mechanics. The travel time of the light was responsible for the discrepancies.

An **occultation** occurs when the Moon passes in front of a star or planet, or when any other solar system object passes in front of a star. The fact that objects do not fade slowly when occulted by the Moon but wink out of view instantly was proof to early telescopic observers that the Moon lacks an atmosphere. (Modern high-speed detectors can actually measure the small amount of time it takes for stars to disappear behind the Moon and, in so doing, can determine the diameters of some stars.) An important result from occultation studies is measurement of the sizes and shapes of solar system bodies by timing their occultations of stars. The time taken for an asteroid, for example, to pass in front of a star can be measured by several observers, and the diameter and shape of the object can be learned from the length of time each observer sees the star occulted. In 1977, unexpected occultation events just before and after Uranus passed in front of a star revealed the existence of its ring system.

The planets Mercury and Venus occasionally pass directly between the Earth and the Sun, becoming visible as black dots moving across its face. These events, called **transits,** occur rather frequently for Mercury but are rare for Venus (so rare that none occur between 1882 and 2004). Timing of transit events was at one time the most accurate method for determining the length of the astronomical unit, but this technique has been supplanted by bouncing radar beams from the surfaces of other planets.

■ Chapter Summary

Most early models of the solar system proposed a geocentric configuration. In the 1500s, Copernicus became the first modern proponent of the heliocentric system. Using Tycho's detailed planetary observations, Kepler determined that planets orbit the Sun and derived his three great laws of orbital motion. Galileo, using the newly invented telescope, provided observational evidence for the heliocentric theory, whereas Newton developed three laws of motion and the law of gravity, which provide the theoretical basis for understanding orbital dynamics.

Orbits can be described using a set of quantities called orbital elements, and the location of a planet at any time can be calculated. A peculiar relationship, Bode's law, predicts the spacing of the orbits of many of the planets. A number of specific terms are applied to various locations and alignments of planets with respect to the Earth and the Sun.

■ Chapter Vocabulary

aphelion
argument of perihe-
    lion
Aristarchus
Aristotle
Bode, Johann
Bode's law
celestial mechanics
centripetal force
conic section
conjunction
Copernicus, Nicholas
deferent
eccentricity
eclipse
ellipse
elongation

epicycle
escape velocity
focus point
Galileo
Gauss, Karl F.
geocentric concept
geosynchronous
    satellite
gravity
greatest elongation
heliocentric concept
hyperbola
inclination
inertia
inferior planet
Kepler, Johannes
Kepler's laws

kinetic energy
Lagrange, J.L.
Lagrangian points
line of nodes
longitude of ascend-
    ing node
Newton, Isaac
Newton's laws of
    motion
node
occultation
opposition
orbital velocity
orbital elements
parabola
perihelion

perturbations
phases
potential energy
Ptolemy
quadrature
retrograde loop
semi-major axis
sidereal period
superior planet
synodic period
time of perihelion
    passage
Titius, Johann
transit
Tycho
vernal equinox

■ Review Questions

1. Describe the two major theories of solar system configuration and orbital motion.
2. If you could go back in time and talk to Ptolemy, how would you try to convince him that the geocentric concept was incorrect?
3. Explain the contribution of each of the following astronomers to the development of the heliocentric concept: Copernicus, Galileo, Tycho, Kepler.
4. Sketch an ellipse and label its important properties. Describe what a conic section is.
5. State Kepler's three laws of planetary motion.
6. Explain the importance of kinetic and potential energy in understanding orbits.
7. What procedure did Newton use to prove that Kepler's laws were correct? How did he refine these laws?
8. Summarize Newton's laws of motion and the law of gravitation.
9. Explain what is meant by orbital velocity and escape velocity.

10. List the seven orbital elements, and explain what each describes.
11. What is a perturbation, and how is it caused?
12. Describe Lagrangian points and how they are important in the solar system.
13. State Bode's "law," and explain its historical consequences.
14. Distinguish between sidereal and synodic periods.
15. What phenomena are visible in the Earth's skies as a result of the orbital motions of the Earth and the other bodies of the solar system?

■ For Further Reading

Berry, Arthur. *A Short History of Astronomy.* Dover, New York, 1961. (Reprint of 1898 edition.)

Dreyer, J.L.E. *A History of Astronomy from Thales to Kepler.* Dover, New York, 1953. (Reprint of 1906 edition.)

Durham, Frank and Robert D. Purrington. *Frame of the Universe.* Columbia University Press, New York, 1983.

4

Observing the Solar System: Light, Spectroscopy, Telescopes

CHAPTER OBJECTIVES

After studying this chapter, you should be able to

1. Understand the particle, wave, and photon concepts of light.
2. Describe the basic properties of light, such as brightness, wavelength, and frequency.
3. Describe the forms of light found in the electromagnetic spectrum and their contributions to observational astronomy.
4. Understand Wien's law, the mathematical relationship between the temperature of an object and the wavelength at which it emits the most light.
5. Describe the parts of an atom and the role electrons play in spectroscopy.
6. Distinguish among continuous, absorption, and emission spectra, and explain their causes.
7. Understand the Doppler effect.
8. Explain the functions performed by telescopes, and describe basic telescope types.

To an astronomer, "light" means not only the visible form, but invisible forms such as radio. Here, the 140-foot radio telescope at the National Radio Astronomy Observatory in Green Bank, West Virginia, studies emission from Comet IRAS-Araki-Alcock in May, 1983. Radio telescopes can be used day and night, unlike optical ones. (Photograph by the author.)

As astronomers survey celestial objects from the nearby Moon to the farthest galaxies, the "raw material" reaching them from space is light. With the exception of the spacecraft that have landed on the Moon, Mars, and Venus, all astronomical study of the solar system has been done by **remote sensing**, or observation from a distance. It is light that carries information about celestial objects to Earth, so a thorough understanding of light, its properties, and how it is studied using telescopes and spectroscopes is essential in astronomy.

Most of the light that we see in the solar system comes from the Sun, either directly or indirectly by reflection. (There are a few exceptions, such as lightning, auroras, and the glow of meteors.) Because we view many objects by reflected light, this chapter also discusses the reflection process.

■ What Is Light?

Although light is certainly familiar, it is somewhat difficult to describe and fully understand. Light may be described as radiant energy produced by a source, but that is only part of the story. We are used to thinking of light as something we can see, but there are many forms of "light" that our eyes cannot detect.

If we flip a switch and turn on the lights in a room, what is happening that was not occurring before? Can we catch and examine the light and determine its composition? Until the 20th century, physicists were divided in their description of light. Some felt that it was composed of particles and pointed to some of light's observed properties to support this idea. Others felt that light was a type of wave motion, similar to sound traveling through air or earthquake waves traveling through the Earth. Some of light's properties are best explained using wave theory.

The modern idea of light, developed by physicists in the first part of the 20th century, emphasizes its dual nature. Light is now considered to be composed of **photons**, which can be visualized as packets of wave energy with some particle-like properties. Accordingly, it still makes sense to discuss many of light's properties using wave terminology. The number of photons in light is exceedingly large. For example, at a distance of 1 centimeter from a 100-watt light bulb, there are 1 billion photons in every cubic centimeter! Light travels at the exceedingly fast speed of 300,000 kilometers per second (186,000 miles per second) in a vacuum. (When light travels through a medium such as air, water, or glass, its speed decreases.) The **speed of light** is an important physical quantity and is designated as c.

■ Properties of Light

Anyone who has thrown a rock into water is familiar with waves, which in this case travel outward from the point where the impact occurs. Let us use the wave analogy to explain light (Figure 4.1). Although some of our discussion will refer specifically to visible light, the same concepts apply to other forms.

The most noticeable property of light is its brightness, or intensity. The brighter the light, the greater the **amplitude** (height) of the light waves. The reason brightness variation occurs is that some light sources are more energetic than others and produce higher-amplitude waves, just as larger rocks thrown into a pond might be expected to make larger waves than smaller rocks would.

Astronomers measure the brightness of celestial objects using a quantity called **magnitude.** The magnitude scale was developed in the second century B.C. by the Greek astronomer **Hipparchus,** who called the brightest stars magnitude 1 and the faintest, magnitude 6. Although this early classification was arbitrary, physicists invented devices in the 19th century that could accurately measure the amount of light energy detected by a telescope. These detectors showed that the light of a first-magnitude star was about 100 times more intense than that of a sixth-magnitude star. The magnitude scale was therefore refined so that a five-magnitude difference corresponds exactly to a hundredfold increase or decrease in light intensity. This allowed the magnitude scale to be extended beyond the original limits of 1 and 6. Faint objects, visible only telescopically, have magnitudes beyond 6. For example, a magnitude-11 star is 100 times fainter than a magnitude-6 star, or 10,000 (100 times 100) times fainter than one of magnitude 1. The faintest stars visible through the largest telescopes are fainter than magnitude 20! Conversely, objects brighter than the brightest stars are given magnitudes with

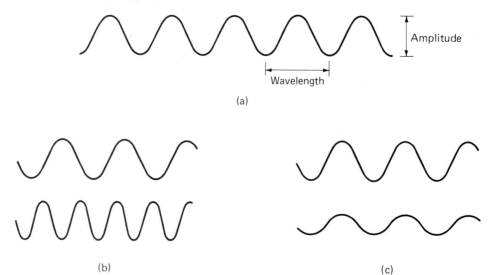

(a)

(b)   (c)

■ Figure 4.1 Wave properties. (a) Wavelength and amplitude. (b) Waves with the same amplitude but different wavelengths. (c) Waves with the same wavelength but different amplitudes.

zero and negative values. Venus at its brightest is about magnitude −4; the full Moon, −12; and the Sun, −27.

An important characteristic of visible light is its color, which is determined by **wavelength.** Visible light ranges from violet, with wavelengths about 400 nanometers, to red, with wavelengths near 700 nanometers. (The nanometer, frequently used to describe light wavelength, is 10^{-9} meters.) There are many forms of light of longer or shorter wavelengths we cannot see. These include ultraviolet, X-rays, and gamma rays (shorter than violet) and infrared, microwave, and radio (longer than red).

A quantity directly related to wavelength is **frequency**, the number of waves passing a given point each second. (Frequency is given in units of **hertz** [Hz] or waves per second.) Frequency (v, the lower-case Greek letter nu) and wavelength (λ, the lower-case Greek letter lambda) are related by the equation

$$c = \lambda v.$$

This means that if either frequency or wavelength is known, the other can be calculated. For example, yellow light with wavelength of 600 nanometers has a frequency of 5×10^{14} Hz. Longer-wavelength light has lower frequency, and shorter-wavelength light has higher frequency.

Every photon has a specific energy that is determined by its wavelength or frequency and given by the equation

$$E = hv = \frac{hc}{\lambda},$$

where h is a numerical constant. Therefore, the shorter the wavelength of a photon (or the higher its frequency), the higher its energy. The photons of ultraviolet light, for example, are able to burn the skin because they have more energy than those of visible light.

Another property of light is its **polarization**, which is the degree to which the direction of wave motion in a beam of light is aligned. Water oscillates up and down in its waves, but when light travels through air or space, vibration may be up and down, side to side, or anywhere in between. Some physical processes (reflection from a surface is an example) remove light of all vibration directions but one, producing polarized light. Polarized sunglasses reduce glare from horizontal surfaces, which polarize light reflected from them in the horizontal direction. Because these sunglasses transmit only light that is polarized vertically, they eliminate glare. Examination of the polarization of light reflected from a planetary surface can give insight into the nature of the surface, because the amount of polarization will depend upon the composition and roughness of the surface.

■ The Electromagnetic Spectrum

As mentioned earlier, there are forms of light we cannot see with our eyes. The full range of light, illustrated in Figure 4.2, is called the **electromagnetic spectrum** because the propagation of light waves entails both electric and magnetic effects. The term spectrum refers to the breakdown of light into its component parts and is used in describing the rainbow or any other situation in which light is resolved into its colors. The term **electromagnetic radiation** is often used to describe both visible and invisible light when no distinction between them is being made.

A light-emitting object will typically produce several types of light. Some celestial objects that are completely invisible in normal light emit copious amounts of other light that can be detected and studied. With the exception of some ultraviolet and infrared light, and most radio/microwave, the Earth's atmosphere blocks nonvisible light, preventing astronomical observation in these regions (Figure 4.3). Since the start of the space age in the late 1950s, a number of astronomical observatory satellites have been placed into orbit above the atmosphere, allowing us to study the universe using these new wavelengths. The results have greatly increased our understanding of the universe.

The most extensive gamma-ray and X-ray observations were conducted by the three High Energy Astrophysical Observatories (HEAOs) orbited during the 1970s. The majority of the high-energy sources that they observed were hot or unusual objects in our galaxy or beyond. The Sun emits X-rays, although not as many as some other

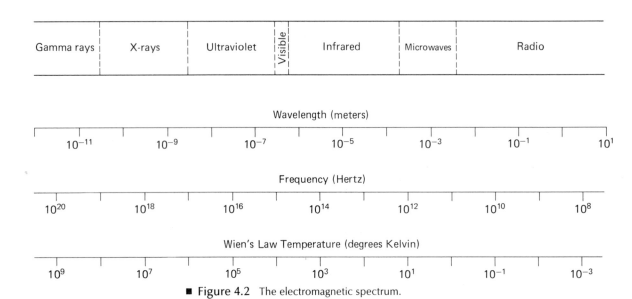

| Gamma rays | X-rays | Ultraviolet | Visible | Infrared | Microwaves | Radio |

Wavelength (meters)

| 10^{-11} | 10^{-9} | 10^{-7} | 10^{-5} | 10^{-3} | 10^{-1} | 10^{1} |

Frequency (Hertz)

| 10^{20} | 10^{18} | 10^{16} | 10^{14} | 10^{12} | 10^{10} | 10^{8} |

Wien's Law Temperature (degrees Kelvin)

| 10^{9} | 10^{7} | 10^{5} | 10^{3} | 10^{1} | 10^{-1} | 10^{-3} |

■ **Figure 4.2**   The electromagnetic spectrum.

stars do, and the solar observatory that was part of the Skylab manned mission in 1973–1974 extensively studied the Sun in X-ray light. When solar X-rays hit planets, secondary X-rays are produced. Secondary X-rays from the Moon were analyzed from orbit during two Apollo lunar missions. These missions also detected natural X-rays produced on the Moon by the decay of radioactive elements.

The detection and study of far-ultraviolet radiation have been conducted by several spacecraft, most notably the International Ultraviolet Observer (IUE), launched in 1978. Most of the objects studied by IUE are beyond the solar system, because the Sun is too bright to be examined and many small solar system objects are too faint. The IUE observatory has been used to observe the Moon, planets, and comets. Ultraviolet re-

■ **Figure 4.3**   Atmospheric transmission. The *arrows* indicate whether light from various portions of the spectrum passes through the atmosphere to the Earth's surface. (Length of *arrow* indicates degree of transmission. The *dotted line* represents the top of the atmosphere.)

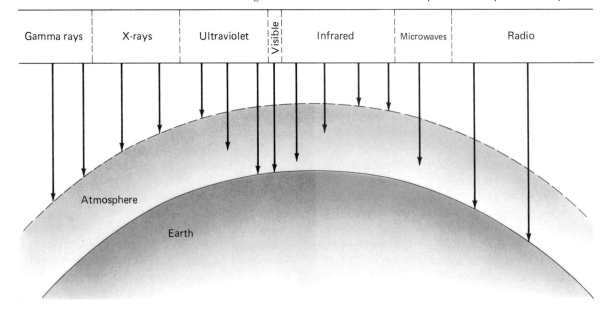

| Gamma rays | X-rays | Ultraviolet | Visible | Infrared | Microwaves | Radio |

Atmosphere

Earth

search is particularly useful in the study of planetary atmospheres, but it also gives information about the composition of solid planetary bodies.

Although significant infrared observations can be made from Earth-based observatories, water vapor in the atmosphere hinders observations and blocks some infrared wavelength areas entirely. In 1983, the Infrared Astronomy Satellite (IRAS) was launched, and it spent nearly one year surveying the entire sky to detect new infrared sources and study those already known. The Infrared Astronomy Satellite discovered and studied several new comets, asteroids, and other solar system objects.

Radio astronomy was born in the 1930s, when **Karl Jansky** (1905–1950) detected radio interference coming from the Milky Way. Although most radio sources are beyond the solar system, the Sun and Jupiter are both powerful sources. In addition to passive studies of solar system sources, active studies have been made using **radar**, a process by which radio radiation is beamed toward a solid object and the reflected signal studied. (Radar, originally developed to detect aircraft during World War II, is an acronym for "radio detection and ranging.") Radar yields much information about solar system objects; it is used to measure distance by timing its outward and return times, because the speed at which it travels is accurately known. This is the most accurate way to measure the length of the astronomical unit. When radar is bounced from a planet, other information can be obtained, most notably rate of rotation. This is because radar allows detection of speed by means of the Doppler effect, which we discuss later in the chapter. This technique is useful for objects whose surface is either obscured or difficult to see clearly. Radar also allows the characterization of the surface from which it has been reflected, because a rough surface and a smooth surface will reflect differently. In 1976, radar measurements were used to select landing sites for the Viking spacecraft on Mars after the first choices were seen from orbit to be too rough. Finally, radar can map surface features of a cloud-covered planet such as Venus, because radio waves can easily penetrate thick cloud cover. Such mapping of Venus has been done both from radio telescopes on the Earth and from orbit around the planet.

Although an object may emit many of the forms of light mentioned previously, it will emit more of one form than any other. For example, the Sun emits more visible light than other kinds. All colors are emitted, as evident from a rainbow, yet the Sun appears yellow in the sky. The reason for this is somewhat complicated, because the Sun's light output is actually greatest in the green portion of the spectrum. However, the lesser amounts of blue and red light emitted combine with the green light, and the Sun's color is further modified by the passage of sunlight through the atmosphere, which subtracts blue light.

There is a mathematical relationship between the temperature of an object, T, and the wavelength at which it emits the most light, λ_{MAX}. Called **Wien's law**, after **Wilhelm Wien** (1864–1928), it states that

$$\lambda_{MAX} = \frac{C}{T},$$

where C is a numerical constant. (If λ_{MAX} is expressed in nanometers and T in degrees Kelvin, C has the value 2.898×10^6 K-nm.) In other words, the hotter the light-emitting object, the shorter the wavelength at which it emits the most energy (bluer, if visible light). This means that a star cooler than the Sun will appear orange or red, whereas a hotter one will be blue. The Earth, which is even cooler, emits light, but the maximum wavelength is in the infrared region.

■ Atoms and Spectroscopy

Although we have discussed light in terms of stars and planets, to understand many of the processes involving light we need to descend to the smallest level and examine the atom itself. Although atoms and their components are complicated and a complete description would be rather involved, we can briefly summarize atomic properties pertinent to understanding light.

An **atom** can be visualized as a central region (the **nucleus**) around which a swarm of **electrons** are orbiting, somewhat like a miniature solar system (Figure 4.4). The nucleus is the most massive portion of the atom and is composed of particles with positive charges (**protons**) and neutral particles (**neutrons**). The nucleus is very small compared to the electron orbits around it. Electrons

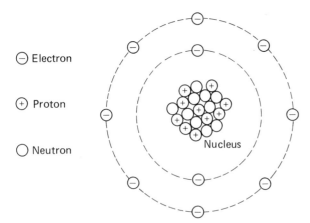

⊖ Electron

⊕ Proton

○ Neutron

Nucleus

■ **Figure 4.4** Parts of the atom. Protons and neutrons are located in the nucleus, whereas electrons orbit it.

have negative charges, and in a neutral atom, their number will be equal to that of the protons. (Some atoms, called **ions**, have either lost or gained electrons and, therefore, have a net electric charge.) Although we speak of electrons "orbiting" the nucleus, there are major differences from planetary orbital motion. First of all, although the planets can theoretically orbit the Sun at any distance, electron orbital levels are fixed and separated by gaps at which no electrons occur. Secondly, more than one electron can occupy a given orbital level. Finally, an electron can "jump" to a higher orbital level if it absorbs outside energy and fall to a lower level if it emits energy. An electron that is occupying a higher orbital level than normal is said to be excited. Although these orbital concepts are helpful in understanding activity within atoms, modern physics describes electrons as having a probability of being in a given location at a given time, rather than predicts exact locations, as in celestial mechanics. However, the details of this probability study are not essential to a basic understanding of light and the atom.

Although there are many sources of the various forms of electromagnetic radiation, all of them involve processes at the atomic or nuclear level. Let us consider just a few of the ways by which light can be produced. First are interactions among atoms, or chemical reactions, exemplified by fires, fireworks, and fireflies. Light can also be produced by electronic interactions, such as electricity, which is electron flow. In an incan-

descent lamp, the electric energy passing through the tungsten filament causes the electrons in the tungsten to jump into higher energy levels, from which they drop, emitting light. A fluorescent bulb operates similarly, but instead of exciting atoms in a filament, the gaseous material filling the tube is made to emit ultraviolet light, which excites the electrons in the atoms of the substance, called a phosphor, which coats the inside of the tube. When the excited electrons in the phosphor return to their normal states, the light emitted causes the tube to glow.

Most of the starlight we see in the night sky was originally produced by reactions involving nuclei of atoms. These take place primarily at the centers of the stars, where nuclear fusion occurs as hydrogen nuclei combine to form helium nuclei. Light can be produced in several other ways, but because they are only important outside the solar system, we will not discuss them.

As discussed earlier, atoms are made of protons, neutrons, and electrons. The various types of atoms, distinguished by their characteristic numbers of protons (and electrons, in neutral atoms), are called **elements**, and about 92 of them occur in nature. One of the goals of astronomical study is determining the composition of celestial objects, and **spectroscopy**, the study of the light that they either emit or reflect, is the main tool for compositional research. The basis for spectroscopy is the fact that electrons in atoms can absorb energy and jump up to a higher orbital level. This energy typically is obtained from light that the atoms intercept. All elements have different numbers of electrons, and, more importantly, the spacing between electronic energy levels is different for each element.

When an electron absorbs light and jumps to a higher energy level, exactly the right energy is needed. The situation is similar to climbing a ladder, where you must raise your foot precisely the distance to the next rung; therefore, the electrons will absorb only those photons with the exact energy to raise them to higher levels (Figure 4.5). By studying various materials in the laboratory, chemists have determined which wavelengths of light are absorbed by each of the elements.

The Sun was one of the first celestial objects studied by spectroscopy. If a glass prism or a diffraction grating (a piece of glass or plastic with

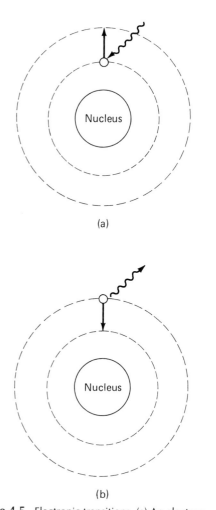

(a)

(b)

■ **Figure 4.5** Electronic transitions. (a) An electron absorbs incoming light and jumps to a higher orbital level. (b) An excited electron emits light and drops to a lower level.

inscribed lines spaced a few thousandths of an inch apart, which is about the wavelength of visible light) is placed in a beam of sunlight, it will disperse the light into its component colors. This experiment was done by Newton, but in the early 19th century, experimental apparatus became more sophisticated and the solar spectrum was seen to be more complicated. Numerous black lines occurred as gaps in the continuous "rainbow" of sunlight. **Joseph Fraunhofer** (1787–1826) cataloged these lines in 1815, and they are now called **Fraunhofer lines.** The explanation of the lines (Figure 4.6) is in accordance with what we discussed earlier: A **continuous spectrum** of light containing light of all wavelengths is produced in the Sun's interior. However, the various elements present in the Sun's partially opaque outer atmospheric layers absorb some of the light by electronic transitions, producing the dark lines. For example, electrons in hydrogen atoms will produce an absorption line at 122 nanometers, if they jump from the first to the second orbital levels, or at 656 nanometers, if they jump from the second to the third. Our modern knowledge of the Sun's composition comes from the identification of the elements responsible for these lines (Figure 4.7). The abundance of various elements can also be determined, because wider lines mean that more absorbing atoms are present. This type of spectrum is called an **absorption spectrum** because intervening material has absorbed some light between the source and the observer.

■ **Figure 4.6** Absorption spectrum. Light passing through the solar atmosphere is subject to the absorption process, and its spectrum will contain dark lines where light was removed.

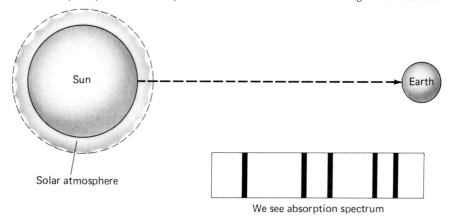

We see absorption spectrum

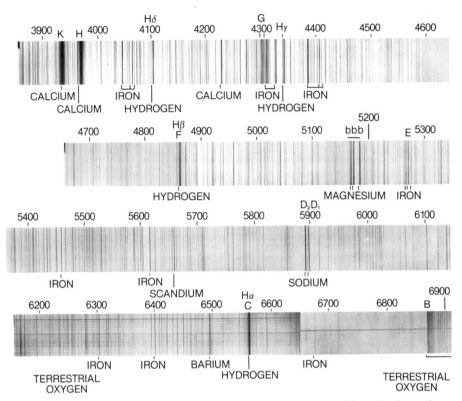

■ **Figure 4.7** Example of an absorption spectrum: the spectrum of the solar photosphere. Elements responsible for the major absorption lines are identified. (Photograph courtesy of the Observatories of the Carnegie Institution of Washington.)

If a cloud of gas exists near a star, many of the electrons in its atoms will be excited as they absorb light from the star. If we look at the cloud, or nebula, it will often appear bright, even away from the line of sight to the star. Examination of the gas's spectrum will reveal a completely different type of phenomenon (Figure 4.8): The spec-trum will be mostly dark, with some bright lines at separated intervals. This **emission spectrum**, as it is called, is caused by excited electrons dropping down to lower energy levels. When they do so, the light that they previously absorbed from the star is re-emitted in all directions. Emission spectra are visible from some solar system objects,

■ **Figure 4.8** Emission spectrum. When the Sun is covered during an eclipse, only its outer atmosphere can be seen. The light coming from the outer atmosphere is produced by the emission process, and its spectrum will consist of bright lines.

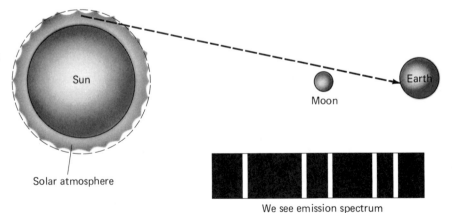

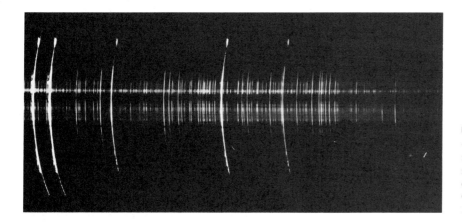

particularly comets, and the outermost atmosphere of the Sun, when the Sun itself is covered during an eclipse (Figure 4.9).

When visible light illuminates a planet or other object, the various light-absorbing processes that occur in atoms near its surface permit only a fraction of the incident light to be reflected. The percentage of the incident light that a surface reflects is called its **albedo**; for example, black objects have very low albedos, whereas white ones have high albedos. Many people prefer to wear light-colored clothing in the summer because it absorbs less light, making the wearer feel cooler. The albedos of solar system objects vary widely, from a few percent for the darkest asteroids and planetary satellites to well over 90% for bright, ice-covered objects.

Because solar system objects other than the Sun shine only by reflected sunlight, they are studied by a technique known as reflectance spectroscopy, which is the analysis of the sunlight reflected from their surfaces. Electrons in the rocks or other surface material present absorb incoming sunlight of certain wavelengths but not of others. As a result, an object will reflect a larger percentage of the light striking it at some wavelengths than at others. By measuring and graphing the percentage of light reflected at each wavelength, reflectance spectra for various solar system objects can be obtained. Compositional studies can be conducted by comparing these spectra with those obtained in the laboratory of materials of known composition. Absorption during reflection is responsible for the colors of most of the objects we see in everyday life. Chemical dyes in clothing, for example, absorb light of certain colors and reflect others. (Color is determined by temperature, as given by Wien's law, only for objects hot enough that their light output peaks in the visible wavelength region.)

■ The Doppler Effect

Composition is not the only information that can be determined using spectroscopy. An object's light also reveals the velocity of its relative motion toward or away from the Earth. The basis for this observation is a phenomenon known as the **Doppler effect**, after **Christian Doppler** (1803–1853), who first described it.

The Doppler effect states that if a source emits waves of a given wavelength, an observer will detect waves of changed wavelength if either the source or the observer is moving (Figure 4.10). In practice, the shift is detected by looking at spectral absorption features whose normal wavelengths are known. If the source and the observer are approaching, the observed wavelength will be shorter than the emitted wavelength, and visible light will appear "bluer" than it actually is. This result is called a blue shift. If the source and the observer are separating, the wavelength will appear longer (redder) than it was when emitted, yielding a red shift. (The terms red shift and blue shift are still used even if nonvisible light is being studied.) The Doppler effect works the same whether the source is fixed and the observer is moving, or vice versa. The Doppler effect holds for forms of wave phenomena other

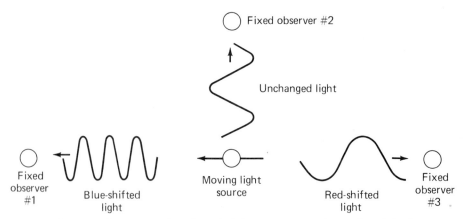

■ **Figure 4.10**   The Doppler effect. If a light source moves from right to left, oberver #1 will see shortened (blue-shifted) waves. Observer #3 will see lengthened (red-shifted) waves. Observer #2 will see unchanged waves, because the motion is neither toward or away from him.

than light, and Doppler shifting of sound waves is a familiar phenomenon. If we hear a train whistle or a siren as it approaches us, passes by, and then recedes, the pitch of the sound will noticeably change. The sound emitted by the whistle or siren is constant, but the change in relative motion affects the pitch we hear.

The magnitude of the observed wavelength shift is directly proportional to the relative motion of the source and the observer. If λ is the normal, nonmoving wavelength, $\Delta\lambda$ is the shift in wavelength (equal to the observed wavelength minus the normal wavelength, v is the velocity of relative motion, and c is the velocity of light, then

$$\frac{\Delta\lambda}{\lambda} = \frac{v}{c}.$$

(This form of the equation is valid in solar system study, but a more complicated relation must be used when the value of v becomes a significant fraction of the velocity of light.)

The most important astronomical application of the Doppler effect is the study of stars and galaxies to ascertain their radial velocities relative to the Sun and Milky Way, respectively. These velocity studies have revealed the stellar orbital motions within our galaxy; they have also enabled us to determine that the entire universe is expanding and at what rate this expansion is taking place. In the solar system, an important application of the Doppler shift is the determination of rates of planetary rotation. If a radio beam is bounced from a planet, the portion of the beam reflected from the side of the planet turning toward us will be blue-shifted, and the part from the side turning away will be red-shifted. The rotational velocity of the planet can be obtained from the magnitude of these shifts (Figure 4.11).

■ Telescopes

Although photons from celestial objects are constantly hitting the Earth, study of them is complicated by two problems. First, very few photons are typically received from celestial objects, because even bright objects appear faint at great distances. Second, the distances of these objects cause their apparent sizes in the sky to be very small. The **telescope** is one of the most important tools of the astronomer, as it overcomes both of these difficulties. Although most of our discussion will deal specifically with **optical telescopes** (those observing visible light), telescopes using other wavelength regions are also used and are based upon the same principles.

A telescope's first function is to gather much more light than the unaided human eye. This is done using either a lens, a mirror, or both. The diameter of the lens or mirror is called the **aperture** of the telescope, and the larger the aperture, the more light is gathered. The main light-gathering lens or mirror of a telescope is called its **objective**. (The term **primary** is often used if the objective is a mirror.) The largest optical telescope in the world, located in the Soviet Union, has an aperture of 6 meters (236 inches). Amateur

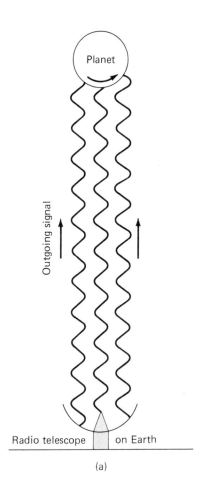

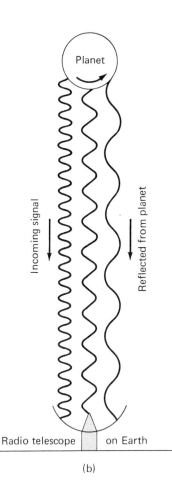

■ **Figure 4.11** Illustration of how the Doppler effect is used to measure the rate of rotation of a planet. In (a), a terrestrial radio telescope sends a signal toward the planet. The reflected signal, shown in (b), contains a blue-shifted component from the side of the planet rotating toward Earth, and a red-shifted component from the side moving away from Earth.

astronomers often use telescopes with 20-centimeter (8-inch) mirrors, which gather 512 times more light than the human eye, whose pupil aperture is less than 1 centimeter. Obviously, telescopes reveal countless stars and other objects too faint to be seen with the naked eye.

The second function of a telescope is the magnification of distant objects. The ability to distinguish two close objects, or two features on the same object, is called **resolution.** The minimum angle that can be resolved by the human eye is about 1 minute of angular measure, whereas telescope resolution can be 1 second or less. (Note that a telescope is said to have *higher* resolution than the human eye, even though the minimum angle that it can resolve is *smaller.*) The actual distance on a planetary surface to which this angular resolution corresponds depends on the object's distance. A telescope's magnification is varied by using different eyepieces, but magnification cannot be increased without limit, be-

cause atmospheric turbulence and distortion are also magnified, and high magnification yields dim images because the incident light is spread over a larger area.

The primary hindrance to telescopic observation is the Earth's atmosphere. Obviously, clouds and bad weather prevent visual observations, but there are problems even on clear nights. (Radio observations, however, can be done during the day and even during light rain.) Two criteria are used in describing the sky's quality. **Transparency** is the amount of light that the atmosphere transmits. Good transparency is essential for studying faint objects, but it is not as important for planets, which are brighter than most other astronomical objects. For bright objects, the major criterion is **seeing**, which describes the steadiness of the atmosphere. Bad seeing, caused primarily by wind and turbulence in the upper atmosphere, distorts images of objects and causes telescopic views to lack sharp-

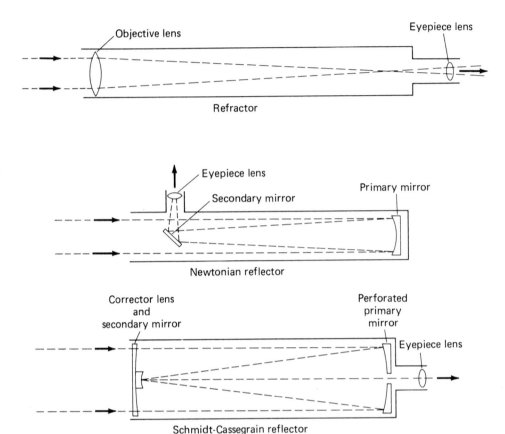

■ **Figure 4.12**   Telescope types. The *broken lines* indicate the paths of the light.

ness. Ideally, an observer hopes for both high transparency and good seeing. The Hubble Space Telescope (HST), with an aperture of 2.4 meters (96 inches), has been placed into orbit far above the atmosphere by the Space Shuttle. Even though it is smaller than the largest-aperture telescopes on the ground, the HST will not suffer from effects of bad seeing and transparency and should outperform them.

Optical telescopes, as mentioned earlier, rely on lenses and mirrors to gather light. (Diagrams of the optical configurations of various types of telescopes are shown in Figure 4.12.) A telescope using a lens to gather light is called a **refractor.** The largest refractor in the world is at Yerkes Observatory in Wisconsin, and it has a 1-meter (40-inch) objective (Figure 4.13). Larger refractors are impractical because the lens, which is supported only at the edge, would sag under its own weight. The largest optical telescopes are **reflectors,** which rely on large mirrors to gather

light. Because the mirror can be supported from behind as well as at its edge, large reflectors such as the 5-meter (200-inch) telescope at Mt. Palomar in California and the 6-meter (236-inch) telescope in the Soviet Union are possible (Figure 4.14). There are several different styles of reflectors. The design problem is that the light is reflected by the primary mirror back in the direction from which it came. A secondary mirror must deflect this final image to a place where it can be seen. A Newtonian reflector has a diagonal mirror near the front of the telescope to reflect the image to the side. A Cassegrain reflector has a secondary mirror that reflects the image back toward the primary mirror, which has a hole cut in the center through which the image passes.

Recent years have seen the development of a number of compound (or catadioptric) telescope designs, which use both mirrors and lenses. The lens is usually placed in front of the telescope, where it bends the light slightly to

■ **Figure 4.13** The 1-meter (40-inch) refractor at Yerkes Observatory, the world's largest refractor. (Yerkes Observatory Photograph.)

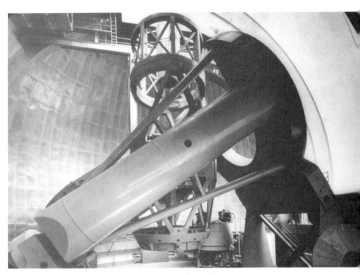

■ **Figure 4.14** The 5-meter (200-inch) reflector at Mt. Palomar Observatory, the world's second largest reflector. (Palomar Observatory Photograph.)

Recent years have seen the development of improved methods for recording telescopic images, including films that record only specific wavelength regions of light and electronic devices that record light directly and greatly amplify its brightness. Electronic detectors do not produce images directly but yield an output consisting of numerical signals. These numbers can be

allow the mirror to produce a better image. The best-known compound telescope is the Schmidt–Cassegrain telescope (Figure 4.15).

From the time of Galileo until the middle of the 19th century, telescopic observations were recorded by sketching what was seen. Although many impressive drawings resulted, the artistic skill of the observer played a role in the results. The development of photography revolutionized astronomy. Not only could an image be recorded independently of any bias on the part of the observer, but a photographic film or plate has a big advantage over the human eye: Light builds up cumulatively, allowing a time exposure to reveal details too faint for the eye to record. This is not always significant in solar system research, because many solar system objects are relatively bright, but it is essential in photographing fainter objects, such as asteroids and comets, as well as most objects beyond the solar system.

■ **Figure 4.15** A 20-centimeter (8-inch) Schmidt–Cassegrain reflector. This type of telescope is commonly used by amateur astronomers. (Photograph by the author.)

■ **Figure 4.16**   The 100-meter (300-foot) radio telescope at the National Radio Astronomy Observatory in Green Bank, West Virginia. This telescope collapsed in 1988 as a result of metal fatigue, but a replacement is planned. (Photograph by the author.)

manipulated by computers, producing enhanced images that reveal details invisible to the naked eye and even photographic film.

Telescopes that observe nonvisible light are nearly all reflectors, because there are no materials that transmit other wavelengths and allow lenses to be constructed. **Radio telescopes** are a good example, and the principle of their operation should be familiar to most people, as they are the "big brothers" of the satellite dishes many people use to receive television broadcasts. The antenna is a large parabolic dish that focuses the incoming signal onto a device that sends it to an amplifier. Whereas satellite receivers look for broadcast programs, radio telescopes look for natural radio signals emitted by celestial objects, most of which sound like static when converted to audio signals.

Because radio waves have much longer wavelengths than visible light, the laws of optics require radio telescopes to have relatively large apertures in order to provide resolution comparable to that of optical telescopes. The largest radio telescope is in Arecibo, Puerto Rico, and it was constructed in a natural bowl-shaped valley about 300 meters (1000 feet) across. Unfortunately, the construction of the Arecibo telescope only allows it to observe approximately straight up. The largest steerable dishes are about 100 meters (300 feet) in diameter (Figure 4.16).

■ Chapter Summary

Light is the raw material for most astronomical observations, and it consists of photons of many forms, collectively called the electromagnetic spectrum. Light can be characterized by brightness, wavelength, frequency, and other properties. Although the Earth's atmosphere transmits mostly visible light, satellites now explore the universe using all wavelength regions. Earth-based radio observations have been especially important in solar system studies.

Spectroscopy allows compositional studies of objects using light received from them. Most spectroscopic features are produced by electronic transitions. Wien's law allows the temperature of a light-emitting object to be determined based upon the wavelength at which it emits the most energy. The Doppler effect allows radial velocity determination from spectroscopy. Astronomers use telescopes to gather light and to magnify the images of distant objects. The basic types of telescopes include refractors, which gather light with lenses, and reflectors, which use mirrors.

■ Chapter Vocabulary

absorption spectrum
albedo
amplitude
aperture
atom
continuous spectrum
Doppler, Christian
Doppler effect
electromagnetic
    radiation
electromagnetic
    spectrum

electron
elements
emission spectrum
Fraunhofer, Joseph
Fraunhofer lines
frequency
hertz
Hipparchus
ions
Jansky, Karl
magnitude

neutron
nucleus
objective
photon
polarization
primary
proton
radar
reflector
refractor

remote sensing
resolution
seeing
spectroscopy
speed of light
telescope
transparency
wavelength
Wien, Wilhelm
Wien's law

■ Review Questions

1. Why is a thorough understanding of light and its properties important in astronomy?
2. What is the definition of light?
3. Describe the basic properties of light, and explain the phenomena responsible for them.
4. What is the electromagnetic spectrum, and what forms of light does it contain?
5. Why has the placement of satellites in Earth orbit greatly increased our knowledge of the universe? Summarize some of the discoveries made by these satellites.
6. What information can radar yield about solar system objects?
7. Explain Wien's law.
8. Summarize the various ways in which light is produced.
9. Explain the formation of absorption and emission spectral lines in space.
10. Explain the Doppler effect and its use in astronomy.
11. What two observational limitations does a telescope overcome? How does the telescope function to do this?
12. Describe the basic types of telescopes.
13. Summarize the techniques used to record and analyze astronomical observations.

■ For Further Reading

Adler, Irving. *The Story of Light.* Harvey House, New York, 1971.
Asimov, Isaac. *Eyes on the Universe: A History of the Telescope.* Houghton Mifflin, Boston, 1975.
Augensen, Harry and Jonathan Woodbury. "The Electromagnetic Spectrum." In *Astronomy*, June, 1982.
Cornell, James and John Carr, editors. *Infinite Vistas — New Tools for Astronomy.* Scribners, New York, 1985.
Cornell, James and Paul Gorenstein, editors. *Astronomy from Space.* MIT Press, Cambridge, MA, 1983.
King, Henry C. *The History of the Telescope.* Dover, New York, 1979. (Reprint of 1955 edition)
Stencil, R. et al. "Astronomical Spectroscopy." In *Astronomy*, June, 1978.

Exploring the Solar System: Space Travel

CHAPTER OBJECTIVES

After studying this chapter, you should be able to

1. Describe the physical principles upon which rocket propulsion is based and distinguish between the two types of rocket engines.
2. List the basic types of space exploration missions and which solar system objects have been explored by each.
3. Describe the types of instrumentation carried by spacecraft designed to explore various solar system objects.
4. Summarize arguments for and against manned spaceflight.

One of the greatest stimuli of solar system study has been the development of space exploration in recent decades. Just as increasingly better telescopes seemed to bring the planets closer to Earth-based observers, space technology now allows us to reach out into the solar system and observe planets and other objects firsthand.

■ Rocket Propulsion

Although calculations of orbits and interplanetary trajectories have been possible ever since the time of Newton, and fictional accounts of such travel have existed for many years, space travel did not become a reality until the mid-20th century. This is because all of the techniques of transportation used on the Earth's surface or in the atmosphere require oxygen and will not function in the vacuum of space. Only recently have technological breakthroughs allowed travel in space.

The key to space exploration is the **rocket** engine, which is the only form of propulsion capable of functioning in a vacuum. The rocket was invented nearly a thousand years ago by the Chinese, and it was used primarily as a weapon until this century. All of the rockets used in the early part of the space age were based on military weapons systems, which in turn were derived from the German V2 rocket used during World War II. Following the war, most German rocket scientists were taken to the United States and the Soviet Union, where they provided the expertise upon which these nations' space programs were founded.

The theory of rocket operation is very simple and is based on Newton's third law of motion (Figure 5.1). A chemical reaction shoots a jet of hot gases out through a nozzle at the rear of the rocket. As a result of the force exerted by the rocket's engine, the entire rocket is pushed in the opposite direction. The force exerted by a rocket

■ **Figure 5.1** How a rocket functions, as demonstrated by the liftoff of a small model rocket. The engine forces hot exhaust gas downward, with force F_1. By Newton's third law, the entire rocket moves in the opposite direction, upward, as a result of the reactive force F_2. (Photograph by the author.)

engine is called its **thrust**, which is measured using newtons or pounds. The thrust is typically larger than the weight of the rocket, and its value may vary during the firing of the rocket's engine. (For example, a rocket's thrust will be greater in a vacuum than in the atmosphere, all other conditions being equal.)

Rocket engines are powered in two ways (Figure 5.2). The earliest rockets were **solid-fuel** varieties, in which the fuel burns continuously from the moment of ignition until it is expended. Solid rocket fuels are formulated from chemicals that burn even in a vacuum, because oxygen to support combustion is contained within them. Examples of solid-fuel rockets range from the small ones used in model rocketry to the enormous pair of boosters used to launch the Space Shuttle (Figure 5.3). Each solid booster of the Space Shuttle has a thrust of 11,790,000 newtons (2,650,000 pounds). The second type of engine was devised by **Robert Goddard** (1882–1945) in the 1920s. The **liquid-fuel** engine pumps separate fuel and oxidizer into the combustion chamber. Most modern rockets are of this design, because it has several advantages over solid-fuel engines. The engine can be stopped and restarted and can be throttled according to the amount of propellant reaching it. Examples of liquid-fuel rockets include the Saturn rockets used

■ **Figure 5.2** Types of rocket engines.

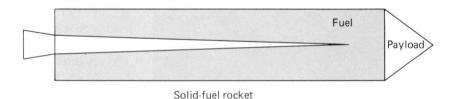

Solid-fuel rocket

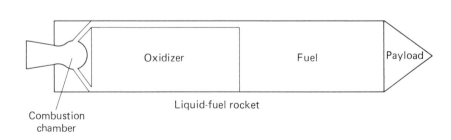

Liquid-fuel rocket

■ **Figure 5.3**   Launch of the Space Shuttle. One of the two solid-fuel rocket boosters can be seen at left, and the three liquid-fuel main engines can be seen at right. (NASA Photograph.)

in the Apollo program and the three Space Shuttle main engines (Figures 5.3 and 5.4). The five F-1 engines of the Saturn 5 rocket used in launching Apollo lunar missions were the most powerful liquid-fuel rockets ever developed. Each F-1 engine, powered by kerosene and liquid oxygen, developed 6,800,000 newtons (1,530,000 pounds) of thrust. Each of the main engines of the Space Shuttle, which use liquid hydrogen and liquid oxygen, develops 1,650,000 newtons (370,000 pounds) of thrust.

A typical liquid-fuel rocket is composed primarily of propellant tanks. As the fuel and oxidizer are depleted, much useless weight would be carried along. For this reason, most rockets are staged, with the first section of tanks and engines dropping off to lighten the remaining load. This may continue several times, as some rockets have three or even four separate stages.

How is the path of a rocket controlled? While still in the atmosphere, a rocket may be steered by means of aerodynamic fins that are usually located at its base. Other guidance methods include rocket nozzles that swivel to change thrust direction; moveable vanes that deflect the rocket's exhaust; and small, separate rocket engines that alter the rocket's direction. The rocket stays on course because it contains one or more gyroscopes, which maintain a stable orientation. Devices sense the current orientation of the rocket in relation to the course fixed in the gyroscope and cause appropriate changes in course direction to be executed.

■ **Figure 5.4**   Launch of a Saturn 5 rocket carrying three Apollo astronauts toward the Moon. The first stage of this three-stage rocket was powered by five liquid-fuel rocket engines. (NASA Photograph.)

■ Table 5.1   Earth Satellites Most Significant in Solar System Studies

Mission	Nation	Launch Date	Accomplishments
Explorer 1	USA	1/31/58	Discovered Earth's Van Allen radiation belts
Pioneer 1	USA	10/11/58	Unsuccessful lunar probe; located outer edge of Van Allen belts
Orbiting Solar Observatory 1–8	USA	3/07/62 to 6/21/75	Eight spacecraft that studied the Sun
Pegasus 1–3	USA	2/16/65 to 7/30/65	Measured density of micrometeoroids in Earth orbit
Skylab	USA	5/14/73	Space station; visited by three crews; conducted numerous solar studies
International Ultraviolet Explorer	USA/ESA*	1/26/78	Obtained ultraviolet spectra of many solar system objects
P78-1/Solwind	USA	2/24/79	Studied the Sun and discovered six Sun-grazing comets
Solar Maximum Mission	USA	2/14/80	Studied the Sun during the peak of its activity cycle
Infrared Astronomy Satellite	USA/ESA	1/25/83	Surveyed sky for infrared-emitting objects and found many new solar system objects while doing so

* ESA = European Space Agency.

■ Types of Missions

The simplest type of rocket mission involves the **sounding rocket**, which carries a payload straight up, so that it is above a large portion of the atmosphere, and then parachutes it safely to Earth. Although much more sophisticated missions have been devised, there is still a role for sounding rockets because their flights are relatively inexpensive and instrumentation can be reused many times. They can also be launched on short notice if an unexpected celestial event occurs.

The majority of space missions to date have been orbital, and thousands of objects have been placed into Earth orbit, beginning with the Soviet Sputnik 1 satellite on October 4, 1957. Orbital missions most important in solar system studies are listed in Table 5.1. Recall from Chapter 3 that most satellites are positioned about 160 kilometers (100 miles) high in what is called **low Earth orbit**. Rockets are usually launched from west to east to take advantage of Earth's rotational motion, which is nearly 1600 kilometers per hour (1000 miles per hour) at the equator. (This value decreases away from the equator, which is why most launch facilities are as close to it as possible). Some spacecraft are placed into polar orbit, which allows them to view the entire surface of the Earth.

As mentioned in Chapter 3, satellites in equatorial orbit 35,700 kilometers (22,200 miles) above the surface will orbit once every 24 hours, remaining stationary above an equatorial point. Most communications satellites use these **geo-synchronous orbits**, because antennas pointed at them can remain fixed in place after initial adjustment. A few spacecraft designed to probe the outer reaches of the environment around Earth have orbited at even greater distances. Obviously, reaching higher orbits requires more powerful rockets. Most satellites intended for geosynchronous orbits or beyond are first placed into low Earth orbit, where a small rocket engine is fired to boost them higher. This is because most satellite-launching rockets lack sufficient power to reach these high orbits directly. Objects in higher orbits have the advantage of experiencing minimal atmospheric drag. Even though satellites in low Earth orbit are above most of the atmosphere, there is still enough material present to slow them gradually so that they eventually re-enter the atmosphere.

After Earth orbit was reached in 1957, the Moon presented a tempting target. How different the history of space exploration would have been if the Earth lacked a natural satellite! The first lunar spacecraft simply flew past the moon, returning photographs and other information, and this same type of **flyby mission** has since been accomplished for all planets of the solar system except Pluto. (Unfortunately, there are no plans for any missions to Pluto. Major planetary missions, including flybys, are listed in Table 5.2.) To reach another planet from Earth, a spacecraft must be placed into an elliptical orbit, called a **transfer orbit.** Although variations in transfer orbits occur, the one most commonly used, called a **least-energy orbit**, requires the minimum expenditure of energy. To go from the Earth to a superior planet, the rocket leaves Earth at perihelion of the transfer orbit and arrives at the planet at the orbit's aphelion (Figure 5.5). To go to an inferior planet, the rocket leaves the Earth at the transfer orbit's aphelion and arrives at the planet at perihelion of the transfer orbit. Of course, in either case, launches must be timed so that the planet will be in the proper location when the spacecraft arrives. The obvious limitation of a flyby mission is that only a short time is available to study the object.

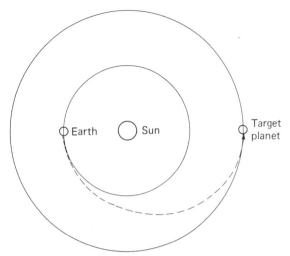

■ **Figure 5.5** Transfer orbit used to travel from the Earth to another planet. The planets' orbits are nearly circular, whereas the transfer orbit is half of a more eccentric ellipse.

The next step was actually landing spacecraft on the Moon and planets. Hard landings were accomplished first, with spacecraft designed to transmit information and photographs before hitting the surface. Soft landings, which used retro-rockets to slow the spacecraft just before it reached the surface, were the next step and were accomplished on Venus and Mars as well as the Moon. Two even more sophisticated types of unmanned missions were accomplished only on the Moon. One of these missions involved automatically gathering lunar material and launching it back to Earth for recovery. Another involved a mobile exploratory craft, which left the landing craft and traveled many miles over the surface of the Moon, returning more information than would have been available at a single site. Plans for Martian missions involving automatic sample return and mobile landers are under consideration.

Some spacecraft have been placed into orbit around the Sun at various distances, returning information on conditions observed. Spacecraft have also been placed into orbit around the Moon, Venus, and Mars for long-term observation, and a combined orbital and atmospheric entry probe mission is currently en route to Jupiter. In order to go into orbit around another planet after traveling to it from Earth, a spacecraft

■ Table 5.2 Examples of Missions to Solar System Objects

Object	Mission	Nation	Launch Date	Accomplishments
Sun	Pioneer 5	USA	3/02/60	Placed into solar orbit to study interplanetary space
Sun	Helios 1	USA	12/10/74	Passed within 30 million miles of Sun
Moon	Luna 2	USSR	9/12/59	First lunar impact
Moon	Luna 3	USSR	10/04/59	First photographs of lunar far side
Moon	Ranger 7 Ranger 8 Ranger 9	USA	7/28/64 2/17/65 3/21/65	Photographed surface before impacting
Moon	Luna 9	USSR	1/31/66	First soft landing
Moon	Luna 10	USSR	3/31/66	First lunar satellite
Moon	Apollo 11	USA	7/16/69	First manned landing
Moon	Luna 16	USSR	9/12/70	First automatic sample return
Moon	Luna 17	USSR	11/10/70	First automatic lunar roving vehicle
Mercury	Mariner 10	USA	11/03/73	Three flybys
Venus	Mariner 2	USA	8/27/62	First successful flyby mission
Venus	Venera 7	USSR	8/17/70	First soft landing
Venus	Mariner 10	USA	11/03/73	First photographs during flyby
Venus	Venera 9	USSR	6/08/75	First surface photographs and first orbiter
Venus	Pioneer–Venus Orbiter	USA	5/20/78	First radar mapping from orbit
Mars	Mariner 4	USA	11/28/64	First successful flyby mission
Mars	Mariner 9	USA	5/30/71	First spacecraft to orbit another planet
Mars	Viking 1 Viking 2	USA	8/20/75 9/09/75	First soft landings
Jupiter	Pioneer 10	USA	3/03/72	First Jupiter flyby
Jupiter Saturn	Pioneer 11	USA	4/06/73	Jupiter flyby, then first Saturn flyby
Jupiter Saturn	Voyager 1	USA	9/05/77	Jupiter and Saturn flyby missions
Jupiter Saturn Uranus Neptune	Voyager 2	USA	8/20/77	Jupiter and Saturn flybys, then first Uranus and Neptune flybys

■ Table 5.2   *(continued)*

Object	Mission	Nation	Launch Date	Accomplishments
Comet Giacobini–Zinner	International Comet Explorer	USA and ESA*	8/12/78	Originally in Earth orbit, retargeted to comet flyby in 1985
Comet Halley	Vega 1 Vega 2	USSR	12/15/84 12/21/84	Halley flybys, after flybys of Venus
Comet Halley	Sakigake	Japan	1/08/85	Halley flyby
Comet Halley	Giotto	ESA	7/02/85	Halley flyby
Comet Halley	Suisei	Japan	8/19/85	Halley flyby

* ESA = European Space Agency.

must be slowed enough by a rocket engine to allow capture by that object.

One of the most exciting aspects of planetary exploration is that several planets can be examined during one mission. As a spacecraft passes a planet, its trajectory is altered by gravitational forces from the planet. With careful planning of the trajectory, the gravitational boost can be used to speed the craft along on its journey. For example, the Voyager 2 spacecraft was launched from Earth in 1977 and passed Jupiter in 1979, Saturn in 1981, and Uranus in 1987. Neptune was reached in 1989, only 12 years after launch. A direct flight to Neptune in a least-energy orbit would have required nearly 30 years! This speeding up of a spacecraft, often called the **slingshot effect**, involves energy transfer from a planet to the spacecraft. When Voyager 2 flew by Jupiter, for example, it accelerated by gaining kinetic energy at the expense of the planet, which lost potential energy. As a result of losing potential energy, Jupiter now orbits the Sun a bit closer than before, but the change is far too small to be detected. Voyager 2, whose mass is much smaller, was greatly accelerated. The total combined energy of Jupiter and Voyager 2 was conserved during this encounter.

Manned exploration has so far been limited to Earth orbit and the surface of the Moon. Mars will probably be the next target for human exploration, but manned Martian missions are probably decades away from reality. Meanwhile, the success of the Soviet Salyut and Mir space stations has brought us to a point where humans are almost constantly in Earth orbit.

■ Spacecraft Instrumentation

Although the design of a spacecraft varies according to its specific mission, we can identify certain types of devices that are often used to study solar system objects (Figure 5.6). (We will describe specific spacecraft sent to various solar system bodies in detail in the chapters covering those bodies.) Two basic factors contribute to spacecraft design. First, the equipment must be as light as possible due to rocket payload capacity limitations. In addition, the equipment must be reliable, because once launched, repairs are impossible (except for satellites in low Earth orbit). Essential systems are usually redundant, with backups standing by to take over should any major systems fail.

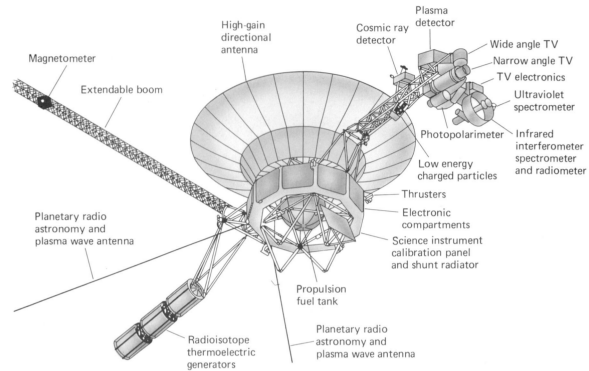

Magnetometer

Extendable boom

High-gain
directional
antenna

Cosmic ray
detector

Plasma
detector

Wide angle TV

Narrow angle TV

TV electronics

Ultraviolet
spectrometer

Photopolarimeter

Infrared
interferometer
spectrometer
and radiometer

Low energy
charged particles

Planetary radio
astronomy and
plasma wave antenna

Thrusters

Electronic
compartments

Science instrument
calibration panel
and shunt radiator

Propulsion
fuel tank

Radioisotope
thermoelectric
generators

Planetary radio
astronomy and
plasma wave antenna

■ **Figure 5.6** The Voyager spacecraft. The instruments shown here are representative of those carried by interplanetary spacecraft. For scale, the large high-gain directional antenna is 3.7 meters (12 feet) across. (Redrawn after a NASA photograph.)

A spacecraft needs a source of power for its instruments. For anything longer than a short-duration sounding rocket flight, batteries are insufficient, and two alternate sources are used. Solar cells convert sunlight into electrical power and are used for most Earth satellites and interplanetary probes in the inner solar system. For probes to the outer solar system, where the intensity of sunlight is much lower, or for instruments spending considerable time in darkness, such as the Apollo experimental packages left on the Moon, a radioisotope generator is used. This device generates electricity from the heat produced by the decay of radioactive materials.

Radio communication with the Earth is essential. Antennas for sending and receiving the signals are needed, as well as electronic instruments to transmit signals and process those received from Earth. The signals from probes to distant planets are very weak by the time they reach Earth, and large radio telescopes are used to receive the faint incoming data. Radio signals are studied closely if they pass through a planetary atmosphere or ring system during a flyby mission. These occultation studies, as they are called, can reveal significant information about atmospheric properties or ring structure.

A spacecraft requires a navigational system to determine its location and orientation. Because gyroscopes require considerable power to remain spinning and may fail during a long mission, interplanetary probes and some Earth satellites navigate by sensing the location of the Sun and a bright star in the sky. Small thruster rockets are used to change the spacecraft's orientation, such as when a satellite observatory is targeted toward another celestial object or when a planetary probe needs to change course.

On-board computers coordinate spacecraft activity and translate commands sent from Earth into detailed spacecraft instructions. For a craft so far from Earth that there is a long delay time for radio contact, the computer will store many instructions in advance so that the craft may function properly at all times. Space exploration would have been impossible without the digital computer, and many advances in computers were spurred by the need to miniaturize them in order to decrease the weight of the spacecraft.

Nearly all spacecraft carry cameras of some sort to return images to Earth. No other method has yet equaled photographic film for the amount of detail that can be recorded, but it is impossible to return film for processing from the distant reaches of the solar system! Although some Earth satellites did return exposed film to Earth in a re-entry capsule, and the Lunar Orbiter satellites carried on-board processing equipment for film and transmitted the processed pictures back to Earth by television, other techniques are more common. Usually the images returned to Earth are television pictures in digital form. To photograph scenes in color, the black-and-white camera photographs the scene through several colored filters, and the colored image is reconstructed back on Earth. Extensive techniques, called **image processing**, are used to enhance images once they have been received on Earth. This image enhancement is often used to increase contrast, allowing nearly invisible features to be recognized and studied. One must realize that, when looking at extremely colorful views of other planets, these colors have often been greatly exaggerated and do not picture what the human eye would actually see.

In addition to visible-light cameras, detectors have been devised to study virtually all wavelengths of the electromagnetic spectrum. Some of these can form images, whereas others are **spectrophotometers**, instruments that measure only the amounts of radiation at various wavelengths. One confusing aspect of this type of work is that colorful ''photographs'' are often made using this data. In such cases, the colors are used to indicate the relative intensities of light of different wavelengths, and these images may have little relationship to the visual appearance of the object.

A very common spacecraft instrument is the magnetometer, which detects and measures magnetic fields. Other instruments may detect cosmic rays (mysterious high-energy particles that originate beyond the solar system) and impacts on the spacecraft from small meteoroids. Spacecraft that either enter the atmosphere of another planet or land on its surface include instruments specifically designed to explore their immediate environment. Basic atmospheric properties such as composition, temperature, pressure, and wind velocity need to be known in order to understand global weather and climate. A descending craft can obtain considerable information on the vertical structure of the atmosphere in a short period of time. Surface craft often contain a seismograph to measure earthquake activity, because the wave energy recorded is very informative in revealing the interior structure of the object. A mechanical scoop can be used to pick up samples and place them in automated laboratories, where their compositions can be determined. Similar studies can be carried out to detect whether living organisms are present.

■ The Human Factor

The first manned spaceflights occurred in 1961, and since then, manned missions have been a spectacular part of space exploration. Admittedly, manned spaceflight is very expensive and has been accomplished only with some loss of human life. For these reasons, it has become somewhat controversial.

Many space scientists have argued that manned flights are unnecessary, at least in terms of scientific studies, and their point is certainly valid. After the Space Shuttle program began in 1972, nearly all American space funding was diverted to it, and scientific satellites and space probes became very infrequent. Military interest in space has also intensified in recent years, and many scientists fear that space may become a battlefield during future conflicts, rather than continuing as an area of research and exploration. For example, many scientists were understandably upset when an American antisatellite weapon test in 1985 destroyed a still-functional military research satellite that had made significant discoveries about the Sun!

On the other hand, there are clear arguments supporting manned flights. Many balky experimental devices have been fixed by human ingenuity in space, and many Earth satellites are now being designed to be serviced by crews in orbit, or returned to Earth for extensive overhaul.

Humans are by nature curious and love to explore, and space is the next natural stepping stone for our species. We are shocked and saddened by such tragedies as the explosion of the Space Shuttle *Challenger,* but loss of life has occurred in all exploratory ventures, and we can hardly expect spaceflight to be an exception.

Although both points of view have merit, it is likely that manned spaceflight will continue. It is hoped that it will not be at the expense of the typically far less expensive unmanned scientific missions.

■ Chapter Summary

Knowledge of the solar system has increased greatly in recent years as a result of space travel. The basis for this travel is the rocket engine, which operates because of Newton's third law. Both solid- and liquid-fuel rockets are used in space exploration.

Space missions include sounding rockets, satellites in both low and high Earth orbit, landings on other solar system objects, and craft placed into orbit around them, and flyby missions, many of which use gravity boosts. Spacecraft are designed for studying the specific objects toward which they are sent, and their equipment must be lightweight and reliable.

Manned spaceflight is the most dramatic area of space exploration, but it is somewhat controversial because of its high cost and the fact that human lives have been lost.

■ Chapter Vocabulary

flyby mission	least-energy orbit	slingshot effect	spectrophotometer
geosynchronous orbit	liquid-fuel rocket	solid-fuel rocket	thrust
Goddard, Robert	low Earth orbit	sounding rocket	transfer orbit
imaging processing	rocket		

■ Review Questions

1. How does a rocket engine propel the rocket?
2. What are the two types of rocket engines? How does each work? What are the advantages and disadvantages of each?
3. Why are rockets staged during flight?
4. How is a rocket's flight path controlled?
5. List the various types of space missions that have been conducted and what kind of information each can give us.
6. What factors contribute to the design of a spacecraft?
7. What sources of power are used for spacecraft?
8. How do spacecraft obtain images of solar system objects and return them to Earth?
9. What are the arguments for and against manned space flight? What is your personal feeling on the matter?

■ For Further Reading

Baker, David. *The History of Manned Space Flight.* Crown, New York, 1981.

Baker, David. *The Rocket — The History and Development of Rocket and Missile Technology.* Crown, New York, 1978.

Gatland, Kenneth, editor. *The Illustrated Encyclopedia of Space Technology.* Crown, New York, 1981.

Yenne, Bill. *The Encyclopedia of U.S. Spacecraft.* Exeter Books, New York, 1985.

6

The Sun

The Sun, eclipsed by the Earth, as viewed from the vicinity of the Moon during the Apollo 12 mission in November, 1969. (NASA Photograph.)

The Sun is the "star" of our solar system in more ways than one. Its gravitational force holds the system together, and it provides all of the system's visible light. To an inhabitant of a planet orbiting another star, the Sun would be the only member of the solar system visible.

■ What Is a Star?

Of the more than 200 billion stars in the Milky Way galaxy, the Sun is the one about which we know most. There are several reasons for this, the most important being that the Sun is the only star close enough to be seen as a disk (Figure 6.1). All other stars appear as only pinpoints of light, so their surface details can't be studied. (Indirect methods are now allowing stellar surface features to be analyzed, even though they are not observed directly.) The light received from many stars is so faint that the finer details of their spectra cannot be studied. Because there is no such shortage of light from the Sun, even the finest details of its spectrum are extremely well known.

Although the Sun's appearance in the sky is familiar to everyone, we must list several characteristics that define what a star is. A **star** is composed of gaseous material, and no liquid or solid is present, even in its interior. As we will see, the high temperature and internal pressure of the Sun cause its gas to have properties that are quite different from those of gases found on Earth. Most stars are also spherical, a shape that results from gravitational forces. (The only exceptions are extremely fast rotating stars or those deformed gravitationally by close stellar companions.) A star emits tremendous amounts of energy, which is produced by nuclear fusion reactions occurring in its core. Finally, a star (at least one like the Sun) is in equilibrium, a state of balance. This means that most stars have constant radii, because the inward pull of gravity is matched by the outward pressure of the hot gases

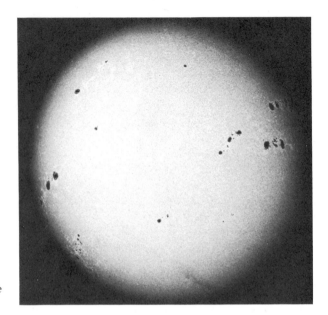

in their interiors, as well as by pressure from the outward flow of radiation within the star.

As we will learn in Chapter 29, scientists have determined that the age of the Earth is about 4.6 billion years and believe that the Sun, as well as the rest of the solar system, probably formed at the same time. Calculations using the rate at which nuclear fusion occurs inside the Sun and the total amount of fuel available suggest a lifetime of 10 billion years for the Sun, making it now "middle-aged." Although we will discuss the origin of the Sun and solar system in detail in Chapter 29, current theories suggest that stars form when areas within concentrations of gas and dust, called nebulae, collapse and condense into smaller objects. After forming, a star spends the majority of its life with a fairly unchanging set of physical characteristics. The most important quantity influencing a star's properties is its mass; the more massive the star, the more luminous it is and the shorter its lifetime will be, because it uses fuel at a much faster rate than a smaller, less luminous star. The mass of a star also influences the events of the final stages of its life and its ultimate fate. The most massive stars, which are relatively rare, expand near the ends of their lives to become red supergiant stars, after which they explode as supernovae, leaving behind remnant neutron stars or black holes. Less massive stars, including the Sun, have much less spectacular

deaths: late in life, they expand to become red giants, whose outer atmospheres slowly dissipate into space, leaving behind the stars' exposed cores as gradually fading objects known as white dwarfs.

■ Physical Properties and Composition

The average distance between the Earth and the Sun is, by definition, 1 astronomical unit, about 149.6 million kilometers (92.9 million miles). This quantity is best determined by bouncing radar beams from the surfaces of other planets. Because their distances from the Earth in astronomical units can be calculated from celestial mechanics, accurate measurement of planetary distances by radar yields the value of the astronomical unit. (Prior to the development of radar, the value of the astronomical unit was determined by observations of transits and triangulation of planetary distances.) The Sun's distance is such that 8.3 minutes are required for its light to travel to the Earth.

From Earth, the apparent angular diameter of the Sun is about 32 minutes. This value varies slightly over the course of the year due to Earth's orbital eccentricity. Given the distance to the Sun, this yields an actual solar diameter of 1.39

million kilometers (865,000 miles), which is about 109 times that of Earth. When the volume of the Sun is calculated, it is found that a "hollow Sun" could contain about 1.3 million Earths.

The mass of the Sun can be determined using Kepler's third law, and it is equal to that of 333,400 Earths, or about 2×10^{30} kilograms (4.4×10^{30} pounds). The Sun is so huge that it contains over 99% of the mass of the solar system! Even though the Sun is so massive, its density is only 1.4 grams per cubic centimeter. (This is an average figure, as the density is much lower at the surface and much higher at the center.)

The Sun, like every other object in the solar system, rotates on its axis. Unlike the solid terrestrial planets, the gaseous Sun rotates at different rates in different latitudes, a phenomenon known as **differential rotation.** The rotational rate varies from about 26 days near the equator to about 36 days near the poles. These values are obtained by timing the rate of sunspot motion and by measuring the Doppler shift of light emitted from the Sun's surface (Figure 6.2). Because the Sun rotates differentially at a relatively slow rate, it has no interior stresses to cause polar flattening. This phenomenon, a decrease in an object's polar diameter compared to its equatorial, is observed in the planets, especially in the gas giants, which rotate differentially but much more rapidly than the Sun. The Sun's axis is nearly perpendicular to the ecliptic plane, but it is tilted about 7° from true perpendicularity.

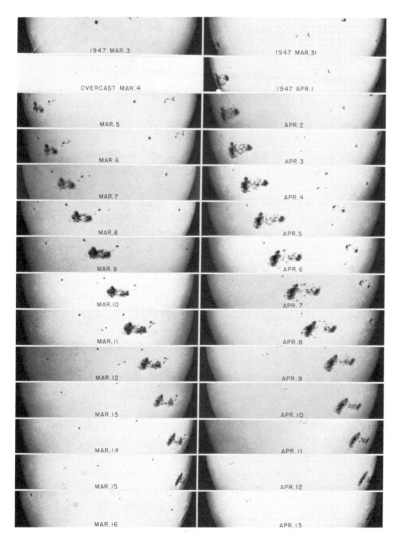

■ **Figure 6.2** How sunspots are used to time solar rotation. These photographs, taken in 1947, show the passage of a large sunspot group across the face of the Sun. (Photographs courtesy of the Observatories of the Carnegie Institution of Washington.)

The composition of the Sun has been determined using spectroscopy (see Figure 4.7). Although it is believed that composition is uniform throughout the interior of a newly formed star, significant variations are expected in an older star like the Sun. This is because the nuclear fusion that takes place at the Sun's core changes its internal composition, and there is little vertical mixing to spread the heavier elements formed by fusion through the rest of the star. Because spectroscopy studies light emitted from the solar surface, it is this part of the Sun whose composition is revealed by it. From the surface composition and theories about the solar interior, the overall composition can be estimated.

The Sun, like all stars, is composed primarily of hydrogen, with helium second in abundance. (This is the reason for its low average density.) There are two ways to express composition. The first is by number of atoms present. About 89% of the Sun's atoms are hydrogen and about 11% are helium. Composition may also be expressed by mass fraction. Because a helium atom is more massive than a hydrogen one, the Sun is about 76.4% hydrogen and 21.8% helium by mass. Oxygen is a distant third (0.8% by mass), followed by carbon, nitrogen, neon, silicon, magnesium, sulfur, and iron. All naturally occurring elements are believed to exist in the Sun, although only about 70 have actually been detected spectroscopically. (In addition, about 20 molecules have been identified in the Sun's spectrum.) For example, Egyptian astronomer Nahed Youssef has found that there are nine gold atoms in the Sun for every trillion hydrogen ones. This means that there are 10^{16} tons of gold in the Sun, whose value on Earth would be truly astronomical! Of course, this "solar gold" is safer than it would be in any bank vault on Earth!

Atoms in the Sun are nearly all ionized, and many have even lost all of their electrons due to the enormous temperature, which increases from 5800 K at the visible surface to a staggering 15 million K at the core. Gas consisting of ionized atoms is called **plasma**, which is sometimes described as the fourth state of matter.

As mentioned in Chapter 4, the wavelength at which the Sun emits the most energy is about 500 nanometers, corresponding to a Wien's law temperature of about 5800 K (about 10,000°F).

This wavelength is located in the green spectral region. The Sun, of course, does not appear green, because it also emits enough red and blue light to "neutralize" the green light and make sunlight appear almost pure white. (The color of the Sun is slightly yellow, because the colors emitted do not balance perfectly into white light. This yellowness is increased by the passage of sunlight through the Earth's atmosphere. This effect is enhanced to red at sunrise or sunset, when sunlight travels a longer distance through the atmosphere.)

■ Energy Transport

In order to understand many of the phenomena occurring within the Sun, it is necessary to be familiar with how thermal energy is transported. One of the fundamental properties of heat energy is that it always travels from an area of higher temperature to one of lower temperature. There are three distinct ways by which thermal energy can be transferred (Figure 6.3). As we will see, two of them (convection and radiation) are important inside the Sun.

One method of heat transport is **conduction**, which involves atom-to-atom contact and occurs primarily in solids. Temperature is a measurement of the rate of atomic motion, and the higher the temperature, the faster the atoms in a solid will vibrate. A fast-moving atom next to a slower-moving one will transfer some of its vibrational kinetic energy to its neighbors. The transfer then continues to the next atoms, and so on. A good example of this involves a serving container of hot gravy and a silver serving spoon. If the gravy sits for a few minutes before being served, the end of the spoon sticking out of the gravy will be surprisingly hot! This is because the thermal energy of the gravy was transferred to the silver atoms in the submerged part of the spoon and then carried along the stem of the spoon. Some materials are better conductors than others, with the electronic structure of the atoms being the determining criterion. Conduction is most important in solids, because their atoms are generally in closest contact.

A second method of heat transfer, **convection**, involves circulation in a fluid (liquid or gas).

(a)

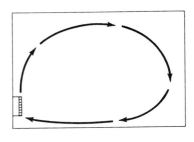

(b)

Sun

Earth

(c)

■ **Figure 6.3** Heat transport. (a) Conduction is illustrated by a spoon in hot gravy. (b) Convection is illustrated by air circulation in a room. (c) Radiation is illustrated by photons moving from the Sun to the Earth.

Several everyday examples are familiar. Electric baseboard heat relies on convection to heat a room, as there are no fans or blowers. The air heated by the unit near floor level rises, displacing cooler air, which is forced to sink. Some of the sinking air reaches the heat unit, where it is warmed, causing the cycle to repeat. Similarly, liquids being heated do not remain motionless as they cook, but circulate in their containers. This causes an entire pan of soup to be heated, not just the bottom layer close to the burner. Stirring, of course, accelerates the process, but it would still occur by itself.

The final method of energy transport is **radiation**, which is transferred by photons as they travel from the heat source to the recipient. There is nothing between the Sun and the Earth, so convection and conduction cannot transport energy. Radiation is the only transport method that works in the vacuum of space, but it also occurs elsewhere. When you hold your hands in front of a fire, for example, you feel warmth, because the photons from the fire are traveling directly to your skin.

■ Magnetism and the Sun

Magnetism, a force generated by electrical interactions, is familiar to most people, as it is encountered frequently in our everyday lives. Many solar system objects, including the Sun, generate magnetic fields internally, causing them to behave like giant magnets. (**Field** is the term applied to the invisible pattern of force generated by a magnet.) Because magnetism is very important in many forms of solar phenomena, we will discuss it at this point. (Magnetism as it relates to planets is discussed in Chapter 9.)

Magnetism and electricity are closely related physical phenomena. An electric current will generate a magnetic field; this is the principle behind the electromagnet. At the same time, moving magnetic fields can be used to produce electric currents, as demonstrated by an electric generator, which contains moving magnets. There is one major difference between electricity and magnetism. Electric charge exists in positive and negative forms, and it is possible to have an isolated charge of either type. A magnet, however, lacks charge but has polarity, a term derived from the fact that a compass needle points toward the poles of Earth. (One end of a magnet is the north pole; the other, the south pole.) If a magnet is cut in half, each smaller magnet will also have a north and south pole. Any magnetized object will have both a north and south pole, and it is not possible to have an isolated pole of either type. Figure 6.4 illustrates how a magnet generates a magnetic field.

Because a magnet has two poles, the field it produces is called a **dipole field.** The fields of the Earth, other planets, and the Sun are all dipole fields. The exact cause of these fields is not yet known, but the Earth's field is probably caused by electrical currents produced by circulation of hot, molten nickel-iron metal in the Earth's outer core. The Sun's field probably results from the electrically conductive plasma within it. The Sun's internal field strength is much greater than the Earth's, but its surface is so far from the central source of the field that the average magnetic in-

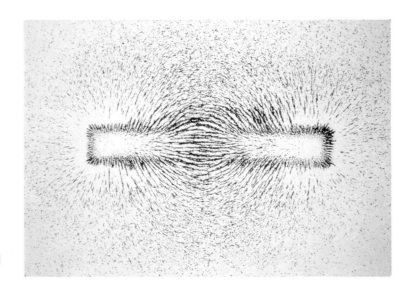

■ **Figure 6.4** A magnetic field. If a piece of paper is placed atop a bar-shaped magnet and iron filings are then sprinkled on the paper, the filings will align themselves along the magnetic field lines. (Photograph by the author.)

tensity at the surface is only about twice as strong as Earth's. The Sun's field possesses much more small-scale variation than Earth's, as there are some places where the Sun's surface field intensity is 4000 times greater than its average value. The differential rotation of the Sun is believed to distort its magnetic field lines, promoting the development of this magnetic variation.

The Sun's magnetic field is studied using spectroscopy. On an atomic level, magnetism can change electron energy levels, affecting electronic transitions and, hence, spectroscopic lines. The effect on spectroscopic lines (known as the Zeeman effect, in honor of its discoverer) is that a single spectral line is split into three closely spaced lines. The two additional lines are caused by electronic transitions involving new electron energy levels produced by the magnetic field. Both intensity and polarity of the magnetic field at any area on the Sun's surface can be determined by studying Zeeman triplets in its spectrum.

■ The Solar Atmosphere

When we look at the Sun, we see a yellow-white orb with a clearly defined boundary. When we think of the nature of gas on Earth, this appearance is difficult to understand. It is hard to imagine gas having a sharp, distinct boundary, and because most gases are transparent, we wonder why we can't see through the Sun. How can we explain these properties?

The solution to our dilemma is found to be a layer in the Sun known as the **photosphere** (literally, "sphere of light"). Compared to the Sun as a whole, the photosphere is rather small, as it is only about 200 to 400 kilometers (120 to 240 miles) thick. However, it contains hydronium (H^-) ions, which cause the layer to be opaque. The opacity of the photosphere causes this to be the layer from which sunlight emanates, in much the same way that the light of a frosted light bulb emanates from the frosted area, even though it is ultimately produced by the glowing filament. Although the photosphere appears to be a distinct surface of the Sun, no such surface exists. Instead, the Sun decreases in density with increasing distance from its center. The density at the top of the photosphere is only 2.8×10^{-8} grams per cubic centimeter, about 10^{-5} Earth's surface atmospheric pressure, and the density decreases above this point.

When the Sun's disk is viewed, it is immediately apparent that the photosphere appears darker at its edge (see Figure 6.1). This effect, called **limb darkening**, is another result of the photosphere's opacity. As we look at the center of the Sun, our line of sight travels straight down into the photosphere, allowing us to see light coming from a deeper, hotter region. When we

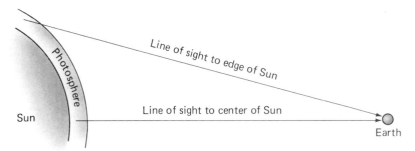

■ **Figure 6.5** The cause of solar limb darkening. When we look at the center of the Sun, we see light from deeper in the photosphere, where higher temperatures produce brighter light. When we look at the edge of the Sun, we see fainter light from higher, cooler areas of the photosphere. In both cases, our line of sight penetrates the same linear distance into the photosphere.

look near the limb of the Sun, our line of sight travels a longer linear distance through the photosphere, but because of the oblique viewing angle, we see a higher, cooler region of the photosphere, which emits less light (Figure 6.5).

The temperature of the photosphere, obtained using Wien's law, is about 5800 K. Underneath the photosphere, energy is being brought to the surface by convection of hot gases, and the photosphere has a mottled appearance when viewed through a telescope due to the numerous regions of convection, called convection cells, which come and go in a shimmering fashion (Figure 6.6). Each of these cells, called **granules**, lasts about 10 minutes and is about 1000 to 2000 kilometers (600 to 1200 miles) across. Granules are separated by dark lanes that are 400 K cooler than the granules themselves. The gas in granules is rising at the rate of 1600 kilometers per hour (1000 miles per hour), and this upward movement shoots material up to 100 kilometers (60 miles) above the photosphere's normal surface. This eruptive activity produces sound waves, which propagate upward into the atmospheric layers above the photosphere, where they become shock waves similar to "sonic booms."

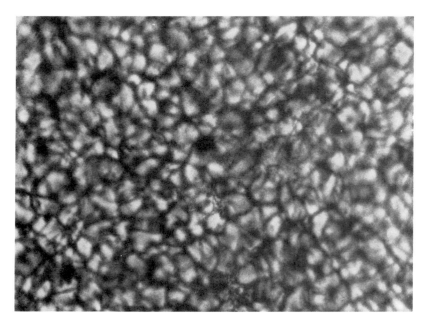

■ **Figure 6.6** A highly magnified view of granulation in the Sun's photosphere. Each granulation cell is about 1000 to 2000 kilometers (600 to 1200 miles) across. (National Solar Observatory.)

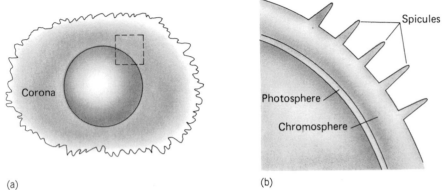

(a)   (b)

■ **Figure 6.7**   The solar atmosphere. (a) The corona is shown surrounding the photosphere and chromosphere, which can't be seen at this scale. (b) An enlarged view of the dotted region in (a) reveals the photosphere and the chromosphere. Note the spicules in the chromosphere.

The photosphere and the layers above it are known collectively as the solar atmosphere (Figure 6.7). The next layer, the **chromosphere** ("sphere of color"), is located directly above the photosphere. It emits light, but the light is so faint that it is normally invisible, overwhelmed by the brilliance of the photosphere. The chromosphere's pink light is visible to the unaided eye only during a solar eclipse, when the Moon blocks the photosphere's light. Special filters allow the chromosphere to be observed and photographed through a telescope at any time

■ **Figure 6.8**   The solar chromosphere. This photograph, taken with a filter that allowed only hydrogen alpha light to pass, shows light produced within the chromosphere by the emission process. (National Solar Observatory.)

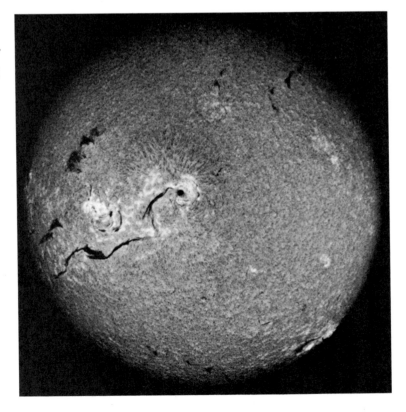

■ **Figure 6.9**   Spicules viewed at the edge of the Sun. These short-lived features extend above the chromosphere and are associated with the supergranulation pattern. (National Solar Observatory.)

(Figure 6.8). The light of the chromosphere is pink in color because of emission by an electronic transition of hydrogen, called the hydrogen alpha transition, which occurs at a wavelength of 656 nanometers (see Figure 4.9). The temperature of the chromosphere ranges from 8000 to 10,000 K, with the high temperature possibly due to energy transmitted by the sound waves propagating from below. The density of the chromosphere is only 0.1% of that of the photosphere.

The chromosphere is generally 2500 kilometers (1500 miles) thick, but it has regions, called **spicules**, that can extend upward as far as 15,000 kilometers (9000 miles). Spicules, which last about 10 minutes, produce a flickering pattern somewhat similar to a flame (Figure 6.9). They are believed to be associated with **supergranulation**, a large-scale effect of convection from deeper within the Sun than that causing photospheric granulation. Unlike granulation, supergranulation is invisible in white light, but it can be seen in light

of the hydrogen alpha transition. The supergranulation pattern is about 30,000 kilometers (19,000 miles) in size and has a lifetime of about one day. During its lifetime, a supergranulation region can generate numerous short-lived spicules.

Immediately above the chromosphere is a thin transition region only a few tens of kilometers thick that is marked by an extreme rise in temperature and a decrease in density. Above this is the **corona** ("crown"), the outermost part of the solar atmosphere. This region is visible during a total solar eclipse as a spectacular glowing region surrounding the eclipsed Sun (Figure 6.10). A special telescope called a coronagraph can be used to block the photospheric light and reveal the corona without waiting for an eclipse to occur, but the effect is not nearly as spectacular. Astronomers generally considered the corona to be an optical illusion until it was photographed in the 1800s, and spectroscopy was later used to study

■ **Figure 6.10**   The solor corona, as photographed during total solar eclipses. (a) This eclipse occurred near solar maximum, when the corona appears nearly circular. (b) This eclipse occurred near solar minimum, when the corona appears somewhat elongated. (Lick Observatory Photographs.)

(a)

(b)

the region. The spectral lines seen were very unusual and did not match those associated with any terrestrial elements, suggesting the presence of unknown materials. In 1940, however, a simple explanation of the lines was revealed; they are caused by highly ionized atoms that have lost many electrons, not just the one or two normally lost during ionization. For example, coronal iron has lost 13 of its normal 26 electrons, a process that requires a tremendous amount of energy! In order to ionize iron and other elements present in the corona, such as nickel, neon, and calcium, in this way, a temperature of 2 million K is needed! Because the density of the corona is only about 10^{-10} that of Earth's atmosphere, there is very little thermal energy, because even though the ions are moving very rapidly, there aren't many of them present.

What is responsible for the high temperature of the corona? It was once believed that shock waves from the photosphere were responsible, but it is now thought that all of their energy is dissipated in heating the chromosphere. Current theories suggest that solar magnetic fields generate the energy that heats the corona. In addition to heating the corona, magnetic fields are important in shaping the corona. The corona is not uniform; it has an irregular shape and contains streamers associated with solar magnetic field lines.

■ The Solar Wind

Although we can consider the corona to be the outermost layer of the Sun, it has no definite outer edge but gradually extends into interplanetary space. This is because coronal gas is so hot that fast-moving particles within it have enough energy to escape from the corona. This material flowing outward from the Sun into interplanetary space is called the **solar wind** and is composed primarily of protons (hydrogen nuclei), electrons, and ions of other elements found in the corona.

The existence of the solar wind was first suggested by Ludwig Biermann in 1951. Biermann noted that the classical explanation of the source of the form of comet tails, the radiation pressure of sunlight, was by itself inadequate to account for all of the features observed in comet tails. He suggested that, in addition to this pressure, an outward streaming of particles from the Sun was responsible for shaping comet tails. In 1958, Eugene Parker theorized that the solar wind was material derived from the corona itself.

The existence of the solar wind was verified by detectors carried on space probes early in the space age. The velocity of the wind is about 400 kilometers per second (250 miles per second) in the vicinity of the Earth, and it takes about five days for solar wind particles to reach our planet from the Sun. The average density of the solar wind near the Earth is from 5 to 80 protons per cubic centimeter. (Despite this continuous outflow of material, the Sun loses only 1 part in 10^{14} of its mass annually to the solar wind.)

The Mariner 2 spacecraft discovered that variation in solar wind velocity occurs, and that the timing of this variation corresponds to the Sun's rotational period. Studies of the Sun using X-ray and far-ultraviolet light during the Skylab manned orbital missions revealed that "holes" exist in the corona, and that these **coronal holes** are the sources of especially fast solar wind. In coronal holes, the density and temperature are only 7% and 75%, respectively, of their normal values in the rest of the corona. In the normal portion of the corona, magnetic field lines loop outward and back, whereas in the holes, they extend outward and do not return. Although solar wind particles travel straight outward from the Sun, the rotation of the Sun (to which they are still attached) causes these extended magnetic field lines to form a spiral pattern (Figure 6.11). For example, they cross Earth's orbit at an angle of about 60°. Because the solar wind carries the Sun's magnetic field, it interacts with planetary magnetic fields. Because the solar wind is fairly well understood, observing its interactions with other planets is a useful tool for gaining understanding of their magnetic fields. (We will discuss planetary magnetic fields in Chapter 9.)

Eventually, somewhere beyond Pluto, the solar wind ends as it collides with, and is probably stopped by, the similar material that is believed to pervade the space between stars in our galaxy. The huge region containing the solar wind is called the **heliosphere**, and its outer boundary is

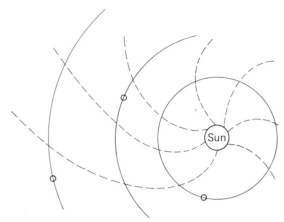

■ **Figure 6.11** How the solar wind carries the Sun's magnetic field outward, forming the interplanetary magnetic field. Whereas solar wind particles (not shown) travel straight outward from the Sun, the magnetic field lines form a spiral pattern. The three large *circles* represent planetary orbits.

called the **heliopause.** (It is hoped that the Pioneer 10 and 11 and Voyager 1 and 2 spacecraft will reach the heliopause before contact with them is lost. It is thought that the heliopause may be located between 50 and 100 astronomical units from the Sun, but the actual value is unknown.)

■ The Solar Interior

Below the photosphere, our Sun becomes a much more mysterious object. There are no techniques for directly observing stellar interiors, so our knowledge is based on theoretical computations verified by indirect measurements.

The mathematics used to "explore" stellar interiors is rather complicated and can best be described as an attempt to "model" a star by breaking it down into numerous layers from the surface to the center. Equations describe five physical properties — mass, pressure, luminosity, density, and temperature — as functions of radius. Solutions to these equations, laboriously derived in the past, are now obtained fairly quickly using computers. The product of such calculations is a page filled with columns of numbers describing the conditions at any point within the star.

The results of studies of the Sun's interior are rather startling. We must travel one-tenth of the way from the photosphere to the center of the Sun before the density reaches that of Earth's atmosphere, and halfway to the center before the Sun becomes as dense as water. The central density is an incredible 150 times that of water! The Sun is so centrally concentrated that the inner 40% of its radius contains 80% of its mass, and the inner 60% of its radius contains 95% of its mass. The pressure at the center of the Sun is about 2.2×10^{11} atmospheres.

Temperature rise is also dramatic, reaching 1 million K 15% of the way toward the center and 10 million K 80% of the way. The Sun's central temperature is estimated to be almost 15 million K. Because the process of nuclear fusion that powers the Sun works only at temperatures in excess of 10 million K, the majority of the Sun's energy (90%) is generated in the core, the inner 20% of its radius.

Theoretical studies indicate that the method of energy transport inside the Sun is not the same throughout (Figure 6.12). Thermal energy first travels outward from the core by radiation, but the solar interior becomes convective about 80% of the way to the surface. The change is due to the variations in physical conditions at different depths.

Activity occurring within the convective zone allows verification of theoretical predictions of conditions there. The method is similar to the

■ **Figure 6.12** The solar interior. The numbers represent the fraction of the Sun's total radius. See text for details.

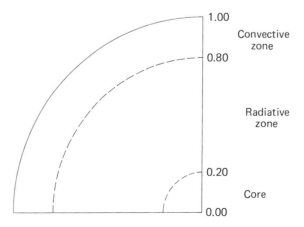

way geologists study the Earth's interior by examining the behavior of earthquake waves as they pass through it. Wave motion occurs inside the Sun, as its interior is actually a very good conductor of sound waves. Careful Doppler shift observations of the solar photosphere reveal that it oscillates up and down with a period averaging 5 minutes. These oscillations are caused by waves that travel upward from the base of the convective zone and are then reflected back downward from layers near the surface. Models of the structure of the convective zone were used to predict the frequencies of oscillation that would be expected to be seen, and then the observed frequencies were used to refine the interior models. Additionally, Doppler shift comparisons of oscillation frequencies on the approaching and receding hemispheres of the Sun allow the rotational rate of the convective layer to be measured. Although results are still inconclusive, it appears that the Sun's interior rotates at a faster rate than does the photosphere.

■ The Solar Activity Cycle

As we look at other stars in our galaxy, we notice that many of them are variable, emitting more light at some times, less at others. If our Sun varied its light output, there would be dire consequences for Earth and the life on it. Just how stable is our Sun?

The amount of radiation received at the Earth from the Sun is called the **solar constant** and amounts to 1.36×10^6 ergs per square centimeter per second. The most accurate values of this quantity are obtained by spacecraft in Earth orbit. The Solar Maximum Mission satellite has found the variation in the solar constant to be only about 0.1% of the average value, which is fortunately a very small figure.

What is responsible for the variability in solar output? Although the full story is not yet known, part of the answer must lie in the solar activity cycle, which has been observed for many centuries. As we will see, this cycle, like so many other solar phenomena, is closely linked to the Sun's magnetic field.

The most conspicuous indicators of solar activity are **sunspots**, which were discovered by Galileo with his simple telescopes in 1609. (Large sunspots were seen with the naked eye before that, dating back to Greek observations in the fourth century B.C., but no scientific study of them had been performed due to their rarity.) Sunspots are regions of the photosphere that appear dark because of their temperatures. Unlike the rest of the photosphere, whose temperature is nearly 6000 K, sunspots have temperatures of only about 4500 K. Although still hot by terrestrial standards, they emit so much less light than the rest of the photosphere that they appear dark by comparison. Sunspots are a direct result of local magnetic variation, as the intense magnetic field in the area of a sunspot stops convection from occurring, thereby reducing the amount of energy emitted.

As shown in Figure 6.13, a sunspot has a darker inner region (**umbra**) and a lighter outer region (**penumbra**). Sunspots often occur in pairs or in larger groups that contain as many as 100 members. When sunspot pairs occur, the leading and trailing spots (with respect to the direction of solar rotation) in each solar hemisphere (northern and southern) will have opposite magnetic polarity. To further complicate the situation, this polarity will be reversed in the other hemisphere. For example, if leading spots in the northern hemisphere have south magnetic polarity, the leading ones in the southern hemisphere will have north magnetic polarity. An individual spot may last from hours up to months, and spots have long been used to time the rotation rate of the Sun. Eventually, sunspots are broken apart by convective activity occurring in the surrounding photosphere.

Heinrich Schwabe (1789–1875) discovered in 1843 that both the number of sunspots and their location vary over an 11-year period now called the **sunspot cycle**. At the start of the cycle, there are only a few sunspots, and they are located at solar latitudes of 20° to 30°. As the cycle progresses, sunspots increase in number and tend to occur at latitudes of about 15°, with the maximum number occurring about 5.5 years after the start of the cycle. At the end of the cycle,

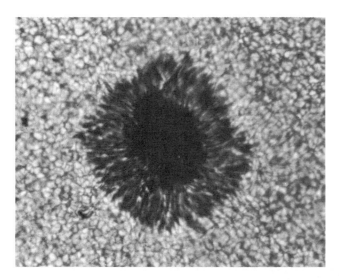

■ **Figure 6.13** A sunspot. Note the dark, inner region (the umbra) and the lighter, outer region (the penumbra). (National Solar Observatory.)

there are once again only a few sunspots, which are located at latitudes of 5° to 10°. Meanwhile, a new cycle begins with a few spots at higher latitudes (Figures 6.14 and 6.15). The shape of the corona also changes, from symmetrical at maximum to elongated in the Sun's equatorial plane at minimum. During alternate 11-year cycles, the magnetic polarity of the leading sunspots in each hemisphere's sunspot pairs reverses, leading some researchers to believe that the entire sunspot cycle should actually be considered to last 22 years.

The sunspot cycle is much more complex than even the preceding discussion suggests. The number of spots varies greatly from maximum to maximum, and the exact length of the cycle is not constant. The reason for the length of the cycle is unknown; it is also not known why it even exists in the first place. While checking past sunspot records in 1890, **E. W. Maunder** (1851–1928) determined that practically no sunspots had been seen from 1645 to 1715, a period now called the **Maunder minimum.** Apparently, this was a short enough time after the discovery of sunspots that their behavior was not understood and their absence unnoticed. Records indicate that the climate of Europe was unusually cool during this period, with a "little ice age" marked by glacial advance. This is only one example of the possible influence of the solar cycle on our climate, but the relationship is not fully understood.

Because sunspots reduce the emission of sunlight, it might be expected that the "missing" energy must come out elsewhere. Careful studies of the solar constant, however, indicate that this is not the case, and it is believed that the energy is stored internally by a mechanism that is not yet understood.

■ **Figure 6.14** The sunspot cycle. (a) Just after minimum, there are a few spots at high latitudes, and the corona is elongated. (b) At maximum, there are many spots at middle latitudes, and the corona is rounded. (c) As the cycle concludes at the next minimum, there are a few spots at low latitudes, and the corona is again elongated.

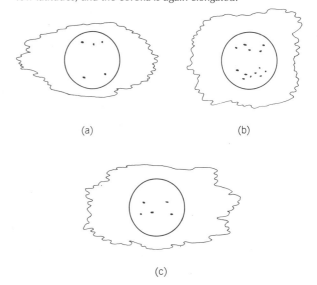

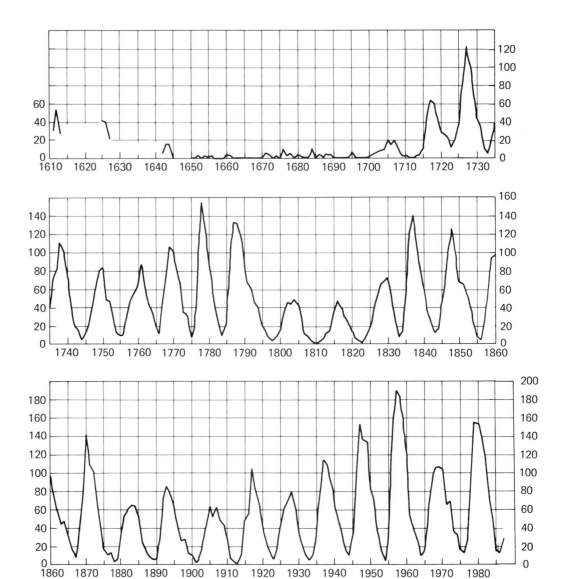

■ **Figure 6.15** Graphs of sunspot numbers from 1610 until the 1980s. Note the 11-year cycle and the Maunder Minimum during the 1600s. (Courtesy of John A. Eddy and the University Corporation for Atmospheric Research.)

In contrast to sunspots, several phenomena above the Sun's surface emit greater-than-normal amounts of energy. One example is the phenomenon of **faculae**, which are small, bright emission areas that are seen above the photosphere. These are visible from Earth only when they are near the edge of the Sun and are believed to be associated with magnetically active regions that are independent of sunspots.

The most spectacular examples of solar activity are **flares**, which dramatically release energy stored in solar magnetic fields (Figure 6.16). Flares are eruptive events that begin abruptly and rise to peak intensity in only a few minutes, then usually dissipate within an hour. Flares are generally associated with sunspot regions. Flares exhibit a wide range of intensities, and most of their energy is emitted as nonvisible light. Occasion-

■ **Figure 6.16** A solar flare, as photographed using hydrogen alpha light. (Courtesy of Big Bear Solar Observatory, California Institute of Technology.)

ally, however, there are visible-light flares, observable in optical telescopes with special filters. Most flare energy is released above the photosphere, in the chromosphere and the corona. The largest flares have energies of 10^{32} ergs (equivalent to 2% of the Sun's average luminosity) and extremely high temperatures. In fact, spectroscopy of powerful flares has shown iron ions that have lost all but one of their 26 electrons, indicating that temperatures of 10 million K are present!

Except for the fact that flare activity is greatest near sunspot maximum, there is no way to predict specific flares or their location and intensity. Recent research has indicated that flare activity may be somewhat cyclical, with a much shorter period than the sunspots, about five months. Additional observations will be needed before such periodicity is verified. Flares send large amounts of high-energy radiation toward the Earth, as well as contribute an unusually high number of particles to the solar wind. Whereas the radiation reaches the Earth in only a few minutes, the arrival of the solar wind takes several days. When it arrives, long-range communications are disrupted because of changes in the Earth's ionosphere, and brilliant displays of **au-**rora may be seen in the sky. The aurora borealis ("northern lights") and aurora australis ("southern lights") occur in the northern and southern sky because the orientation of the Earth's magnetic field lines near the poles (perpendicular to the surface) allows the aurora-producing particles to enter there. Near the equator and mid-latitudes, the Earth's magnetic field lines are oriented parallel to the surface, and they block the entrance of charged particles. The glow of light seen in an aurora is caused by emission of light from excited atoms in the upper atmosphere, similar to a fluorescent light bulb, but on a much larger scale!

Another form of solar activity consists of **prominences**, which are dense loops of gas suspended above the photosphere by magnetic fields (Figure 6.17). Prominences are cooler and denser than the coronal gas around them, and can last for days or weeks. At the end of this time, the material generally falls down to the lower chromosphere. In some cases, however, a sudden release of magnetic energy ejects the prominence upward, occasionally causing it to escape the Sun completely. The source of the material in prominences, as well as the mechanism by which

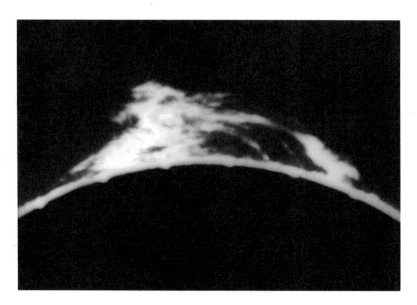

■ **Figure 6.17**   A solar prominence. This photograph was taken using a coronagraph, which blocks the light from the photosphere, allowing activity in the upper solar atmosphere to be seen. (Photograph courtesy of the Observatories of the Carnegie Institution of Washington.)

magnetic fields suspend and isolate them, is not fully understood. Except during eclipses, prominences can be seen only with the aid of special filters.

■ Solar Energy Generation

The Sun releases an enormous amount of energy, 3.9×10^{33} ergs per second. Even more remarkable than this output is the fact that it has been occurring for about 4.6 billion years! Over the years, several mechanisms have been proposed to account for the Sun's energy generation. The correct explanation of the Sun's energy generation must account for its extremely high productivity.

The earliest suggestion of how the Sun obtains its energy was that it is on fire. Hydrogen, the Sun's main constituent, is extremely combustible, but even if the Sun were on fire, the chemical reactions associated with fire are so inefficient that there would only be enough energy to last about 30 million years.

In the 19th century, the extreme age of the Earth (and consequently the Sun) came to be understood, and scientists realized that a better process was required. At that time, the process of **gravitational contraction** was proposed. A very slight contraction of the Sun, imperceptible from Earth except over long time periods, would re-

lease enough energy to account for the Sun's light output. This **Kelvin–Helmholtz contraction**, named for the physicists who proposed it, would provide a solar lifetime of at least 100 million years, but we now know that even this is insufficient. Although studies of star formation reveal that gravitational contraction is important when a star is first developing, it is superceded by another process (nuclear fusion) once the star has fully formed. Over the last century, considerable observational work has been done in an attempt to measure any decrease in the average size of the Sun. Although the evidence is inconclusive, the consensus seems to be that the Sun's diameter is probably not decreasing in size. (Theoretical models of stellar evolution actually predict that stars like the Sun slightly increase in size during their lifetime as a normal star, with an associated increase in temperature and luminosity. For example, such models predict that the Sun is now 40% brighter than immediately after its formation. Climatologists, however, argue that there is no evidence that the warming effect that this would have had on the Earth has actually occurred. Astrophysicists respond by suggesting that the atmosphere's heat-trapping ability has diminished through time at a rate that has effectively balanced the luminosity increase.)

Another suggested energy source was that energy is released from the Sun as meteors fall from elsewhere in the solar system into it. The main argument against this theory is that the mass increase of the Sun, which this process would cause, is not observed. Although a few comets have been observed to collide with the Sun, the total amount of material doing so is low.

The 20th century has seen the understanding of the processes that occur within atomic nuclei. **Nuclear fission** involves the splitting of nuclei, accompanied by the release of energy. Uranium is a common radioactive substance that spontaneously splits and releases energy in this way. Much of the internal heat of the Earth has been produced by fission of uranium. Could fission account for the Sun's energy output? If the Sun were pure uranium, fission could account for its total energy output. However, the Sun contains only one atom of uranium for every 100 million atoms of hydrogen, so fission is not the answer. (It was not until the 1920s that astronomers learned that the Sun is primarily hydrogen.)

After the Sun was found to be mostly hydrogen and helium, and as our understanding of nuclear processes increased, we learned that the answer to the question of stellar energy generation was **nuclear fusion**, or the combination of light nuclei to form a heavier one. In the Sun, this process involves the combination of four hydrogen nuclei to form one helium nucleus, which weighs slightly less than four hydrogen nuclei. The difference in mass is important, because the mass loss is responsible for the energy production.

Prior to the 20th century, mass and energy were believed to be entirely separate quantities. The German physicist **Albert Einstein** (1879–1955), however, shocked the scientific world by proving that they are equivalent and related by the famous equation

$$E = mc^2,$$

where E is energy, m is mass, and c is the speed of light. When four hydrogen nuclei fuse to form a helium nucleus, 0.007 of the mass is converted to energy. For 1 gram of hydrogen being fused to helium, 6.3×10^{18} ergs are produced. To yield the Sun's total energy output, 6×10^{14} grams of hydrogen must undergo fusion each second. An-

other way of stating what happens is that every second in the Sun, 657 million tons of hydrogen become 653 million tons of helium. Even at that prodigious rate of conversion, the Sun is so massive that fusion will last 10 billion years.

In order for fusion to occur at a rate sufficient to sustain the Sun's energy output, temperatures in excess of 10 million K are required. This is because the positively charged hydrogen nuclei are unable to overcome the electrical repulsion of their neighbors and combine with them unless they have the kinetic energy associated with such high temperatures. Consequently, fusion occurs only in the hot central core of the Sun.

Fusion occurs in a number of ways in stars. The exact process depends on core temperature and the stage in the star's life. The fusion process most important in the Sun is the **proton–proton chain**, in which two protons (hydrogen nuclei) join to form a nucleus of "heavy" hydrogen (deuterium). One of these nuclei combines with another proton, yielding a nucleus of "light" helium. Two of the light helium nuclei combine, giving a normal helium nucleus as well as two normal hydrogen nuclei. The net result is that four hydrogen nuclei become a helium nucleus. In addition, gamma-ray photons and **neutrinos**, which are massless subatomic particles that travel at the speed of light, are produced. We might expect the gamma-ray photons to reach the Sun's surface quickly, but this is not the case. Some energy from the Sun's core travels outward, but the material inside the Sun does not conduct heat very well, and relatively little of its total thermal energy escapes. (This is a good thing, because if more heat escaped, the Sun's core would soon cool to a point where fusion would no longer be possible!) The thermal energy of the gamma rays instead is transmitted by the less-efficient process of radiation. As the gamma rays travel outward, they are absorbed and re-emitted many times. Each time this happens, a slight amount of energy is lost, and some photons are re-emitted in an inward, rather than an outward, direction. As energy gradually moves outward, it travels through succeedingly cooler areas, and photons of longer and longer wavelength are produced. (Near the surface of the Sun, as mentioned previously, conditions change and the energy is transported by convection instead of radiation. The main reason for this is that the rate of temperature change with depth in the convective region is greater than the rate of change in the radiative region.) It is estimated that photons require an average of 1 to 10 million years to reach the surface. Neutrinos, however, travel out of the Sun in seconds, because they rarely interact with matter and are therefore not slowed.

■ The Neutrino Experiment

Because we have no way to observe the center of the Sun directly, astronomers have sought ways to verify their theories of what is occurring there. One of the major experimental efforts in this area has been the attempt to detect neutrinos produced in the center of the Sun during the fusion process. Our understanding of the fusion cycle in the Sun leads us to expect that a certain quantity of neutrinos is being produced in the Sun's core. The number of neutrinos produced and the number expected to hit the Earth can be calculated easily. The major difficulty in studying neutrinos is that they are hard to detect because they do not interact with matter to any great extent. Neutrinos travel through the Earth almost as though it were not there, so stopping them and counting them is no easy task. (Approximately 10 trillion solar neutrinos pass through the average person's hand every second!)

In the 1960s, Raymond Davis devised an experiment that could detect neutrinos. A giant tank containing 100,000 gallons of cleaning fluid was placed nearly 1 mile underground in a gold mine in South Dakota (Figure 6.18). (The apparatus was placed deep underground so that only neutrinos could enter the fluid, with other forms of radiation and particles being blocked by the intervening rock.) Neutrinos passing through the tank have a probability of interacting with chlorine atoms in the cleaning fluid and converting them to argon, eventually producing a measurable quantity of argon gas. Calculation indicated that six such interactions would occur every day. Several times during the year, helium gas is purged through the tank, collecting the radioactive argon atoms produced by neutrino interactions. Al-though the actual amount of gas produced is low, the quantity can be measured because the argon is radioactive.

The experiment has been going on since the 1960s. Throughout that time, the results have been disturbing, as evidence has indicated that only about one-third of the expected number of neutrinos are being detected. At first, the discrepancy was attributed to experimental difficulties, but improvements have convinced the experimenters that all neutrinos expected to cause interactions as they pass through the tank are being counted.

What is responsible for the discrepancy? There are a number of possibilities, but the answer is not yet known. Several variations of the proton–proton chain process occur, and each produces neutrinos with different energies, depending upon the specific sequence of reactions followed. Unfortunately, the current experiment can only detect the highest-energy neutrinos, which the Sun is thought to be producing in far less abundance than lower-energy ones. Experiments that could detect lower-energy neutrinos are possible, but they are more complicated and require detecting material that is much more expensive than cleaning fluid. Another complication is uncertainty about the nature of the neutrino itself. Although current theory predicts the neutrino to be massless, alternate theories suggest that it may have a small mass after all. If the neutrino indeed turns out to possess mass, there will be major implications for both theory and experiment.

The most disturbing aspect of the neutrino detection experiment is that it may be telling us that we do not understand the Sun as well as we thought we did. Because neutrinos leave the Sun's core instantly, but photons require 1 million years to reach the photosphere, the lack of neutrinos may mean that something unusual is currently occurring in the Sun's core. The relationship, if any, between the neutrino deficiency and such phenomena as the solar cycle and the Maunder minimum is also unknown. This is certainly an area of science in which future breakthroughs in understanding are likely and necessary.

■ **Figure 6.18** The underground neutrino detector, a tank containing 100,000 gallons of cleaning fluid. Neutrinos penetrating rock and reaching this tank may interact with chlorine atoms in the fluid, after which evidence of the interactions can be analyzed. (Brookhaven National Laboratory/Raymond Davis.)

■ Solar Spacecraft

Since the beginning of the space age, many spacecraft have studied the Sun. Because of the Sun's extreme brightness, normal astronomical observatory satellites cannot observe it, so special spacecraft have been sent into space to conduct solar studies. The United States launched a series of Orbiting Solar Observatories (OSOs) early in the 1960s. The OSOs had several advantages over ground-based observations in that their view was not hindered by atmospheric distortion and wavelength limitations, nor was observation limited by long hours of darkness. The most sophisticated solar observations from space were conducted by the Skylab manned orbital station in 1973 and 1974. A solar telescope with X-ray and ultraviolet detectors revealed much new information about the solar atmosphere and its interactions with interplanetary space.

In the 1970s, the Helios spacecraft was placed into a solar orbit, taking it within 0.1 astronomical units of the Sun for measurement of atomic particles and magnetic fields. The Solar Maximum Mission, as mentioned earlier, was launched in 1980 to monitor the Sun during the sunspot maximum at that time. It blew several fuses early in its mission, but it was repaired by a Space Shuttle crew in 1984.

The next major solar mission planned is Ulysses, which will observe the Sun from far above and far below the ecliptic plane. It will be launched from the Space Shuttle toward Jupiter, whose gravity will place it in an orbit that takes it above and below the Sun, exploring previously inaccessible regions of the solar system. Study of these regions will be informative, because coronal holes are most often found near the poles of the Sun, indicating that solar wind conditions above and below the Sun may be far different from those in its equatorial plane, which roughly corresponds to the ecliptic plane.

Because many wavelength regions of solar light output cannot be studied from the Earth, new orbiting observatories for solar study are planned for the 1990s. These new spacecraft, along with refined Earth-based observations, improved neutrino detection experiments, and advances in theoretical studies, should greatly increase our knowledge of "our star."

■ Chapter Summary

The Sun is a star composed largely of hydrogen and helium. It is a gaseous object that lacks a distinct surface and decreases in density gradually from the center out. What appears to be the visible surface of the Sun is the photosphere; above that lie the chromosphere and corona, from which a constant stream of material, the solar wind, flows.

The interior of the Sun can't be studied directly, but it is modeled theoretically. The core of the Sun, where its energy is produced, is extremely hot and dense. Energy flows outward by radiation and then convection.

Magnetism is important in the Sun, which undergoes cyclic changes over an 11-year period. Sunspots are dark areas in the photosphere produced where local magnetic field variations reduce the amount of energy released. Other solar activity includes faculae, flares, and prominences.

The Sun's energy is produced by nuclear fusion, the combination of hydrogen into helium. Terrestrial experiments seek to detect neutrinos, a by-product of solar fusion, but the numbers of them detected to date have been lower than predicted.

■ Chapter Vocabulary

aurora	flare	Maunder minimum	radiative zone
chromosphere	granulation	neutrinos	Schwabe, Heinrich
conduction	gravitational contrac-	nuclear fission	solar constant
convection	tion	nuclear fusion	solar wind
convective zone	heliopause	penumbra	spicule
corona	heliosphere	photosphere	star
coronal hole	Kelvin–Helmholtz	plasma	sunspot
differential rotation	contraction	prominence	sunspot cycle
dipole field	limb darkening	proton–proton chain	supergranulation
Einstein, Albert	magnetic field	radiation	umbra
faculae	Maunder, E.W.		

■ Review Questions

1. In what ways does the Sun affect the solar system and its various objects?

2. Why do we know more about the Sun than other stars?

3. What is a star?
4. What is the age of the Sun, and what is its expected lifetime?
5. What is the current theory of how a star like the Sun is born and dies?
6. Give the following basic physical properties of the Sun: distance from Earth, diameter, mass, rotational rate, composition.
7. What forms the Sun's apparent outer boundary?
8. Describe the Sun's three atmospheric layers.
9. Explain both granulation and supergranulation.
10. Explain why the chromosphere and corona are so hot.
11. What is the solar wind, and what effect does it have on Earth?
12. How is the Sun's interior studied? Describe the major interior layers of the Sun.

13. Describe the conditions at the center of the Sun.
14. Describe the three mechanisms for transporting heat energy. Which are important inside the Sun, and where?
15. Describe the solar activity cycle, and list the solar phenomena associated with it.
16. What sources of the Sun's energy have been suggested over the years?
17. How does the process of nuclear fusion work?
18. How long does it take protons and neutrinos to travel from the core of the Sun to its surface? Why do they travel at different speeds?
19. Why is the ongoing solar neutrino detection experiment important?
20. Summarize the work of spacecraft designed to study the Sun.

■ For Further Reading

Eddy, John A. *A New Sun: The Solar Results from Skylab.* NASA, Washington, DC, 1979.

Frazier, K. *Our Turbulent Sun.* Prentice-Hall, Englewood Cliffs, NJ, 1983.

Giovanelli, Ronald. *Secrets of the Sun.* Cambridge University Press, Cambridge, England, 1984.

Nicholson, Ian. *The Sun.* Rand McNally, Chicago, 1982.

Noyes, Robert W. *The Sun, Our Star.* Harvard University Press, Cambridge, MA, 1982.

Wentzel, Donat G. *The Restless Sun.* Smithsonian Institution Press, Washington, D.C., 1989.

Zirin, Harold. *Astrophysics of the Sun.* Cambridge University Press, Cambridge, England, 1988.

7

EARTH

MARS

MERCURY

MOON

IO

EUROPA

GANYMEDE

CALLISTO

VENUS

TITAN

Introduction to the Planets and Their Satellites

CHAPTER OBJECTIVES

After studying this chapter, you should be able to

1. Realize that the nine planets are a diverse group of objects and that each has distinctive characteristics.
2. List the three basic types of planets, and describe the characteristics of each.
3. Understand tidal forces and their results.
4. Describe the phenomenon of spin–orbit coupling.
5. Understand phenomena associated with rotation, such as obliquity, precession, and oblateness.

Several planets and satellites, reproduced to scale. (NASA Photographs. Composite prepared by Stephen Paul Meszaros.)

One of the many important results of the space age is that the planets and their satellites have changed from being simply "lights in the sky" to worlds whose atmospheres, surface features, internal structures, and histories are beginning to be understood. Objects about which very little was known a decade ago, such as the satellites of Uranus, are now being discussed in detail in numerous scientific papers. At the same time, new satellites have been discovered, and the discovery of more satellites and even a planet or two beyond Pluto may occur in the future.

No two of the nine planets are exactly alike, and the same thing can be said of the many satellites as well. At the same time, similarities exist between various bodies. It helps in understanding the diversity among planets and satellites if we categorize them into groups of objects with somewhat similar properties. As mentioned in Chapter 2, planets and satellites can be divided into three types — terrestrial, gas giant, and icy. This is an oversimplification (for example, some satellites are intermediate between terrestrial and icy), but it is a good starting point for further discussion.

Recall also from Chapter 2 that there is no physical difference between planets and satellites, except that a satellite will always be smaller than the planet around which it orbits. However, seven satellites are larger than the smallest planet, Pluto, and two satellites are larger than the second smallest planet, Mercury. On many occasions, therefore, we will discuss planets and satellites collectively. For want of a better term, we will generally use the terms object and body to denote a planet or satellite when no distinction is made.

Appendix 2 contains planetary data, including a table of the planets' orbital and physical properties, a diagram of their relative diameters, and maps of their orbits. Appendix 3 contains similar data for the planetary satellites.

■ Planets

The nine planets together comprise the most interesting and diverse group of objects in the solar system, ranging from huge, gaseous behemoths like Jupiter and Saturn, to Earth, our familiar home in space, to cold, small, icy Pluto. All of them except Pluto have been explored by flyby spacecraft, with both orbital missions and landings accomplished at Venus and Mars.

Why are there so many differences among the planets? The main reason is clearly distance from the Sun, because the three types—terrestrial, gas giant, and icy—occur in that order from the Sun. This is underscored by the fact that three pairs of "twin" planets, with very similar bulk properties, exist: Venus and Earth, Jupiter and Saturn, and Uranus and Neptune. In each pair, the planets are adjacent to each other in distance from the Sun. While we will discuss the formation of the planets in detail in Chapter 29, the modern theory can be summarized as follows: the Sun and planets formed from a nebular region that contained gas and dust. As the Sun formed in the center of the solar system, where the temperature, pressure, and density of the nebula were highest, solid grains of material condensed from the remainder of the nebula as it cooled. Temperatures were low enough in the outer solar system that the condensation process formed ices, whereas only rocks and metals were produced in the warm inner solar system. Grains of these materials subsequently joined together, or accreted, and eventually formed planetary objects. Gas giants and icy objects formed in the cool outer solar system, whereas terrestrial planets formed in the warm inner solar system, where ices and light gases could not survive. Because objects that formed at different distances from the Sun experienced different conditions and incorporated different materials, adjacent planets might be expected to be similar to each other.

However, distance from the Sun is not the only factor responsible for influencing planetary properties, as considerable differences exist even between members of these adjacent pairs. Venus and Earth, for example, have atmospheres that are totally different, and their surface geologies are dissimilar as well. Although there is no one simple explanation for these differences, planetary geologists are beginning to understand some of the reasons for them. As we examine the planets individually in the coming chapters, an underlying goal will be to learn why each planet is the way it is.

■ Planetary Satellites

Every planet except Mercury and Venus is accompanied by at least one satellite. Sixty planetary satellites are now known, 56 of which are associated with the four gas giant planets. Although each satellite is unique, several statements can be made about planetary satellites in general. Two terrestrial planets (Earth and Mars) have satellites, and their satellites can also be considered terrestrial objects. Each gas giant planet has numerous satellites, most of which are probably intermediate between terrestrial and icy bodies. Both Pluto and its satellite, Charon, are icy objects.

One important property of satellites is that they are generally very small in relation to the planet around which they orbit. There are only two exceptions to this rule: Charon is about one-half the diameter of Pluto, and the Moon is about one-fourth the diameter of Earth. Satellites generally orbit in the equatorial planes of their planets, although some orbit in the ecliptic plane. (The Moon's orbit, for example, lies nearly in the ecliptic plane, not in the Earth's equatorial plane.) Most satellites orbit in the prograde direction (counterclockwise as seen from north of the ecliptic.) Those with retrograde orbits are considered likely to have been gravitationally captured by their planets. (Incidentally, there is no known case of a planetary satellite having a satellite of its own.)

Modern theories of solar system formation propose that most satellites accreted from small particles orbiting the planets as they themselves were forming. (The exceptions are those objects captured gravitationally by the planet around which they now orbit.) In the case of Jupiter's satellite system, there is variation in the physical properties of the major satellites based upon distance from Jupiter, with the inner satellites more rocky and the outer ones more icy. This indicates that as the satellites formed, heat emitted by Jupiter caused conditions to vary according to distance from the planet, just as with planetary formation around the Sun.

■ The Three Groups of Planets

The Earth and its nearest neighbors provide a convenient starting point for discussing the three basic planetary types. The four innermost planets (Mercury, Venus, Earth, and Mars) are called **terrestrial,** or Earth-like, since Earth is the best-known example and serves as a prototype for the group. (Many planetary satellites can be considered terrestrial objects as well.) Terrestrial objects are dense and composed primarily of rock and metal. They have definite surfaces and are solid throughout, except for a molten core, which the Earth now has and some of the others may have had in the past. The larger of these objects have atmospheres whose compositions are different from those of the solid portions of the planets. Terrestrial planets have either few or no satellites (Earth and Mars, collectively, have three), and none have systems of rings.

Terrestrial planets have a wide range of diameters: Earth is the largest (almost 13,000 kilometers, or 8000 miles), and Mercury is the smallest (about 5000 kilometers, or 3000 miles). Satellites that can be considered terrestrial objects range in diameter from slightly larger than Mercury to less than 30 kilometers (20 miles). An interesting property of terrestrial objects is that the smallest (those no more than a few tens of kilometers across) are generally not spherical. The spherical shape of solar system objects is produced by internal gravitational attraction, which in the smallest objects is too weak to produce sphericity.

Terrestrial planets are the densest solar system objects. The Earth's density is highest at 5.52 grams per cubic centimeter. Mercury and Venus also have densities above 5 grams per cubic centimeter, whereas those of Mars, the Moon, and some satellites of outer planets are between 3 and 4 grams per cubic centimeter. (It is estimated,

however, that if Earth's interior were not compressed by gravitation, its average density would only be 4.03 grams per cubic centimeter, whereas Mercury's uncompressed density would be a higher 5.31 grams per cubic centimeter.) The high density of terrestrial planets is in contrast to that of the gas giants and icy objects, whose densities range from about 2 to as low as 0.7 grams per cubic centimeter, the latter density being less than that of water! Density is a clear indicator of composition, and as we will learn in Chapter 9, the range of values for terrestrial objects gives clues to internal structure, particularly whether a metallic core is present and what its extent may be.

The **gas giant planets** (Jupiter, Saturn, Uranus, and Neptune) are located in the outer solar system. (These objects are sometimes called the **Jovian planets** because Jupiter, or Jove, is the largest example. There are no planetary satellites of this type.) As the name implies, these objects are large and composed mainly of hydrogen and helium gas, causing them to have low average densities. Inside these planets, hydrogen and helium probably liquify as a result of the high pressures found there. Uranus and Neptune may each contain a layer of liquid water, ammonia, and methane near their centers, whereas all four of the gas giants probably contain some solid materials (rock, metal, or ice) in their cores. While we speak of these objects as having atmospheres, they are simply the outermost layers of the planets, not separate regions. Like the Sun, gas giants lack distinct surfaces, and there is no distinct planet–atmosphere demarcation as with terrestrial planets. The gas giants have numerous (8 to 17) satellites each and possess ring systems composed of countless small orbiting icy particles.

There is a considerable size gap between the largest terrestrial planet, Earth, which is 13,000 kilometers (8000 miles) across, and the smallest gas giant, Neptune, which is 50,500 kilometers (31,000 miles) across. The reason why our solar system has no objects of intermediate diameter is not yet known.

Although Pluto is the only icy planet, many satellites in the outer solar system are also composed primarily of ices. (When we use the term ice for these objects, we mean not only water ice,

■ **Figure 7.1** Differential gravitational forces. Each of the three objects (A, B, and C) at left feel a gravitational attraction from the planet at right. The length of each *arrow* represents the relative magnitude of the attractive force. Note that as a result of the differences in attraction, the separations between the three objects increase.

but any frozen material that would be liquid or gaseous under normal room-temperature conditions. Besides water [H_2O] other common ices in the solar system include methane [CH_4], ammonia [NH_3], and carbon dioxide [CO_2].) In addition to ices, rocky material is probably present inside these objects, and their densities are intermediate between those of the gas giants and the terrestrial planets. Icy bodies are solid throughout and generally lack atmospheres, although some of Pluto's outer layers vaporize at perihelion and form a temporary atmosphere.

It is very easy to distinguish terrestrial objects from gas giants, and there are no intermediate forms. However, the distinction between terrestrial and icy objects is less distinct. There is no clear-cut boundary between them, and overall density is the main criterion used to identify the proportion of ice present. As mentioned previously, many satellites of the outer planets are made of mixtures of rocky and icy materials. Even though predominantly icy objects may contain rocky cores, their overall densities are generally lower than 2 grams per cubic centimeter.

■ Tidal Interactions

As mentioned in Chapter 3, planets influence one another gravitationally, with the main result being orbital perturbations. On the other hand, the separation between planets is great enough that, for the most part, the gravitation of one planet is too small to influence the physical properties, such as shape and rotational rate, of another planet. (As we will see, the Sun's stronger gravitation has influenced Mercury, the planet closest to it.) The situation changes, however, when we consider planets and their satellites. Because smaller separations are involved, gravitational forces are able to cause physical interactions through the mechanism of tides.

Tidal forces are **differential gravitational forces** caused by variation in gravitation with distance. Suppose a massive object attracts three smaller bodies, labeled A, B, and C in Figure 7.1. (We imagine a hypothetical case of fixed bodies with no orbital motion involved.) Because A is farthest from the attracting object and C is closest, A will feel the least attraction and C the most, as shown by the relative length of the arrows. Note that as a result of the displacements caused by differences in gravitational attraction, the distances separating A, B, and C increase.

Let us change the diagram, replacing the three separate bodies with a single, larger body (Figure 7.2). The same effect happens at points A, B, and C in the larger body, with the net effect being a "stretching" of the body along the line connecting points A, B, and C with the attracting body. (In practice, of course, as one body is stretched tidally by another, it will in turn have a similar effect on the other body.) Tidal deformations of solid bodies are generally very small. The Earth's liquid oceans, however, deform easily and produce the well-known ocean tides. Figure 7.3 shows two views of Earth's oceans, one as they would appear with no tidal deformation and another as they are deformed, primarily by the Moon. The rotation of the solid Earth under the ocean bulges causes coastal areas to experience high tide twice daily when at points 1 and 3 and low tide twice daily when at points 2 and 4. Tides do not occur at the same time every day, because the Moon's orbital motion around the Earth causes the position of the bulges to move around the Earth once a month.

As was discussed in Chapter 3, the force of gravitational attraction decreases with the inverse

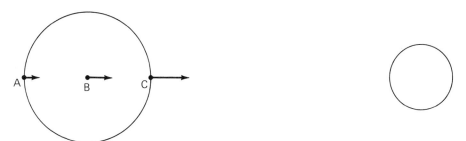

■ **Figure 7.2** Differential gravitational forces in a larger body. Each of the three points (*A, B,* and *C*) in the planet at left feel a gravitational attraction from the satellite at right. A slight "stretching" of the planet is the result, similar to the effect on the three separate objects shown in Figure 7.1.

square of the distance between objects. (We call this a 1/R² dependence.) Calculations show, however, that tidal forces decrease with the inverse cube of the distance between objects, or a 1/R³ dependence. This means, for example, that although the Sun exerts a much greater gravitational attraction on the Earth than does the Moon, its distance causes its tidal influence on the Earth to be only about half that of the Moon. (The precise value is 45%.) At small separations, tidal forces increase dramatically. We will see in Chapter 18 that if two approaching bodies come close enough together, tidal stretching can actually cause the smaller of them to be broken apart.

■ **Figure 7.3** Tidal deformations of Earth's oceans. (a) The appearance of the Earth and its oceans if no tidal influences were present. (b) The Moon causes the oceans to be deeper than normal at points 1 and 3 and shallower than normal at points 2 and 4. (For simplicity, the Sun's contribution is ignored. The effect shown here is greatly exaggerated.)

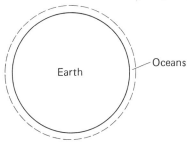

(a)

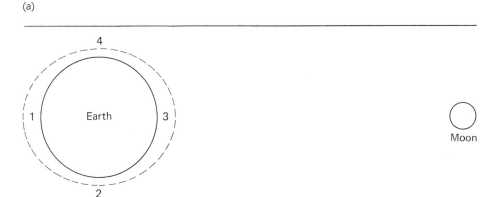

(b)

■ The Rotation of Planets and Satellites

In addition to their orbital motions, each of the planets and satellites rotates around an axis. An object's rate of rotation can be measured by looking at its surface features and timing how long it takes them to complete a circuit around the planet, or by the use of Doppler effect radar techniques, as described in Chapter 4. Just as most objects orbit in the prograde direction, the rotational direction of most objects is also in this prograde sense. Recall from Chapter 3 that a planet's orbital period can be measured either with respect to the stars or to the Earth and Sun. In much the same way, **sidereal day** is measured with respect to the stars and represents the actual time required for the planet to rotate. The **solar day** is the time between successive passages of the Sun at its highest point in the sky, as at noontime on Earth. The lengths of these two periods differ because the planet is orbiting the Sun at the same time it is rotating.

The "natural" rotation period for most solar system objects seems to be about 10 to 12 hours. In other words, some researchers believe that most of the solar system's objects rotated at that rate immediately after they were formed. Obviously, many objects, including the Sun, Earth, and Moon, rotate much more slowly than this today, so some force must have been responsible for slowing their rotational rates. Tidal forces are believed to have caused these slowings of rotation.

When one object orbits another, the orbital period and the rotational periods of both objects are, in general, three different quantities. However, the gravity of each object slightly deforms the other, producing the tidal bulges described earlier. As a satellite orbits a planet, their mutual gravitational forces will attempt to hold their tidal bulges in line, even though both objects may be rotating faster than the orbital period. As a result, frictional forces are produced that heat both objects internally and also change both rotational and orbital periods.

The effects of tidal forces influence all planet–satellite systems. The Earth currently rotates faster than the Moon orbits the Earth (24 hours versus 27.3 days). This transfers energy from the Earth into the Moon's orbit. As a result, the Earth's rotation is gradually slowing, and the Moon is moving farther away from the Earth. In the case of Mars and its satellite Phobos, the satellite orbits faster than Mars rotates (7.7 hours versus 24.6 hours). In this case, the energy transfer works in the opposite direction, and Phobos is losing orbital energy and slowly spiraling toward the planet, whose rotational rate is gradually increasing.

As a result of tidal forces, many solar system objects have rates of rotation that are identical to their orbital periods, or else are related by some simple numerical ratio. This phenomenon is called **spin–orbit coupling.** The Moon is a good example, as its orbital and rotational periods are both 27.3 days. (This is why we only see one side of the Moon from the Earth.) The Earth's and Moon's rotational rates were probably once about 10 to 12 hours, as mentioned previously. The reason that the Moon's rate has been slowed so much more than the Earth's is that the Earth's greater mass causes more tidal braking effects on the Moon's rotation than the Moon causes on the Earth. It is theorized, however, that given sufficient time, both of the rotations and the orbital period in the Earth–Moon system will slow further and eventually become equal. The majority of the satellites in the solar system also have this 1 : 1 coupling. (Pluto and its satellite Charon are a special case, because both of them rotate in the same amount of time as Charon's orbital period.) Mercury is close enough to the Sun that solar tidal forces have caused it to display spin–orbit coupling, although not at a perfect 1 : 1 ratio. Mercury orbits the Sun in 88 days and rotates in 59 days, which is a 3 : 2 coupling. (This means that Mercury rotates three times during two orbits of the Sun.)

The line on a planet marking the halfway point between its rotational poles is called its equator. For most planets, the plane of the equator is close to the plane of the planet's orbit, making the rotational axis nearly perpendicular to the orbital plane. The angle between the axis and true perpendicularity is called the **obliquity** of the planet (Figure 7.4). The obliquity of the Earth's axis is 23.5° and is responsible for seasons, as we will see in Chapter 11. The gravitational forces of

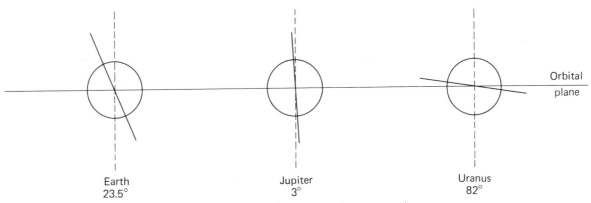

■ **Figure 7.4**   Obliquity. The axial tilts of the planets Earth, Jupiter, and Uranus are shown relative to the orbital plane. The *vertical dotted lines* represent zero-obliquity rotational axes, whereas the *solid lines* represent the planets' actual rotational axes.

other solar system bodies can combine to cause small periodic variations in planetary obliquities over long times. For example, Earth's obliquity is believed to vary through a range of about 1° over a period of 10^4 to 10^5 years.

If we examine a planet over the course of several rotations, we note that its axis points toward a particular direction in space. The star closest to this fixed point in the sky is called the **pole star** of that planet. For example, the Earth's northern pole star, **Polaris,** is a fairly conspicuous star now located within 1° of the true pole position. Polaris therefore appears to remain virtually motionless during the course of a night, whereas other stars rise and set due to the Earth's rotational motion (Figure 7.5).

■ **Figure 7.5**   Time exposure of the northern sky showing the motion of the stars around the north celestial pole as a result of the Earth's rotation. Polaris is the bright star near the top edge of the photograph, just left of center. Because of its proximity to the pole, Polaris' motion was insignificant during this 10-minute exposure. (Photograph by the author.)

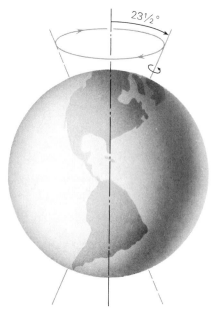

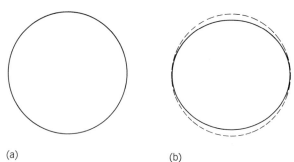

(a)   (b)

■ **Figure 7.7** Oblateness. (a) A circle. A figure with the obliquity of the Earth could not be distinguished from a circle at this scale. (b) A figure with an oblateness equal to that of Saturn. Using a ruler, you can easily verify that it is not a perfect circle.

■ **Figure 7.6** Precession. The Earth's present rotational axis (*solid line*) gradually moves, tracing out a circle over a period of 26,000 years. In 13,000 years, the axis will point in the direction indicated by the *broken line*. (From Abell, George, David Morrison, and Sidney Wolff. *Exploration of the Universe,* 5th ed. Saunders College Publishing, Philadelphia, 1987.)

Polaris, however, has not always been the pole star. This is because tidal effects cause the orientation of the Earth's axis to change slowly, somewhat like the wobble of a child's top. This process, called **precession**, occurs because the Earth's rotation causes it to bulge slightly at the equator. Because of the Earth's obliquity, its equatorial bulge is not perfectly aligned with the Sun and Moon. The Sun and Moon try to pull the bulge into the plane of the ecliptic and the plane of the Moon's orbit, respectively. Instead of decreasing obliquity, however, the result is a precession of the axis. Precession is extremely slow, and a complete circuit requires about 26,000 years (Figure 7.6). (Other planets also precess;

Mars' precessional period is about 175,000 years.)

As planets rotate on their axes, the forces that result from their rotational motion cause them to deform somewhat from perfect sphericity. Although this effect is most noticeable in the gas giant planets, it also occurs in the solid ones. As a result, there is a net loss of material from polar regions and a net gain in equatorial regions. This causes the diameter of the planet, measured through the poles, to be smaller than the diameter at the equator (Figure 7.7). The quantity that describes this polar flattening is called **oblateness**. Oblateness is calculated by dividing the difference between a planet's equatorial and polar diameters by its equatorial diameter. For example, Earth's polar diameter is 12,713.5 kilometers (7900.2 miles) and its equatorial diameter is 12,756.3 kilometers (7926.8 miles), a difference of 43.0 kilometers (26.7 miles). This means that Earth's oblateness is 1 part in 297, or about 0.003. Jupiter's oblateness is much larger (0.065), whereas Saturn, the most oblate planet, has a value of 0.108. Mercury and Venus, the slowest-rotating planets in the solar system, have oblateness values of 0.

■ Chapter Summary

The planets and their satellites compose an interesting and diverse group of solar system objects. Much of the diversity probably results from the fact that each planet formed in a different portion of the solar nebula.

The planets and satellites can be divided into the terrestrial bodies, which are largely rock and metal; the gas giants, which are primarily hydrogen and helium; and objects composed of ice and some rock.

Planets and satellites are generally close enough that they can deform one another by tidal interactions. Tidal effects cause slowing of planet and satellite rotation in many cases, often producing the phenomenon of spin–orbit coupling. Planetary rotation is also complicated by precession, a long-term variation in axial orientation.

■ Chapter Vocabulary

differential gravita-
    tional force
gas giant planets
ice

icy planets
Jovian planets
oblateness
obliquity

Polaris
pole star
precession
sidereal day

solar day
spin–orbit coupling
terrestrial planets
tidal forces

■ Review Questions

1. List the basic properties of each of the three types of planets.
2. Describe the differences and similarities between planets and satellites.
3. What is the normal relationship between planet and satellite diameter? List any significant exceptions.
4. How can the rotational rate of another object be measured from Earth?
5. Explain differential gravitational forces.
6. Explain how tides work in the Earth's oceans.
7. Why are many solar system objects rotating slower than the rapid rate assumed to be "natural" for them?
8. Explain what is meant by spin–orbit coupling.
9. What is obliquity, and with respect to what is it measured?
10. What is precession, and what is responsible for it?
11. What is oblateness, and why are planets oblate?

■ For Further Reading

Scientific American. *The Planets*. Freeman, San Francisco, 1983.

Scientific American. *The Solar System*. Freeman, San Francisco, 1975.

8

Constituents of Terrestrial Planets: Elements, Minerals, and Rocks

CHAPTER OBJECTIVES

After studying this chapter, you should be able to

1. Understand basic concepts involving atoms, such as ionization, bonding, isotopes, and radioactive decay.
2. Describe the distribution of chemical elements in the solar system and within a typical terrestrial planet.
3. Summarize the basic properties of minerals, and explain the difference between a mineral and a rock.
4. Describe each of the three types of rocks and how each is formed.
5. Explain how the ages of rocks are determined by the use of radiometric age dating.

Although most people rarely think about the rocks beneath their feet, the Earth contains many natural wonders, such as this giant crystal of the mineral beryl at the Palermo Pegmatite near North Groton, New Hampshire. What mineral wonders await geologists on Mars, Mercury, and other worlds? (Photograph by the author.)

Although there are many differences among the various terrestrial planets, all of these worlds have some features in common, as we have seen in Chapter 7. One of their similarities is composition, as all of them contain many of the same major constituents. In this chapter, we describe the materials of which the terrestrial planets are composed.

The compositions of terrestrial planets can be examined on several levels. When we look at planetary surfaces, we see rocks. Upon closer examination, however, we find that rocks are made of simpler components, minerals, which are naturally occurring chemical compounds. We can further examine minerals and describe the atomic elements of which they are composed. In this chapter, we examine the components of the terrestrial planets at each of these three levels.

■ Atoms and Elements

As discussed in Chapter 4, the atom is the basic building block of all matter. Recall that the atom is made of three simple components (see Figure 4.4): protons, positively charged particles found in the nucleus; neutrons, chargeless particles found in the nucleus; and electrons, negatively charged particles orbiting the nucleus. Electrons are much less massive than protons and neutrons, which both have about the same mass. In a neutral atom, the number of protons (a quantity called the **atomic number**) is equal to the number of electrons. The number of neutrons is generally equal to, or greater than, the atomic number.

Specific varieties of atoms, called **elements**, are distinguished by their atomic numbers. For example, the atomic number of oxygen is eight. All oxygen atoms have eight protons, but the number of both electrons and neutrons may vary. Variations in the number of neutrons give rise to **isotopes**, or varieties of elements with different masses. Most oxygen has eight neutrons, produc-

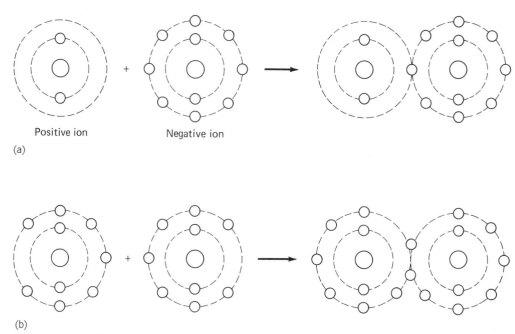

Positive ion   Negative ion

(a)

(b)

■ **Figure 8.1** Atomic bonding. (a) A positive ion and a negative ion join by ionic bonding to form a molecule. (b) Two neutral atoms join by covalent bonding to form a molecule.

ing oxygen-16. (The number 16 is the sum of the number of protons and neutrons and represents the relative weight of the nucleus.) Other isotopes of oxygen include oxygen-17 and oxygen-18, with 9 and 10 neutrons, respectively. Atoms often occur without the normal number of electrons and are called **ions** when in this form. Unlike neutral atoms, ions have net electrical charges, because their numbers of protons and electrons are not equal. Positive ions are produced when electrons are lost, and negative ions, when electrons are gained. The most common ion of oxygen, for example, has two extra electrons, giving a charge of −2. Why does oxygen commonly occur as a doubly charged negative ion? The reason is oxygen's electronic configuration. A neutral oxygen atom has six electrons in its outer electron shell, or orbital level. This level has a capacity for eight electrons, and by gaining two additional electrons, the ion formed is in a filled orbital configuration, which is very stable. Many ions are found to have gained or lost just enough electrons to produce this stable, filled orbital configuration.

Bonding, the process of combining atoms to form molecules and compounds, involves inter-

actions among atoms on the electronic level. (A **molecule** is any combination of atoms, but a **compound** contains at least two different elements.) Two forms of bonding are important in forming minerals (Figure 8.1). **Ionic bonding** involves oppositely charged ions joining together because of charge difference. Ordinary table salt forms in this fashion. A sodium ion with a +1 charge and a chlorine ion with a −1 charge join to form neutral salt. (Both of these ions have the stable configuration of eight electrons in their outermost orbital levels.) **Covalent bonding** involves sharing of electrons between atoms so that both seem to have eight in their outer levels. The oxygen (O_2) molecule forms in this way from two atoms of oxygen. (Some bonding is intermediate between strictly ionic or covalent.)

A total of about 90 stable elements occur in nature and are found in terrestrial planets. In addition to these stable elements, there are a number that are **radioactive** and can be produced in the laboratory but are never found in nature. If an element is radioactive, its nucleus spontaneously changes into another type of nucleus by emitting particles or radiation. Given the age of the solar system, most such elements once present in the planets are gone, having decayed into other elements. In addition to elements that are always radioactive, most common elements have at least one radioactive isotope that decays by some process. Carbon-12, for example, is stable, but carbon-14 undergoes radioactive decay. Another example is aluminum and its isotopes. Although aluminum is a major component of terrestrial planets, the radioactive isotope aluminum-26 does not exist in planets today. It was present at the time of the solar system's formation but quickly decayed into stable magnesium-26, which survives to the present.

■ The Distribution and Origin of Elements

The two most abundant elements in the universe are hydrogen and helium, which are the major components of stars. Hydrogen and helium are the two lightest elements, and for this reason, they are rare in planets like the Earth. The velocity with which an atom or molecule moves in a planetary atmosphere is inversely proportional to its mass. Hydrogen and helium are so light that they move fast enough to escape from the gravity of even the most massive terrestrial planet, Earth. (Considerable hydrogen exists in the water molecules on Earth, of course, and some helium is trapped inside the Earth, having formed there as a product of the radioactive decay of uranium.)

Because terrestrial planets contain so little hydrogen and helium, they consist primarily of the heavier elements, which are relatively scarce in the universe as a whole. For example, the most abundant elements in the Earth's outermost layer, or crust, are oxygen, silicon, aluminum, iron, calcium, potassium, sodium, and magnesium (Table 8.1). These eight constitute the majority of the crust, and none of the other elements constitute even 1% of the total. (As we shall see, the abundances of elements in the Earth as a whole are not the same as their crustal abundances.) Many of the scarce elements in the Earth's crust are economically important natural resources. Elements like gold, silver, and copper have very low average abundances and can only be recovered economically from areas of the crust where some geological process has greatly concentrated them compared to their average abundances.

Careful study of the solar system reveals many peculiarities in element distribution. First of all, different elements are found in different parts of the solar system. We have already seen that the Sun is mostly hydrogen and helium, whereas terrestrial planets are made of heavier elements. The gas giants are essentially identical to the Sun in composition, and icy objects contain frozen gases, which are not very abundant elsewhere in the solar system. These differences are due primarily to conditions in the solar system while the objects in it were forming. (Temperature variations were probably most important.) Another factor is mass of the objects, because large planets have gravity that is strong enough to retain lighter, fast-moving elements.

At the same time, we find that there are different elements at different places within each planet. Although we have no way to sample material directly deep inside planets, we know from indirect evidence (discussed in Chapter 9) that the core of the Earth has a density of about 13 or 14 grams per cubic centimeter. A region this dense could not be made of the same materials as the crust, whose average density is under 3 grams per cubic centimeter; instead, the core is believed to be mainly metallic iron and nickel. The process responsible for this difference is **differ-**

■ Table 8.1   Composition of the Earth's Crust*

Element	Symbol	Abundance by Mass	Abundance by Number of Atoms
Oxygen	O	46.60%	62.55%
Silicon	Si	27.72%	21.22%
Aluminum	Al	8.13%	6.47%
Iron	Fe	5.00%	1.92%
Calcium	Ca	3.63%	1.94%
Sodium	Na	2.83%	2.64%
Potassium	K	2.59%	1.42%
Magnesium	Mg	2.09%	1.84%
All others		1.41%	†

* Data from Mason, Brian. *Principles of Geochemistry.* Copyright(©) 1966 by John Wiley and Sons, Inc.
† Values in this column include only the eight most abundant elements.

entiation, the separation of a newly formed planet into layers as heavy materials sink to the core and lighter ones rise to the surface. We examine the differentiation process in more detail in Chapter 9.

Where did the elements in the terrestrial planets originate? According to present theory, conditions during the Big Bang allowed only hydrogen, helium, and small amounts of lithium to form. (That is why hydrogen and helium are the most common elements in the universe.) The first stars to form after the Big Bang, therefore, contained only these light elements. However, nuclear reactions inside stars produce heavier elements, particularly during the brief but spectacular supernova explosion that marks the end of the life of a very massive star. This production of elements, called **nucleosynthesis**, has been occurring ever since the Big Bang and was the source of the heavier elements now contained in the solar system. In addition to providing the setting for considerable nucleosynthesis, supernova explosions also spew newly formed elements into the interstellar medium, from which they can be incorporated into newer stars and solar systems as they form. For these reasons, we believe that the terrestrial planets contain elements originally formed inside stars older than the Sun.

■ Minerals

Of the approximately 90 nonradioactive elements found in the Earth's crust, only about two dozen occur in pure, solid form. Some examples are carbon (which occurs as the minerals graphite and diamond), gold, sulfur, silver, and copper. Most elements occur in minerals that are compounds of two or more elements. A **mineral** is defined as a naturally occurring solid substance (either a pure element or a compound) with a crystalline structure and a definite chemical formula. Because the atoms in minerals are arranged in an orderly structure called a crystal lattice, the external appearance of most minerals depends in large part on this internal arrangement of atoms, as well as on which specific types of elements the mineral contains (Figure 8.2). It is because of the crystalline arrangement of the atoms inside minerals that many of them occur as beautiful **crystals,** or solid forms bounded by flat surfaces.

Well over 3000 terrestrial minerals have been identified, but fewer than two dozen of these, called the **rock-forming minerals,** constitute the majority of the Earth's crust. These minerals are also important in lunar samples and meteorites. (The names and chemical formulas of all minerals mentioned in this book are listed in Appendix 4.) Examples of common rock-forming

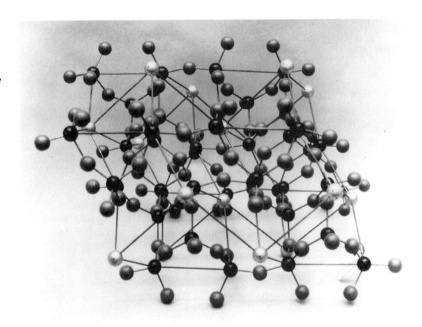

■ **Figure 8.2** The crystalline structure of a mineral. This model of the atomic arrangement of the mineral feldspar shows the regularity of that arrangement. The physical properties of minerals depend in large part upon their crystalline structures. (Photograph by the author.)

■ **Figure 8.3** Examples of common rock-forming minerals. (Top) Left to right: olivine, pyroxene, amphibole, and biotite. (Bottom) Left to right: muscovite, orthoclase feldspar, plagioclase feldspar, and quartz. (Photograph by the author.)

minerals are shown in Figure 8.3. Minerals are characterized by a number of important physical properties, such as color of bulk specimens, color of powdered material, hardness, density, surface luster, crystal shape, and whether breakage occurs along a cleavage plane (a flat surface caused by a weakness in the crystal structure) or irregularly. With practice, identification of the major minerals is not extremely difficult. A major difficulty is that minerals rarely occur in the crust in pure crystalline form but generally occur in mineral mixtures called rocks. Individual minerals in rocks are often difficult to identify.

Minerals are divided into eight groups based on chemical composition (Table 8.2). The most abundant of these, constituting over 90% of the Earth's crust by mass, is the **silicate** group. All silicates contain a structural unit made of one silicon atom surrounded by four oxygen atoms (Figure 8.4). This group of atoms forms a unit known

■ **Table 8.2   The Eight Mineral Groups**

Name	Description	Examples	Formulas
Native elements	Pure element	Diamond	C
		Graphite	C
Sulfides	Metal and sulfur	Troilite	FeS
		Galena	PbS
Oxides	Metal and oxygen	Hematite	Fe_2O_3
		Corundum	Al_2O_3
Halides	Metal and halogen	Halite	NaCl
		Fluorite	CaF_2
Carbonate	Metal and carbonate (CO_3)	Calcite	$CaCO_3$
		Dolomite	$CaMg(CO_3)_2$
Sulfates	Metal and sulfate (SO_4)	Barite	$BaSO_4$
		Gypsum	$CaSO_4 \cdot 2H_2O$
Phosphates	Metal and phosphate (PO_4)	Apatite	$Ca_5(PO_4)_3F$
		Brabanite	$CaTh(PO_4)_2$
Silicates	Metal and silicate (SiO_4)	Quartz	SiO_2
		Olivine	$(Fe,Mg)_2SiO_4$

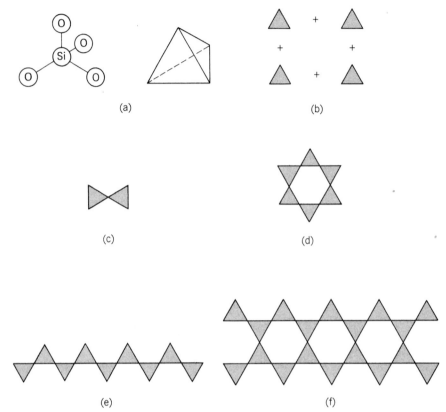

■ **Figure 8.4** Silicates. (a) The arrangement of one silicon atom and four oxygen atoms to form the silicate tetrahedron. A tetrahedron is a four-sided figure, each side of which is an equilateral triangle. (b) Some silicates contain isolated tetrahedra linked by positive metal ions. (c) Pairs of tetrahedra are the basis for forming other silicates. (d) Rings of six tetrahedra provide the basis for forming some silicates, including the large six-sided crystal shown in this chapter's opening photograph. (e) Many silicates contain single chains of tetrahedra. (f) An even more complex form of silicate is based on double chains of tetrahedra.

as the silicate tetrahedron, because of its shape. Silicate minerals contain these tetrahedra (either singly or joined in rings, chains, sheets, or networks) as well as metal ions linking them together. Familiar silicates include quartz, feldspar, emerald, and garnet.

Although the Moon is the only other solar system object whose surface rocks have been examined on Earth, indications are that silicates are important components of most terrestrial objects. Lunar rocks were found to contain silicates similar to those found on Earth, and silicates are also important components of many meteorites. Although individual minerals could not be identified by the spacecraft that landed on Venus and Mars, these spacecraft were able to identify

which elements were present in the surface rocks and soil on these planets. The results obtained support the theory that their surfaces are composed of silicates.

For other solar system objects, such as Mercury, asteroids, and planetary satellites, evidence obtained by reflectance spectroscopy indicates that silicates are present at their surfaces. Recall from Chapter 4 that reflectance spectroscopy is the analysis of sunlight reflected from an object. For a given mineral, the percentages of incoming light reflected at various wavelengths are often unique, allowing mineral identification by remote sensing. In practice, spectra of terrestrial minerals are obtained in laboratories and compared with those of other solar system objects obtained by

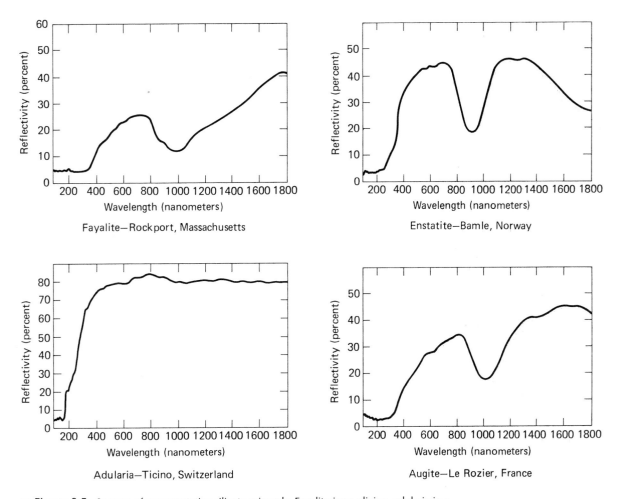

■ **Figure 8.5** Spectra of representative silicate minerals. Fayalite is an olivine, adularia is a feldspar, enstatite is an orthopyroxene, and augite is a clinopyroxene. (Spectra measured by the author.)

remote sensing. This process allows the identification of specific minerals in solar system objects. (Examples of reflectance spectra of terrestrial examples of important silicates are shown in Figure 8.5).

■ Rocks

As mentioned earlier, minerals rarely occur as large, pure samples but are more likely to occur in mixtures called **rocks.** Some rocks contain individual mineral grains that are large enough to be identified, whereas others are so homogeneous that individual minerals can no more be recognized in them than individual ingredients can be

identified in a baked cake! Rocks are divided into three types according to how they were formed. **Igneous rocks** are those formed by the solidification of hot, molten rock, called **magma** or **lava.** These are the most abundant rocks in the Earth's crust and presumably also in the crusts of other terrestrial objects. **Sedimentary rocks** are those formed from the accumulation of **sediment,** fragments of pre-existing rocks that have been transported and deposited. The formation of sedimentary rocks usually requires water, and this type of rock is therefore probably limited to Earth. (It is possible that Mars may have some sedimentary rocks, because indications are that water may have been present briefly in the past.) The final type of rock is **metamorphic,** the result of a rock's experiencing **metamorphism,** which is a change in mineralogy and texture due to exposure to heat and pressure. It is important to note that a rock cycle occurs on Earth, where rocks of one type are transformed into rocks of other types by crustal activity. (One result of the terrestrial rock cycle is that extremely old rocks are rare on the Earth, because most of the crustal rocks formed in the first billion years or so of the Earth's history have been changed into something new. This complicates matters for geologists seeking to unravel the secrets of Earth's early history!) Most terrestrial planets and satellites are less active geologically than Earth, and rock cycles may not be occurring on them.

Many rocks form in layers called **strata.** This is particularly the case for terrestrial sedimentary rocks (Figure 8.6), but it is also true for other rock types, on both the Earth and the other planets. The fact that these strata form sequentially is the basis for interpreting the geological history of the Earth and other planets. Geologists are able to study layered rocks and place them into sequence, from oldest to youngest. (Old layers are on the bottom, and young layers are at the top, unless geological processes have inverted the rock layers.) Examination of the various rock layers can lead to understanding of geological conditions at the times when the rocks were forming.

When geologists study a sequence of rock strata, the individual layers that were formed at specific times are called **units** or **formations.** Individual formations that have distinctive properties are generally given names, such as the Cayley formation on the Moon. In many cases, a formation can be traced over a wide area, but this is not always possible, as it may be too deeply buried by younger formations, or it could have been removed by erosional activity since the time when it formed.

A very important tool of the geologist is the **geologic map,** which shows which rock formation is exposed at each point on a planetary surface. Geologic maps exist for most areas of the Earth, and geologic mapping of the Moon and planets has been an important accomplishment during the space age. Geologic maps usually include as a key the stratigraphic column, or sequential arrangement of formations, and a cross section illustrating surface topography and orientation of the rock layers (Figure 8.7).

■ **Figure 8.6** A photograph of sedimentary rock layers, or strata. Geologists use layered rocks to help decipher the Earth's past geological history. (Photograph by the author.)

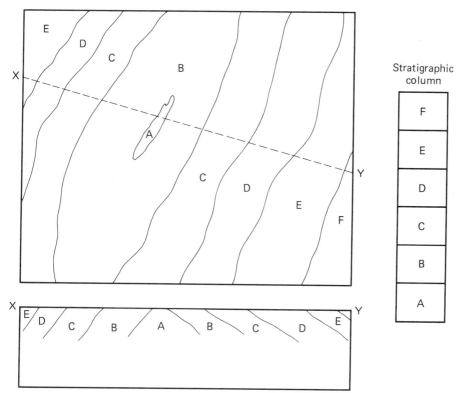

■ **Figure 8.7** Simplified geologic map of a hypothetical region. The map of the region is shown at upper left, whereas the cross section from point *X* to point *Y* on the map is shown at bottom left. The stratigraphic column for the area is shown at right.

■ Igneous Rocks

Igneous rocks are probably the most abundant type in the crusts of all terrestrial objects. Formed from the cooling of molten rock, they are subdivided into two groups based on mode of formation. **Extrusive** igneous rocks form when magma cools at the surface. Due to their more rapid rate of cooling, there is insufficient time for large crystals to develop and extrusive rocks have a homogeneous appearance. **Intrusive** rocks, on the other hand, form beneath the surface, where the surrounding rock acts as an insulator, allowing slow cooling. Gradual cooling promotes the formation of larger crystals that can be seen individually and identified in intrusive rocks.

Magma is formed when heat melts pre-existing rocks. Although there is a layer of molten material in the Earth's core, this is not the direct source of magma. Instead, convection within the

Earth brings heat up toward the crust, with some areas receiving more heat than others. The areas where concentrations of heat develop are where magma forms and igneous activity occurs. As we will see in Chapter 9, the Earth's internal heat ultimately results from radioactive decay occurring deep within the planet.

Magma varies in composition, depending primarily on the type of rock melted during its formation. Although a wide variety of magmas occurs, terrestrial magmas can be divided into acidic and basic varieties. The former have a higher quartz content and are lighter in color, more viscous, and somewhat cooler. Basic magmas have low quartz content and are darker in color, less viscous, and hotter. Indications are that igneous rocks derived from acidic magmas are absent from, or at least are very rare on, other planets. Basic igneous activity, however, has occurred on numerous solar system objects, includ-

■ **Figure 8.8** The common igneous rocks basalt *(left)* and granite *(right)*. (Photograph by the author.)

ing the Moon, Mercury, and Mars. The chemical composition of a magma has a significant effect on its behavior when it reaches the surface during a volcanic eruption and also on the types of volcanic features that will be formed. (Volcanic activity will be discussed in Chapter 10.)

Igneous rocks are classified on the basis of mineralogy (which, in practice, is often identified largely on the basis of color) and texture. For example, one of the most important igneous rocks is basalt, which is dark colored and fine grained. These characteristics show that basalt is an extrusive rock formed from basic magma. Another important igneous rock is granite, which is light colored and coarse grained, being an intrusive rock formed from acidic magma. (Photographs of samples of basalt and granite are shown in Figure 8.8). Some igneous rocks, especially lunar ones, have been fragmented, with the fragments subsequently welded together. Such rocks are said to be **brecciated,** and lunar brecciation, at least, seems to have been primarily the result of fragmentation by impacting bodies.

■ Sedimentary Rocks

Sedimentary rocks are very common at the surface of the Earth's continents but constitute only about 8% of its total crustal volume. This is because most of the steps involved in the formation of these rocks occur at the surface, rather than at depth. As mentioned earlier, sedimentary rocks

are probably limited to the Earth because water is so important in their formation. A special property of terrestrial sedimentary rocks that makes them especially valuable to geologists is that they may contain **fossils,** the preserved remnants of ancient life.

The first step in the formation of sedimentary rocks is the weathering of rocks exposed at the Earth's surface. **Weathering** is a general term applied to all processes that break rocks down into fragments, called weathering products. Weathering may be due to physical processes, such as expansion of water as it freezes in cracks in rock, or biological activity, such as plant roots helping to break rocks. These forms of weathering yield products that are chemically unchanged from the original material. Even more important is chemical weathering, or reactions that change the minerals in rocks to new and different forms, as well as dissolving some of the minerals in water. (Clay minerals are examples of new minerals formed from old ones during weathering.) Rocks differ in their susceptibility to the various forms of weathering, and within a given rock, the different minerals present will not all weather at the same rate and may produce a variety of weathering products. Weathering can certainly occur on planets other than Earth, but biological weathering would be impossible, and reactions involving water would be less common. Nevertheless, even the Moon has a surface layer of powdered rock, largely due to the fragmenting of rocks by meteorite impacts.

Some material produced by weathering remains in place as soil. (The term soil is normally used only in reference to Earth, because it implies a biological component from decay of organic material and wastes from earthworms and other soil-dwelling creatures. A more general term, applicable to all planets, is **regolith.**) Other weathering products are removed from their point of origin, transported by some mechanism, and deposited. The most important transport mechanism on Earth is running water, although wind and glacial ice also transport some sediment.

When rivers carrying weathering products reach standing water, such as lakes and oceans, the particulate material quickly settles, forming a deposit called sediment, whereas the dissolved material remains in solution for a time. Sedimentary rocks that are derived from the sediment are called **clastic,** whereas those formed from the dissolved material are called **nonclastic.** Clastic rocks can also form from particulate material deposited by wind or glacial ice.

Clastic rocks form primarily by one of two methods. The water that surrounds individual mineral grains in a sediment deposit contains abundant dissolved minerals. Crystals of these dissolved minerals may precipitate (form as solids directly from solution) and, in so doing, act as cement to joint the sediment grains together.

Sandstone is an example of a rock formed in this manner. (Figure 8.9 shows photographs of common sedimentary rocks.) The second method of clastic rock formation involves small, plate-like clay mineral grains. Pressure from the weight of overlying material causes these flat grains to align and adhere to one another, forming the rock shale.

Most terrestrial nonclastic rocks form by chemical precipitation of mineral grains in the ocean. Such precipitation occurs because the capacity of ocean water for holding dissolved minerals such as calcite and quartz is limited. New dissolved materials are constantly being added to the oceans as rivers bring them in. As various dissolved minerals reach the level of maximum concentration, called saturation, crystals of them begin to precipitate, settle to the bottom, and adhere to one another, forming rock. Limestone forms by the precipitation of dissolved calcite, and chert, by the precipitation of quartz.

■ Metamorphic Rocks

Metamorphic rocks originate when rocks experience conditions unlike those under which they formed. These new conditions disturb the chemical and physical equilibria existing within rocks

■ **Figure 8.9**   The common sedimentary rocks sandstone *(left),* shale *(center),* and limestone *(right).* (Photograph by the author.)

■ **Figure 8.10** Common metamorphic rocks. (Top) Left to right: slate, schist, and gneiss. (Bottom) Left to right: quartzite and marble. (Photograph by the author.)

since their formation and create new textures and mineralogies. There are three main causes of metamorphism. One cause is heat, which produces metamorphism because increased temperature causes more rapid atomic motion, promoting rearrangement of crystal structures. A second cause is pressure, particularly directed pressure (stress), which causes realignment of mineral grains. Pressures of up to several thousand atmospheres are required to cause metamorphism. Finally, some metamorphism is caused by chemical reactions between a rock and its surroundings.

On Earth, metamorphism most often occurs in areas of mountain building. Most mountains are formed by the tremendous compressional forces that the horizontal movement of crustal rocks generates, and considerable stress results. Rocks undergoing **regional metamorphism,** as this process is called, usually exhibit **foliation,** the alignment of mineral grains in a specific direction in response to the stress. Common examples of foliated metamorphic rocks include slate, schist, and gneiss (Figure 8.10). The type of foliated rock produced depends upon what type of rock was originally present, as well as the amount of stress experienced. As we will learn in Chapter 10, because horizontal crustal movement appears to be mainly a terrestrial phenomenon, foliated rocks may be limited to our planet.

Metamorphism also occurs in rocks adjacent to bodies of magma located beneath the surface, a process known as **contact metamorphism.** Rocks may be "baked" in this way for a considerable distance away from the boundary of the magma body. These rocks do not exhibit foliation, because there is no stress. Common nonfoliated metamorphic rocks include quartzite and marble, which are derived from sandstone and limestone, respectively. Contact metamorphic rocks undoubtedly occur on other terrestrial bodies, because igneous activity is not limited to Earth.

Metamorphism may also be produced by instantaneous shock pressure, in addition to the more gradual forms mentioned previously. Terrestrial rocks rarely experience such shock, but airless bodies such as the Moon have experienced numerous impacts by meteorites. The shock produced by these impacts has been sufficient to metamorphose rocks in the areas of these impacts.

■ Radiometric Age Dating

When we pick up a rock, either in our backyard or on the surface of another world, an obvious question to ask is how long ago the rock formed. One of the important achievements of 20th-century geology has been the ability to determine the ages of many rocks by **radiometric age dating,** which is based on radioactive decay. Radioactive decay occurs in a number of ways. Some radioactive isotopes, or radioisotopes, undergo a simple, one-step decay process, as is the case with potassium-40, which decays directly into argon-40. Other radioisotopes follow a complicated decay sequence, as exemplified by uranium-235, which decays into lead-207 in a process requiring several intermediate steps. At each intermediate step, another unstable isotope is produced and then undergoes further decay.

For a given atomic nucleus, the likelihood of undergoing decay can best be expressed as a probability. In other words, no one can specify exactly when a single radioisotope nucleus will decay. If 1000 nuclei are present, not all will decay simultaneously. Some will decay quickly, whereas others will decay only after a longer time. The most commonly used way to measure the rate of radioactive decay is with a quantity called the **half-life,** which is the average time required for half of the radioactive substance to undergo decay. In the example of the 1000 nuclei used above, if the half-life of that particular radioisotope were 1 million years, it would mean that 500 of the 1000 nuclei would have decayed by 1

million years after the start of the experiment. After 2 million years, half of the remaining 500 radioisotopes would also have decayed, leaving only 250. In other words, for every time equal to the half-life that passes, the amount of original material remaining is once again halved.

A wide range of half-lives of radioisotopes is used in age determination. Carbon-14 has a half-life of 5760 years; aluminum-26, 720,000 years; and uranium-238, 4.5 billion years. Calculations indicate that a radioisotope never completely disappears, at least in theory. From a practical standpoint, however, after several half-lives, the percentage of the original isotope remaining is so low that it cannot be measured.

Although the foregoing theory is straightforward, the actual measurement of the age of a rock is a difficult laboratory procedure. First, the rock must be crushed and its component minerals separated, as only certain minerals can contain radioisotopes in their crystal lattices. Then, the amount of the "parent" radioisotope present in each mineral containing it is measured, as is the amount of the "daughter" decay product present. The latter determination is complicated by the fact that some of the material may not be the result of decay. For example, a mineral that contains both uranium and lead may have some lead that was originally present, as well as other lead that is a decay product. (Lead produced by the decay process is said to be **radiogenic**.) After the true amounts of the parent and daughter materials are determined, the age of the rock can be calculated. The more of the radioisotope present in relation to the decay product, the younger the rock.

Such radioactive age dating determines the time since cooling for an igneous rock and the time since metamorphism for a metamorphic rock. Age dating is generally not done for sedimentary rocks, because they are composed of fragments of many pre-existing rocks that could have a wide variety of ages. If a rock has experienced shock, as would be the case if it were a meteorite that had collided with another meteorite before reaching Earth, the age obtained could be erroneous. This is because shock can "reset" the age clock by allowing gaseous decay products to escape from the rock. Fortunately, shocked rocks can be recognized by their textural features, warning of such problems in dating them. Although many experimental uncertainties complicate the age-dating process, different minerals in the same rock, or different rocks from the same locality, can be sampled in the search for consistent results. If such consistent results are obtained, the age determined is said to be **concordant.**

■ Chapter Summary

Although hydrogen and helium are the most abundant elements in the universe, terrestrial planets and satellites tend to be made of heavier elements. These elements were produced inside stars long before the solar system itself formed. The terrestrial objects contain the largest concentrations of heavy elements in the solar system.

Planetary crusts contain a variety of chemical elements, but they rarely occur in pure form. They are more likely to occur as minerals, which are compounds in crystalline form. Minerals, in turn, usually occur in mixtures called rocks. Igneous rocks are those formed by the solidification of hot, molten rock; sedimentary rocks are those formed by the accumulation of fragments of pre-existing rocks; and metamorphic rocks are those that have been changed by heat and pressure.

One of the significant accomplishments of modern geology has been the development of the technique of radiometric age dating for determining the ages of many rocks. This is a major tool for determining the history of individual solar system objects as well as the solar system as a whole.

■ Chapter Vocabulary

atomic number	element	isotope	regional metamor-
bonding	extrusive	magma	phism
brecciation	foliation	metamorphic rock	regolith
clastic	formation	metamorphism	rock
compound	fossil	mineral	rock-forming mineral
concordant age	geologic map	molecule	sediment
contact metamor-	half-life	nonclastic	sedimentary rock
phism	igneous rock	nucleosynthesis	silicate
covalent bonding	intrusive	radioactive	strata
crystal	ion	radiogenic	weathering
differentiation	ionic bonding	radiometric age dating	

■ Review Questions

1. Describe the atom, the ways in which atoms bond, and the difference between stable and radioactive substances.
2. Describe the distribution of elements within the Earth.
3. How did the elements originate?
4. Define what a mineral is and list the properties used in identifying them.

5. What is the most important of the eight groups of minerals?
6. Define what a rock is and describe the three types of rocks. List one or two rocks representative of each type.
7. Describe the two groups of igneous rocks.
8. What is weathering, and how does it occur?
9. How are sedimentary rocks formed?

10. How are metamorphic rocks formed?
11. List several examples of radioisotopes and their half-lives.

12. Explain how radiometric age dating works.

■ For Further Reading

Mason, Brian. *Principles of Geochemistry.* Wiley, New York, 1966.

Simpson, Brian. *Rocks and Minerals.* Pergamon Press, Oxford, England, 1966.

Tennissen, Anthony C. *Nature of Earth Materials.* Prentice-Hall, Englewood Cliffs, NJ, 1983.

9

Interiors of Terrestrial Planets

An aurora, as photographed from Minnesota. Although it is a sky phenomenon, the aurora is related to the Earth's magnetic field, generated deep in its core. (Photograph by Sherman Schultz, courtesy of Dennis Milon.)

F or the most part, it is relatively easy to study the surfaces of the planets and the many interesting features found on them. (Venus is an exception, because its cloud-filled atmosphere prevents direct observation of the surface.) However, the surface is actually only a limited part of the planet, and surface rocks and features are usually not representative of the entire planet. The vast majority of the material in a planet is hidden from view, and the interior of a planet is generally much different in composition and structure than its surface. In addition, many important processes occur inside planets, where they are obviously rather difficult to study. In order to understand a planet fully, its interior regions must be explored. In this chapter, we examine how we study the interiors of terrestrial planets and what we learn when we do so.

■ Studying Planetary Interiors

How do we study the interiors of Earth and other planets? It is obvious that direct observation is very difficult, if not impossible. The deepest holes that have been drilled into the Earth are only a few kilometers in depth, and the remainder of its interior will probably remain unreachable in the future. The depths that unmanned spacecraft and astronauts can reach on other planets are even more limited. It is obvious that indirect methods of study are required.

The first step in studying a planet's interior is determining the planet's overall average density, a straightforward calculation once its mass and diameter are known. For example, the density of the Earth is 5.52 grams per cubic centimeter. (Recall that this figure is so high partly because pressure greatly compacts Earth's interior. The Earth's uncompressed density is somewhat lower — 4.03 grams per cubic centimeter.) What does this tell us? Because the densities of most rocks found at the surface range from 2.5 to 3.5 grams per cubic centimeter, it is clear that there must be some much denser material inside the Earth in order for the average density to be so high.

This idea is verified by measuring the moment of inertia of the Earth. The **moment of inertia** (*I*) of a rotating object is a quantity controlled by the distribution of mass within it, and is given by

$$I = C M R^2,$$

where *M* is the mass of the object, *R* is its radius, and *C* is a quantity known as the moment-of-inertia coefficient. This coefficient varies according to the distribution of material within the rotating object. Imagine three spheres of equal mass, one with most of its mass concentrated toward the center, one with most of its mass concentrated toward the surface, and one with a uniform mass distribution. Even if all three rotated at the same rate of speed, their moments of inertia would be different and the three could be distinguished. A uniform body has a moment-of-inertia coefficient of 0.4. The coefficient becomes higher if mass is concentrated toward the surface of the sphere, reaching 0.67 if all of the mass is concentrated at the surface, forming a hollow sphere. The coefficient becomes lower if the sphere's mass is concentrated toward its center. This is the case for the Sun, Earth, Mars, and Moon, whose coefficients are 0.06, 0.33, 0.365, and 0.39, respectively.

In practice, the moment of inertia of a planet or other object is measured by examining how it responds to gravitational forces applied to it by an object in orbit around it. For example, the Earth's moment of inertia is calculated by looking at the amount of precession caused by the Moon's tidal attractions.

Because we know from its moment-of-inertia coefficient that the Earth's density increases toward the center, the next step is to learn whether the increase is gradual or due to a series of abrupt changes. The only way to get such detailed information is by studying earthquake waves. **Earthquakes** are the most destructive natural phenomena on our planet and consist of shock waves, called **seismic waves,** which are produced when rocks break and slide along a surface known as a **fault.** Faults are caused by either tensional or compressional forces within our planet's crust. When rocks slide along each other at the fault plane, the shock waves generated can travel throughout the Earth and be detected by instruments known as **seismographs.**

Seismic waves reveal Earth's interior structure in two ways. The first is based on the fact that there are several distinct types of seismic waves, each of which travels at different velocities. P (primary) waves, for example, are faster than S (secondary) waves, but there are also other differences between them. Whereas P waves can propagate through both liquids and solids, S waves can travel only through solids. When earthquakes occur, S waves are not observed at regions more than 103° away from the earthquake site (Figure 9.1). This observation reveals that there is a liquid region within the Earth's core and allows its size to be determined. A second characteristic of seismic waves is that their velocity increases as they travel through denser material, because atoms there are closer together and, thus, better able to transmit the wave motion. As a result of their changing velocities, waves travel curved, rather than straight, paths through the

■ **Figure 9.1** How seismic waves show that the Earth has a liquid layer within it. When an earthquake occurs, seismographs more than 103° away from it are unable to detect S (secondary) waves. These waves travel only through solids, and the liquid portion of the Earth's core blocks their passage.

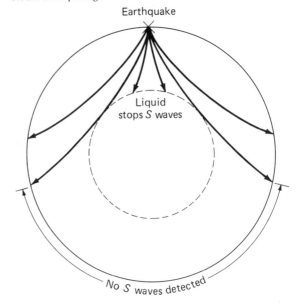

Earth. More importantly, abrupt changes in density can be detected by examination of seismic data. Studies of seismic waves have revealed several **discontinuities,** as these changes are called, inside the Earth, indicating that density increases in several abrupt steps, causing discrete internal layers to be present.

Seismographs were left on the Moon by the Apollo expeditions and have revealed much about its interior. "Moonquakes" are much less frequent and energetic than their terrestrial counterparts, but they do occur. As an aid in understanding the lunar interior, spent rocket stages and used lunar modules were sent crashing into the Moon's surface to produce moonquakes of known intensity for study. Although both Viking Mars landers carried seismographs, one failed and the other gave readings that may have been influenced by wind vibrations of the spacecraft. Moreover, one seismic station alone is insufficient to reveal much about internal structure. At least one "Marsquake," however, was definitely observed. (Detailed descriptions of the interiors of the Earth, Moon, and Mars are given in Chapters 14, 15, and 16, respectively.)

To enhance the information gained by seismic wave investigations, considerable study of planetary interiors is done by theoretical calculations similar to those described for the Sun in Chapter 6. These give temperature, density, pressure, and other properties as a function of depth. Based on these quantities, researchers can speculate what materials might be present, because they know which rocks and other substances would have the observed physical properties at the required temperatures and pressures. For example, one line of evidence that supports the theory that the liquid portion of the Earth's core is composed of nickel and iron metal is that this material would be liquid at the temperature and pressure conditions believed to exist there.

■ Differentiation

Because we know that terrestrial planets are not uniform throughout and that discrete layers are present, we need to explain why their interiors are in this condition. Modern theories of plane-

tary formation state that planets formed by a process known as **accretion**, whereby smaller objects called **planetesimals** collided and were held together gravitationally, forming the present planets. Assuming that the planetesimals were all of about the same composition, the planets probably were homogeneous immediately after their formation.

After the planets formed, however, their interiors became hot due to the gravitational energy released during accretion and to the decay of the radioactive elements present in them. It is not certain whether the terrestrial planets were ever completely molten inside, but there was enough melting that the lighter materials in them could float to the surface and the denser materials could sink to the center. This process, called **differentiation,** is responsible for the layered structure within terrestrial planets. Because interior heating also drove off volatile materials trapped inside the planets as they accreted, differentiation was also responsible for the formation of atmospheres and (at least in the case of Earth) oceans. Only planets with sufficiently strong gravity were able to retain these volatiles.

Although details vary from object to object, three basic layers have formed within terrestrial planets as a result of differentiation (Figure 9.2).

■ **Figure 9.2** The interior of a typical terrestrial planet contains a thin crust, a mantle, and a core.

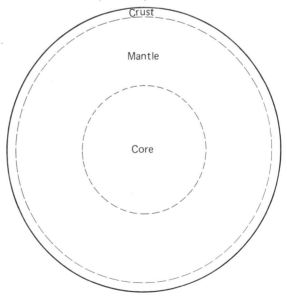

The **crust** is the thin, outermost layer containing rocks of low density. Crusts are subject to the numerous external processes of erosion, which produce a wide variety of landform features. Beneath the crust lies the **mantle,** which generally contains the largest volume of material within the planet. The density of the mantle is higher than that of the crust, and rock composition is different. Innermost is the **core,** which is often metallic and may be at least partially molten. The smaller the terrestrial object, the less likely it is to have a molten core. The core of a small object is generally not molten at present, because its small size has allowed sufficient cooling to occur since the object formed so that the object is now entirely solid.

■ Heat Within Terrestrial Planets

Geologists have long known that there are sources of heat within our planet, as attested to by the frequent volcanic eruptions that occur. Prior to the 20th century, most believed that the Earth was entirely molten at the time of its formation and has been cooling ever since. In the 19th century, the estimated rate of cooling was used to calculate the age of the Earth and yielded estimates of about 100 million years. Modern ideas are quite different. Although few dispute that more of the Earth was molten in the past than in the present, it is considered unlikely that our planet was ever completely in liquid form. The change in ideas was brought about by the discovery of a new heat source, radioactive decay.

Although much of the Earth's initial heat was produced by the gravitational potential energy released during the infalling of material to make our planet, radioactivity allowed heat production to continue after accretion had ended. The main source of radioactive heat is the gradual decay of uranium, thorium, and potassium in rocks. (Although these elements occur in greatest abundance in crustal rocks, they are distributed throughout the Earth's interior. Radioactive decay rates are independent of the changes in temperature and pressure that occur within our planet.) Obviously, as these materials decay, the amount of them remaining decreases, so heat production has decreased through time. Therefore, the Earth and other terrestrial planets are slowly cooling as heat escapes and as the heat source (radioactive materials) is depleted. Enough radioactive materials are present in Earth that its interior is still hot even after about 4.6 billion years!

Another source of initial heating was the vast quantity of short-lived radioactive isotopes that were incorporated into solar system objects at the time of their formation but decayed so quickly that they are "extinct" today. Examples include iodine-129 and aluminum-26, whose rapid decay caused them to be depleted within a few million years. The burst of heating from these materials may have been responsible for the differentiation of the planets.

Investigation of certain kinds of meteorites has shown that they originated in bodies that were once molten. As we will see in Chapter 28, the source of most meteorites is believed to be asteroids, which would have been too small to be heated significantly by gravitational potential energy or to accumulate enough radioisotopes for heating. What could have caused internal heating of the solar system's smaller objects? It has been suggested that a process called **ohmic heating** was responsible. It is likely that shortly after its formation, the Sun produced a stronger interplanetary magnetic field than is present today. Calculations show that if an asteroid contained conducting materials, such as metals, its orbital motion through the magnetic field would have induced electrical currents to flow through the asteroid's interior, heating and perhaps partially melting it.

Today, internal heat is studied by means of **heat-flow** experiments, which measure the actual amount of energy reaching a planetary surface from below. Although numerous heat-flow measurements have been made on Earth, the Moon is the only other object whose heat flow has been measured. Heat-flow experiments on Earth show that the release of thermal energy varies significantly according to surface geology, but the average heat flow is 62 ergs per square centimeter of the Earth's surface every second, which means that the total amount of energy released worldwide is about 10^{28} ergs annually. (Although this may seem like a large figure, recall that the Sun emits approximately 10^{33} ergs of energy every *second*!) The corresponding figures for the Moon are 17 ergs per square centimeter per second and about 10^{26} ergs per year. (Even though the Moon's heat flow is over one-quarter that of the Earth, its total surface area is much smaller, giving a far smaller annual total.) The smaller quantity of heat present within the Moon probably indicates that it no longer can generate magma for igneous activity, and this is verified by the fact that none of the Moon's igneous rocks are very young. It is likely that the core of the Moon is completely solid.

Unfortunately, heat-flow measurements require that probes be placed down holes drilled at least several meters below the surface, and this has not yet been attempted by unmanned spacecraft. The Apollo astronauts found that drilling heat-flow holes was one of the most strenuous and difficult portions of their lunar explorations. Future measurements of heat flow on other planets should reveal considerable information about their internal workings. For example, some surface features on Mercury appear to be the result of shrinkage of that planet's crust. Crustal shrinkage would be occurring if the planet were contracting, a possibility if Mercury's interior is cooling.

■ Planetary Magnetic Fields

Everyone who has used a compass to find directions has experienced the effect that the Earth acts like a giant magnet, interacting in some way with the little magnet turning inside the compass. Although this observation is an easy one to make, understanding why the Earth is a magnet and exactly how it works if far more difficult.

As described in Chapter 6, both a small laboratory bar magnet and the Earth generate dipole (two-pole) magnetic fields comprised of invisible lines of magnetic force. This field can be mapped by sprinkling iron filings on a piece of paper over a bar magnet, or by measuring the horizontal and vertical orientations of a compass needle at different points on the Earth (see Figures 6.4 and 9.3). By doing this on the Earth, researchers have learned that our planet's magnetic poles are close

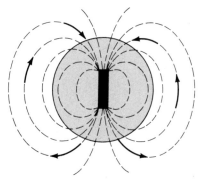

■ **Figure 9.3** Planetary magnetic field. Compare with Figure 6.4. (From Levin, Harold L. *Contemporary Physical Geology*, 3rd ed. Saunders College Publishing, Philadelphia, 1990.)

to, but not identical to, the poles defined by the axis of rotation. (There is an angle of 11.4° between the rotational and magnetic axes.) This similarity, even though it is not exact, leads scientists to believe that the magnetic field is somehow related to the Earth's rotation.

The Earth's present magnetic field could not be a result of the planet's magnetization when it formed, as such a field would have decayed long ago. Instead, some internal process must continually be generating a magnetic field. The modern explanation for the Earth's magnetic field is that our planet acts like a giant generator, or dynamo. **Dynamo theory** proposes that the magnetic field is generated within the liquid outer core of the Earth, which is probably composed of highly conductive nickel and iron. The rotation of the Earth presumably causes electric currents in this circulating liquid metal, which in turn produce the magnetic field.

There is one major complication to our understanding of the Earth's field. Certain igneous rocks contain information about the condition of the Earth's magnetic field at the time of their formation, because the magnetic field aligns susceptible mineral grains that are forming as the magma cools so that they are parallel to the field, and they retain this orientation after the rock has solidified. This **remanent magnetism**, as it is called, can be used to determine the location of the rock with respect to the magnetic field at the time of the rock's formation. This allows us to study the changing positions of Earth's continents, as will be discussed in the next chapter. However, the surprising fact revealed by studies of remanent magnetism in rocks is that the Earth's magnetic field periodically reverses polarity, and has been doing so at intervals of about 1 million years as far back as can be determined. Because this has not happened during historic times, we do not know how long it takes for such a reversal to occur, but presumably the field fades away, then returns with polarity reversed. If this were to happen now, our compasses would point south instead of north! Although some versions of dynamo theory attempt to explain this phenomenon, there is not yet a consensus regarding its origin.

After looking at the Earth's magnetic field, we may conclude that a reasonable set of requirements for a planet to have a magnetic field would be a liquid metallic core and a rapid rotation rate. Unfortunately, magnetism is much more complicated than this, as revealed by magnetic studies of the planets. (Planetary magnetic fields are studied by magnetometers carried by flyby and orbital spacecraft. All planets except Pluto have been studied by spacecraft with magnetometers, although there has been insufficient study of Mars to characterize adequately the nature of its magnetic field or even determine with certainty if it has one. In addition, indirect evidence about planetary magnetic fields can be obtained remotely from Earth. This is done by studying radio waves emitted by the radiation belts of planets with magnetic fields, which will be discussed in the next section.) One of the biggest problems in planetary magnetism studies is trying to understand Mercury. Mercury is so small that it may lack a molten core, and it also requires 59 days to rotate. When the Mariner 10 spacecraft flew by the planet in 1974, one of the most surprising results was the discovery of a magnetic field. Mercury shows that the generation of magnetic fields is not a well-understood process.

The magnetic fields of planets and other solar system objects can be measured using two units — total magnetic moment and surface field strength. The **total magnetic moment** is the overall strength of a magnetic field; it is measured in units of gauss-cubic centimeters. Because planets vary in diameter, however, two planets with identical total moments might not have the same magnetic field strength at their surfaces. This is because the surface of a larger planet will be farther away from the internal source of the field than the surface of a smaller planet would be. The average **surface field strength** is determined by dividing the total magnetic moment by the cube of the planet's radius and is expressed in units of gauss or gammas (1 gauss is 10^5 gammas). For example, the total magnetic moment of the Earth's field is 7.98×10^{25} gauss-cubic centimeters, whereas the average surface field is 0.308 gauss, or 30,800 gammas.

■ Effects of Magnetic Fields

Planetary magnetic fields have a number of important effects, many involving the solar wind and the interplanetary magnetic field that pervades our solar system. The interaction between the solar wind and a planet depends on whether the planet has a magnetic field. If a planet has a magnetic field, the field will interact with the charged particles in the solar wind and keep them from impinging directly upon the planet. An object with no magnetic field, such as the Moon, will have no protection from the solar wind, and solar wind particles will strike its surface.

As described in Chapter 6, energetic particles in the solar wind interact with the Earth's magnetic field, with their entrance blocked by magnetic field lines at the equator and middle latitudes. At the poles, where magnetic field lines are approximately perpendicular to the surface, incoming particles bombard the atmospheric molecules present and cause discharges of light visible from the ground as **auroras**. Most auroral displays occur between 80 and 160 kilometers (50 and 100 miles) above the surface. Auroras can best be seen from northern Canada and Alaska, but displays can occasionally be seen from the northern United States, especially during sunspot maxima. The aurora is an eerie light that can have many colors and change appearance over a time span of a few minutes (Figure 9.4). Terrestrial auroral displays have even been photographed from space (Figure 9.5). Auroras occur around other planets that have trapped radiation belts; Jupiter's auroras were photographed by the Voyager spacecraft.

■ **Figure 9.4**   The aurora, as it commonly appears in the sky. (National Research Council of Canada.)

One effect of a planetary magnetic field is that it can trap electrically charged particles within it, forming what is called a radiation belt. Earth satellites in the late 1950s revealed the presence of two large doughnut-shaped radiation belts in space around the Earth that were called the **Van Allen belts**, after the astronomer (**James A. Van Allen**) who interpreted the satellite data revealing their presence (Figure 9.6). The inner belt is centered about 3000 kilometers (1800 miles) above the surface over the equator and has a thickness of at least 5000 kilometers (3100 miles). The outer belt, which extends farther north and south, is about 15,000 to 20,000 kilometers (9000 to 12,000 miles) above the surface and 6000 to 10,000 kilometers (4000 to 6000 miles) thick. These regions lack sharply defined boundaries and are composed mainly of electrons and protons from the solar wind trapped by the Earth's magnetic field. (Some of the particles in the inner belt have probably been produced by interactions between cosmic rays and the molecules in the Earth's outer atmosphere.) Radiation belts of other planets cannot be "seen" with visible light but can be detected and mapped using the radio portion of the spectrum. This is because the charged particles within radiation belts emit radio waves, most notably in the case of Jupiter.

■ **Figure 9.5**   An aurora, as viewed from Earth orbit during a Space Shuttle mission in May, 1985. (NASA Photograph.)

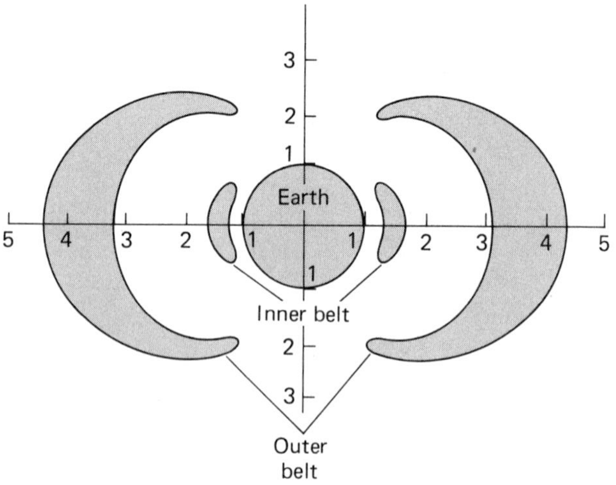

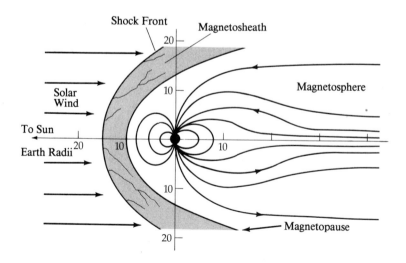

■ **Figure 9.6** The Van Allen radiation belts, showing a cross section through the Earth and the inner and outer belts. The *vertical line* represents Earth's magnetic axis, and the *horizontal line* represents Earth's magnetic equator. Both lines are marked in units of Earth radii. (From Zeilik, Michael and Elske v.P. Smith. *Introductory Astronomy and Astrophysics,* 2nd ed. Saunders College Publishing, Philadelphia, 1987.)

Containing the Van Allen belts and extending outward far beyond them is the **magnetosphere**, the outermost region where the Earth's magnetic field is important. It fends off much of the incoming solar wind, producing an area of interaction called the bow-shock region. The shape of the magnetosphere is elongated, extending much further in the direction away from the Sun (Figure 9.7). Beyond the boundary of the magnetosphere (the **magnetopause**) and a transition area (the **magnetosheath**), the magnetic field carried by the solar wind is dominant.

In addition to studying the magnetic fields of planets such as Mercury and the gas giants, spacecraft have examined the interactions between the solar wind and their magnetospheres.

Some planets, such as Venus and Mars, lack appreciable magnetic fields but still interact with the solar wind. Such interactions can occur around planets without magnetic fields, provided they possess atmospheres whose outer layer is electrically conducting, producing what is called an **ionosphere**. The ionospheres of these planets also produce magnetopause effects.

An object like the Moon, on the other hand, which has neither magnetic field nor atmosphere, is struck directly by the solar wind. (There is, of course, a solar wind "shadow" region on the side of the Moon directed away from the Sun.) During the time they were on the lunar surface, Apollo astronauts deployed foil material to collect solar wind particles for return to the Earth for study.

■ **Figure 9.7** Earth's magnetosphere. The Earth is shown at the center, as well as the magnetosphere, magnetopause, magnetosheath, and incoming solar wind. (From Zeilik, Michael and Elske v.P. Smith. *Introductory Astronomy and Astrophysics,* 2nd ed. Saunders College Publishing, Philadelphia, 1987.)

■ Chapter Summary

Unlike their surfaces, the interiors of solid planets can be studied only by using indirect techniques. Calculation of density and moment of inertia, as well as seismic and theoretical investigations, allows us to determine planetary interior structure. Planets are believed to have differentiated into distinct crust, mantle, and core regions.

Planetary interiors were initially heated by the release of potential energy during accretion and, subsequently, by radioactive decay. The

Earth, and presumably the other terrestrial planets, generate dipole magnetic fields by internal dynamos. Planets with magnetic fields have radiation belts and magnetospheres.

Magnetic effects clearly show that the influence of events occurring deep inside planets extends outward to a greater distance than might initially be expected. In the next chapter, we will see that internal activity plays important roles in shaping planetary surfaces as well.

■ Chapter Vocabulary

accretion	earthquake	mantle	seismograph
aurora	fault	moment of inertia	surface field strength
core	heat flow	ohmic heating	total magnetic
crust	ionosphere	planetesimal	moment
differentiation	magnetopause	remanent magnetism	Van Allen, James A.
discontinuity	magnetosheath	seismic wave	Van Allen belts
dynamo theory	magnetosphere		

■ Review Questions

1. List the techniques used for gaining information about planetary interiors.
2. Arrange the following in order of increasing moment-of-inertia coefficient: a normal bowling ball, a bowling ball that has been hollowed out, a bowling ball that is not uniform internally but has a core made of lead.
3. Describe the three basic layers of terrestrial planet interiors, and explain why this layering occurs.
4. What is the source of planetary internal heat? How is this heat measured?
5. What is the explanation of the source of the Earth's magnetic field?
6. Explain how we are able to study the Earth's past magnetic field.
7. What did the discovery of Mercury's magnetic field tell us about planetary magnetism?
8. How is planetary magnetism measured?
9. What are the effects of a planet's magnetic field on the space around it? Use the Earth for an example.

■ For Further Reading

Cole, G.H. *The Physics of Planetary Interiors*. Adam Hilger, Bristol, England, 1984.

Cook, A.H. *Interiors of the Planets*. Cambridge University Press, Cambridge, England, 1980.

Howell, Benjamin F. *Introduction to Geophysics*. Krieger, Melbourne, FL, 1978.

Hubbard, William B. *Planetary Interiors*. Van Nostrand Reinhold, New York, 1984.

Seymour, Perry. *Cosmic Magnetism*. Adam Hilger, Bristol, England, 1986.

Surfaces of Terrestrial Planets

The average person associates the word "landscape" with a view containing hills, trees, and sky. This stunning photograph of the crater Copernicus on the lunar surface, taken by Lunar Orbiter II in 1966, is more typical of what the landscapes of terrestrial planets and satellites are like: stark, cratered, and lifeless. (NASA Photograph, courtesy of the National Space Science Data Center and the Lunar Orbiter II Principal Investigator, Leon J. Kosofsky.)

One of the most interesting aspects of terrestrial objects is the diversity of landforms that their surfaces contain. Some planetary landforms are large enough to be visible even from Earth, whereas others remained undetected until viewed firsthand by spacecraft. Most landforms are the result of **vertical relief,** or elevation differences. No solid planet is a perfect sphere, and all planetary surfaces contain numerous high and low areas. On the Earth, elevations are measured in relation to sea level, a convenient reference level. Because other planets lack oceans, elevations on them are generally measured with respect to an arbitrary level called the planet's **average radius.** This radius is obtained by calculating the average distance from the planet's center to all the points on its surface. On planets with atmospheres, an elevation having a specific atmospheric pressure is sometimes used as a starting point for elevation measurements. Pressure decreases above this elevation and increases below it.

■ Internal and External Processes

As we look around us at the geological processes occurring on the Earth, it is clear that some of them tend to build up the surface, producing mountains and other areas of high elevation. As we examine this building up of the Earth's surface, we find that it generally results from processes occurring beneath the surface of the Earth. In other words, **internal processes** lead to development of elevated surface features. As we will see, the heat generated within a planet is ultimately responsible for most of these processes.

If only internal processes were active, the Earth's surface would be high and rough everywhere. This is not the case, because there are numerous processes that also wear down, or erode, the surface. Nearly all of these processes

127

are **external,** occurring at or very near the surface. If only they were active, the Earth's surface would be eroded down to sea level in only a few million years. On our planet, a delicate balance is maintained between internal and external processes, resulting in a constantly changing landscape. Most other terrestrial objects appear to be much less active internally and are therefore not currently building up their surfaces. At the same time, airless bodies are not now experiencing the same external processes that are shaping the Earth. For this reason, their elevated features are still in existence, even if they lack ongoing internal processes.

■ Crustal Types

The processes described previously act primarily on the outer layer, or crust, of a planet. As we examine Earth and other worlds, we find that most have more than one type of crustal material. At the same time, we see that the surfaces of many objects can be divided into two hemispheres that have radically different properties.

As an example, let us consider the Earth. The two crustal types (shown in Figure 10.1) are **continental crust,** which comprises the continents, and **oceanic crust,** which underlies the oceans. Continental crust is thicker and of lower density. Although it contains rocks of all types, its average composition is similar to the igneous rock granite.

The elevation of the surface of the continental crust averages 92 meters (300 feet) above sea level. Oceanic crust is thinner and of higher density. It is composed of the igneous rock basalt, and its surface lies 3 to 5 kilometers (2 to 3 miles) below sea level. A bit of experimentation with a globe will reveal that one can orient it so that one hemisphere is primarily land and the other ocean. The implication of this is unclear, but it matches observations of other planets.

The Moon has light-colored crust comprising highland regions and dark-colored regions of basaltic lava flows, called **maria.** (Maria is the plural of the Latin word mare, meaning "sea," which is what Galileo and other early observers believed these smooth, dark areas were.) Most of the maria are on the hemisphere facing Earth. (As we will see in Chapter 15, the reason for this is probably related to Earth's gravity.) Radar studies of Venus show several continental regions rising above the lowland plains that cover most of the planet. Their distribution, however, appears to be random. Mars has one hemisphere that is mostly old, cratered terrain and another that contains volcanoes and other evidence of more recent internal activity.

Just as mapping of surface rock units, described in Chapter 8, is a helpful tool in understanding compositional differences resulting from the geological history of an object, we can also map planetary surfaces according to crustal type and landforms. We can define a topo-

■ Figure 10.1 Terrestrial crustal types. Oceanic crust (*right*) is thin and composed of basalt. Continental crust (*left*) is thicker. It is made of granite and other igneous rocks, which are generally overlain by a veneer of sedimentary rocks.

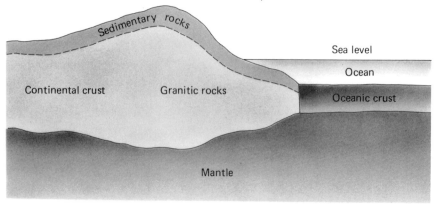

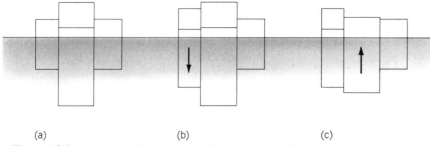

(a)   (b)   (c)

■ **Figure 10.2**   Isostasy. (a) Blocks of wood float in water, with part submerged and part above the surface. (b) If wood is added to a block, it sinks lower. (c) If wood is removed from a block, it rises higher.

graphic province as an area of a planetary surface containing specific landforms, all produced in the same way at about the same time. For example, the Tharsis volcanic province on Mars is a large area containing several huge volcanoes. Although a geologic map of relatively small-scale variations in rock type and a topographic province map may not appear similar at first, upon closer examination, the information each contains is seen to complement the other considerably.

■ Isostasy

As described in Chapter 9, the density of a planet increases with depth due to the vertical layering resulting from differentiation. The crust of the Earth or another planet can be thought of as "floating" on a deeper, denser layer. Although our planet underwent differentiation a long time ago, there is evidence that some vertical motion still occurs within the Earth's crust. The theory that planetary surface layers are in vertical balance as a result of this motion is called **isostasy.**

A simple analogy of how isostasy works is a block of wood floating in water, as illustrated in Figure 10.2. Because wood is lighter than water, just as a planet's crust is lighter than its mantle, the wood floats. However, part of the wood block rises above the water's surface, whereas the rest is below. If material were added to the top of the wood block, the entire block would sink lower; if material were taken away, it would rise. Similarly, on the Earth, mountains extend into the air but also have "roots" extending down into the man-

tle. Mountains, like the wood, can also move vertically if material is added to them or removed from them.

Why does isostasy occur? Suppose a mountain lacked roots of low-density material descending into the mantle (Figure 10.3). If you flew over the mountain, the "extra" material extending above the surface would cause a greater-than-normal gravitational attraction. However, due to isostatic adjustment, the low-density material under a mountain peak reduces the gravitational contribution from beneath it, effectively compensating for the presence of the raised material. The Earth and other planets are generally well compensated in this respect. However, **gravity anomalies,** as they are called, occur

■ **Figure 10.3**   Isostasy in mountains. (a) If mountains lacked roots, the extra crust would cause a gravity anomaly. (b) In actuality, a "root" of low-density crust extends downward into the mantle, compensating for the elevated material.

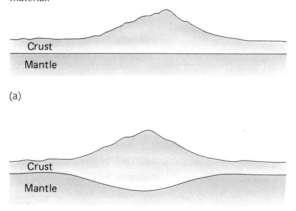

(a)

(b)

where isostatic compensation is not complete. One example occurs on the Moon beneath some maria, which are large igneous rock bodies formed when magma flooded large basins caused by meteorite impacts. Some maria are not isostatically adjusted, because by the time that the flooding of them occurred, the lunar crust and mantle had become rigid enough that vertical adjustment could not take place. As a result, gravity anomalies are associated with these uncompensated maria. Gravity anomalies are detected by close observation of the path of a spacecraft as it orbits a given body, because the anomalies slightly vary the gravitational acceleration that the spacecraft feels. Lunar gravity anomalies were discovered in this way in 1968.

There are several important results of terrestrial isostatic adjustment. As mountains are eroded away, they slowly rise as the material at their peaks is removed. This causes mountains to exist longer than might otherwise be expected. During the ice age, which occurred during the past 2 million years, the enormous weight of the glaciers covering certain areas of the Earth caused the crust in those areas to sag. Now that the ice has melted, isostatic rebound is occurring at the rate of a few inches per century as areas such as Scandinavia and the Great Lakes region slowly rise. Finally, when volcanic islands are formed in the ocean, they often sink below the surface due to sagging of the oceanic crust as a result of the weight added to it when the volcanoes were formed.

■ Plate Tectonics

The most spectacular surface features on the Earth are the huge chains of mountains that occur on the world's land masses. Many of these mountain ranges, such as the Rocky Mountains in North America and the Andes Mountains in South America, occur near the edges of continents. Others, such as the Ural and Himalaya Mountains in Asia, occur within continental interiors. In addition to these, exploration of the ocean floor has revealed huge chains of undersea mountains called the mid-oceanic ridges, which extend for

almost 66,000 kilometers (40,000 miles) and constitute the largest landform on Earth. What is responsible for these large-scale chains of mountains? In recent years, geologists have developed a revolutionary way of looking at the Earth's crust, and this new theory, called **plate tectonics**, adequately explains the origin of most of our planet's large-scale surface features. A plate is a section of the Earth's crust and upper mantle (collectively known as **lithosphere**) that moves as a distinct unit, and tectonics is a term that refers to the large-scale movement of rocks. In other words, the theory of plate tectonics states that the Earth's crust is not one solid unit but a number of discrete sections that move around atop the underlying mantle material. In contrast to isostasy, which involves vertical crustal motion, plate tectonics involves horizontal crustal movement. On the Earth, horizontal crustal movement is responsible for the formation of many important landforms.

As a result of plate motions, the Earth's continents move with respect to each other, and occasionally even collide. Although this was long suspected on the basis of similarities between the western coastline of Africa and the eastern coastline of South America, suggesting that they were once linked together, evidence proving this was not obtained until the middle of the 20th century. Important evidence comes from the study of fossils. Although widely separated continents such as South America and Africa now have completely different plants and animals, fossils show that 200 million years ago, these continents had similar plant and animal populations, indicating that there was once close contact between them. In addition, fossils from Antarctica show that tropical plants and animals once lived there. Because tropical conditions cannot exist at the poles, this indicates that Antarctica has not always been at the South Pole.

The most important evidence for continental motion comes from **paleomagnetism**, the study of the Earth's ancient magnetic field. When magmas cool, magnetic minerals in them are aligned along magnetic field lines. When samples of igneous rocks are obtained, the magnetic field information preserved in them can be detected. Because the ages of many igneous rocks can be measured by studying the radioactive elements

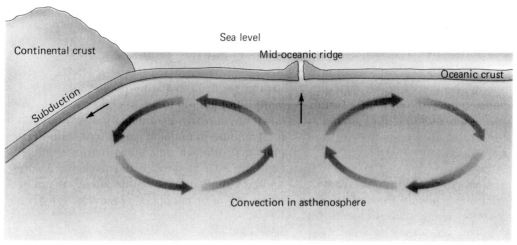

■ **Figure 10.4** Plate tectonics. Convection in the asthenosphere brings up heat, producing magma and resulting in volcanic activity that forms a mid-oceanic ridge. Downward convective movement causes subduction as oceanic crust is pulled down below the continental crust.

and decay products they contain, the rock's location relative to the Earth's magnetic field can be determined for the time of its formation. Because the Earth's magnetic axis is approximately parallel to its rotational axis and has probably always been aligned in this way, studying the remanent magnetism preserved in rocks allows the location of continents to be traced back into the past for hundreds of millions of years.

There is also evidence for plate tectonics that has been obtained by studying the rocks of the ocean floor. This evidence, which shows that a symmetrical pattern of paleomagnetism and rock ages exists on either side of the mid-oceanic ridges, reveals that new oceanic crustal material is formed at these ridges. These mountains therefore result from the accumulation of new crustal material before it spreads outward from the ridge in both directions. Because sea-floor spreading, as this process is called, occurs at these ridges, oceanic crust must be destroyed elsewhere in order to maintain the same total area of ocean floor. Oceanic crust has been found to be destroyed by a process known as **subduction,** which can occur at places where crustal plates are converging. During subduction, one crustal plate is slowly pushed down into the mantle, underneath the other plate. Continental crust has such a low density that it cannot be subducted, so sub-

duction always involves oceanic crust. Most sites of subduction are marked by especially deep areas in the ocean, called trenches. (The formation and destruction of oceanic crust is shown in Figure 10.4).

The Earth's crust is composed of about seven major plates and a few dozen smaller ones (Figure 10.5). Plates typically contain both oceanic and continental crust and move at rates of several inches per year. The North American plate, for example, contains most of North America, as well as the floor of the North Atlantic Ocean west of the Mid-Atlantic Ridge. Plate motion is believed to be caused by convective motion in the deeper mantle, or **asthenosphere,** but the exact mechanism is poorly understood. Evidence of past tectonic activity indicates that plate motion can change rate and direction, or even stop and reverse, due to changes within the mantle.

Most major landforms are produced at plate boundaries, and the interiors of plates are relatively free of earthquakes, volcanoes, mountain building, and other effects of plate motion. Which process will occur at a plate boundary is determined by whether the plates are converging or diverging. Diverging plate motion occurs where upward convective motion in the mantle brings heat to the surface, producing magma. This magma forms new oceanic crustal material,

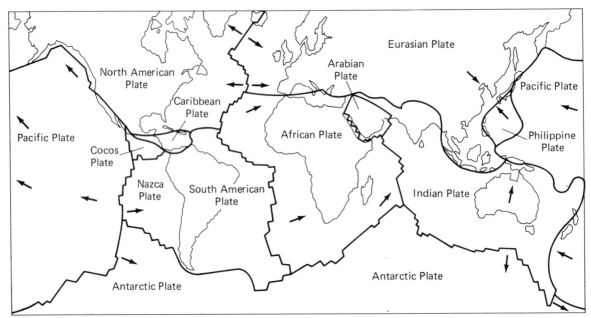

■ Figure 10.5 A map of the Earth's major crustal plates. The *arrows* represent the present directions of plate motion, and plate boundaries are indicated. (Some minor plates are omitted from this diagram.) (From Levin, Harold L. *Contemporary Physical Geology*, 3rd ed. Saunders College Publishing, Philadelphia, 1990.)

which is pushed away from its birthplace as more material forms. A mid-oceanic ridge marks the actual site of crustal formation. (Divergent motion can also occur within a continent; however, oceanic crust, not continental crust, forms in this case, too. As a result, the continent splits into two portions that are separated by an ever-widening ocean.)

Converging plate motion occurs where convection brings plates together and also pulls material downward. The precise result of such activity depends upon which types of crust are present. If one of the two converging plates contains oceanic crust, it will be subducted under the other plate. When a plate containing oceanic crust collides with a continent, the force of the collision, along with the magma generated as the subducted material melts, will produce mountains on the continent; this activity has produced the Andes Mountains in South America. Because continental crust is too light to be subducted, continents will collide after the oceanic crust between them has been subducted. The result of such a collision is a huge mountain range, such as

the Himalayas, formed when India and Asia collided.

How can we tell whether plate tectonic activity is taking place on other planets? The only way is to look for features characteristic of plate tectonic activity, such as spreading centers (the equivalent of Earth's mid-oceanic ridges), trenches, and long chains of mountains. Recently, researchers have interpreted several surface features on Venus as possibly being the result of divergent plate boundary activity. With this exception, plate tectonic activity seems to be unique to Earth. This is not too surprising in view of the fact that the driving force for plate tectonics is convection caused by the Earth's internal heat, which as we have already seen is greater than that of the other terrestrial planets. Another factor whose significance is not fully understood is the relationship between the Earth's ocean water and plate tectonics. It is not known if the process would occur in exactly the same way if the oceans were not present. It is hoped that these and other questions will be answered as we learn more about both Earth and other planets.

■ Rock Deformations

As we examine the crusts of the Earth and other terrestrial planets, we find evidence that crustal material has experienced severe deformations. A **deformation** may be defined as a change in structure, position, or orientation of a rock body. As described in Chapter 8, many rocks on the Earth and other planets form in horizontal layers called strata. This is particularly true for sedimentary rocks on Earth. If sedimentary rocks are found in deformed layers, it is evidence that past geological activity has occurred within the Earth. Most terrestrial deformations occur as a result of horizontal crustal movement caused by plate tectonics and involve either compressional or tensional forces within the crust. However, we see evidence of crustal deformation on other worlds where Earth-style plate tectonics may not be operating, so deformations may be caused by other processes as well. These may include crustal expansion or contraction due to internal heating or cooling, as well as forces resulting from impact activity.

There are three types of crustal deformations (Figure 10.6). The first is **folding,** which produces regular bends in rock. Folding is produced by compressional forces acting on rock layers. Assuming that older layers are on the bottom and younger layers are on the top, the fold can either bend downward, with the older rocks in the middle, or upward, with the younger rocks in the middle. These two types of folds are called anticlines and synclines, respectively, and are very common in terrestrial mountain belts.

The second type of deformation is **jointing,** or cracking in rocks. Joints occur in all types of rocks and seem to be caused by the weight of overlying material or by cooling in the case of igneous rocks. Joints, which often occur in parallel sets, play an important role in weathering on Earth, because moisture seeps into joints and freezes, causing expansion.

Related to jointing but somewhat more spectacular is **faulting,** or displacement occurring along a crack. Fault motion can occur in either the horizontal or vertical direction. Compressional forces result in reverse faults, marked by the "uphill" movement of rock along the fault itself. Tensional forces result in normal faults, marked

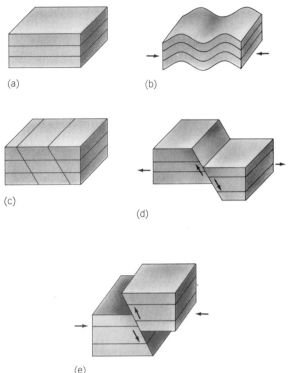

■ **Figure 10.6**   Rock deformations. Layered rocks (a) subjected to compression become folded (b). Joints (c) are cracks in rocks. Normal faults (d) and reverse faults (e) are produced by displacement along breaks in rocks. Tensional forces cause normal faulting, whereas compressional forces cause reverse faulting.

by "downhill" movement of a block of rock along the fault. Both reverse and normal fault motion can cause vertical walls, called **scarps,** to appear at the surface. A down-dropped block of rock, bounded by normal fault scarps on either side, is called a **graben.** A strike-slip fault occurs when motion at the fault is from side to side, often as a result of crustal plates moving laterally past one another. The San Andreas fault in California is a well-known example.

When fault motion occurs, the rough nature of the fault surface produces vibration, which manifests itself as shock waves. These waves travel outward in all directions, producing an event known as an earthquake. The **focus,** or point at which the fault motion occurs, may be at the surface or many kilometers below it. As mentioned in Chapter 9, study of earthquake waves as

■ **Figure 10.7** A batholith exposed above the Earth's surface. This feature, located in Yosemite National Park in California, is made of granite that was formed below the surface and subsequently exposed as the rock surrounding it was eroded. (From Levin, Harold L. *Contemporary Physical Geology,* 3rd ed. Saunders College Publishing, Philadelphia, 1990.)

they pass through our planet's interior has enabled us to learn much of what we know about its interior structure, and seismic wave studies of the Moon and Mars have also been conducted.

■ Igneous Landforms

Numerous planetary surface features are produced by igneous activity occurring above and below the surface. This is especially the case on objects other than Earth, because they lack sedimentary rocks. Even igneous formations produced underground (**intrusions**) can become surface features on Earth, because igneous rock resists erosion better than most sedimentary rocks, and the intrusion may eventually stand out above the Earth's surface (Figure 10.7).

Intrusive activity occurs on both large and small scales. Magma may spread apart pre-existing layers of rock, forming a thin intrusion, called a **sill,** which is parallel to them. Magma can also cut across older layers of rocks, forming a **dike,** so named because it may look somewhat like a wall when exposed at the surface after the surrounding material is eroded. On a much larger scale, an intrusion called a **batholith** may be many square miles in area and replace large quantities of older rock.

Extrusive activity occurs when magma reaches the surface and produces a landform called a **volcano.** A wide variety of volcanic eruptions can occur, and several distinct types of landforms may result, depending on the type of material erupted. Magmas reaching the Earth's surface have a wide range of compositions that lead to diverse physical properties, the most important of which is viscosity. Basic magmas are runny and will not accumulate at the point of eruption and form a high mountain. On the other hand, acidic magmas are very viscous and form high volcanoes, rather than flow over large land areas. The eruptions involving viscous magma also tend to be more violent, because gases cannot escape from the magma as easily.

In addition to the magma itself, other materials are released during a volcanic eruption. A number of gases are emitted after having been under pressure deep within the Earth. These gases can cause explosive events if they are prevented from escaping freely, as was the case when Mt. St. Helens exploded in 1980. The pressure from gas can also shoot droplets of magma into the air, where they solidify while falling. These fragments, as well as pieces of rock ejected from the volcano, are called **pyroclastic material.** Pyroclastic material ranges in size from fine ash to large boulders.

If we examine volcanic features on the Earth and other solar system objects, we can characterize four major types of eruptions (Figure 10.8). **Lava flows** occur when low-viscosity basaltic magma reaches the surface and covers large surface areas. Repeated lava flows in an area can accumulate large thicknesses of material, as in the Columbia Plateau in the northwestern United States. Lava flows are the most common volcanic feature on the Moon and have filled many large basins caused by meteorite impacts, forming the maria (Figure 10.9). These large, usually circular features have surfaces that are marked by ripples and other markings indicative of flow. Landforms similar in appearance to stream channels can also be produced by flowing lava. This can occur at the surface, but the features more commonly form when the surfaces of lava tunnels inside so-

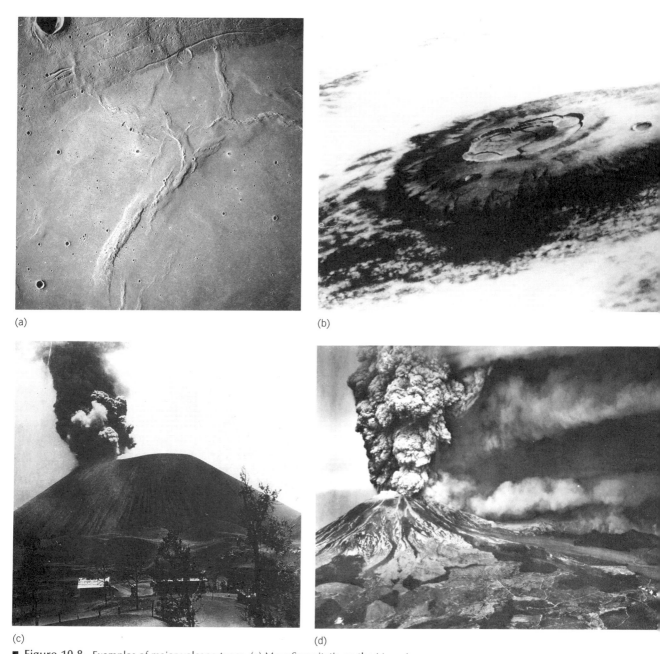

(a)

(b)

(c)

(d)

■ **Figure 10.8**   Examples of major volcano types. (a) Mare Serenitatis on the Moon is a lava flow. (NASA Photograph, courtesy of the National Space Science Data Center and the Apollo 17 Principal Investigator, Frederick J. Doyle.) (b) Olympus Mons on Mars is a shield volcano. (NASA/JPL Photograph.) (c) Paricutin in Mexico is a cinder cone. (U.S. Geological Survey Photograph.) (d) Mount St. Helens in Washington is a composite cone. (U.S. Geological Survey Photograph by Austin Post.)

(a)

■ **Figure 10.9** Mare formation. (a) A large impacting body has formed a huge basin with shattered rock inside. Magma is present at depth. (b) The magma has reached the surface and formed a lava flow.

(b)

lidifying rock collapse (Figure 10.10). Such features are found on both the Earth and the Moon. Many such lunar features are of considerable size and are called rilles.

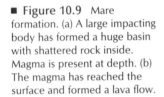

■ **Figure 10.10** Hadley Rille, a lunar feature that is probably a collapsed lava tunnel or related structure. The Apollo 15 mission landed in this area. (NASA Photograph, courtesy of the National Space Science Data Center and the Apollo 15 Principal Investigator, Frederick J. Doyle.)

A second type of volcano is the **shield volcano**, so named because a cross-sectional view of one resembles a shield lying on its back. These are produced by basaltic magma with viscosity high enough that it accumulates to form a mountain. Shield volcanoes have gently sloping sides but are generally very large. The best-known examples of terrestrial shield volcanoes are the Hawaiian islands, which are over 9 kilometers (5 miles) in height, rising from the 5-kilometer (3-mile) deep ocean floor to an elevation of 4 kilometers (2.5 miles) above sea level. Mars has several large shield volcanoes, including Olympus Mons, the largest known shield volcano in the solar system. Olympus Mons is about 550 kilometers (340 miles) across and 25 kilometers (15 miles) high. A shield volcano has a large summit **caldera**, the opening where magma has poured out; the caldera on Olympus Mons is 70 kilometers (45 miles) across.

A third type of volcano is the **cinder cone**, an accumulation of pyroclastic material around the site of the eruption. Because of the nature of cinder cones, they erode quickly and are not durable, at least on Earth. Although the main volcano is composed of pyroclastics, some lava typically escapes from the base of the mountain but is not incorporated into the volcanic cone itself.

Cinder cones are generally less than 500 meters (1600 feet) high and are therefore difficult to detect on other planets. There are indications of some cinder cones in the Marius Hills region on the Moon, but they apparently are not common lunar features. Paricutin in Mexico is an example of a terrestrial cinder cone.

The final type of volcano is the most spectacular. The **composite cone,** or **stratovolcano,** has very steep sides and forms a conspicuous mountain peak. These volcanoes are produced by high-viscosity magma and are composed of layers of rock from both lava and pyroclastic materials, giving them high strength and resistance to erosion. They may be as large as 8 kilometers (5 miles) across and 3 kilometers (2 miles) high. Mount St. Helens is a well-known example, as are Mt. Fuji in Japan and Mt. Kilimanjaro in Africa. Acidic magma seems to be rare on other planets, so composite-cone volcanoes are probably limited to Earth.

In addition to these types of eruptions, all based on molten silicate magmas, there is volcanism on two planetary satellites based on entirely different heat sources and magma chemistry. The heat for the volcanism on Jupiter's satellite Io appears to come from tidal interactions between Io and other satellites of Jupiter, and the material erupted is unlike material in terrestrial eruptions. This magma contains abundant sulfur in a variety of compounds, and the amount of silicates it contains is unknown. We will discuss the details of Io's volcanism when we discuss Jupiter's satellites in Chapter 20. Neptune's satellite Triton has a form of volcanism in which liquid nitrogen below the surface, apparently heated by the absorption of sunlight, breaks through to the surface and spews icy material into Triton's tenuous atmosphere. Triton will be discussed in more detail in Chapter 24.

Several terms are used to describe the stages of a volcano's life cycle. An **active** volcano is one that is erupting material. Earth contains 500 to 600 active volcanoes, and Io had seven when photographed by Voyager. The total number of active volcanoes on Triton is not yet known. (No other solar system objects have currently active volcanoes.) Active volcanoes do not erupt continuously, but the magma source is present, and eruptions can occur at any time. A **dormant** volcano is one whose eruptive activity has ceased but one that has the capability to resume activity if its magma source is replenished. A volcano may be dormant for decades or centuries before renewed activity occurs. Mount St. Helens had been dormant for over 150 years prior to its 1980 eruption. An **extinct** volcano is one that is no longer capable of erupting. This can result from the solidification of all of the source magma, with no further opportunities for magma production. The Moon and Mars appear to have cooled sufficiently so that no further magma generation will ever occur. Therefore, all of their volcanoes are extinct. On Earth, extinct volcanoes are eventually eroded away.

■ Surface Erosion

At the same time that internal processes are at work building up the surface of Earth and other worlds, external processes are acting to wear them away. Although some planets and satellites are essentially "dead" and now lack internal activity, none of them are immune to surface erosion.

In order for erosion to occur, outside agents must act upon planetary surfaces. The nature of the object will determine which of several processes will act upon its surface. The most important consideration is whether the planet has an atmosphere. If an atmosphere is lacking, impact cratering will be the most important form of surface erosion. The only other forms of erosion possible will be mass wasting (the downhill motion of material as a result of gravity) and minor changes caused by the impinging solar wind. If an atmosphere is present, it will greatly reduce the importance of impact phenomena, because most incoming bodies will be burned up before hitting the surface. However, the atmosphere (depending upon composition and pressure) will allow other erosive phenomena to occur. On Earth, these include running water, flowing ice in the form of glaciers, ground water, wind, and waves eroding the shores of large bodies of water. We will examine these erosive processes in detail later.

Geologists have made extensive studies of the surface erosive processes on Earth. Certain landscape features can easily be recognized as being the result of a specific process. Therefore, features can be recognized even after a specific process is no longer active. For example, glacial ice was an important agent of erosion in northern North America during the ice age of the past few million years. Even though this ice has long since retreated, numerous glacial features exist on land that was once covered by ice. Similarly, many features on Mars appear to have been formed by running-water erosion, even though liquid water is now absent from its surface.

When surface processes act on a landscape, three distinct actions can be identified. The first, **erosion**, includes both the breaking down of rock into smaller particles by weathering and the removal of both rocks and regolith from their place of origin. The various agents of erosion each produce characteristic features as a result of removing material. Next, the removed material is transported by a medium such as wind, water, or air. Finally, material may be deposited in some area, again producing distinctive landforms.

■ Impact Cratering

There can be no doubt that the process of one object impacting upon another is responsible for the majority of planetary surface features in the solar system. However, the fact that the Earth has comparatively few impact features prevented this from being realized until relatively recently. Two factors are responsible for the scarcity of terrestrial impact features: the Earth's atmosphere shields us from most impacting bodies, and most craters that do form are quickly weathered away by our planet's active surface geology. (Many old terrestrial craters are visible only as faint outlines from aircraft or satellite photographs.) Even the importance of cratering on the Moon was not unequivocally realized until the 1960s. Although impact cratering had been suggested by astronomers as early as the late 1800s as being the main agent of lunar erosion, some geologists disagreed and claimed that lunar craters were volcanic. The matter was not settled until the lunar exploratory missions of the 1960s. This firsthand exploration of the Moon and spacecraft photography of other solar system bodies have revealed that impact cratering is a ubiquitous phenomenon.

Even though solar system objects tend to be relatively far apart, their paths sometimes bring them close enough that collisions can occur. Most collisions involve small objects, because there are proportionally more of them than larger ones. What is the source of these colliding objects? The solar system's large bodies (planets and satellites) formed primarily by accretion, in which small objects in similar orbits stuck together as electrostatic and then gravitational forces held them together to form larger masses. This accretion was incomplete, and many fragments remained after the process stopped. Because of this, collisions were extremely frequent when the solar system was young. Today, 4.6 billion years later, relatively few fragments remain, the rest having already smashed into other bodies.

What happens when solar system objects collide? Several factors are involved. One factor is the relative diameters of the objects. If two objects of nearly equal diameter collide, both are likely to be fragmented. (This process occurs even today in the asteroid belt. As a result, some newly generated fragments are placed into orbits that cause them to collide eventually with a planet.) A more common occurrence is that a smaller object strikes a larger one, because, as mentioned earlier, there are more smaller objects than planets and satellites. In fact, many particles that strike the Moon are microscopic and produce "craters" consisting of small pits on the faces of rocks that they strike. Another factor is the relative velocities of the objects and the angle at which the collision occurs. The results of a head-on collision will be different from what happens if one body gradually overtakes another while both are moving in nearly the same orbit. Also, the hardness and strength of the materials involved are important.

Depending on these factors, several things can occur as a result of collision. The objects may fuse together, as in the accretion process. The objects could be broken apart by the force of the impact, with the fragments either dispersing completely or reassembling due to their gravitational attraction. The most common result, however, is an **impact crater**. Craters range in size from al-

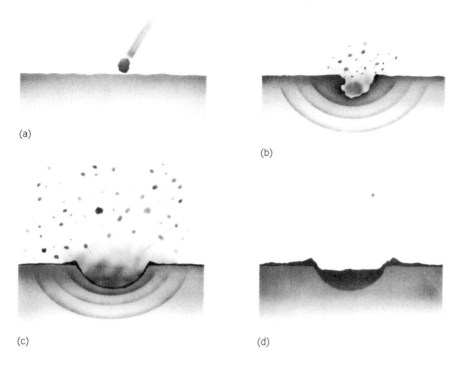

(a)

(b)

(c)

(d)

■ **Figure 10.11** The impact process. A meteorite approaches (a) and vaporizes as it hits (b). The shock produces a crater and ejects large amounts of material (c). The final crater floor is covered by a layer of shattered rock (d). (From Abell, George, David Morrison, and Sidney Wolff. *Exploration of the Universe,* 5th ed. Saunders College Publishing, Philadelphia, 1987.)

most microscopic to hundreds of miles across. Mimas, a satellite of Saturn with a 390-kilometer (240-mile) diameter, has a crater 100 kilometers (60 miles) across. This is thought to be nearly the largest crater size possible on that world, because a larger impact would have shattered it.

Figure 10.11 illustrates what occurs during the cratering process. Orbital velocities in the vicinity of Earth are about 30 kilometers per second (20 miles per second), so an object entering our atmosphere will be traveling at speeds that most of Earth's residents would find difficult to comprehend. Smaller meteoroids are slowed by atmospheric friction to only a few hundred kilometers per hour, but larger objects have so much kinetic energy that they are slowed very little. Accordingly, the impact process happens quickly, and incredibly large amounts of energy are released.

When a large meteorite hits the surface, it produces a shock wave that melts the surrounding rock. The impacting fragment is likely to be completely vaporized, along with some of the surface rock. In the late 1800s, researchers drilled into the bottom of Meteor Crater in Arizona, looking for the remains of the meteorite thought to be responsible for the impact. The failure to find anything was considered evidence that the crater was actually a volcanic feature. However, we realize today that the energy released upon impact would have been sufficient to vaporize the impacting meteorite, as well as to knock observers as far as 30 kilometers (20 miles) away off their feet! The object that produced this crater is estimated to have had a mass of 27,000 kilograms (30 tons) and to have struck the surface at a speed of about 11 kilometers per second (7 miles per second).

Some of the rock at the impact site is pushed upward and outward into the air. Any of this material that is molten cools and solidifies during flight, forming glassy objects called **tektites** (Figure 10.12). Large numbers of tektites have been found near several major impact sites on Earth. Most material, however, remains solid (although fragmented) and ends up near the impact site, piling up around the sides of the hole excavated by the blast. This **ejecta**, as it is called, may contain layers of rock that were removed from inside the crater and literally flipped upside down.

This scenario of crater formation has been verified by laboratory experiments in which high-

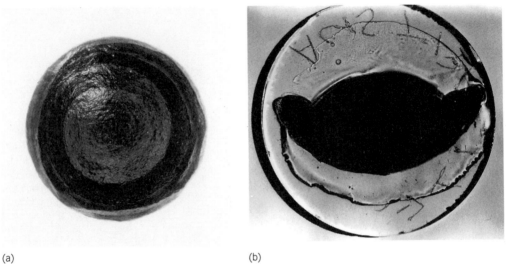

(a)   (b)

■ **Figure 10.12** Tektites from Australia. These objects, which are about 2.5 centimeters (1 inch) across, formed from molten material generated by a meteorite impact. Their aerodynamic shape indicates that they solidified while in flight. (a) The surface of a tektite. (Smithsonian Institution Photograph, courtesy of Brian Mason.) (b) Cross section of a tektite, showing its aerodynamic shape. (Photograph by the author.)

speed projectiles are shot into various materials. In addition, field examination of craters on the Earth and Moon has shown that our understanding of the cratering process is now fairly complete. Because other features on Earth may resemble impact craters, care must be taken in interpreting circular "holes in the ground." Fortunately, the shock wave passing through rocks creates characteristic features called **shatter cones,** which are never produced by any other type of activity (Figure 10.13). These shatter cones have been used to identify positively some circular features on Earth as impact sites, because they remain in the rock even after the crater itself has been eroded away. Another line of evidence for impacts is the presence of high-pressure forms of minerals near the surface, where they are not ordinarily found. Over 100 impact sites are now recognized on our planet (Figure 10.14).

Impact craters, whether small or large, share several important properties (Figure 10.15). Nearly all craters are circular, even if the impacting body struck the surface at an angle. This is because the crater is excavated by the explosive vaporization that travels outward from the point of impact, rather than because the object itself

■ **Figure 10.13** Shatter cone found in rock near an ancient impact structure in Australia. The specimen is 20 inches long. (Smithsonian Institution Photograph, courtesy of Brian Mason.)

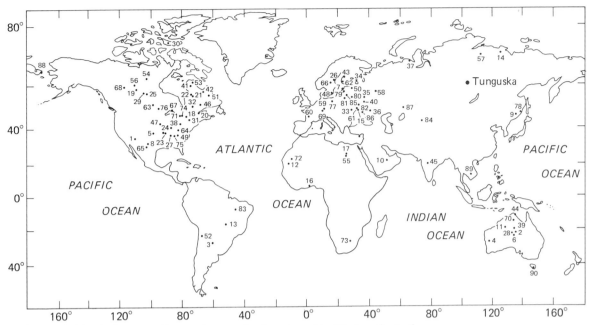

■ **Figure 10.14**   The location of impact features that have been identified on the Earth. (Map by R.A.F. Grieve, reproduced from Taylor, Stuart Ross. *Planetary Science: A Lunar Perspective.* Copyright 1982, Lunar and Planetary Institute, Houston.)

"plows" the surface. Some craters are squarish or hexagonal, but this is a result of joint structures in the rock that was impacted. An example is Meteor Crater, which appears somewhat squarish due to the underlying rock structure (Figure 10.16). Craters have rims of ejecta that are raised above the original level of the surface. The crater floor lies below the original level of the surface and is covered with shocked and shattered rock. Many craters, particularly larger ones, have **central peaks.** These are especially prominent in many lunar craters and can be seen through small telescopes. It is thought that these peaks may form as a result of isostatic rebound of the center of the crater or from an accumulation of material that slides down from the inside of the crater's rim. The main crater is often surrounded by **secondary craters** produced by large fragments of ejecta. On the Moon, long streaks of ejecta, called **rays,** extend outward in a radiating pattern for hundreds of miles around several fairly young craters. They are lighter in color than the underlying material and are therefore very conspicuous (Figure 10.17). They are visible only around

young craters because they eventually darken, presumably because of exposure to solar wind hitting the lunar surface and also because impacts of small objects into the ray material mix it into the underlying regolith. The various ejecta-related phenomena vary significantly from object to object, because ejecta travels farther on small worlds with weak gravity than on larger worlds with stronger gravity.

The largest impact structures have slightly different forms and are called **multiringed basins.** They may be hundreds of kilometers or more in diameter and contain several concentric mountain ridges that form an overall bull's-eye pattern. The Orientale basin on the Moon is 900 kilometers (550 miles) across and contains four mountainous rings formed from faulting associated with the impact event (Figure 10.18). As mentioned earlier, the lunar maria were formed when magma flooded such basins. Mare Orientale, however, was formed later in the history of the Moon, and only a small amount of magma remained to enter it, allowing much of its internal structure to remain visible.

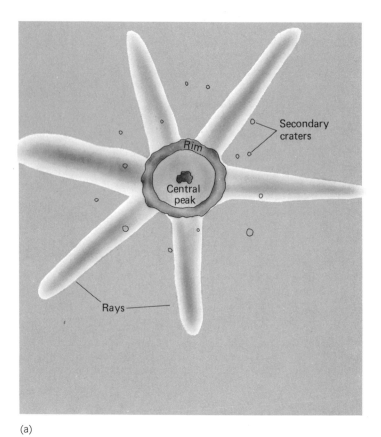

(a)

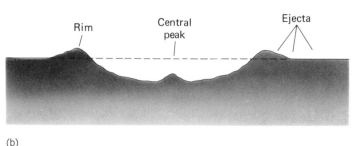

(b)

■ **Figure 10.15** Crater features as shown in a map view (a) and a cross section (b). A crater is round and has a raised rim. It may have a central peak, and a ray system and secondary craters are often present. (In the cross section, the *dotted line* represents the original surface.)

Craters have become important tools in understanding the history of solar system objects. Many studies of planet and satellite surfaces involve **crater counting** to estimate age. The basis for this technique is the fact that most impacts occurred very early in the solar system, when there were abundant fragments left over from the accretion process. In general, if we see a heavily cratered surface, we can conclude that the craters were formed long ago and that nothing has since happened to destroy them and renew the surface. If we see a crater-free surface, we can conclude that it is young, renewed by either surface or interior processes, because no solid surface was able to escape the early bombardment period. Much of the Moon is an example of an old, heavily cratered surface, whereas the Earth's surface, which has few craters, is very young. Earth has been reworked constantly since the long-ago era of heavy bombardment during which its original surface was probably heavily cratered. Enough solar system objects have been

■ **Figure 10.16** An aerial view of Meteor Crater, located near Winslow, Arizona. Note the squarish shape of the crater, the result of structural features in the rock present before the impact occurred. (U.S. Geological Survey Photograph by David J. Roddy and Karl Zeller.)

■ **Figure 10.17** Lunar crater rays. This photograph was taken at full moon, when the rays are most prominent. Some of the rays from the crater Tycho *(top center)* extend for a considerable distance. (Lick Observatory Photograph.)

■ **Figure 10.18** Mare Orientale, one of the most impressive lunar surface features. This large multiringed impact basin is 900 kilometers (550 miles) across. (NASA Photograph, courtesy of the National Space Science Data Center and the Lunar Orbiter IV Principal Investigator, Leon J. Kosofsky.)

studied using crater-counting techniques that researchers are even able to determine ages in years for various planetary and satellite surfaces. One assumption inherent in such work is that impact rates were essentially identical throughout the solar system, from Mercury to the gas giants, during the early bombardment period.

■ Mass Wasting

Mass wasting, the downhill movement of rock and regolith caused by gravity, is an important erosional process that occurs on both the Earth and airless bodies. On Earth, mass wasting is a very important process, but the average person is unaware of its existence unless a spectacular, rapid form of the process, such as a landslide, occurs. Material moved by mass wasting is called **colluvium** and can be identified in photographs of the surfaces of other solar system objects, even if we have never witnessed the mass-wasting motion. Colluvium can be identified by its appearance, as it is usually jumbled and has an irregular surface, and it may clearly look as though it has flowed from a higher elevation. In addition, steep areas near the colluvium may bear scars formed when regolith or rock broke free.

Several factors control the occurrence and rate of mass wasting at a given locality. Gravity is the cause of mass wasting, and the process will be more important on planets with higher surface gravity. Another example of the role of gravity is that a large boulder will be more difficult to dislodge than a small grain of sand, but if the material supporting the boulder is eroded away and it starts moving, it will travel farther than a smaller particle would. Because mass wasting involves downhill movement, the process will only occur where there is a slope, and it will be most prevalent in areas of highest slope. It will also be higher in areas covered by regolith than in areas where bedrock is exposed at the surface, because in the latter case, a fragment of rock will need to break free before it can move downhill.

One of the most important factors promoting terrestrial mass wasting is the presence of water in the soil. Water adds weight and also lubricates the soil particles so that they flow more easily.

Terrestrial occurrences of mass wasting often take place during the rainy season of a given area. Vegetation tends to hold soil in place, preventing mass wasting of soil, but tree roots also help break apart rocks on a cliff face, promoting mass wasting in that environment.

The factors that control mass wasting in a given area determine the steepest slope possible in that area. As an example of slope stability, imagine a child playing in a sandbox. It is impossible to make a pile of dry sand with vertical sides. If the child attempted to make such a pile, it would immediately mass waste material from the top, making a pile with sloping sides. The same thing occurs in nature; the slopes of a new mountainous area undergo mass wasting until they are stable. On the Moon, many cases of mass wasting have occurred in the rims of craters. The cratering process forms steep slopes that are unstable, and mass wasting therefore results. Boulders that have rolled downhill and left tracks in the regolith are often seen, as are large masses of colluvium that have moved as a single unit.

Many types of mass wasting have been identified and named. **Creep** is the slow movement of soil downhill. Although barely noticeable, it will eventually cause stone walls to bulge, fences to move, and posts to tilt downhill. **Rockfall** is the rapid falling of a single rock down a slope. **Landslides** are similar, except that a huge mass of material moves all at once. Landslides on Earth are often triggered by earthquakes, and huge masses of colluvium have traveled for miles, at speeds of hundreds of miles per hour, and done considerable damage. Geologists once theorized that large landslide masses travel on a cushion of trapped air, but indications are that similar events happen on Mars, where the atmospheric pressure is considerably less. **Slumping** is a process in which a coherent mass of material slides downhill along a curved surface of rupture. Slumps are very often seen in the rim regions of craters.

■ Surface Running Water

Because liquid water can exist only on the surfaces of planets with atmospheres, its geological importance in the solar system is severely limited.

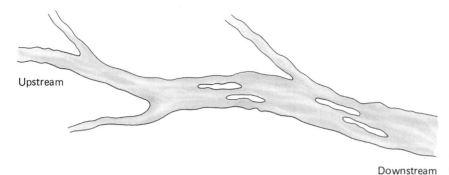

■ **Figure 10.19** Stream characteristics. Notice that the stream's path is not perfectly straight; that it gets wider downstream; that tributaries enter, making small angles with the upstream direction; and that braiding occurs.

However, the effects of surface water on the two worlds (Earth and Mars) where it has played a role in shaping the surfaces are very dramatic.

It can be argued that surface running water is the most effective agent of landform change on the Earth. Even in deserts, where rainfall is rare, the lack of surface vegetation to hold soil in place allows extensive erosion when rain does fall. In normal climates where rainfall is abundant, numerous streams and rivers collectively sculpt the land, wearing away high areas and decreasing overall relief. The main erosional features made by running water are the channels in which the water flows and the larger valleys that contain these channels. Widened by mass wasting and deepened by long periods of downward erosion, many terrestrial valleys attain enormous dimensions. Running water is also a very effective medium for transporting sediments. Both the maximum particle size that can be carried and the overall sediment capacity increase as water velocity rises. Stream depositional features are formed when something causes water velocity to drop rapidly. When a stream descending a steep slope reaches a level area, a semicircular deposit known as an **alluvial fan** may form. When a river enters a lake or an ocean, a large deposit known as a **delta** may form, unless currents in the water disperse the sediment.

There currently seems to be no liquid water present on the surface of Mars, yet many of its features are considered to be fluvial (stream formed). How can we tell that some Martian features were formed by running water if there is currently no surface water present? Several characteristics of surface water flow are shown in Fig-

ure 10.19. One clue is that streams are usually curved, unlike faults and grabens, which are usually straight. Streams and rivers also have tributaries entering them. These tributaries normally enter at small angles to the upstream direction, and their additional water causes the channel to widen, thus providing two pieces of evidence of flow direction. A third unique aspect of stream channels is that **braiding** is often present. This means that the channel separates into several paths for water flow, with heavy sediment deposits separating them. Braiding occurs when the amount of water is insufficient to carry all of the sediment present. Another indicator of fluvial features on Mars is that streamlined, teardrop-shaped deposits of sediment appear behind craters and other obstructions that lowered water velocity enough that sediment was deposited.

Because the Martian surface appears to be waterless today, the origin of its fluvial features is difficult to explain. It is possible that Mars' atmospheric pressure was higher at one time, allowing liquid water to exist at the surface. Many researchers believe that Mars has (or had) large amounts of frozen water below its surface. When meteorite impacts occurred, enough heat was released to cause some of the ice to melt. Even though water ordinarily cannot exist in liquid form at the temperature and pressure conditions now present at the Martian surface, the water would have survived long enough before vaporizing to form short-lived streams and rivers.

Several satellites of the outer planets appear to contain large amounts of water ice and may even contain liquid water at some point within them. However, they contain no features indicat-

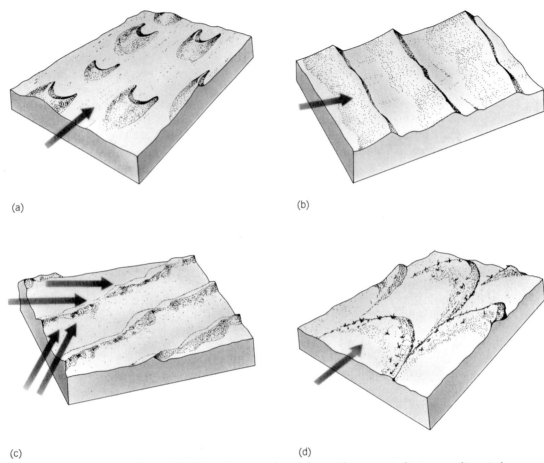

(a)

(b)

(c)

(d)

■ **Figure 10.20** Major types of sand dunes. The *arrows* indicate prevailing wind directions. (a) Barchan. (b) Transverse. (c) Longitudinal. (d) Parabolic. (From McKee, Edwin D. *A Study of Global Sand Seas.* U.S. Geological Survey/Department of the Interior, Washington, D.C., 1979.)

ing that stream flow ever occurred on their surfaces. One confusing point about surface running water is that, occasionally, lava flowing on the surface can create channels that mimic those formed by stream flow. This is the interpretation of certain stream-like features on the Moon, although the release of frozen water beneath the lunar surface by impact heating cannot yet be ruled out completely.

■ Wind

Wind, or horizontal movement within an atmosphere, has been detected on every planet whose atmosphere has been studied. Wind is produced by variations in air pressure, with air moving from regions of high pressure to those of low pressure. These pressure variations are the result of uneven solar heating.

On Earth, wind erosion is not nearly as important as water erosion. This is due mainly to the fact that the vegetation covering most locales holds soil in place, allowing wind erosion to be effective only in areas where little vegetation is present, such as deserts and beaches. Wind is limited to moving only small particles, and the total volume of material that can be carried by a given quantity of air is small. On the other hand, wind is not limited to a particular channel (like a river is, for example), and the amounts of wind-borne material can be enormous, given the right conditions.

The main features associated with wind erosion are **ventifacts**, rocks shaped by the sand-blasting effects of air-carried sediment; **yardangs**, larger hills shaped by wind; and **deflation basins**, low areas from which wind has removed material. The major features of wind deposition are **dunes**, which are mounds of sand. Dunes occur in a wide variety of shapes and sizes, determined by such factors as constancy of wind direction and the amount of sand present (Figure 10.20). Dunes change with time, because sand is blown up one side and redeposited on the other, causing the entire dune to move downwind.

Venus has a dense, viscous atmosphere, but the Soviet Venera landers have indicated the presence of winds that sometimes have sufficient velocity to transport sand. However, no specific wind-caused features have been detected, either by landers or by surface radar studies from orbiting spacecraft.

Mars, like the Earth, has been shaped by wind and contains numerous wind-caused features. For over a century, observers have considered yellow "clouds" sometimes seen in the Martian atmosphere to be dust storms. When the Mariner 9 spacecraft began orbiting Mars in 1971, the entire surface was obscured by wind-blown dust. It and subsequent orbiting spacecraft photographed many dune fields, which are particularly common at high Martian latitudes, as well as numerous yardangs. Many Martian craters have dark or light streaks extending to one side that are apparently the result of wind erosion or deposition; the crater acts as a windbreak, leading to deposition behind it (Figure 10.21).

■ Ice

Frozen water plays several roles in surficial erosion. One is that colder regions of Earth have a layer of permafrost below the surface, and the soil layer at the surface remains totally frozen for a large part of the year. When the surface layer melts, significant mass wasting can occur because the soil is virtually saturated. Another effect is that some frozen areas display an unusual large-scale pattern of frost-induced surface markings whose

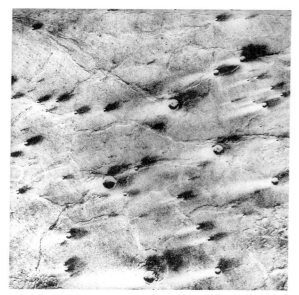

■ **Figure 10.21** Martian crater streaks. These light and dark features are believed to be the result of the craters' acting as "wind shadows" as wind erosion and deposition occur. (NASA/JPL Photograph, courtesy of the National Space Science Data Center.)

origin is not yet fully understood. These areas of patterned ground can be seen in aerial photographs of terrestrial Arctic areas, and similar patterns have been spotted in photographs of the surface of Mars (Figure 10.22).

The most important ice-related feature on Earth is the **glacier**, a large body of ice moving over a land surface. Glaciers are very effective erosional agents; they wear away rock and soil from the areas over which they travel. Glaciers currently cover 10% of the Earth's land surface, but this figure was tripled during much of the last 2 million years, when an ice age occurred for reasons that are not fully understood. An **ice age** is a period of time during which glaciers cover an unusually large portion of the Earth's surface. Evidence indicates that ice ages have occurred at intervals of several hundred million years throughout the history of the Earth.

Two types of glaciers occur on Earth. **Alpine glaciers** are limited by terrain and are most commonly found in valleys (Figure 10.23). Because only part of the land is eroded by Alpine glaciers, they tend to increase an area's relief. The main landform characteristic of Alpine glaciation is the

■ **Figure 10.22** Martian patterned ground. This pattern is similar to that found in places in the Arctic on Earth, but it is much larger. The features here are 5 to 10 kilometers (3 to 6 miles) across. (NASA/JPL Photograph, courtesy of the National Space Science Data Center.)

glacial trough, a U-shaped valley that is quite distinct from the V-shaped valley normally formed by stream erosion. The second type of terrestrial glacier is the **continental glacier,** so named because it covers a very large expanse of land. Continental glaciers in Antarctica today are over 3 kilometers (2 miles) thick in places, so even mountains can be covered and eroded. As a result, areas that have experienced continental glaciation are characterized by an overall lowering of

■ **Figure 10.23** An Alpine glacier, located in a valley in Canada. This view looks down the valley from the higher portion of the glacier. (U.S. Geological Survey Photograph by Austin Post.)

relief. Continental glaciers covered much of the northeastern United States during the most recent ice age.

Although the north polar region of Earth is not covered by a land mass, an ice pack forms at the surface of the Arctic Ocean. Mars has polar caps, but rather than thick masses of ice, as are found in glaciers, they are thought to be thin deposits of water ice and carbon dioxide ice (dry ice) frost. Could glaciers have existed on Mars in the past? Although there is no definite evidence, some researchers have suggested that the "stream" channels on Mars were actually formed by Alpine glaciation, rather than by stream erosion. This is because ice erosion is more efficient than stream erosion, because ice carries larger sediment fragments and can move greater overall quantities of sediment. For this reason, a smaller quantity of solid ice could have formed the Martian channels, compared to the amount of liquid water that would have been required.

■ Ground Water

Tremendous quantities of **ground water** are found in the pore spaces within rock and soil beneath the Earth's surface. Most of this water occurs within about 800 meters (2600 feet) of the surface, because the pressure from overlying material would close any openings below that. It has been estimated that if all of this subsurface water could be brought to the surface, its average depth would be 30 meters (100 feet)! In addition to providing a source of water for plants and humans,

ground water performs considerable erosion if the right type of rock is present. Limestone and other rocks made of carbonate minerals are soluble in water that is slightly acidic, as is most terrestrial ground water. Numerous caverns have been formed in areas where the rock is limestone and considerable ground water is present, and some have reached immense size. Solution of limestone at the surface, or the collapse of a cave roof, leads to a feature known as a sinkhole. Terrestrial areas where caves and sinkholes are abundant are said to have **Karst topography**, named after a region in Yugoslavia.

No Karst topography has been seen on any other planetary surface, and it is likely that if extraterrestrial ground water is present, it is actually "ground ice," which would be active in erosion only if melted and released at the surface, as mentioned previously.

■ Waves

The final agent of landform development is one that is undoubtedly limited to Earth. Waves are the result of wind, which causes oscillatory movement at the surface of large bodies of standing water. Waves in Earth's lakes and oceans are effective in eroding rock at shorelines in some areas and in depositing beaches and sand bars in others. Even if water were present at the surface of Mars or other bodies as a result of impact-induced melting of subsurface ice, it is unlikely that the water could persist long enough for any wave activity to occur.

■ Chapter Summary

Terrestrial planets are marked by a wide diversity of surface landforms, some produced by internal processes within the planets, others resulting from external activity. Planetary crusts are affected by vertical motion (isostasy) and by horizontal motion, exemplified by plate tectonics on the Earth. Horizontal motion is responsible for faults and folds, the most important crustal deformations.

Many landforms are produced by intrusive and extrusive igneous activity. On airless

bodies, the most important external process is impact cratering. Other important external processes of erosion include mass wasting, running water, wind, and ice.

As we examine the planets in later chapters, we will see that no other world has quite the diversity of surface features found on the Earth. Nevertheless, all planets have a wide variety of surface features, and many contain very spectacular and interesting landforms.

■ Chapter Vocabulary

alluvial fan
Alpine glacier
asthenosphere
average radius
batholith
caldera
central peaks
cinder cone
colluvium
composite cone
continental crust
continental glacier
crater counting
creep
deflation basin
deformation

delta
dike
dune
ejecta
erosion
external processes
fault
focus
fold
glacier
graben
gravity anomaly
ice age
impact crater
internal processes

intrusion
isostasy
joint
Karst topography
landslide
lava flow
lithosphere
mare
mass wasting
multiringed basin
oceanic crust
paleomagnetism
plate tectonics
pyroclastic material
ray

rockfall
scarp
secondary crater
shatter cone
shield volcano
sill
slump
subduction
tektite
topographic province
ventifact
vertical relief
volcano
wind
yardang

■ Review Questions

1. Compare the method used for making elevation measurements on Earth to those used on other terrestrial planets.
2. What are the differences between the surface features caused by internal and external processes? Summarize the major contributors to each process.
3. List examples of several solar system objects that have significant differences between two hemispheres.
4. Explain the theory of isostasy, and give examples of how it works.
5. What is plate tectonics? Why do we believe that plate tectonic motion occurs on Earth?
6. How does plate tectonics produce major landforms on Earth?
7. How can we determine whether plate tectonics is active on other planets?
8. List and describe the three types of crustal deformations.
9. Describe several features produced by intrusive igneous activity.

10. Describe the four types of volcanic eruptions.
11. Summarize the events that happen during surface erosion.
12. What factors control which types of surface erosion will occur on a particular planet?
13. Summarize the events that occur during the cratering process.
14. Sketch a crater, and label its important features.
15. List the factors that control the occurrence and severity of mass wasting.
16. List some examples of mass wasting that occur on the Earth as well as on other planets.
17. How would one identify features made by surface running water on a totally dry planet?
18. Describe the effects of wind erosion.
19. Explain the two types of glaciers and the results of each type on Earth.

■ For Further Reading

Cox, Allan and Brian R. Hart. *Plate Tectonics: How It Works.* Blackwell Scientific Publications, Oxford, England, 1986.

Decker, Robert and Barbara Decker. *Volcanoes.* Freeman, San Francisco, 1981.

Greeley, Ronald. *Planetary Landscapes.* Allen and Unwin, London, 1985.

Mark, Kathleen. *Meteorite Craters.* University of Arizona Press, Tucson, 1987.

Murray, Bruce et al. *Earthlike Planets.* Freeman, San Francisco, 1981.

Rice, R.J. *Fundamentals of Geomorphology.* Longman, London, 1988.

Thornbury, William D. *Principles of Geomorphology.* Wiley, New York, 1969.

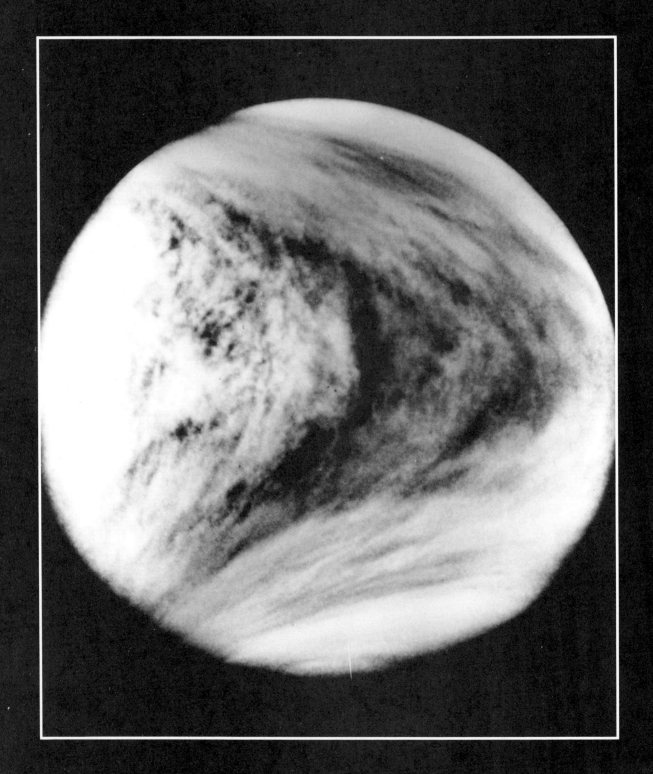

Atmospheres of Terrestrial Planets

Cloud-shrouded Venus, whose atmosphere is the densest of any terrestrial planet. Relatively featureless when viewed in visible light, some markings are revealed when using ultraviolet light. (NASA Photograph.)

Three of the four terrestrial planets and at least two of the many planetary satellites are surrounded by atmospheres. An **atmosphere** may be defined as a blanket of gases held in place around a planet by gravity. Although certain features are common to all atmospheres, the precise nature of an atmosphere depends upon a number of factors, and no two atmospheres are alike.

■ Factors Controlling Planetary Atmospheres

Whether a planet will have an atmosphere and what the composition of the atmosphere will be depend mainly on two factors, the temperature of the planet and the strength of its gravity. A planet's temperature depends, in turn, upon its distance from the Sun and its albedo. (Another factor, the greenhouse effect, may be involved, but we will defer discussion of it until later in the chapter.) The larger the semimajor axis of a planet's orbit, the less energy it receives from the Sun. The amount of energy striking an area of 1 square centimeter above the Earth's atmosphere every second (a quantity called the **solar constant**) is 1.37×10^6 ergs. For comparison, the values for Mercury and Pluto are 9.16×10^6 and 882 ergs, respectively. What happens to the energy that is incident upon a planet? Some of it is absorbed, whereas the rest is reflected back into space. The fraction of incident light reflected is called the albedo of the planet and is controlled by the nature of the planet's surface or atmosphere. For example, the Earth's albedo is about 37%, meaning that only 63% of the light incident upon the Earth actually contributes toward warming our world. There is a wide variation in albedo values among the terrestrial planets, with dark Mercury having a value of 11% and cloud-covered Venus having a value of 65%.

Why is temperature so important to atmospheres? Recall that temperature is a measure of

153

the rate of molecular motion. The higher the temperature at a planet's surface, the faster any gas molecules located there will be moving. The average velocity, V_G, of a gas atom or molecule is given by the equation

$$V_G = \sqrt{\frac{3kT}{m}},$$

where k is a numerical constant, T is the temperature, and m is the mass of the atom or molecule. (A listing of atoms and molecules commonly found in planetary atmospheres, and their relative masses, is contained in Table 11.1.) This equation tells us that more massive molecules travel more slowly than light ones. A complication arises when considering gas velocity, however. Not all of the atoms or molecules in a given quantity of gas at a certain temperature will be moving at this average velocity; some will be moving faster, others slower. The fastest-moving atoms or molecules will be capable of escaping from the planet's atmosphere. Recall that a planet's escape velocity, V_E, is given by the equation

$$V_E = \sqrt{\frac{2GM}{R}},$$

where G is the gravitational constant, M is the mass of the planet, and R is its radius. For a particular gaseous component to remain stable in an atmosphere, its average velocity must be no greater than one-sixth of the escape velocity, or

$$V_G < \tfrac{1}{6}V_E.$$

(The reason for the 1/6 in the equation is that some atoms or molecules with greater than average velocity would be lost if V_G were just slightly less than V_E. If V_G is less than one-sixth the value of V_E, very few of even the fastest-moving atoms or molecules will have escape velocity, and the atmosphere will be retained indefinitely.)

As an example of the role of temperature and gravity, consider the Earth and Moon, which have about the same average surface temperatures, although the Moon's temperature values vary much more widely. The Moon's escape velocity is so low that it has no atmosphere at all. The Earth has retained heavier gases such as oxygen, nitrogen, and carbon dioxide, but the lightweight (and, therefore, faster-moving) gases hydrogen and helium have escaped into space. Only cold, massive planets such as the gas giants can retain hydrogen and helium. In general, the larger the planet and the more distant from the Sun it is, the more likely it will be to have an atmosphere, and the lighter the gases that will be present.

Composition, temperature, and pressure vary widely in terrestrial planet atmospheres. Table 11.2 summarizes the properties of atmospheres of terrestrial and icy solar system objects. Each object's atmosphere will be described in detail in the chapter discussing that object.

■ Incident Sunlight and the Greenhouse Effect

When sunlight hits an airless planet or satellite, the entire spectral range of solar energy strikes its surface. Some of this energy may be reflected directly back into space, and some is reradiated at other wavelengths after heating the rocks and regolith of the planet's surface. The presence of an atmosphere greatly complicates the interactions between the Sun and a planet.

An important process occurring in atmospheres is the absorption of certain wavelengths of light by atmospheric gases. In fact, certain wavelength regions are completely blocked in this way. For example, Earth's atmosphere pre-

■ Table 11.1   Gases Commonly Found in Planetary Atmospheres

Component	Symbol	Relative Mass
Hydrogen (atomic)	H	1
Hydrogen (molecular)	H_2	2
Helium	He	4
Methane	CH_4	16
Ammonia	NH_3	17
Water	H_2O	18
Neon	Ne	20
Carbon monoxide	CO	28
Nitrogen (molecular)	N_2	28
Oxygen (molecular)	O_2	32
Argon	Ar	40
Carbon dioxide	CO_2	44
Ozone	O_3	48
Ammonium hydrosulfide	NH_4SH	51

■ Table 11.2   Atmospheres of Selected Terrestrial Objects

	Venus	Earth	Mars	Titan
Average surface temperature	730 K 855°F	288 K 59°F	218 K −67°F	92 K −294°F
Average surface pressure (atmospheres)	90	1	0.007	1.6
Atmospheric composition (by volume)	CO_2 96% N_2 4%	N_2 78.08% O_2 20.95% Ar 0.93% CO_2 0.034% H_2O 1–4%	CO_2 95.3% N_2 2.7% Ar 1.6% O_2 0.13% CO 0.08%	N_2 92–98% CH_4 2–8%

Data from Pollack, James B. "Atmospheres of the Terrestrial Planets," in *The New Solar System*, Beatty, J. Kelly and Andrew Chaikin, editors. Sky Publishing Corporation, Cambridge, MA, 1990. Reproduced with permission.

vents gamma rays, X-rays, and some ultraviolet and infrared light from reaching the surface. These forms of light are absorbed at different regions in the atmosphere, and, as a result, atmospheric heating occurs in these regions, as will be discussed more fully later. Ultraviolet radiation, in addition, can break apart atmospheric molecules into their component atoms. This **photodissociation**, as it is called, can produce atoms light enough to escape, unlike the heavier molecules from which they were derived. Because most other wavelength regions interact with higher areas within the atmosphere, visible light plays the greatest role in heating the Earth's surface and producing weather phenomena in the troposphere, the lowest atmospheric layer.

As the surface of a planet is warmed by incident sunlight, it will, in turn, emit radiation according to Wien's law. The Earth's average surface temperature, for example, is such that the wavelength of maximum light emission is in the infrared region. The infrared radiation leaving the surface is mainly responsible for cooling at night, and if this radiation did not leave the Earth, our planet would gradually become hotter and hotter.

All other conditions being equal, a terrestrial cloudy night is generally warmer than a clear one. This is because the water vapor in clouds blocks the infrared light, preventing some heat from escaping. Carbon dioxide also blocks infrared light, but this effect does not depend upon the degree of cloudiness. Trapping of heat in this way is called the **greenhouse effect**, because it is somewhat similar to the heat-trapping design of glass windows in a greenhouse (Figure 11.1). The

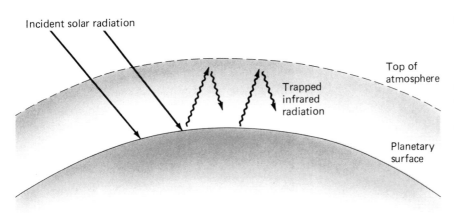

Incident solar radiation

Top of atmosphere

Trapped infrared radiation

Planetary surface

■ Figure 11.1   The greenhouse effect. Visible sunlight enters the atmosphere and heats the surface. On Venus, there is enough carbon dioxide in the atmosphere that infrared light is blocked from escaping the atmosphere, causing extreme heating.

greenhouse effect on Earth is fairly small, but the situation is much different on Venus. Because of the huge amount of carbon dioxide in Venus' atmosphere, very little infrared light can escape from that planet. For this reason, Venus' surface is incredibly hot, averaging 730 K (855°F), nearly 300 K hotter than Mercury, despite the latter's proximity to the Sun. The condition of Venus' surface has raised concerns about the amount of carbon dioxide that is being added to Earth's atmosphere because of the burning of fossil fuels (coal, oil, and natural gas). Even a slight increase in the Earth's average surface temperature would have disastrous effects, as the polar ice subsequently melted would raise sea level, inundating the major coastal cities of the world! In addition, global warming would change the climates of inland regions, such as making the central United States' grain-producing regions much warmer and drier, reducing their agricultural productivity.

■ Energy Transfer, Clouds, and Precipitation

Because of the temperature variations that occur in planetary atmospheres, many of the normally gaseous components of these atmospheres may also occur in the liquid or solid state. Perhaps the best example is water in Earth's atmosphere, which can occur in all three states, depending upon temperature. Both liquid and solid water also occur on the surface of the Earth. Changes of state play an important role in transferring energy within atmospheres. This is because the melting of solids or the evaporation of liquids requires energy, and the opposite processes (freezing and condensation) release energy.

As an example of atmospheric energy transfer, consider water in the Earth's troposphere. The troposphere receives mostly visible light, because shorter wavelengths are absorbed higher in the atmosphere, heating the gas in its upper layers. Because the troposphere is transparent to visible light, few absorptions occur within it, allowing most of the light to reach the surface. How, then, is the troposphere warmed? The process most responsible is the **evaporation** of ocean water. As ocean water evaporates, the

immediate effect is that the oceans themselves are cooled, because heat energy is required for evaporation. Later, much of this atmospheric water vapor condenses, forming clouds composed of small water droplets or ice crystals. **Clouds** form when the atmosphere can hold no more of a vaporized substance, a condition known as **saturation.** When saturation is reached, liquid droplets or solid crystals begin to condense. Energy is released during the **condensation** process and warms the atmosphere.

In addition to Earth's familiar clouds of water vapor, haze and clouds are found in the atmospheres of other planets. (Cloud droplets are generally about 10^{-6} meters in diameter, whereas **hazes** are composed of droplets at least one order of magnitude smaller.) Venus has clouds of sulfuric acid and elemental sulfur that are dense enough that the surface is completely obscured from view. The planet Mars has haze and clouds comprised of both water and carbon dioxide ice. Saturn's satellite Titan has a surface completely obscured by haze and clouds of methane and other hydrocarbons. The atmosphere of Neptune's satellite Triton contains a haze layer that is not substantial enough to obscure Triton's surface.

Within terrestrial clouds, water droplets and ice crystals coalesce and eventually form units too heavy to remain suspended. At this point, precipitation—rain and snow—results. It has been estimated that approximately 1 million cloud droplets are necessary to form a raindrop. Raindrops are typically at least one-half millimeter across. It is considered likely that precipitation of liquid and solid methane occurs on Titan.

■ Vertical Structure

The characteristics of a planetary atmosphere change dramatically with height. As one travels upward from the surface of a planet, the most notable effect is that pressure decreases as more gas is below rather than pushing down from above. Closely related to the decrease in pressure is decrease in density with altitude. In addition, compositional changes occur with altitude. For example, there is very little water vapor in the regions of the Earth's atmosphere above the tro-

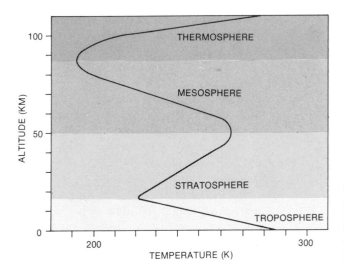

■ **Figure 11.2** Vertical structure of Earth's atmosphere. Altitude in kilometers is plotted against temperature in degrees Kelvin. The four atmospheric layers are labeled. (From Pasachoff, Jay M. *Contemporary Astronomy,* 4th ed. Saunders College Publishing, Philadelphia, 1989.)

posphere. Temperature also changes with height; its rate of change is called the **lapse rate.** Temperature normally decreases with height, yielding a lapse rate that is negative. The lapse rate near the Earth's surface is such that a rise of 300 meters (1000 feet) yields 1.9 K (3.5°F) of cooling. Prior to the 20th century, it was assumed that all of the Earth's atmosphere had this lapse rate. However, balloon and airplane investigations revealed that this was not the case. In certain atmospheric layers, the lapse rate is such that warming occurs with increasing altitude. An atmospheric region with this property is said to be in **inversion.**

The **vertical structure** of an atmosphere may be displayed graphically by plotting temperature (or another physical property) against altitude. (Vertical structure for the Earth's atmosphere is shown in Figure 11.2.) In this way, areas of inversion can easily be identified. These areas present challenges to meteorologists, because the reason for the warming occurring in each inversion layer must be determined. Atmospheres are generally divided into several distinct layers with boundaries marked by changeovers between normal cooling and inversion. For example, Earth's atmosphere has four layers, two with normal cooling and two with inversion. Above the troposphere lies the stratosphere. The heat that causes it to be in inversion is generated by the absorption of solar ultraviolet light. This light is absorbed by a stratospheric layer containing a form of oxygen called ozone (O_3). Above the stratosphere lies the

mesosphere; the uppermost layer, the thermosphere, is also an inversion layer. The thermosphere is heated primarily by gaseous absorption of incoming X-rays and extremely short wavelength ultraviolet radiation. The outer edge of the thermosphere, marking the boundary between the Earth's atmosphere and interplanetary space, is called the exosphere. A planet's exosphere is a particularly important region, as it is at this level that gas molecules may escape from the atmosphere, depending upon the conditions discussed earlier.

As mentioned in Chapter 9, many planets, including the Earth, have atmospheric layers called ionospheres. (The Earth's ionosphere lies within the thermosphere.) Because they contain electrically charged atoms, ionospheres interact with incoming solar wind particles to produce auroras. Above planetary atmospheres lie radiation belts and magnetospheres, if the planet has a magnetic field.

■ Winds and Global Circulation

One characteristic that is shared by all atmospheres is that they are never at rest. **Wind,** the horizontal movement of atmospheric gas, has been observed in all atmospheres of terrestrial objects. Wind is ultimately produced by varia-

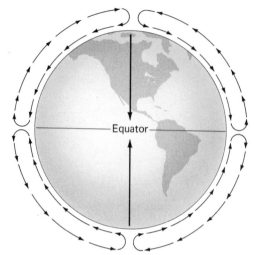

■ **Figure 11.3** Hadley circulation. Warm air rises at the equator and sinks at the poles. Surface wind flow is from north to south in the northern hemisphere. (From Turk, Amos and Jonathan Turk. *Physical Science.* W.B. Saunders, Philadelphia, 1977.)

tions in solar heating, which in turn produce variations in atmospheric pressure. Wind is the movement of air from areas of high pressure to areas of low pressure.

At the same time that horizontal movement occurs, there are also vertical air motions caused by the rising of warm air and the sinking of cool air. Warm equatorial regions have rising air, whereas air sinks over cool polar regions. Of course, when air rises in one area, a vacuum would form there if other air did not move in to replace it, and an area of sinking air would have very high pressure if the air remained at that point. Obviously, these things do not occur, because horizontal and vertical movements are interrelated, yielding a global system of atmospheric circulation. On Earth, for example, we would expect rising warm air at the equator to travel north and south to the poles, where, having been cooled, it would sink. Subsequently, the air would travel at the surface from pole to equator, where the cycle would repeat (Figure 11.3). This **Hadley circulation,** as it is called, occurs on Venus and Mars, but the actual situation on Earth is not this simple.

The Earth's rotation is rapid enough that the Coriolis force affects its atmospheric circulation. **Coriolis forces** are forces experienced on a rotat-

ing object that cause deflections of moving objects in the rotating frame of reference. An example would be rolling a ball while on a moving merry-go-round. Because of the rotational motion, the ball would not travel in a straight line. Objects moving on the Earth and through its atmosphere are similarly affected. For example, missiles and long-range guns must take the Coriolis force into account when aiming at targets. Because of the Coriolis force, gunners in the northern hemisphere aim to the left of their targets, because the Coriolis force causes the projectile to be deflected to the right. (Coriolis forces have the opposite direction in the southern hemisphere.) The Coriolis force disrupts the Earth's global atmospheric circulation so that a single Hadley cell in each hemisphere is impossible; instead, three smaller cells occur in each hemisphere. Air rises at 60° latitude as well as at the equator, and sinks at 30° latitude as well as at the poles (Figure 11.4). In addition, within each of these circulation cells, surface flow is changed from being in a strict north-to-south or south-to-north direction. Between the equator and 30° north latitude, pre-

■ **Figure 11.4** Earth's global circulation. Air rises at the equator and at 60° latitude, and air sinks at the poles and at 30° latitude. Coriolis forces cause surface wind flow to be offset to the directions shown. (From Zeilik, Michael, and Elske v.P. Smith, *Introductory Astronomy and Astrophysics,* 2nd ed. Saunders College Publishing, Philadelphia, 1987.)

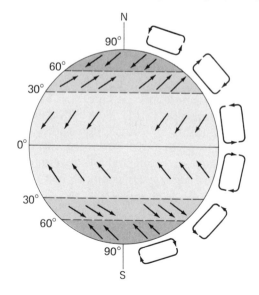

vailing winds are from the northeast. Between 30° and 60° north latitude, prevailing winds are from the southwest. (This is why most storms travel across the United States from west to east.) Between 60° north and the north pole, prevailing winds are from the northeast.

Because Mars rotates nearly as fast as the Earth, its atmospheric circulation might also be expected to be modified by the Coriolis effect. However, its circulation is different from the Earth's, primarily because of the lack of Martian oceans. Because the planet lacks water in which to store thermal energy, Martian temperatures are more directly dependent upon incident solar radiation, and the global circulation pattern changes seasonally.

Other factors contribute to large-scale atmospheric circulation. On Earth, temperature differences between oceans and continents produce standing waves within the atmosphere near shoreline regions. Certain features observed in clouds in these areas are related to the resulting atmospheric wave pattern. On Mars, the seasonal evaporation of a polar ice cap at one pole, followed by the condensation of the other at the opposite pole, contributes to the global circulation. Mars also has significant thermal tides, which are winds caused by the extreme temperature differences between daytime and nighttime.

In addition to large-scale circulation, wind patterns exist on a smaller scale. On the Earth, the most important manifestations of this are high- and low-pressure systems (Figure 11.5). Air flows outward from a high, but the circulation direction depends upon the hemisphere, due to changes in the Coriolis force. In the northern hemisphere, circulation around highs is clockwise as seen from above; in the southern hemisphere, the direction is counterclockwise. Winds blow toward a low, circulating in a counterclockwise direction in the

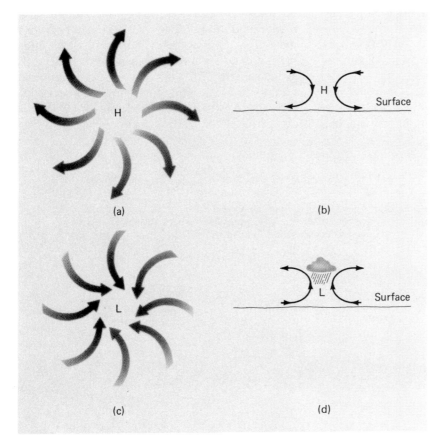

■ **Figure 11.5** High- and low-pressure systems in the northern hemisphere. (a) Winds flow outward from the high's center and circulate clockwise. (b) In addition, air descends at the high's center. (c) Winds flow inward toward a low's center and circulate counterclockwise. (d) In addition, air rises at the low's center, producing clouds and precipitation. Circulation directions in (a) and (c) would be reversed in the southern hemisphere.

northern hemisphere and in a clockwise direction in the southern. Many of the Earth's major storms, including hurricanes and tornadoes, are associated with low-pressure systems, which are also called cyclones. (Highs are also known as anticyclones.)

■ Weather, Climate, and Seasons

Atmospheric conditions on a planet's surface vary from place to place and from time to time. Variations in atmospheric conditions are called **weather,** which manifests itself on Earth as periodic variation in temperature, pressure, and other conditions. Precipitation and severe storms also add variety to the Earth's surface weather. Over a longer period of time, we find that certain weather conditions can be expected in particular areas of the Earth. We use the term **climate** to describe the expected average weather conditions of an area. Climate of terrestrial areas is controlled by such factors as latitude, elevation, surface topography, and proximity to major bodies of water.

On Earth, as well as on other planets, we find that several factors cause weather conditions to vary on a periodic basis. Conditions vary due to a planet's rotation, and temperatures generally rise during the daylight hours to a maximum value

and then cool to a minimum after the Sun has set. If the Earth's rotational period were shorter, we would experience less difference between day and night temperatures. If days and nights were longer, the difference would be greater, due to longer periods of heating and cooling. Conditions may also vary on an annual basis, but only if a planet has a highly eccentric orbit. If the orbit is nearly circular, which is the case for Earth, the change in the amount of incident sunlight as our distance from the Sun slightly varies is too small to produce a noticeable effect.

The most important cause of variation during the course of a planet's year is seasonal variation. **Seasons** occur on planets whose axial obliquities are more than just a few degrees. For example, Earth's obliquity is 23.5°, whereas that of Mars is 25°. Because axial direction is constant (other than for effects of precession), orbital motion causes the tilt of a planet's axis to vary with respect to the Sun. Four specific events are defined during the course of a planet's year (Figure 11.6). Summer solstice occurs in the hemisphere of a planet when the pole in that hemisphere has its maximum tilt toward the Sun, whereas winter solstice occurs when the pole has its maximum tilt away from the Sun. Vernal (spring) and autumnal equinoxes occur when the tilting direction is intermediate, with the axis tilted neither toward nor away from the Sun. The reason seasonal changes occur is that the amount of sunlight incident upon a given location on the Earth's surface varies. At

■ **Figure 11.6** The cause of seasons. Summer solstice occurs when the northern hemisphere is tilted toward the Sun, and winter solstice occurs when it is tilted away. The equinoxes occur when the axis is tilted in an intermediate direction.

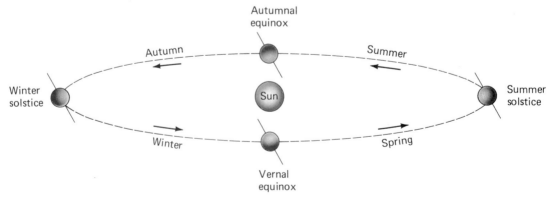

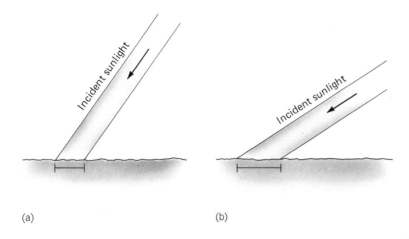

■ **Figure 11.7** How solar heating varies due to seasonal change. (a) During spring and summer, the Sun is higher in the sky, and sunlight is more concentrated on the Earth's surface and warms it more effectively. (b) During autumn and winter, the Sun is lower in the sky, and sunlight is less concentrated and less effective at heating the surface.

the equinoxes, the hours of daylight and night are equal. After the beginning of spring and during summer, early sunrise and late sunset combine to give long hours of daylight, and the incident sunlight is more concentrated on the Earth's surface. After the beginning of autumn and during winter, daylight is shorter, and sunlight strikes the surface obliquely, making the heating less efficient (Figure 11.7). (On the Earth, the hottest and coldest times do not occur precisely at the summer and winter solstices, respectively, but a month or so afterward. This thermal lag, as it is called, occurs because considerable time is required for the large mass of gas in the atmosphere to be heated or cooled.) As we will learn in Chapter 23, the planet Uranus has such a high obliquity (82.14°) and a long orbital period (84 years) that it has the most extreme seasons of any planet.

On the Earth, perihelion occurs during northern hemisphere winter and southern hemisphere summer, whereas aphelion occurs during northern hemisphere summer and southern hemisphere winter. This effect causes southern hemisphere seasons to be a bit more extreme, but the effect is very minor. The effect on Mars, which has a more eccentric orbit, is much more pronounced. The Martian southern hemisphere has winter and summer corresponding to aphelion and perihelion, respectively, and its seasonal variation is more pronounced than that observed in the northern. Venus, with a nearly circular orbit and low obliquity (3°), has very little long-term variation in weather.

■ Sources of Atmospheres

There are several theories regarding the initial source of planetary atmospheres. One suggests that as planets formed from the accretion of solid rock particles, volatile materials were trapped inside them. Later, as the planetary interiors warmed, this material **outgassed**, primarily at the sites of volcanic eruptions. On the Earth, volcanoes emit large amounts of gas even today, indicating that such a mechanism is reasonable. Alternatively, it has been proposed that atmospheres were added to planets from the outside after they had formed. There could have been several possible external sources for atmospheric gases: they may have been derived from the solar nebula itself, from which the solid particulate matter of the solar system had already condensed; they may have come from the solar wind, which would have been much more substantial during the youth of the Sun; or they may have been introduced as volatile-rich bodies impacted upon the newly formed planets. (The origin of the solar system will be discussed in more detail in Chapter 29.)

The topic of atmospheric origin is still rather controversial, and it is likely that it will be a while before a consensus is reached. Regardless of how the planets' atmospheres originally formed, it is certain that numerous changes have occurred in them since that time. It has been suggested that the first atmospheres (or else portions of them) were stripped away from the planets by the more powerful solar wind of that time. Subsequent outgassing from within the planets replaced the lost gas, producing new atmospheric compositions in some cases.

It can be assumed that Venus, Earth, and Mars, having formed adjacent to one another, might have originally possessed very similar atmospheres. However, these three worlds have entirely different atmospheres today, and these differences contribute significantly to the variations among their surface features. How can we account for these differences? Although the three planets probably started with atmospheres of mostly nitrogen and carbon dioxide, and had both surface and atmospheric water, changes involving water have altered each world. Earth had temperatures that allowed liquid water to continue to exist, and life eventually appeared, probably in the oceans. Photosynthesis by plants produced Earth's atmospheric oxygen, whereas carbon dioxide (in the form of calcium carbonate) was precipitated from the oceans to form carbonate rocks. Venus lost its original water, which probably photodissociated and escaped into space. The carbon dioxide remained in the atmosphere and heated Venus by the greenhouse effect. Mars, which is cooler and has weaker gravity, lacks liquid water, probably because most of it escaped or was trapped under the surface as ice. Because both Venus and Mars lacked oceans and life, they lacked the mechanism for atmospheric carbon dioxide removal and oxygen generation.

Additional factors, such as the release of gaseous decay products from radioactive elements in rocks, have also caused changes in atmospheres. (We will discuss the history of each planet's atmosphere in more detail in the following chapters.) On the Earth, concern exists at present that human activity is increasing the amount of atmospheric carbon dioxide and destroying the ozone layer in the stratosphere. When we consider our lifeless neighbors Venus and Mars, we realize just how special our atmosphere is and that we should make every effort to preserve its present characteristics!

■ Chapter Summary

Whether a planet has an atmosphere is controlled largely by its gravity and temperature. The cooler the planet and the stronger its gravity, the more likely it will be to have a substantial atmosphere.

Some incident sunlight is absorbed by atmospheric gases, whereas other light reaches the surface and heats it. Planetary surfaces reradiate energy at infrared wavelengths; some of this energy may be trapped by the greenhouse effect.

Evaporation and condensation of moisture in the Earth's atmosphere transfers thermal energy, in addition to producing clouds and precipitation. Temperature changes with altitude define divisions of atmospheres into discrete layers. Temperature and pressure differences produce winds and atmospheric circulation on both local and global scales.

Atmospheric conditions vary on a short term due to weather effects, as well as on a longer term due to seasonal changes. Modern theories suggest that atmospheres either outgassed from within planets or were added to planets from the outside after the planets had formed.

■ Chapter Vocabulary

atmosphere	evaporation	lapse rate	seasons
climate	greenhouse effect	outgassing	solar constant
clouds	Hadley circulation	photodissociation	vertical structure
condensation	haze	precipitation	weather
Coriolis force	inversion	saturation	wind

■ Review Questions

1. What factors determine whether a planet will have an atmosphere and what its composition will be?
2. One effect of urbanization on Earth is a proliferation of dark surfaces such as roads and parking lots. What effect would this have on the temperature of urban areas compared to other areas?
3. How does a planet's atmosphere affect incoming sunlight?
4. How does the greenhouse effect affect a planet's temperature?
5. Explain how energy transport helps warm the Earth's atmosphere.
6. What is meant by the vertical structure of an atmosphere? What changes occur in the Earth's atmosphere with increasing height?
7. What causes horizontal and vertical atmospheric motion?
8. Explain the difference between idealized Hadley circulation and the circulation that actually occurs in the Earth's atmosphere.
9. Describe the difference between weather and climate.
10. Describe the climate of your locality.
11. Explain the origin of seasons on Earth.
12. Describe the theories explaining the origin of planetary atmospheres.

■ For Further Reading

Barbato, J.P. and E. Ayer. *Atmospheres*. Pergamon Press, New York, 1981.

Cole, Franklyn W. *Introduction to Meteorology*. Wiley, New York, 1980.

Eagleman, Joe R. *Severe and Unusual Weather*. Van Nostrand Reinhold, New York, 1983.

Lockwood, John G. *World Climatic Systems*. Edward Arnold, London, 1985.

Lutgens, Frederick K. and Edward J. Tarbuck. *The Atmosphere: An Introduction to Meteorology*. Prentice Hall, Englewood Cliffs, NJ, 1989.

Wells, Neil. *The Atmosphere and Ocean*. Taylor and Francis, Philadelphia, 1986.

Mercury

Mercury is the planet closest to the Sun, and it is the smallest of the four terrestrial planets. It is also noteworthy because it has a rather unusual orbit, with high inclination and eccentricity. In fact, of the nine planets, only Pluto has greater inclination and eccentricity values. Mercury was named for the Roman messenger of the gods because of its rapid motion through the sky. Its 116-day synodic period is the shortest of any planet. (Mercury's major orbital and physical properties are summarized in Table 12.1.)

One unusual aspect of Mercury's orbital motion once led astronomers to believe that the solar system had another planet and later played an important role in the development of Albert Einstein's theory of relativity. It has been known for several centuries that the major axis of Mercury's orbit does not maintain a permanent orientation in space but gradually advances around the Sun at a rate of about 5600 seconds of arc per century. Part of this advance was found to be the result of perturbations by other planets, but by the 19th century, it had become obvious that a small part of the advance (43″ per century) could not be explained in this way. During the 19th century, astronomers believed that an undiscovered planet closer to the Sun was responsible for the remaining advance, and they confidently gave it the name Vulcan. Searches for Vulcan were conducted during solar eclipses, when it would presumably be visible, but it was never found. The mystery of the 43″ advance was solved early in the 20th century, when Einstein published his theory of general relativity. One of its predictions is that powerful gravitational sources, such as the Sun, can actually bend or warp space itself so that nearby objects do not obey the usual Newtonian laws of motion. Relativity predicts an advance of Mercury's orbit of the exact quantity observed. As a result, it is now considered unlikely that any substantial body orbits the Sun any closer than Mercury does.

The cratered surface of Mercury. Note that both old, degraded craters and young, fresh craters are visible. (NASA/JPL Photograph.)

■ Table 12.1   Summary of Mercury's Basic
              Properties*

Semi-major axis: 0.387 astronomical units
Semi-major axis: 57.9 million kilometers
Semi-major axis: 35.9 million miles
Orbital eccentricity: 0.206
Orbital inclination: 7.0°
Orbital period: 88 days
Diameter: 4878 kilometers
Diameter: 3031 miles
Mass (Earth = 1): 0.055
Density: 5.43 grams per cubic centimeter
Sidereal rotational period: 58.6 days
Axial obliquity: 0.0°
Number of satellites: 0

* For a complete table of planetary data, see Table A2.1 in
  Appendix 2.

■ Appearance and Basic Properties

Even though it is bright enough to be a naked-eye object, Mercury is notoriously hard to see due to its proximity to the Sun. (Tradition has it that Copernicus himself never saw this elusive object!) Recall that Mercury and Venus are **inferior planets** and are therefore never located very far from the Sun in the sky. Mercury is always within 28° of the Sun and is visible only in twilight very soon after sunset or before sunrise. Times when Mercury is seen to be farthest from the Sun are called greatest elongations. Eastern elongation occurs when Mercury is visible after sunset, and western elongation occurs when Mercury is visible before sunrise. In the Northern hemisphere, the orientation of the ecliptic in the sky causes Mercury to be best seen in the evening sky during eastern elongations that occur in the spring, and in the morning sky during western elongations that occur in the fall.

Because Mercury is an inferior planet, its orbit sometimes carries it across the face of the Sun as seen from the Earth, producing a transit. During a transit, Mercury somewhat resembles a sunspot (Figure 12.1). Because of Mercury's orbital inclination, it is aligned perfectly with the Sun and Earth to do this only about a dozen times per century. The first transit of Mercury was ob-

served in 1631, after having been predicted by Kepler four years earlier. Because of the orbital elements of the Earth and Mercury, transits occur in either May or November. Upcoming transits include ones on November 6, 1993; November 15, 1999; May 7, 2003; and November 8, 2006.

Because it is visible only in twilight and at very low elevations in the sky, telescopic observations of Mercury are difficult. Its small physical size leads to a small angular diameter when viewed telescopically, further complicating observations. Although Galileo discovered Venus' phases shortly after he began observing in 1609, Mercury's were not seen until 1639, when improved telescope optics revealed them. As telescopes continued to improve, observers were able to measure Mercury's diameter with moderate accuracy, revealing it as the smallest planet then known. By the 1800s, various observers were mapping the planet's vague surface markings. Determining Mercury's rotational period from these markings, however, was difficult, as very little change could be seen from day to day. In the 1880s, **Giovanni Schiaparelli** (1835–1910) proposed that Mercury had 1:1 spin-orbit coupling, with an 88-day rotational period equal to its orbital period. Most astronomers agreed that this was a reasonable conclusion, because Mercury's proximity to the Sun should have allowed the Sun's gravity to lock Mercury into 1:1 coupling. This meant that Mercury would have an incredibly hot side in constant light and a perpetually dark side with temperatures near absolute zero.

As late as the 1960s, visual mapping of Mercury's surface features supported an 88-day rotational period; however, evidence provided by radio astronomers soon led to new conclusions. In 1962, thermal radio emission was detected from both the day and night sides of the planet, and the emission from the dark side indicated that its temperature was far warmer than expected if it never received sunlight. This surprising result led some researchers to suggest that Mercury had an atmosphere that transported thermal energy from the day to night side. In 1965, Doppler effect measurements of radar beams reflected from the surface of Mercury revealed that the planet's sidereal rotational period is only 59 days, not 88. Earlier observers had apparently been biased by the prediction of perfect

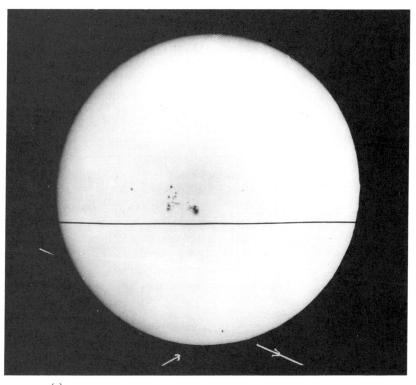

(a)

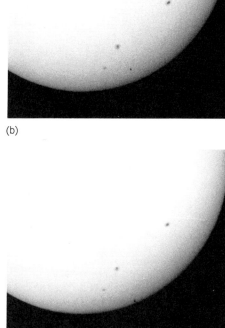

(b)

(c)

■ **Figure 12.1** Transits of Mercury. (a) November 14, 1907. (Yerkes Observatory Photograph.) (b and c) May 9, 1970. The two views show the movement of Mercury across the Sun. (Photographs courtesy of Geoffrey Chester.)

spin-orbit coupling and had "seen" an 88-day rotation in surface features that were actually too vague to yield useful information. It was soon realized that because 59 is two-thirds of 88, an imperfect form of spin-orbit coupling occurs: Mercury rotates three times during two complete orbits of the Sun (Figure 12.2). Such a form of spin-orbit coupling meant that the same face of Mercury was frequently presented toward the Earth, accounting for the mappings that indicated an 88-day rotation.

Why does Mercury have 3:2 spin-orbit coupling instead of 1:1 coupling? Although the precise answer is not known, it is believed that Mercury, like other solar system objects, rotated in 10 hours or less shortly after its formation. Theoretical studies suggest that tidal interactions with the Sun would have slowed Mercury to its present rotational rate by about 1 billion years after the planet formed. These studies indicate that the 3:2

coupling that resulted has a high long-term stability and is capable of lasting indefinitely, rather than eventually slowing further to 1:1 coupling.

■ Spacecraft Exploration

Mercury, despite its relative proximity to the Earth, has been largely neglected during the space age. Only one spacecraft has visited Mercury—**Mariner 10**, which flew past the planet a total of three times in 1974 and 1975. Mariner 10 was launched on November 3, 1973 and photographed portions of the Moon as it passed by. Shortly afterward, it became the first spacecraft to photograph the Earth from farther away than the Moon. Mariner 10 flew by Venus on February 5, 1974. This encounter gravitationally altered its course so that it could reach Mercury, the first instance of a spacecraft's trajectory

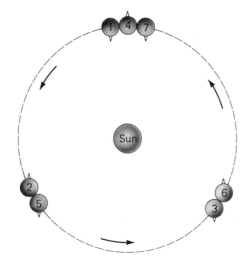

■ **Figure 12.2** Mercury's spin-orbit coupling. At *1*, a locality on the surface (marked by a triangle) points toward the Sun. At *2*, one-half rotation has been completed, and at *3*, one entire rotation is done. At *4*, one and one-half rotations and an entire orbit have been completed. Two rotations have been completed at *5*, and two and one-half rotations have been completed at *6*. At *7*, three rotations and two orbits have been completed, with conditions identical to *1* again. (Note that sidereal rotation, not solar rotation, is being discussed.)

being changed by a planetary encounter. The spacecraft went into a 176-day orbit around the Sun, exactly twice Mercury's orbital period, and it flew by Mercury every other time the planet returned to the same point in its orbit. Photographs returned by Mariner 10 showed that during each of its three flybys past Mercury, the same side of the planet was illuminated. This verified the 3:2 spin-orbit coupling proposed following the radar studies, but, unfortunately, it meant that half of Mercury's surface could not be photographed.

Mariner 10's first Mercury encounter was on March 29, 1974, when the spacecraft flew 1000 kilometers (600 miles) above the planet's night side. During the second encounter, on September 21, 1974, the craft flew 50,000 kilometers (30,000 miles) above the day side, marking the first time that a flyby spacecraft returned to the same planet. The final encounter, on March 16, 1975, saw Mariner 10 pass 300 kilometers (180 miles) above the night side. After this encounter, Mariner 10 depleted its attitude control system fuel and could no longer function. Mariner 10's scientific discoveries about Mercury were nu-

merous. The planet's mass, radius, moment-of-inertia coefficient, and other bulk properties were measured with high accuracy. Most of the illuminated hemisphere was photographed with surface resolutions ranging from 5 kilometers (3 miles) to as high as 100 meters (300 feet) in some areas. Over 2700 images were obtained, allowing Mercury's surface geology to be thoroughly investigated. (As we discuss Mercury's surface geology in the next section, we must bear in mind that the hemisphere not photographed by Mariner 10 may hold some surprises. Unfortunately, neither the United States nor other countries have plans for further Mercury spaceflights, so the remainder of Mercury will be a mystery for quite some time.) Mariner 10 also discovered Mercury's magnetic field, studied its magnetosphere, and verified that it has no satellites.

■ Surface Features

Although only half of Mercury was studied by Mariner 10, the photographs returned revealed a world that looks remarkably like Earth's Moon at first glance (Figure 12.3). This is not surprising, because impacts have greatly shaped both of these airless bodies and have produced abundant craters. However, there are some major differences between these worlds. There is no correlation between topography and albedo on Mercury, whereas rough areas on the Moon are light and smooth areas are dark. Mercury is somewhat brighter than the Moon, having an average albedo of 11% as opposed to under 10% for the Moon. Spectroscopic evidence reveals that Mercury's surface appears to have less iron and titanium than does the Moon's, and there are important differences in their surface geologies. For example, Mercury has no areas completely saturated with craters and has a smaller proportion of smaller craters than does the Moon. Mercury also has several types of landforms (e.g., intercrater plains and lobate scarps, which will be discussed later) that are not found on the Moon.

After planetary geologists examined the Mariner 10 photographs of Mercury, they established four major topographic provinces on the visible hemisphere: two plains units (smooth and inter-

■ **Figure 12.3**   Photomosaics of Mercury made from Mariner 10 images. (a) The "outgoing" side of Mercury. (b) The "incoming" side. (NASA/JPL Photographs, courtesy of the National Space Science Data Center and the Mariner 10 Principal Investigator, Bruce C. Murray.)

crater), heavily cratered terrain, and hilly and lineated terrain (Figure 12.4). In addition, Mercury contains numerous craters, plus several multi-ringed impact basins and fault-related features. Craters on Mercury are generally named for famous authors, composers, and artists.

The **smooth plains** cover about 15% of the surface photographed by Mariner 10 (Figure 12.5). Most of the smooth plains are in Mercury's northern hemisphere, suggesting that their distribution may be asymmetrical, as is the distribution of the lunar maria that they resemble. The smooth plains appear to be the youngest topographic province on Mercury, because they contain relatively few large craters. Because Mariner 10 did not photograph unequivocally volcanic features such as vent and flow structures on the smooth plains, their origin is not as clear cut as that of the lunar maria. One theory is that they are blankets of ejecta from some of the larger impacts on the planet. Another theory is that they are indeed basaltic lava flows. Most researchers support the latter view, because the plains units seem to be much younger than the basin impacts and there seems to be too much plains unit material present to be accounted for by just ejecta. The mystery of their origin will probably not be solved until further exploration of the planet is conducted.

About 45% of the visible surface is composed of topographic provinces called **intercrater plains,** the most widespread type of feature on Mercury (Figure 12.6). Intercrater plains are smooth or gently rolling areas that contain numerous craters, most of them 5 to 10 kilometers (3 to 6 miles) across. Some intercrater plains bury old craters and their ejecta, whereas other craters are superimposed on them. Although this variation in cratering indicates that not all of the intercrater plains are of the same age, they are thought to contain some of Mercury's oldest terrain. As with the smooth plains, researchers have suggested both volcanic and ejecta origins for the

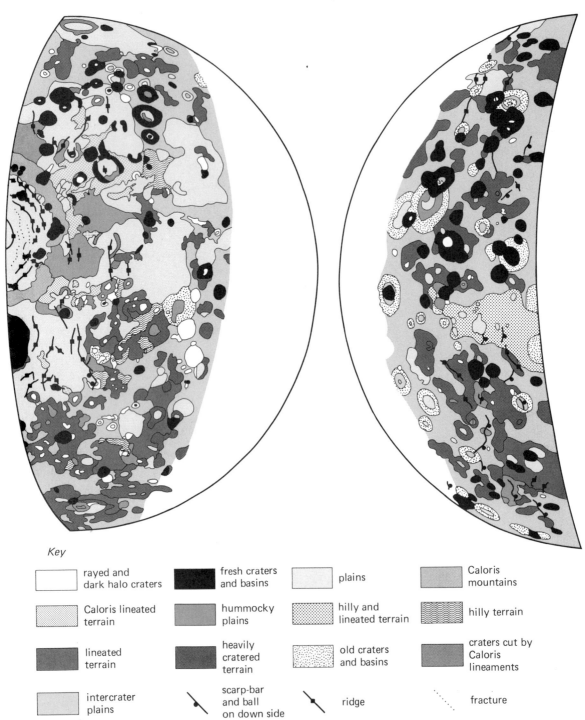

Key

rayed and dark halo craters	fresh craters and basins	plains	Caloris mountains
Caloris lineated terrain	hummocky plains	hilly and lineated terrain	hilly terrain
lineated terrain	heavily cratered terrain	old craters and basins	craters cut by Caloris lineaments
intercrater plains	scarp-bar and ball on down side	ridge	fracture

■ **Figure 12.4** Geologic map indicating the distribution of terrain types on Mercury. (Originally appeared in color in an article by Trask, N.J. and J.E. Guest. In *Journal of Geophysical Research,* volume 80, page 2461. Copyright 1975 by the American Geophysical Union. This version taken from Greeley, Ronald, *Planetary Landscapes.* Allen and Unwin, London, 1985. Reproduced by kind permission of Unwin Hyman Ltd.)

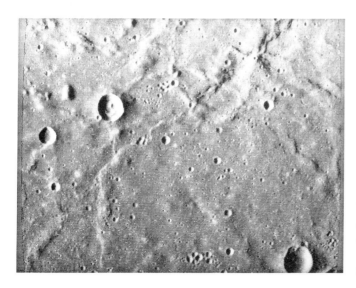

■ **Figure 12.5** Example of a smooth plains area. This region is located southeast of Caloris basin. The area shown is about 200 kilometers (120 miles) wide. (NASA/JPL Photograph, courtesy of the National Space Science Data Center and the Mariner 10 Principal Investigator, Bruce C. Murray.)

intercrater plains. It has also been proposed that they may represent some of Mercury's original crust, but given their range of ages, this is probably unlikely.

A third surface unit is **heavily cratered terrain,** which is similar to highland areas on the Moon. These areas contain numerous craters that are in close contact or even overlap (Figure 12.7). These craters tend to be larger than those in the plains, ranging from about 30 kilometers (20 miles) across to over ten times that large. Mercury's craters have similar morphology to those of the Moon, but there are small differences in the form of their ejecta. Mercury's craters have smaller ejecta blankets, because the planet's higher gravity prevents ejecta from spreading as far as it does on the Moon. Additionally, Mercury has a greater proportion of craters with bright, noticeable rays.

Mariner 10 photographed two multiringed basins, although there appear to be other, older ones buried under plains. Tolstoy is 350 kilometers (215 miles) across, but **Caloris,** at 1300 kilometers (800 miles) in diameter, is the single most impressive feature on Mercury (Figure 12.8). Caloris means "heat," an appropriate

■ **Figure 12.6** Example of an intercrater plains area. The area shown is about 400 kilometers (250 miles) wide. (NASA/JPL Photograph, courtesy of the National Space Science Data Center and the Mariner 10 Principal Investigator, Bruce C. Murray.)

■ **Figure 12.7** A region of heavily cratered terrain on Mercury. A fault scarp can be seen near the top of the photograph. (NASA/JPL Photograph.)

name, because the basin is aligned directly under the Sun when Mercury is at perihelion every other orbit. Caloris basin contains several rings of mountains up to 2000 meters (6600 feet) high, and its floor is covered with ridges and fractures. Many smooth plains regions are near Caloris, probably the result of magma extruded after the impact.

The fourth topographic type on Mercury, **hilly and lineated terrain**, occurs in the region of Mercury on the opposite side of the planet from the Caloris basin. Most researchers believe that it was formed by the disruptive force of seismic waves generated by the Caloris impact (Figure 12.9). Hilly and lineated terrain consists of hills 5 to 10 kilometers (3 to 6 miles) across and as high as 2000 meters (6600 feet), as well as intersecting linear valleys. The hills are the remains of pre-existing craters, whereas the valleys represent large breaks in the surface. This type of landscape was so unlike anything previously seen that the Mari-

ner 10 scientists originally called it "weird terrain."

Mercury also contains many features that are undoubtedly the result of faulting. Two types of fault scarps mark the surface, and many of the scarps are of impressive size. **Lobate scarps** are arcuate (rounded) and appear to be the result of reverse faulting due to compressional forces (Figure 12.10). Lobate scarps range from 20 to 500 kilometers (12 to 300 miles) in length and up to 3000 meters (9800 feet) in height, and they cut across craters and other large-scale features. (These scarps are so large that they can even be detected by Earth-based radar.) **Linear ridges** are straight scarps possibly associated with normal faulting (Figure 12.11). They range from 50 to 300 kilometers (30 to 190 miles) long and are as high as 1000 meters (3300 feet). Other tectonic features on Mercury appear to be grabens formed by normal faulting. **Linear troughs** are grabens that range from 40 to 150 kilometers (25 to 90 miles)

■ **Figure 12.8** Photomosaic of the Caloris basin. (NASA/JPL Photograph, courtesy of the National Space Science Data Center and the Mariner 10 Principal Investigator, Bruce C. Murray.)

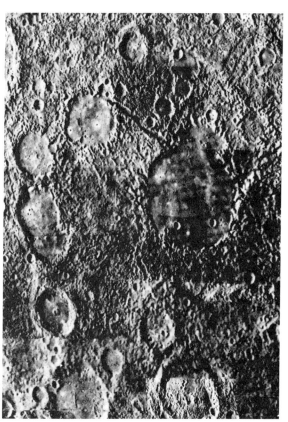

■ **Figure 12.9** A portion of the hilly and lineated terrain on Mercury. (NASA/JPL Photograph.)

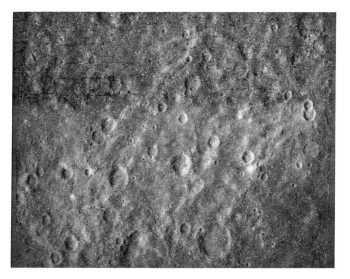

■ **Figure 12.10** A lobate scarp cutting across intercrater plains. The scarp, which extends from the top-center to lower-left portion of the photograph, is about 700 kilometers (430 miles) long. (NASA/JPL Photograph, courtesy of the National Space Science Data Center and the Mariner 10 Principal Investigator, Bruce C. Murray.)

■ **Figure 12.11** A linear ridge (top-center of photograph) known as Mirni Rupes. The area shown is about 400 kilometers (250 miles) wide. (NASA/JPL Photograph, courtesy of the National Space Science Data Center and the Mariner 10 Principal Investigator, Bruce C. Murray.)

long, 5 to 15 kilometers (3 to 9 miles) wide, and 100 to 500 meters (330 to 1600 feet) deep (Figure 12.12).

What is the origin of these numerous fault-related features? Two mechanisms for faulting on Mercury have been proposed, and it is likely that both have contributed toward shaping its surface. One suggestion is that Mercury was relatively oblate shortly after its formation, while its crust was solidifying. As its rotation slowed from less than 10 hours to 59 days, the planet's internal material redistributed itself, producing the perfectly spherical planet of today. Obviously, the solid crust would have been severely deformed during

this process. The evidence pointing to past compressional forces suggests that Mercury has also contracted slightly as a result of its internal cooling as its heat source (radioactive decay) has dissipated. It has been calculated that a decrease in Mercury's radius of only 1000 to 2000 meters (3300 to 6600 feet) could account for the observed fault-related features.

■ Atmosphere

Mercury is too small to have enough gravity to retain a substantial atmosphere. Its temperature ranges from 90 K (−297°F) just before sunrise to

■ **Figure 12.12** A linear trough known as Arecibo Valley, which is located in the hilly and lineated terrain. The valley, located in the lower-left corner of the photograph, is 13 kilometers (8 miles) wide. (NASA/JPL Photograph, courtesy of the National Space Science Data Center and the Mariner 10 Principal Investigator, Bruce C. Murray.)

about 740 K (873°F) at noon. This gives Mercury the greatest temperature range of any planet, and its high average temperature (440 K, or 333°F) would prevent the retention of an atmosphere even if its gravity were stronger. Because Mercury is so close to the Sun, large quantities of solar wind particles impinge upon its surface. Small quantities of hydrogen and helium remain near its surface for short periods of time before escaping into space, giving it a transient "atmosphere" whose pressure (10^{-15} atmospheres) makes it far less substantial than the best laboratory vacuum obtainable on Earth. In addition to hydrogen and helium from the solar wind, Mercury's tenuous atmosphere also contains sodium vapor. This material probably does not come from the solar wind but is produced when either surface rocks or meteorites are vaporized during impact events.

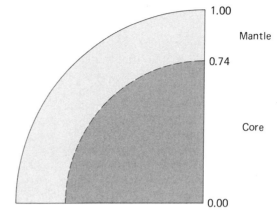

■ **Figure 12.13** Mercury's inner layers. The numbers represent the fraction of the planet's total radius. The thickness of Mercury's crust is presently unknown; therefore, the crust is omitted from this diagram.

■ Interior and Magnetic Field

Because no spacecraft have yet landed on Mercury to probe its interior seismically, our only source of information about its interior is indirect evidence such as density and moment-of-inertia coefficient. Mercury has the distinction of having the highest density of any planet other than Earth. However, the Earth's high density is due in part to compression. Because Mercury is smaller than Earth, it has less pressure to cause internal compression, and its uncompressed density would actually be higher than Earth's (5.31 versus 4.03 grams per cubic centimeter).

Mercury's high density is good evidence that it has a metallic core. It is estimated that Mercury's iron core is proportionally larger than Earth's, occupying the central 74% of the radius, as opposed to 52% for Earth (Figure 12.13). At present, it is not known whether any of Mercury's core is molten, but because the planet is so much smaller than the Earth, it has probably lost enough heat that any formerly molten material has cooled and solidified. Because of the large extent of Mercury's core, it can be estimated that the planet may be as much as 70% iron by mass. How can we account for this iron-rich composition? It is probably the result of Mercury's forming very close to the Sun, in the warmest portion of the solar nebula (the gas and dust from which the planets formed). As Mercury formed, the high

temperatures drove off a greater proportion of volatile materials than were lost from the forming Earth.

In addition to photographing surface features and providing more precise values of mass and radius, Mariner 10 also conducted studies of magnetism in the vicinity of Mercury. Prior to the mission, it was considered unlikely that Mercury would have an internally generated magnetic field, because such fields seem to be generated by currents in a molten metallic core. The currents, in turn, are caused by rapid planetary rotation. One of the most surprising discoveries made by Mariner 10 was that Mercury has a magnetic field about 1% the strength of Earth's. Mercury's field is dipolar and is tilted 11° to the planet's rotational axis. Because Mercury rotates so slowly, the presence of its magnetic field is difficult to explain. Part of Mercury's core may possibly remain liquid, in which case the rate of rotation required for production of a magnetic field must be fairly small. An alternate theory suggests that Mercury had a more normal field long ago, while its interior was completely molten and the planet rotated faster than today. This field then magnetized Mercury's mantle rocks as they cooled and solidified, leaving them with a remanent magnetism that acts as a field source today. Most researchers, however, doubt that Mercury's mantle rocks contain sufficient iron to produce the observed field strength.

Mercury's magnetic field causes it to possess a magnetosphere that interacts with the solar wind. Mercury's magnetosphere, however, is proportionally smaller that the Earth's. Whereas the Earth's magnetosphere extends out to about 11 times the planet's radius, Mercury's extends outward to only 1.5 times its radius. Due to the small size of its magnetosphere, there are no radiation belts of trapped charged particles around Mercury.

■ History

As a result of the photographic studies of Mercury's surface conducted by Mariner 10, planetary geologists have been able to reconstruct a sequence of events describing the planet's history. Because no samples of rocks from Mercury are yet available for radiometric age dating, we cannot describe the actual times when specific events occurred. The best we can do at present is to place events into sequence, a process called **relative dating.** In relative dating, the events in Mercury's history can be discussed in relation to a major event such as the impact that formed the Caloris basin, even though the exact time when the impact occurred remains unknown.

Mercury's history has generally been divided into five periods. During the first, the planet accreted and differentiated into crust, mantle, and core layers. It has been suggested that part of Mercury's original crust may survive in some intercrater plains regions, but it is probably unlikely that these areas could have survived unchanged through the remainder of Mercury's history.

The second period of Mercury's history was the era of heavy meteorite bombardment, which probably lasted from the formation of the planet to about 4 billion years ago. The numerous impacts during this time formed the heavily cratered terrain as well as the planet's older impact basins. It is considered likely that the crustal contraction due to internal cooling that led to the production of the lobate fault scarps occurred during this time. Extrusion of volcanic lava also may have occurred during this period.

The third event in Mercury's history was the Caloris basin impact, which formed both the basin and the hilly and lineated terrain on the opposite side of the planet. The Caloris impact probably occurred near the end of the time of heavy bombardment, because the interior of the Caloris basin contains relatively few craters.

The fourth event was the formation of the smooth plains regions. As mentioned earlier, this was probably the result of lava flow. The final period of Mercury's history, probably including most of the past 4 billion years, has been marked by very little change in surface geology, except for occasional meteorite impacts. The youngest of these craters are marked by systems of bright rays.

■ Prospects for Future Exploration

Although a great deal was learned about Mercury by only one spacecraft, the fact that half of the planet's surface is relatively unknown underscores the fact that more exploration of the planet is certainly needed. Soviet scientists have indicated an interest in planning a future mission to Mercury, but indications are that neither they nor any other country will be sending a spacecraft to Mercury until the next century. In the meantime, our primary source of new data will be radar studies of Mercury's unphotographed side.

■ Chapter Summary

Mercury, the planet closest to the Sun and the smallest terrestrial planet, is fairly difficult to observe. Not until the 1960s was its rotational rate learned, a discovery that revealed its 3:2 spin-orbit coupling.

Mercury was observed firsthand by the Mariner 10 spacecraft, revealing that it is somewhat Moon-like in many respects. Its surface consists of smooth plains, intercrater plains, heavily cratered terrain, and hilly and lineated terrain; its single most impressive feature is the Caloris basin.

Mercury lacks a permanent atmosphere but has a magnetic field, which is surprising for such a slow-rotating object. Most of Mercury's surface features formed early in its history, during the solar system's era of heavy bombardment.

■ Chapter Vocabulary

Caloris basin
heavily cratered
   terrain
hilly and lineated
   terrain

inferior planet
intercrater plains
linear ridge

linear trough
lobate scarps
Mariner 10

relative dating
Schiaparelli, Giovanni
smooth plains

■ Review Questions

1. What is noteworthy about Mercury's size, location, and orbit?
2. Why did Mercury's orbital advance challenge traditional physics?
3. Why is Mercury difficult to observe with a telescope?
4. What observations of Mercury's surface were made by early observers? Were all of their conclusions correct?
5. Describe Mercury's 3:2 spin-orbit coupling.
6. Summarize the major results of Mariner 10.
7. In what ways are the surfaces of Mercury and the Moon similar? In what ways are they different?
8. Summarize the four major topographic types found on Mercury.
9. What types of fault-related features are found on Mercury? What is their probable origin?
10. Describe Mercury's interior and why its magnetic field was unexpected.
11. Summarize the major events in Mercury's history.

■ For Further Reading

Chapman, Clark R. "Mercury's Heart of Iron." In *Astronomy,* November, 1988.

Davies, M. et al., editors. *Atlas of Mercury.* NASA, Washington, D.C., 1978.

Murray, Bruce. "Mercury." In *Scientific American,* September, 1975.

Strom, Robert G. "Mercury." In Carr, Michael H., editor. *The Geology of the Terrestrial Planets.* NASA, Washington, D.C., 1984.

Strom, Robert G. *Mercury, the Elusive Planet.* Smithsonian Institution Press, Washington, D.C., 1987.

13

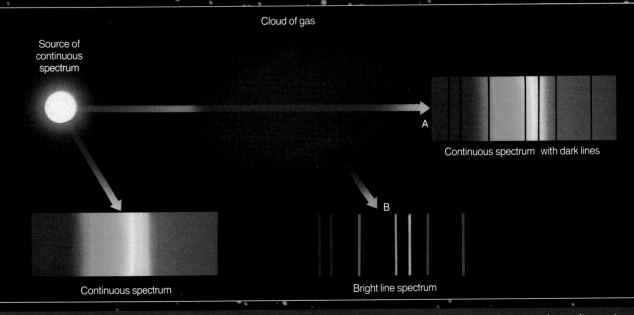

The origin of continuous, absorption (*dark line*), and emission (*bright line*) spectra. See Chapter 4 for a discussion of spectra. (*From* Morrison, David and Sidney Wolff. *Frontiers of Astronomy*. Saunders College Publishing, Philadelphia, 1990.)

The launch of a Titan–Centaur rocket. This rocket was used to launch America's most successful planetary missions: the Vikings to Mars and the Voyagers to the gas giant planets. (NASA Photograph.)

The launch of the first Space Shuttle, *Columbia*, in 1981. The Shuttle will carry many important interplanetary spacecraft into space during the 1990s. (NASA Photograph.)

The Sun, photographed in hydrogen–alpha light. Several sunspots and flares can be seen. (Photograph by George East, courtesy of Dennis Milon.)

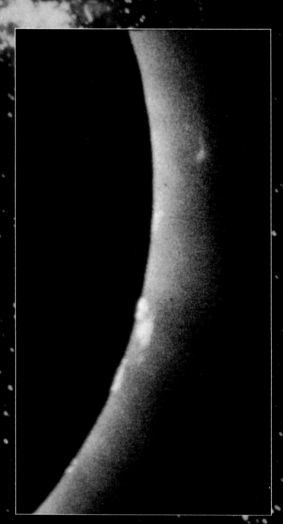

The Sun photographed during the February 26, 1979 total eclipse. Several pink-colored prominences can be seen. (Photograph by Sherman Schultz, courtesy of Dennis Milon.)

The Sun photographed during the February 16, 1980 total eclipse. The corona is prominent in this photograph. (Photograph by George East, courtesy of Dennis Milon.)

The aurora, photographed from Minnesota. The familiar stars of the Big Dipper can be seen at the top of the photograph. (Photograph by Sherman Schultz, courtesy of Dennis Milon.)

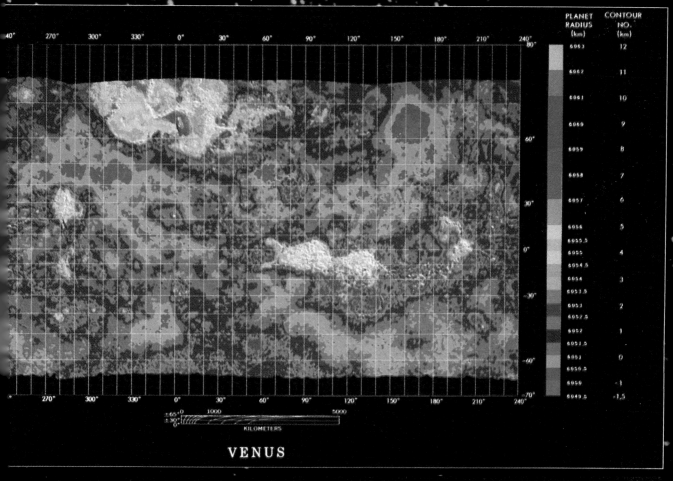

PLANET RADIUS (km)	CONTOUR NO. (km)
6063	12
6062	11
6061	10
6060	9
6059	8
6058	7
6057	6
6056	5
6055.5	
6055	4
6054.5	
6054	3
6053.5	
6053	2
6052.5	
6052	1
6051.5	
6051	0
6050.5	
6050	-1
6049.5	-1.5

KILOMETERS

VENUS

Radar map of Venus' surface derived from Pioneer Venus Orbiter data. The colors represent different elevations, as shown in the scale at right. (NASA/USGS Photograph.)

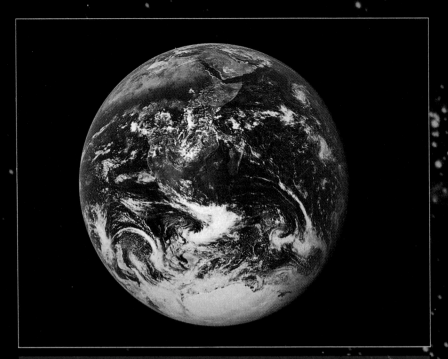

The Earth, as viewed from space during the Apollo 17 mission. (NASA Photograph.)

Thunderstorm activity, as photographed from Earth orbit. (NASA Photograph.)

Venus

CHAPTER OBJECTIVES

After studying this chapter, you should be able to

1. Describe Venus' appearance to the naked eye and through the telescope.
2. Explain how it was learned that Venus has an atmosphere, why the atmosphere hinders the study of Venus' surface, and how this hindrance is circumvented.
3. Summarize the major accomplishments of spacecraft exploration of Venus.
4. Describe the surface geology of Venus, and list the planet's significant surface features.
5. Summarize the properties of Venus' atmosphere, and explain the importance of the greenhouse effect on Venus.
6. Understand why, even though their bulk properties are similar, Venus and the Earth have such different atmospheres and histories.

The Space Shuttle *Atlantis* prepares to deploy the Magellan spacecraft on its voyage to Venus on May 4, 1989. This event marked the first new American planetary mission since Pioneer Venus was launched in 1978. (NASA Photograph.)

Venus is named for the Roman goddess of love and beauty. The name is appropriate, because next to the Sun and Moon, Venus is the brightest object visible in the sky. Venus at its nearest approaches the Earth closer than any other planet, and its overall bulk physical properties are very similar to those of the Earth. (Venus' major orbital and physical properties are summarized in Table 13.1.)

■ Appearance and Basic Properties

Like Mercury, Venus can only be seen before sunrise or after sunset. However, it can reach a distance of 47° from the Sun at greatest elongation and is therefore much less elusive than Mercury. In fact, Venus can be seen in a totally dark sky when at greatest elongation and can even cast shadows when at its brightest. Many ancient cultures had two different names for Venus in the morning and evening sky before they realized both were the same object.

Venus was one of the first objects observed by Galileo, and his observations of the planet were instrumental in convincing him that both Earth and Venus orbit the Sun. He saw two important phenomena — phase changes and variation in the planet's apparent size (Figure 13.1). When Venus is most distant on the far side of the Sun, it is almost fully illuminated and its angular diameter is only 10″. When Venus is close to Earth, a crescent of 64″ diameter can be seen. No other planet attains such a large angular size. Venus is brightest when near the Earth in crescent phase, because its proximity more than compensates for the decrease in the illuminated area. Several observers have claimed to see Venus' crescent with the naked eye, but there is some doubt regarding the validity of their claims.

As telescopes began improving during the 1600s, observers noted that Venus lacked clear

■ Table 13.1 Summary of Venus' Basic Properties*

Semi-major axis: 0.723 astronomical units
Semi-major axis: 108.2 million kilometers
Semi-major axis: 67.2 million miles
Orbital eccentricity: 0.007
Orbital inclination: 3.4°
Orbital period: 224.7 days
Diameter: 12,104 kilometers
Diameter: 7521 miles
Mass (Earth = 1): 0.815
Density: 5.24 grams per cubic centimeter
Sidereal rotational period: 243 days (retrograde)
Axial obliquity: 2.7°
Number of satellites: 0

* For a complete table of planetary data, see Table A2.1 in Appendix 2.

surface markings (Figure 13.2). Attempts to time Venus' rotation using the vague markings that were seen led to estimates that were approximately equal to Earth's rotational period, although Giovanni Schiaparelli, in the late 1800s, suggested a 1:1 spin-orbit coupling with a 225-day rotational period.

Venus occasionally transits the Sun, but such occurrences are much rarer than transits of Mercury, even though Venus' orbital inclination is lower than Mercury's. This is because Venus' greater distance from the Sun and its closer proximity to the Earth combine to give it a greater apparent angular separation from the Sun. Transits of Venus occur in pairs eight years apart, with over a century between pairs. The last transits were in 1874 and 1882; the next will be in 2004 (June 8) and 2012 (June 6). The first transit to be predicted was the one in 1631, by Kepler in 1627, but it occurred during the night in Europe, and nobody could observe it. In late 1639, **Jeremiah Horrocks** (1617–1641) predicted that a transit would occur on December 4, and he (along with some others he was able to notify in time) was the first transit observer. Observations of the four transits during the 18th and 19th centuries were used to measure the value of the astronomical unit, and astronomers traveled all over the globe to be in places where they would be visible. Today, however, radar is used for measuring interplanetary distances, so future transits will simply be interesting curiosities for observers.

Another important observation was made during the 1761 transit by **Mikhail V. Lomonosov** (1711–1765), who noted that Venus' outline was hazy. Lomonosov interpreted this observation as indicating that Venus has an atmosphere at least as substantial as the Earth's. In 1788, **Johann H. Schroeter** (1745–1816) noted additional evidence for an atmosphere when he saw the cusps of Venus' crescent phase extend all around the planet (Figure 13.3). Another effect discovered by Schroeter is that Venus' quarter phases, when exactly half of the hemisphere facing the Earth is illuminated, do not occur exactly when predicted by theory. The observed quarter phase occurs on the average of six days early at evening (eastern) elongation and six days late at morning (western)

■ **Figure 13.1** The appearance of Venus. The planet's angular size and phase change as it orbits the Sun, as seen from Earth through a telescope. (*From* Pasachoff, Jay M. *Contemporary Astronomy,* 4th ed. Saunders College Publishing, Philadelphia, 1989.)

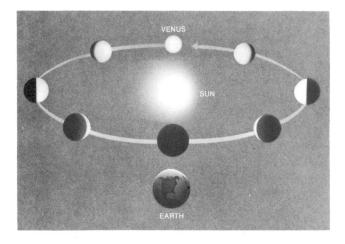

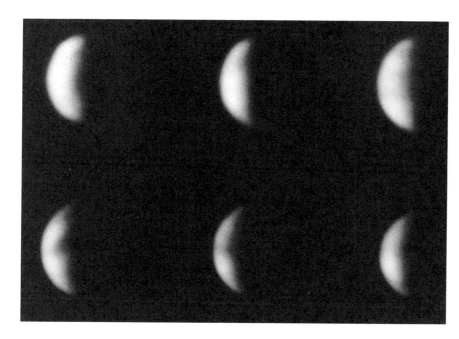

■ **Figure 13.2** Earth-based telescopic views of Venus photographed over a one-month period. Note the vague atmospheric markings and the phase changes. (Photograph courtesy of the Observatories of the Carnegie Institution of Washington.)

elongation. This effect is also the result of Venus' atmosphere. Observers have also seen lights, called **ashen lights,** on Venus' night side, which are possibly lightning or auroras, although the observations may be the result of the extreme contrast between Venus' bright, illuminated side and the dark one.

By the early 20th century, it was realized not only that Venus had an atmosphere, but that it was opaque enough to completely hide the planet's surface. Spectroscopy indicated the presence of carbon dioxide in 1932 and nitrogen in 1960, but the rate of rotation could not be determined from Doppler effect studies of visible light, indicating that the rate was very slow. During the 1950s, motion of atmospheric markings was interpreted as indicating a four-day, retrograde rotational period, which was later verified by spacecraft observation. However, the rotational rate of the solid planet itself remained unknown until 1961, when it was determined using Earth-based radar. The surprising result showed that Venus rotates once every 243 days in the retrograde direction (clockwise as seen from north of the ecliptic). On Venus, the Sun would rise in the west and set in the east, with one "day" equivalent to 116.8 terrestrial days. Not only is Venus the slowest-rotating planet in the solar sys-

tem, but it is one of only three planets with retrograde rotation! What might have been responsible for these unusual rotational properties? One suggestion is that as Venus was accreting, a late impact could have supplied enough energy to change its rotation. However, Venus' obliquity is very low, which would not be expected if such a catastrophic impact had occurred. Another suggestion is that Earth's gravitation may be responsible. If Venus' sidereal rotational period were 243.16 days (the actual value is 243.01), it would rotate four times between intervals when the

■ **Figure 13.3** A sketch of the cusp extension of Venus, when the crescent phase extends all the way around the planet.

Earth and Venus line up on the same side of the Sun. There is no good reason why this would happen, but the times are close enough that it probably isn't a coincidence.

Meanwhile, speculation abounded concerning what might exist below Venus' opaque clouds. Oceans, swamps, tropical rain forests, and deserts were all proposed by researchers and used as settings by science-fiction writers. Thermocouples attached to telescopes recorded upper atmosphere temperatures of about 235 to 240 K (−36 to −27°F) during the 1920s and later, but the first information about the planet itself did not come until 1956, when microwaves from Venus were detected. Analysis of these microwaves indicated that they were produced by a region that had a temperature of about 600 K (621°F). Many researchers felt that Venus' surface could not be this incredibly hot and that the radiation had been produced by Venus' ionosphere. However, the various spacecraft sent to Venus have proved that the microwaves are produced at the surface, verifying that its average surface temperature is about 730 K (855°F), which obviously precludes Earth-like conditions.

■ Spacecraft Exploration

Because Venus approaches the Earth more closely than does any other planet, it was the first planet explored by spacecraft, and more probes have been sent toward it than to any other planet. Although both the United States and the Soviet Union have contributed to the exploration of Venus, America's accomplishments have been somewhat overshadowed by the Soviets' successful soft landings on the planet's surface.

The first successful mission to Venus was the American Mariner 2 spacecraft, which flew within 34,830 kilometers (21,640 miles) of the planet on December 14, 1962. Mariner 2 found that Venus was without a magnetic field or radiation belts, and its instruments determined that the surface temperature of the planet was over 700 K. After several failures, the first successful Soviet mission was Venera 4, which flew past the planet on October 18, 1967 and dropped an at-

mospheric probe that entered the night side of Venus' atmosphere. The probe transmitted data until it was within about 25 kilometers (15.5 miles) of the surface and indicated that the planet's atmosphere is almost entirely carbon dioxide. Mariner 5 passed within 4000 kilometers (2500 miles) of Venus on October 19, 1967, and Veneras 5 and 6 performed missions similar to those of Venera 4 on May 16 and 17, 1969.

On December 15, 1970, Venera 7 became the first spacecraft to reach the surface of Venus. It landed on the night side and reported a pressure of 90 atmospheres. Venera 8 landed on the day side on July 22, 1972 and performed the first analysis of surface materials. Mariner 10 flew by Venus on February 5, 1974 and took the first closeup photographs of the planet. As discussed in the last chapter, Mariner 10 used Venus' gravity to alter its course to provide subsequent encounters with Mercury.

Between 1975 and 1982, the Soviets successfully soft-landed six Venera craft on the planet, four of which returned photographs of the surface and analyses of surface rocks. Because of the severe conditions at the surface, the landers were only able to survive and transmit data for about an hour. Veneras 9 and 10 went into orbit around Venus on October 22 and 25, 1975 (the first spacecraft to do so), after landers had separated from the orbiters, entered the planet's atmosphere, and landed using parachutes and aerodynamic speed brakes. Each lander transmitted photographs from the surface, but the cameras of Veneras 11 and 12, which landed on December 25 and 21, 1978, failed. Veneras 11 and 12 were fitted with microphones and detected numerous thunderclaps while descending through the atmosphere. Veneras 13 and 14 landed on March 1 and 5, 1982, and returned the first color photographs of the surface and also brought surface materials onboard for analysis. (Previous analyses had only been of rocks and regolith in place on the surface.)

In 1978, the United States launched two Pioneer spacecraft toward Venus. Pioneer Venus 1 went into orbit on December 4, and has returned data ever since. The orbiter obtained numerous images of Venus' clouds, performed spectroscopic analysis of the atmosphere in several wavelength regions, and used radar to map much of the planet's surface. It was also used to study Halley's comet in 1986 during the time that the comet was invisible from Earth, lost in the Sun's glare. Pioneer Venus 2 was designed to study the atmosphere; it consisted of a large "bus" that carried four smaller probes. Prior to atmospheric entry, the four probes separated from the bus and entered different parts of the atmosphere on December 9, 1978. The four entry probes, designated night, day, high latitude, and low latitude, along with the bus portion of the craft, returned information on atmospheric conditions until they crashed into the surface. (One entry probe even continued transmitting for about 1 hour after hitting the surface.)

The Soviets placed Veneras 15 and 16 into orbit on October 10 and 14, 1983. These spacecraft, like the Pioneer Venus Orbiter, have mapped much of the surface using radar to penetrate its clouds. In 1984, the Soviet Vega 1 and 2 spacecraft were launched toward Venus, with Halley's comet their final destination. In June, 1985, they flew past Venus, and each spacecraft sent an entry vehicle into the planet's atmosphere. Each entry vehicle contained two devices, a lander and a balloon probe. Together, these four mission components performed well. Although the Vega landers lacked cameras, each contained sophisticated instruments for analyzing surface materials. Both Vegas carried gamma-ray and X-ray spectrometers for surface analysis, although the drill of Vega 1, designed to obtain a soil sample, failed. The balloon probes were the first of their kind, and each consisted of a 4-meter (12-foot) balloon that supported a scientific platform at an altitude of 55 kilometers (33 miles), near the top of the atmosphere's thickest cloud layer. The balloon probes transmitted data for about two days, during which time they were carried nearly one-third of the way around the planet and returned considerable data on atmospheric conditions.

On May 4, 1989, the Space Shuttle *Atlantis* launched the Magellan spacecraft toward Venus. Magellan, which is scheduled to begin orbiting Venus in August, 1990, is designed to use radar to map the planet's surface with the best resolution to date.

■ Large-Scale Surface Features

Even though Venus' clouds cause its surface to be invisible from the Earth, radar can penetrate the clouds and map the planet's large-scale surface features. The first such studies were accomplished using the large radio telescope in Arecibo, Puerto Rico (Figure 13.4). Better results have been obtained from orbital spacecraft (Figure 13.5). The Pioneer Venus 1 Orbiter mapped 93% of the planet's surface with a vertical accuracy of 200 meters (600 feet). (Recall that because Venus has no natural reference point for elevation measurements comparable to sea level on Earth, the planet's average radius is used as the starting point for measuring elevation.) With few exceptions, surface features on Venus are named for famous women in history and from mythologies of various cultures.

The surface topography of Venus is quite different from that of Earth, whose surface is mostly high elevations (continents) and low elevations

■ **Figure 13.4** A radar image of the Maxwell Montes region of Venus obtained by Arecibo Observatory. Bright areas represent rougher surfaces. (Courtesy of D. B. Campbell.)

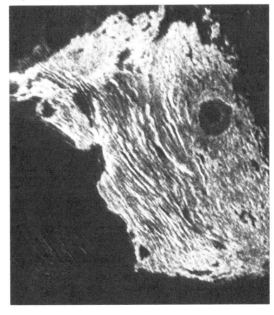

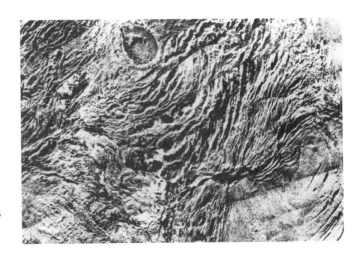

■ **Figure 13.5** A radar image of Maxwell Montes obtained using radar data from Veneras 15 and 16. (Courtesy of the U.S.S.R. Academy of Sciences.)

(ocean basins). (Recall that the bimodal elevation distribution on Earth is the result of its having two types of crust, continental and oceanic.) Venus' surface has areas at three separate elevation levels, and this fact has provided the basis for dividing its surface into three topographic types (Figure 13.6). In addition to having different elevations, it is likely that the three topographic types contain different kinds of rock and have different geological histories. However, it is considered likely that Venus contains only one type of crustal material, whose composition is probably basaltic.

About 65% of Venus' surface consists of **upland rolling plains** whose elevations lie within about 1000 meters (3300 feet) of the average planetary radius. Radar studies reveal that these plains contain many circular features that range from 20 to 800 kilometers (12 to 500 miles) across. Depending upon their precise form as indicated by radar, some have been interpreted as being craters, whereas others appear to be volcanoes. Venus, like other solar system objects, would be expected to have experienced numerous impacts, but the amount of subsequent erosion that any craters and basins produced by them would have undergone is unknown.

The second topographic type on Venus, **depressed lowlands**, covers about 27% of the surface. These areas all lie below the average planetary radius, and some are as far as 1400 meters (4500 feet) below it. They are sparsely cratered and reflect radar poorly, possibly indicating that they are smooth. It is considered likely that these areas are basaltic lava flows that are comparable to the Earth's ocean basins or the Moon's maria, although it has also been suggested that they may contain wind-deposited material. The most extensive lowland region, **Atalanta Planitia**, is about as large as the Gulf of Mexico. Most of it extends 1400 meters (4500 feet) below the average planetary radius, although one portion of it reaches 1600 meters (5200 feet) below.

The third topographic type on Venus, the **highlands**, covers the remaining 8% of the planet. Most of this material is contained in two "continents," Aphrodite Terra and Ishtar Terra, which are comparable in size to Africa and Australia, respectively. A third, smaller area is called Beta Regio, and a few other minor upland areas

■ **Figure 13.6** Maps of Venus. (a) Location of three elevation levels. Upland rolling plains are not shaded, depressed lowlands are shaded light gray, and highlands are shaded a darker gray. (*From* an article by Masursky, Harold et al. In *Journal of Geophysical Research,* volume 85, page 8232. Copyright 1980 by the American Geophysical Union.) (b) Location map of important geological features. (*From* Basilevsky, Alexander T. and James W. Head. "The Geology of Venus." Reproduced, with permission, from the *Annual Review of Earth and Planetary Sciences,* volume 16. Copyright 1988 by Annual Reviews, Inc.)

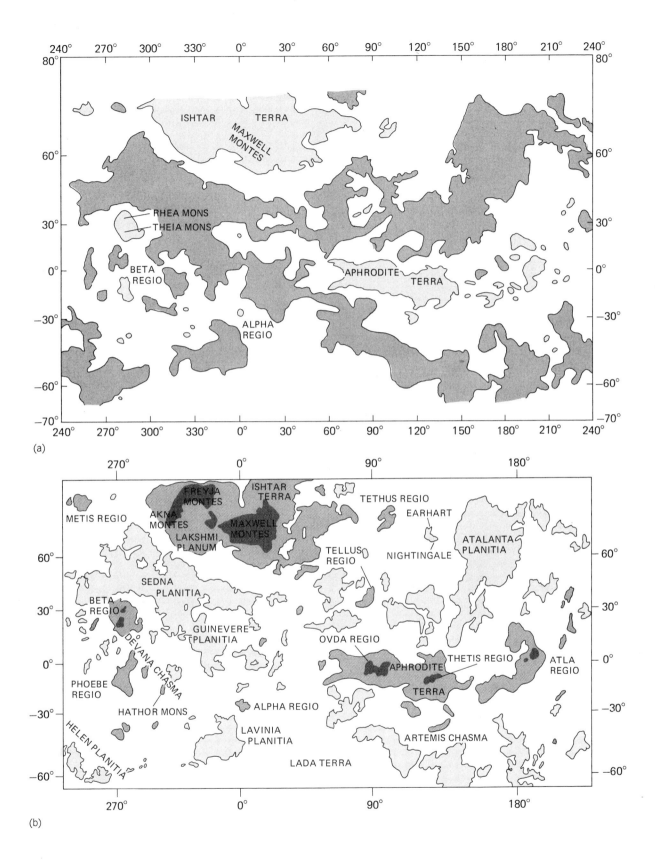

(a)

(b)

185

(a)

(b)

■ **Figure 13.7** Artist's conceptions of two of Venus' highland regions. Vertical relief is highly exaggerated in both views. (a) Aphrodite Terra. (b) Ishtar Terra. (NASA Photographs, courtesy of the National Space Science Data Center.)

also exist. These highlands lack circular radar-reflecting features, suggesting that craters are absent and that the continents are therefore relatively young features. The continents are cooler and at lower atmospheric pressures than the other areas, because Venus' atmospheric conditions vary widely with elevation. For this reason, there may be considerable variation in erosional activity with altitude, and this may cause the sur-

faces of the continents not to be representative of the entire planet. For example, radar measurements have shown that the continents are much rougher than the lower regions.

Aphrodite Terra is the largest "continent" on Venus and is elongated in an east-to-west direction. It contains western, central, and eastern mountain regions, as well as an area of complicated ridges and grabens that separates the central and eastern mountains. **Ishtar Terra** contains the highest point on Venus, **Maxwell Montes,** which reaches 11 kilometers (7 miles) above the average radius (Figure 13.7). East of Maxwell is an area of ridges and grabens that suggests fault-related activity. **Beta Regio** contains two features, **Theia Mons** and **Rhea Mons**, that appear to be 4-kilometer-high (2.5-mile-high) shield volcanoes. It has been suggested that Beta Regio is one area on Venus that may still be active volcanically.

The lowest point on Venus is **Diana Chasma,** which appears to be a fault-related feature. It extends 2000 meters (6600 feet) below the average planetary radius. The fact that this and other linear, graben-like features exist suggests that some sort of tectonic activity has occurred on Venus, and Akna Montes, a mountain belt located in Ishtar Terra, appears to be a result of compressional forces. However, Venus' features appear to be the result of localized tectonic activity rather than the complex style of global plate tectonics found on Earth, because Venus lacks many of the features, such as long mountain belts, associated with terrestrial tectonics. Exactly how much internal activity is now occurring within Venus is unknown, but it is hoped that the improved radar images that the Magellan spacecraft should obtain will add considerable insight to our knowledge of how Venus' surface features have been produced. Spacecraft analysis of gases in Venus' atmosphere initially seemed to indicate that volcanic activity is occurring presently, but this interpretation was later disproved.

■ The Surface of Venus

What types of rock and regolith are present at Venus' surface, and how would the planet's surface appear to someone standing on it? As a result of four Soviet Venera spacecraft, we how have an answer (Figure 13.8). Venus' surface contains both rocks and regolith, indicating that some weathering processes occur there. Because Venus lacks water, weathering could not occur in exactly the same ways as on Earth. Nevertheless, there is sufficient weathering that most of Venus' surface rocks appear to be somewhat smoothed and rounded. In addition, it is likely that mass wasting has moved some material at Venus' surface, and photographs seem to indicate that deposition of wind-blown material has occurred.

Venera 8 landed in an upland plains region. Although it lacked cameras, it analyzed the surface materials on which it rested and indicated that high-potassium gabbros or basalts were present. (Gabbro and basalt are two igneous

■ **Figure 13.8** Photographs of the surface of Venus obtained by Veneras 9, 10, 13, and 14. (USSR Academy of Sciences Photographs. This NASA composite is courtesy of the National Space Science Data Center.)

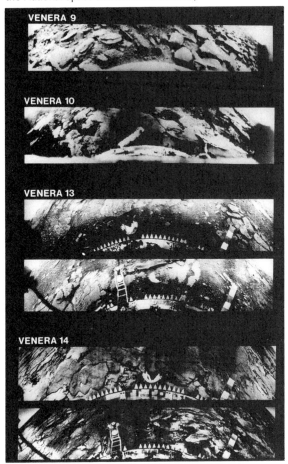

rocks with the same composition. The difference between them is textural: gabbro is intrusive and coarse grained, whereas basalt is extrusive and fine grained. Venera landers studied only composition and were unable to make textural determinations.) Veneras 9 and 10, which landed in upland plains near Beta Regio, transmitted photographs of their surroundings. The images from Venera 9 revealed slabs of rock underlain with regolith, whereas Venera 10 landed in a region of scattered rock outcrops. In both cases, the rocks and regolith had fairly low albedos, and analyses indicated that they were gabbro or basalt. Veneras 13 and 14 landed in upland plains regions containing slabs of flat, reddish-brown rocks that appeared to be layered. At the Venera 13 site, which looked similar to those of Veneras 9 and 10, considerable dark regolith was present between rocks, but regolith was absent at the Venera 14 site. Both Venera 13 and Venera 14 analyses indicated gabbro or basalt rock, but the Venera 13 rocks appeared to have a higher potassium content and may be rich in the mineral leucite. Researchers believe that the rocks at the Venera 13 and 14 sites are **tuff**, which is rock made from compacted volcanic ash rather than directly from cooling magma. Vegas 1 and 2, which unfortunately were not equipped with television cameras, landed near the eastern end of Aphrodite Terra. Vega 1 landed in lowlands north of Aphrodite Terra, and Vega 2 landed in a region of the Aphrodite Terra highlands called Atla Regio. Although some of the instruments on Vega 1 malfunctioned, Vega 2 analyzed surface rocks and found them to be basalt or gabbro, similar to other rocks on Venus' surface. The Soviet scientific team analyzing the results concluded that the Vega 2 rock was of a type called olivine gabbro-norite.

■ Atmosphere

Venus has the densest atmosphere of any of the terrestrial planets. Cloud-filled, and with a surface pressure 90 times that of Earth's, the atmosphere is responsible for Venus' high albedo. In fact, Venus' albedo is so high (about 65%) that if it were not for the greenhouse effect, the planet's surface would be cooler than Earth's!

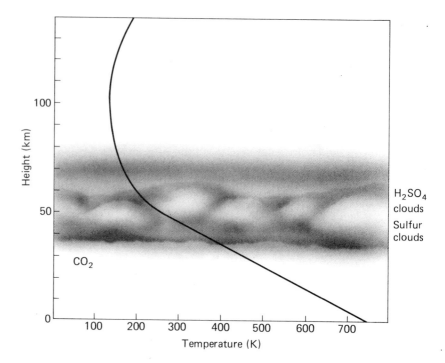

■ Figure 13.9 Vertical structure of Venus' atmosphere. Height in kilometers is plotted against temperature in degrees Kelvin. The location of the cloud layers is marked. (From Abell, George, David Morrison, and Sidney Wolff. *Exploration of the Universe,* 5th ed. Saunders College Publishing, Philadelphia, 1987.)

The atmosphere of Venus is composed of 96% carbon dioxide, 4% nitrogen, and minor amounts of other gases. The atmosphere of Venus is extremely dry because the sulfuric acid, which is the primary cloud component, absorbs water very readily. It is estimated that if all of the remaining water vapor in Venus' atmosphere could be condensed and collected, it would make a layer only about 1 centimeter deep on the planet's surface! Unlike Earth's clouds, which are primarily located within 11 kilometers (7 miles) of the surface, Venus' clouds are much higher and comprise three distinct layers (Figure 13.9). The top layer, ranging from 55 to 70 kilometers (35 to 44 miles) above the surface, is composed of sulfuric acid droplets at a temperature of about 286 K (55°F). The middle layer, between 50 and 55 kilometers (31 and 35 miles), is composed of larger drops of sulfuric acid and particles of solid sulfur at 293 K (68°F). The bottom layer, which is the densest, extends from 45 to 50 kilometers (28 to 31 miles) high and is a mixture of liquid and solid sulfur at a temperature of 475 K (395°F). A haze layer extends from the bottom of the clouds down to a height of about 23 kilometers (14 miles); ''clear'' air exists below this point. The atmosphere's composition below the clouds is somewhat different from that above, as it contains more sulfur dioxide, sulfur trioxide, and water. All of these are believed to be produced by decomposition of sulfuric acid in the clouds. Pioneer Venus 2 discovered that the atmosphere has **polar holes,** meaning that the cloud tops occur at lower altitudes in the polar regions than near the equator.

Venus has atmospheric markings that are vague when viewed in visible light, but spacecraft photographs taken in ultraviolet light clearly reveal a distinct pattern of cloud markings (Figure 13.10). The identity of the atmospheric component that absorbs ultraviolet light and produces this pattern has not yet been determined. These markings reveal that the motion of the atmosphere is much different from that of the planet's surface. The upper layers of the atmosphere, near the cloud tops, rotate much faster than the planet (4 days instead of 243), producing what is called a **super-rotating atmosphere.** The clouds move fastest at the equator, and ultraviolet photographs show a distinctive Y-shaped pattern of atmospheric flow. The winds on Venus have their highest velocity (360 kilometers per hour, or 220 miles per hour) at the cloud tops. Wind velocity drops to half of this value above and below the

   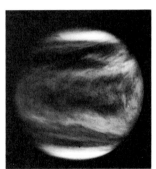

■ **Figure 13.10**   The pattern of atmospheric markings on Venus which is visible using ultraviolet light. These images were obtained by Pioneer Venus 1. (NASA Photograph.)

cloud tops, at elevations of 50 and 100 kilometers (30 and 60 miles).

What is responsible for the high wind speeds in Venus' super-rotating upper atmosphere? The answer appears to be a complex system of Hadley circulation redistributing atmospheric heat. There are several Hadley cells "stacked" atop one another (Figure 13.11). The upward convection supplies energy that drives the super-rotating winds.

Venus' unusual atmosphere would cause surface conditions much different from those on Earth. Surface visibility is only about 2 kilometers (1.2 miles), and refraction from the dense atmosphere would produce distortion of distant views. Although some sunlight reaches the surface, it would only be about one-tenth as bright as a cloudy day on Earth. While surface winds are very light, the rapid upper winds are so effective in transporting heat that even during the long nights, the planet cools by only about 8 K (15°F). Because surface rocks photographed by the Venera landers showed much more evidence of erosion than had been expected, it is considered possible that even light surface winds, when combined with Venus' extremely high surface pressures, can effectively produce erosion.

Venus' atmosphere contains several clues to its origin and history. Venus contains more krypton, argon-36, and argon-38 than the Earth, indicating that these materials may have been more abundant in Venus' region of the solar nebula as the planets were forming. However, Venus has less argon-40 than the Earth. Because argon-40 is mainly derived from the decay of potassium-40 in

rocks, this may indicate that Venus' rocks are deficient in potassium. Venus may originally have contained abundant water, which photodissociated. The hydrogen would have escaped into space, whereas the oxygen would have combined with crustal rocks. Venus has a much higher deuterium–hydrogen ratio than Earth, and this fact is considered evidence of loss of water. Because deuterium is a heavy isotope of hydrogen, it would have been more likely to have been left behind when ordinary, light hydrogen escaped.

One of the biggest differences between the atmospheres of Venus and the Earth is the excess carbon dioxide in Venus' atmosphere. Interestingly, many theorists believe that Venus and the Earth originally had approximately equal

■ **Figure 13.11**   The circulation pattern of Venus' atmosphere. (NASA/ARC Photograph.)

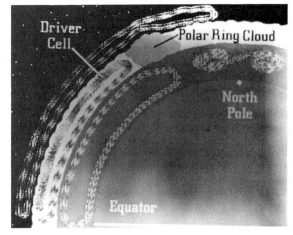

amounts of carbon dioxide. On Venus, the carbon dioxide remained in the atmosphere and contributed to the planet's runaway greenhouse effect. On the Earth, carbon dioxide, in the form of the mineral calcite ($CaCO_3$), was incorporated into carbonate sedimentary rocks and therefore removed from the atmosphere.

■ Interior and Magnetic Field

The interior of Venus is only poorly understood. Unfortunately, the Venera landers had such short lifetimes that no seismic data were obtained, so little knowledge exists about the planet's interior. Venus is slightly smaller and less dense than the Earth, so its internal structure is probably somewhat different. It is likely that the planet differentiated into crust, mantle, and core, while outgassing much of its dense atmosphere at the same time.

Venus is one of only three planets that lack appreciable internally driven magnetic fields. (Mars and Pluto are the others.) (Venus' total magnetic moment is less than 5×10^{-5} that of Earth's.) In view of its slow rotation, this would not be too surprising, except for the fact that the slow-rotating Mercury does possess a field. Furthermore, Venus' Earth-like overall dimensions would make it somewhat likely that a molten core may be present. A comparison of Mercury, Venus, and the Earth shows that our understanding of planetary magnetic fields is far from complete.

Although Venus lacks a magnetic field, it does interact with the solar wind. These interactions take place because it has an ionospheric layer in its atmosphere. The top of the ionosphere generally lies 400 kilometers (250 miles) above the surface, although this changes as a result of variation in the solar wind. The bow shock occurs about 8000 kilometers (5000 miles) above the planet's surface.

■ History

The fact that Venus' surface features are known only moderately well from radar studies prevents our developing an absolute or even relative time scale of the planet's history. Any discussion of Venus' history is based upon assumptions of similarity with the Earth and other well-understood objects. Harold Masursky and other researchers have suggested the following major events in Venus' history: initial accretion, differentiation, and heavy bombardment were followed by crustal redistribution to form lower and higher regions. Next, the continental regions Aphrodite Terra and Ishtar Terra were formed, presumably by extrusion of magma from sources in the mantle. After that, basaltic magmas were extruded in the lowland areas and parts of the rolling plains, after which came the formation of Beta Regio's shield volcanoes. As mentioned previously, Venus may still have internal geological activity today. The loss of Venus' water by photodissociation and the heating of the atmosphere by the greenhouse effect have combined to give Venus a far different atmosphere from that of the Earth.

■ Prospects for Future Exploration

Clearly, many questions remain about Venus, which indicates that further study is needed. Although the Magellan spacecraft should reveal considerable information, the United States has no plans at this time for Venus landers. Presumably the Soviets will continue their efforts in this area, but the technological challenges presented by Venus' surface conditions may make it a long time before significant new information is obtained.

■ Chapter Summary

Venus, the brightest planet in the sky, was found over two centuries ago to have an atmosphere. This atmosphere, being cloud-filled, prevents direct observation of the planet's surface, which led early astronomers to make numerous speculations about its surface. Spacecraft exploration, however, has revealed a surface temperature of 730 K (855°F), ruling out Earth-like surface conditions.

Radar mapping of Venus' surface has revealed three types of topography: upland rolling plains, depressed lowlands, and highlands. Most of Venus' surface is probably basaltic rock, and the importance of tectonic activity on the planet is not yet known with certainty.

Venus' dense atmosphere of carbon dioxide and nitrogen contains clouds of sulfuric acid and sulfur. The atmosphere is opaque, but it has vague markings visible in ultraviolet light. The interior of Venus is poorly understood, but it is believed that it is not generating a magnetic field. The loss of water on Venus and the retention of carbon dioxide in its atmosphere are probably the major factors accounting for its differences from the Earth.

■ Chapter Vocabulary

Aphrodite Terra	Diana Chasma	Maxwell Montes	super-rotating
ashen lights	highlands	polar holes	atmosphere
Atalanta Planitia	Horrocks, Jeremiah	Rhea Mons	Theia Mons
Beta Regio	Ishtar Terra	Schroeter, Johann H.	tuff
depressed lowlands	Lomonosov, Mikhail V.		upland rolling plains

■ Review Questions

1. Why is Venus' name appropriate, and why is the planet less difficult to see than Mercury?
2. What did Galileo see when observing Venus, and why did his observations support the heliocentric system?
3. Why are transits of Venus so rare? Why were they once considered rather important events for astronomers?
4. Why did early observers think that Venus had an atmosphere?
5. Summarize the history of space exploration of Venus.
6. Describe the major surface features of Venus.
7. Why is it considered likely that some features on Venus are volcanic?
8. Describe the atmosphere and clouds of Venus.
9. Why is the surface of Venus so hot?
10. Why do we have problems developing a geological history of Venus?

■ For Further Reading

Burgess, Eric. *Venus, An Errant Twin.* Columbia University Press, New York, 1985.

Hunt, Garry E. and Patrick Moore. *The Planet Venus.* Faber and Faber, London, 1982.

Pettengill, Gordon H. et al. "The Surface of Venus." In *Scientific American*, August, 1980.

Saunders, R. Stephen and Michael H. Carr. "Venus." In Carr, Michael H., editor. *The Geology of the Terrestrial Planets*, NASA, Washington, D.C., 1984.

Schubert, Gerald and Curt Covey. "The Atmosphere of Venus." In *Scientific American*, July, 1981.

Earth

The Hawaiian Islands, photographed from space by astronauts in the Space Shuttle. (NASA Photograph.)

Earth, the home of mankind, is the largest terrestrial planet. For obvious reasons, we know much more about it than all of the other objects in the solar system combined. We have already discussed many aspects of the Earth in previous chapters in which we used our planet as a starting point for comparisons with other bodies. Therefore, this chapter is mainly a summary of the Earth's important properties, as well as an introduction to what we know about its history. (The Earth's major orbital and physical properties are summarized in Table 14.1.)

■ Earth's Motions

Until several centuries ago, the primary question in astronomy was whether the Earth is stationary or orbits the Sun. The work of Copernicus, Galileo, Kepler, and Newton convinced astronomers that the heliocentric theory was correct and that the Earth orbits the Sun. If any doubt remained, it was eliminated in 1729, when **James Bradley** (1693–1762) discovered an effect known as the **aberration of starlight**, which proves that the Earth does indeed orbit the Sun. Aberration of starlight may be defined as the displacement of a star's image on the celestial sphere as a result of the Earth's orbital motion. It manifests itself as a slight change in the apparent direction of a star as the year progresses due to the Earth's motion about the Sun. Aberration effects vary from star to star, as they are greatest for stars lying in a direction perpendicular to the Earth's motion at a given time but do not exist for stars aligned parallel to the Earth's direction of travel. Aberration can be explained using the following analogy (Figure 14.1): Suppose you were standing outside during a rainstorm, holding a large pipe. If you held the pipe vertically, raindrops would pass through from top to bottom. What would happen if you decided to run? In order for raindrops to pass through, you would have to tilt the pipe for-

■ Table 14.1 Summary of Earth's Basic Properties*

Semi-major axis: 1.0 astronomical units
Semi-major axis: 149.6 million kilometers
Semi-major axis: 92.9 million miles
Orbital eccentricity: 0.017
Orbital inclination: 0.0°
Orbital period: 325.256 days
Diameter: 12,756 kilometers
Diameter: 7927 miles
Mass (Earth = 1): 1.0
Density: 5.52 grams per cubic centimeter
Sidereal rotational period: 23.93 hours
Axial obliquity: 23.5°
Number of satellites: 1

* For a complete table of planetary data, see Table A2.1 in Appendix 2.

ward in the direction of motion. Similarly, the Earth's orbital motion requires that to observe stars through a telescope, the telescope tube must be "tilted" by 1 part in 10,000 so that the light of the star passes through without obstruction. (The aberration effect has this value because Earth's orbital velocity is that fraction of the velocity of light.) The extremely small magnitude of aberration explains why it was not discovered until after the telescope had been invented.

The Earth's orbital motion around the Sun also manifests itself in the phenomenon called **trigonometric parallax,** a shift in the location of nearby stars during the course of the year (Figure 14.2). (Bradley was actually trying to detect parallax when he discovered aberration instead.) Parallax causes nearby stars to shift their positions slightly with respect to the distant background stars. The lack of parallax was used by the Greeks as evidence against the heliocentric concept, but stars are so distant that even the closest has a parallax shift much less than 1″ of arc, which can only be detected telescopically. The first parallax measurement was made in 1831 by the German astronomer **Friedrich Bessel** (1784–1846). His measurement of the star designated 61 Cygni was within 6% of the value accepted today. 61 Cygni has a parallax of 0.3″ and is 11.1 light years distant. The nearest star, designated Proxima Centauri, has a parallax of 0.76″ and is 4.3 light years away. The difficulty of making parallax measure-

ments can be appreciated by noting that 0.76″ is the angular size of a postage stamp viewed from 4 miles away. Parallax is a useful technique for measuring the distances of those stars that are close enough to display a shift of 0.05″ or more.

The Earth requires about 365 and one-quarter days to orbit the Sun. Because the orbital period is not an integral number of days, our calendar is designed so that three 365-day years are followed by a 366-day **leap year.** Leap years occur only in years evenly divisible by four (1988, 1992, 1996, and so forth). However, the calendar is slightly more complicated than this. The **sidereal year,** the exact time required for the Earth to orbit the Sun, is 365.2564 days, or 365 days, 6

■ **Figure 14.1** Aberration of starlight. (a) Raindrops pass through a vertical, stationary pipe. (b) If the pipe is moving, as it would be if you were carrying it while running through the rain, the pipe would have to be tilted forward in the direction of motion for raindrops to continue to pass through it. (c) In a similar fashion, a telescope must be deflected slightly in the direction of Earth's orbital motion in order for starlight to pass through it. This causes a small shift in the observed location of the star.

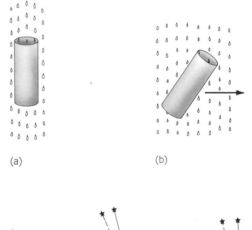

(a)   (b)

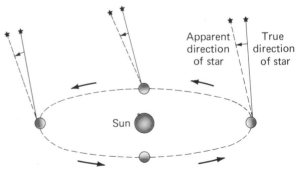

(c)

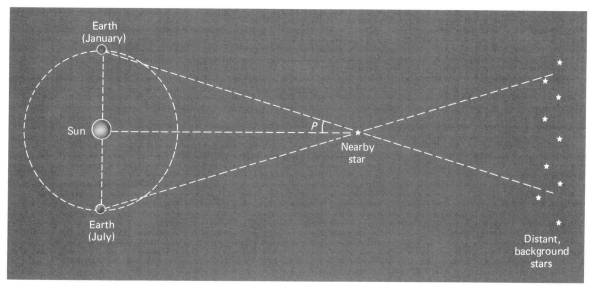

■ **Figure 14.2** Trigonometric parallax. When observed from the Earth when it is on opposite sides of the Sun, a nearby star's position will shift slightly as seen against the distant, background stars. Because the parallax angle, *P*, can be measured, trigonometry can be used to determine the star's distance.

hours, 9 minutes, and 10 seconds. However, the calendar is based on the **tropical year,** which is the time between successive beginnings of spring one year and the next. The tropical year is only 365.242199 days (365 days, 5 hours, 48 minutes, and 46 seconds) because of the effects of precession. Because the tropical year is less than 365 and one-quarter days, not all leap years are needed. This is dealt with by having "century years" (1800, 1900, 2000, and so on) be leap years only if they are evenly divisible by 400, instead of just by 4. Other effects, including the slowing of the Earth's rotation due to tidal interactions, require additional, nonperiodic calendar adjustments. (The adding of "leap seconds" at the ends of several recent years is an example.)

In addition to its orbital motion, the Earth undergoes many additional movements, the most important of which is rotation. The Earth rotates on its axis, and its rotational period defines the day. The 24-hour day with which everyone is familiar is called the **solar day,** and it is defined as the interval between two occurrences of the Sun's being highest in the sky. However, this is not the actual length of time required for the Earth to rotate on its axis. That interval, the **sidereal day,** is measured with respect to the distant

stars and is only 23 hours, 56 minutes, and 4 seconds long. The solar day is nearly 4 minutes longer because while the Earth makes one rotation, it is also traveling 1 part in 365.25 along in its orbit (almost 1°). This means that during the time it takes the Earth to complete one rotation with respect to the distant stars, the Earth has moved enough in its orbit that the Sun's apparent location in the sky will have changed. Therefore, the Earth must turn a little bit extra, a process requiring 4 minutes, in order for the Sun to once again be at the highest point in the sky (Figure 14.3). One result of keeping time by the Sun is that individual stars rise and set 4 minutes earlier every day. (The variations in the Sun's rising and setting times are due to the seasonal effects described in Chapter 11.)

The fact that the Earth rotates was accepted by most Greek astronomers. This was much easier to accept than the alternate explanation for rising and setting of celestial objects — that the entire celestial sphere, which contains numerous objects at different distances, circled the Earth daily. Therefore, a rotating Earth was never as controversial as an orbiting one. Any doubt was removed in 1851, when **J. B. L. Foucault** (1819 – 1868) proved that the Earth rotates. His proof de-

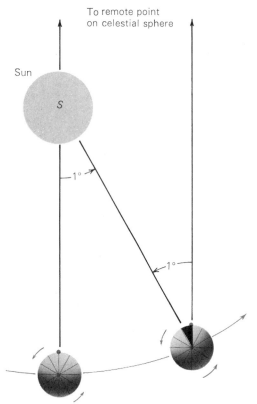

■ **Figure 14.3** Sidereal and solar days. A sidereal day is the time between successive alignments of a point on the Earth with the distant background stars *(parallel solid lines).* A solar day is the slightly longer time between successive alignments of a point on the Earth with the Sun. (From Abell, George, David Morrison, and Sidney Wolff. *Exploration of the Universe,* 5th ed. Saunders College Publishing, Philadelphia, 1987.)

pends upon one of the basic principles of a pendulum (a weight swinging back and forth while suspended on a rope or cable): once a pendulum begins to oscillate, it will continue to oscillate in the same plane. Foucault imagined a pendulum oscillating from the ceiling of a room at the North Pole. Foucault realized that the plane of oscillation of such a pendulum at the North Pole would appear to rotate around the room once each day. In actuality, the room would rotate under the pendulum, whose motion would continue in the same true direction in space (Figure 14.4). Foucault demonstrated such a pendulum in Paris, where complete rotation of the pendulum plane requires about 36 hours, because a pendulum

there is not perfectly aligned with the Earth's axis. Many museums and planetariums have **Foucault pendulums,** and their motion can be seen in less than an hour (Figure 14.5).

As described in Chapter 7, one of the effects of the Earth's rotation is that it is distorted by rotational forces and is not a perfect sphere. The Earth's polar diameter is 43 kilometers (26.7 miles) less than its equatorial diameter, and its oblateness value is 1 part in 297, or 0.003.

As the Earth orbits the Sun, its 23.5° axial obliquity causes it to have seasons, as was discussed in Chapter 11 and illustrated in Figure 11.6. Obviously, the northern and southern hemispheres have seasons that are opposite. In spring and summer, with the appropriate axis tilted toward the Sun, the Sun rises earlier, sets later, is higher at noon, and heats the surface more effectively. With the axis tilted away from the Sun in autumn and winter, the Sun rises later, sets earlier, is lower at noon, and is less effective in heating the surface.

Although the Earth's axis points in a constant direction over the time scale of a human life, observations show that it is variable over longer periods of time, as discussed in Chapter 7 and illustrated in Figure 7.6. The Earth undergoes **precession** with its axis tracing out a circular path in the sky over a 26,000-year cycle. This phenomenon was first discovered in the second century

■ **Figure 14.4** The Foucault pendulum. A hypothetical pendulum at the north pole would oscillate in a fixed plane as the Earth rotates under it. The effect can be observed at all other latitudes as well, except at the equator.

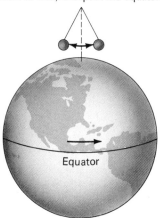

■ **Figure 14.5**   The Foucault pendulum at the Buhl Planetarium in Pittsburgh, Pennsylvania. (Photograph by the author.)

B.C. by the Greek astronomer Hipparchus, who noted a discrepancy between the positions of stars recorded centuries before and those that he himself measured. Precession causes the pole stars to change, as well as the coordinate locations of all other stars. To the ancient Egyptians who built the great pyramids, the star Thuban in the constellation Draco was the north pole star. Currently, however, the pole star is Polaris in Ursa Minor. If calendar adjustments were not made, another effect of precession would be that the dates of the seasons would change. For example, 13,000 years from now, the Earth's axis will tilt in the opposite direction compared to the present, giving us summer in what is now December and winter in what is now July! This illustrates the complexity involved in calendar adjustments.

As described in Chapter 7, precession is caused by two opposing forces: the gravity of the Sun trying to pull the Earth's equatorial bulge into the ecliptic plane, and the Moon's gravity trying to pull it into the plane of the lunar orbit. Precession is complicated by a periodic variation called **nutation,** which is caused by irregularities in the Moon's orbit. Nutation causes the precessional motion of the Earth's axis not to trace out a perfect circle in the sky. Instead, the trace of precessional motion is slightly wavy, with wave motion having an amplitude of 9 arc seconds and a period of about 19 years.

The Earth has many other motions that cannot be discerned by a naked-eye observer and have no effect on time or the calendar. First, the entire solar system orbits the center of the galaxy approximately once every 250 million years. In addition, the Milky Way itself moves with relation to other nearby galaxies, which as a group are moving away from most other galaxy clusters because of the overall expansion of the universe. With all of these dizzying motions taking place, it is probably fortunate that we residents of Earth cannot sense them!

■ Surface Features

It has been remarked that "Earth" is probably a poor name for our planet; "Water" would be more appropriate, as about 70% of our planet's surface is covered by oceans. In space, the most conspicuous feature of our planet is its oceans, but both they and the land may be partially obscured by variable amounts of cloud cover (Figure 14.6). Old science-fiction films usually showed Earth as an unchanging globe hanging in space, but the space age has revealed that the Earth's true appearance is constantly changing as a result of its mantle of clouds.

If we could somehow remove Earth's atmosphere and oceans, we would still note a differ-

■ **Figure 14.6** The Earth from space, as photographed during an Apollo mission. Note how the abundant clouds hide much of the surface. (NASA Photograph.)

mentioned in Chapter 10, the work of running water has been the most important in shaping the surface of most areas of the world. Of the Earth's land surface, about 10% is covered by glacial ice and another 25% is desert.

Many surface features, including rivers, volcanoes, sand dunes, and others, are easily visible from orbit far above the Earth's surface (Figure 14.7). Many manned and unmanned orbital missions have gathered information about the geology of our planet. One might wonder why it would be beneficial to study our planet from space, but many large-scale features and geological trends can be seen better from above, and remote sensing studies enable rock structure and composition to be studied easily. For example, plate tectonic features such as mountains can be discerned easily from space, as they are often long, narrow, and in the form of arc segments. In addition, radar images of a region of the Sahara Desert obtained during a Space Shuttle mission revealed geological surfaces buried under many meters of sand, where they had never been detected by geologists on the surface.

ence between the continental and oceanic areas. About 40% of Earth's surface is **continents,** whose surface elevations average 92 meters (300 feet) above sea level. About one-fourth of all continental rocks are submerged under shallow oceans in areas called the **continental shelves.** Continents contain all three rock types (igneous, sedimentary, and metamorphic), but their average overall composition is similar to that of the igneous rock granite. Continents are old, and indications are that the majority of continental crust formed by about 2.5 billion years ago. As described in Chapter 10, long mountain chains are found at the edges of many continents as a result of continental collisions and subduction of oceanic crust. Earth appears to be the only planet where plate tectonics has produced these types of mountains. The highest point on Earth is Mt. Everest in Asia, whose peak is 8.8 kilometers (5.5 miles) above sea level.

The surfaces of the continents have been shaped by the many geological processes described in Chapter 10. In addition to the internal processes that tend to build up surface features, extensive erosion has worn away rock and sculpted the surfaces that we see around us. As

■ **Figure 14.7** An example of terrestrial surface features photographed from space. This photograph, taken by the Gemini 4 astronauts, shows the Hadramawt Plateau in Saudi Arabia. Note the dry river valleys. (NASA Photograph.)

The remainder of Earth's surface is composed of the **ocean basins,** which average 4.6 kilometers (2.9 miles) deep, although trenches in areas where plate tectonic subduction occurs can be considerably deeper. The deepest trench is the Mariana Trench in the Pacific Ocean, which reaches a depth of 11.0 kilometers (6.8 miles) below sea level. (The total relief of the Earth's crust, from the highest mountain to the deepest trench, is about 19 kilometers, or 12 miles.) In addition to trenches, the ocean floor contains other topographic features, primarily the mid-oceanic ridges caused by the volcanic activity associated with plate tectonics, and volcanic islands, which occur either in isolation or in long chains. The material of the ocean floor is basaltic igneous rock, which is covered by a veneer of sediments that accumulate there from the ocean water. (The thickness of ocean floor sediment varies, depending upon distance from land and other factors.) New oceanic crust is constantly being produced at the mid-oceanic ridges, only to be eventually destroyed by subduction. For this reason, all oceanic crust in existence today is less than 200 million years old, which is fairly young compared to the age of the continents.

■ Oceans

The total amount of water at or near the surface of the Earth is estimated to be about 1.4 billion cubic kilometers (325 million cubic miles). Of this, 97% is in the oceans. The Earth's water undergoes a cycle that takes it from place to place: evaporation takes water from the oceans (and elsewhere) into the atmosphere, condensation forms clouds, and precipitation returns fresh water to the surface (Figure 14.8). Although rivers are considered "fresh" water, they carry small amounts of dissolved material into the oceans. Because these materials remain behind when ocean water evaporates, the oceans have accumulated a significant amount of dissolved materials, 3.45% by weight. The most abundant material in the oceans is, of course, salt (sodium chloride); magnesium chloride, sodium sulfate, and other substances are also present. These materials are precipitated as sedimentary rocks, primarily in arid regions where rapid evaporation of ocean water occurs, and the salinity of the world's oceans appears to be constant, with an equilibrium occurring between processes that add and remove materials.

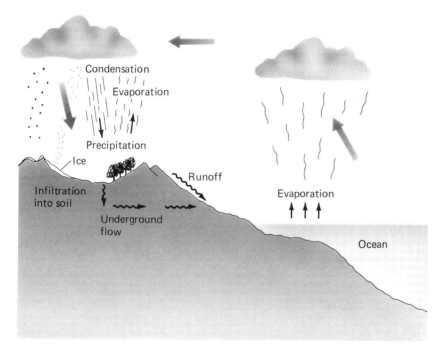

■ **Figure 14.8** The water cycle. Clouds form from water evaporated from the land and oceans and, in turn, precipitate rain and snow onto the surface. Water on the land returns to the oceans, flowing on the land in rivers and beneath the water table as ground water. (From Levin, Harold L. *Contemporary Physical Geology,* 3rd ed. Saunders College Publishing, Philadelphia, 1990.)

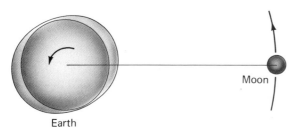

■ **Figure 14.9** Tides. The Moon raises two bulges in Earth's oceans. As the Earth rotates, shoreline areas experience high tide twice daily when they are under the bulge. The Moon's orbital motion causes the time of tides to vary daily. Because of friction, the bulges are not perfectly aligned with the Moon. (From Zeilik, Michael and Elske v.P. Smith, *Introductory Astronomy and Astrophysics*, 2nd ed. Saunders College Publishing, Philadelphia, 1987.)

Ocean water is in constant motion as a result of three processes. First, there are **tides,** the gradual rise and fall of water caused by the gravity of the Moon and, to a lesser degree, the Sun. As described in Chapter 7, the Moon's differential gravitational attraction causes the oceans to "bulge" slightly, with one bulge on the side of the Earth facing the Moon and the other on the opposite side. As the Earth rotates, a coastal area experiences high tide twice daily when it is under the two bulges of the ocean. Low tide occurs when the area is under the shallower parts of the ocean. Tides occur at different times each day because of the Moon's orbital motion (Figure 14.9). When the Sun and Moon are aligned in the sky at new or full moon, tides are more severe than normal and are called spring tides. When the Sun and Moon are at right angles to each other at first and last quarters, tides are milder than normal and are called neap tides. Tidal variation ranges from a few feet to a few tens of feet, with the higher values occurring only along coastlines that have peculiarities in their geology.

The second phenomenon of ocean water movement is the system of large-scale **ocean currents** caused by the Earth's global atmospheric circulation. As described in Chapter 11, prevailing winds over the Earth's surface tend to blow in constant directions. These wind patterns subsequently cause ocean currents to develop, often in the same general directions. These currents carry warm water from equatorial to polar regions and cool water in the opposite direction. A well-known example is the Gulf Stream in the Atlantic Ocean, which carries warm water northward along the east coast of the United States and then eastward toward northern Europe. The third source of movement in the oceans is local wind, which causes **waves.** Waves are oscillatory movements at the water surface and are highly variable; they are most active when storms with high winds occur. Wave activity is the only one of the three motions that shapes the geology of coastal areas.

The Earth's oceans are of extreme importance, particularly because they play a large role in the overall thermal balance of the Earth. The atmosphere is warmed primarily by heat transferred to it from the oceans by evaporation and condensation. The oceans also play an important role in the production of most sedimentary rocks and, for that reason, were probably responsible for removing much of the carbon dioxide that may have once been present in the atmosphere. Life is extremely abundant in the oceans, and many researchers believe that life on Earth may have originated in them.

■ Atmosphere

The Earth's atmosphere is about 78% nitrogen, 21% oxygen, 1% argon, and 0.03% carbon dioxide. Variable amounts of water vapor are also present, ranging from less than 1% to about 4%. The quantity known as specific humidity measures this percentage, whereas the more familiar relative humidity measures the amount of moisture present as a fraction of the atmosphere's total moisture capacity at a given temperature. The Earth's average surface temperature is about 288 K (59°F), and the average surface pressure is, by definition, 1 atmosphere. (Pressure is not constant at the Earth's surface but varies slightly as weather conditions change.)

The Earth's atmosphere has been greatly affected by the life on our planet. The presence of photosynthesizing plants has caused the Earth to be the only known solar system object with any free oxygen in its atmosphere, and it is this oxygen that supports life on our planet today. One of the most important environmental concerns of today is that tremendous quantities of pollutants are being dumped into the atmosphere. Although the atmosphere has absorbed great quantities of such materials while seemingly cleansing itself, problems may await us in the future if the trend is not reversed, particularly if a runaway greenhouse effect begins to occur. As noted in Chapter 11, carbon dioxide is the main cause of the greenhouse effect, and the increasing amounts of this gas being released into the atmosphere could eventually cause global warming.

Recall from Chapter 11 that meteorologists divide the atmosphere into four layers (Figure 11.2). The **troposphere** extends from the surface to a height of about 11 kilometers (7 miles), and, because it contains most of the atmosphere's moisture, it is where nearly all weather occurs. Above the troposphere is the **stratosphere,** which extends to about 48 kilometers (30 miles). It is poorly mixed vertically and is divided into separate layers, or stratified, hence its name. One of the important stratospheric layers is the **ozone** (O_3) **layer,** which blocks out much of the Sun's harmful ultraviolet radiation. Indications are that pollution has been causing deterioration of this layer, which, if continued, could have profound effects on living organisms. Above the stratosphere is the **mesosphere,** which extends to 80 kilometers (50 miles). The uppermost atmospheric layer is the **thermosphere,** which thins until the near vacuum of interplanetary space is reached. (Recall from Chapter 11 that the stratosphere and thermosphere are inversion layers, whereas the troposphere and mesosphere cool with increasing height.)

Most of the gases present in the Earth's thermosphere have been ionized as a result of impacts from incoming high-energy solar particles. For this reason, the thermosphere is sometimes called the **ionosphere.** That ions exist in the upper atmosphere was discovered early in the 20th century, when it was found that long-distance radio communication was possible because radio waves "bounce" from ionospheric layers. Ionospheric layers are variable, especially over the course of the day. (This is why distant AM radio stations can be received at night but not during the daytime.) The two primary ionospheric layers are the E region, between 90 and 120 kilometers (35 and 70 miles), and the F region, between 150 and 300 kilometers (90 and 180 miles).

■ Interior and Magnetic Field

We know more about the interior of the Earth than any other solar system object because we have drilled holes deep into its crust and have also accumulated an enormous amount of seismic data that reveal its inner structure. The Earth has four main interior layers, as illustrated in Figure 14.10. The outermost layer is the crust, which has quite different properties depending upon whether it is oceanic or continental material. Oceanic crust is 5 to 12 kilometers (3 to 7 miles) thick; less than 200 million years old, because it is constantly being created and destroyed by plate tectonic activity; and composed of basaltic rocks. Continental crust has an average thickness of 35 kilometers (22 miles), although some areas are much thicker or thinner; is much older; and is composed of many rocks, the most abundant of which is granite.

Below the crust is the mantle, which is about 2900 kilometers (1800 miles) thick. At least the upper portion of it is probably composed of the ultrabasic igneous rock peridotite, which contains the minerals olivine and pyroxene. Most of the mantle is solid, but the deeper parts of it, where convective motion occurs, are partially molten. This convective motion within the mantle is the ultimate cause of plate tectonics. Some mantle material occasionally breaks through into the crust, where it forms deposits called kimber-

■ **Figure 14.10** The Earth's interior layers. The numbers represent the fraction of the Earth's total radius. See text for detailed descriptions of each layer.

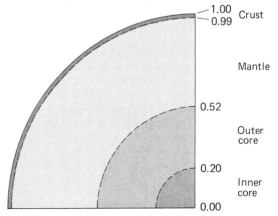

lites. Kimberlites contain diamond, a high-pressure mineral composed of carbon, which can form only at high pressures such as those found in the mantle. At the base of the mantle, the temperature is about 3000 K, the density is about 6 grams per cubic centimeter, and the pressure is nearly 1.4 million atmospheres.

The Earth's core is divided into two layers, both of which are believed to be composed of nickel–iron metal. Evidence for this metallic composition is the presence of the Earth's magnetic field and the abundance of such metal in meteorites, which presumably are fragments of asteroids representative of the smaller bodies that accreted to form the Earth. The outer core is about 2100 kilometers (1300 miles) thick and is molten. We know that it is molten because certain types of earthquake waves that travel only through solids do not penetrate the outer core. The Earth's rotational motion is believed to cause churning currents of the conducting material in this area, which in turn generate electrical currents that produce the magnetic field. The inner core has a radius of 1250 kilometers (775 miles) and is solid, because the great pressure here (about 3.6 million atmospheres) does not allow nickel–iron metal to melt, even though the temperature is about 7000 K, or slightly hotter than the Sun's surface. That the inner core is solid can be verified by study of types of earthquake waves that can penetrate the outer core and travel through the inner core. The density at the center of our planet is estimated to be about 13 grams per cubic centimeter.

The Earth's magnetic field is dipolar, and the magnetic axis is tilted only 11.4° to the rotational axis. This is the strongest line of evidence indicating that the field is the direct result of Earth's rotation. As mentioned in Chapter 9, Earth's magnetic field reverses polarity at irregular intervals, usually of several hundred thousand years. That these magnetic reversals occur can be determined by studying igneous rocks whose magnetic crystals were aligned with the magnetic field present at the time of the rock's solidification. Because the cause of the field itself is not completely understood, the reason for the field reversals is an even bigger mystery.

One of the benefits of the magnetic field is that it blocks the entry of harmful charged parti-cles into the atmosphere, except near the poles, where the magnetic field lines are perpendicular to the surface. Where these incoming particles interact with the ionized gases in the upper atmosphere, they produce the shimmering light of the aurora. Another effect of the field is that it traps charged particles in zones called the Van Allen belts. These two doughnut-shaped belts encircle the Earth above the equator (see Figure 9.6).

■ History

Geologists have been able to use the information contained in rocks to reconstruct the history of the Earth and its life. The basic concept used in reconstructing Earth's history is the **Principle of Uniformity,** which is the idea that the same geological processes we see happening today were operating in the past, probably at about the same rates as today. As we examine rock layers formed long ago, we can usually gain information about the conditions present at the time of their formation. Furthermore, examination of large-scale features, such as mountain ranges, that were formed in the past allows us to see what types of plate tectonic activity occurred long ago. Perhaps most important is the fact that ages of rocks can be determined by the use of radiometric age dating, as described in Chapter 8. Information contained in rocks also allows us to study the history of life on Earth, because many sedimentary rocks contain fossils, which are preserved remnants of ancient life. In most cases, only hard body parts such as bones and shells are fossilized. Not all organisms become fossilized after death, so our knowledge of ancient life is incomplete. Although a detailed summary of geological history is beyond the scope of this book, the following paragraphs present a brief summary of what we know about our planet's past.

The Earth and the rest of the solar system formed about 4.6 billion years ago. (We will examine the formation of the solar system in detail in Chapter 29.) It is thought that this happened as smaller rocky and metallic bodies accreted to form the Earth. The decay of radioactive elements partially melted Earth's interior, producing enough mobility to allow it to differentiate into layers. At this time, the volatile material within the

Earth gradually outgassed, cooled, and condensed, forming the oceans and the atmosphere.

The origin of life is one of science's greatest mysteries, and we will examine theories about it in more detail in Chapter 30. Although we study ancient life by examining fossils preserved in rock, the first organisms were tiny, lacked hard parts, and rarely formed fossils. Therefore, we have no direct evidence of the beginning of terrestrial life. Although there are some differences of opinion, many researchers believe that life may have developed spontaneously in the Earth's oceans from nonliving material.

Geologists divide Earth's history into four eras. The first era, the **Precambrian,** lasted from 4.6 billion years to 570 million years ago. During this era, continents and oceans formed, and plate tectonic activity began. Because there were greater amounts of radioactive elements present at that time, the Earth's interior was presumably warmer, with more convection in the mantle. For this reason, plate tectonic activity would have occurred at a faster rate than today. Even though the Precambrian era comprised over 90% of Earth's history, the geological activity that has occurred since then has erased much of the evidence of Precambrian events, and relatively little is known about them. The earliest fossils, of single-celled organisms called algae, are about 3.8 billion years old. During the Precambrian era, life existed only in the oceans, and although multicelled forms of plants and invertebrate animals developed, there were no organisms with skeletal hard parts. Because hard parts are generally required to form fossils, very few fossils of Precambrian life have been found. One result of the development of Precambrian life was that photosynthesizing organisms had produced an oxygen-rich atmosphere by about 2 billion years ago. (This date can be determined by noting the earliest existence of rocks containing oxidized iron.)

The **Paleozoic** era, which lasted from 570 to 245 million years ago, was marked by numerous continental collisions, and by its end, all of the Earth's continents had combined to form a "supercontinent" we now call Pangaea. This activity produced numerous mountain ranges at the points of collision. For example, the collision of North America and Europe produced the Appalachian Mountains in eastern North America.

During the Paleozoic era, as well as ever since, North America has been moving northward and rotating counterclockwise as a result of plate tectonic motion. The beginning of the Paleozoic era was marked by the rapid proliferation of invertebrate animal types, many of which for the first time had shells and other hard parts that protected them against predators and possibly harsh environmental conditions. The first land plants and animals appeared during the Paleozoic era, as did the first three groups of vertebrates (animals with backbones) — fish, amphibians, and reptiles.

The **Mesozoic** era lasted from 245 to 66 million years ago and was marked by continental separations, as Pangaea broke apart and the continents began to assume their modern appearances. Throughout this time, as North America continued moving northward and westward, the oceanic crust subducting under the west coast continued the development of the Rocky Mountains, a process that is ongoing even today. The final two groups of vertebrates, mammals and birds, first appeared during this era, but the reptiles, particularly the dinosaurs, were the dominant form of life and existed in a wide variety of environments. Dinosaurs were primarily large land reptiles, but other forms of reptiles developed flight, and still others lived in the ocean. Most of these reptiles, as well as numerous other forms of life, became extinct at the end of the Mesozoic era for reasons that are not understood, although global climate change is one possible mechanism. One of the more exciting suggestions of why so many organisms became extinct at this time is that the impact of an extraterrestrial object disrupted the Earth's environment for a time. Although there is evidence indicating that one or more meteorite impacts probably occurred at the end of the Mesozoic era, their role in causing these extinctions is still being hotly debated.

The **Cenozoic** era began 66 million years ago and continues today. Both separations and collisions of continents have occurred during this time. For example, Arabia has separated from Africa, forming the Red Sea, and India has collided with Asia, forming the Himalayan Mountains. One of the most interesting Cenozoic events was the ice age that occurred during the past two million years. During this ice age, almost one-third of the Earth's land was covered by thick, glacial ice. Ice ages are rather infrequent in Earth history, and their cause is not known. Plate tectonic motions may be responsible, because the changing continental locations that this phenomenon causes can alter the flow of ocean currents and, hence, ocean temperatures. Birds and especially mammals have been the dominant forms of life during the Cenozoic era, and humans have now become the first form of life to venture outward from our planet to explore the rest of the solar system directly.

■ Chapter Summary

The Earth orbits the Sun, a fact that can be proven by observing the phenomenon of aberration of starlight. The orbital motion produces trigonometric parallax, which allows stellar distances to be measured.

The Earth rotates on its axis, a fact that can be demonstrated using the Foucault pendulum. Rotation is responsible for the Earth's oblateness.

The Earth's surface consists of continents and basins containing oceans, which cover about 70% of our planet. Continents and oceans are both dynamic regions where many forms of geological activity occur. The Earth's atmosphere is unique in that the presence of free oxygen allows life to exist on our planet.

The Earth's interior contains four distinct layers: the crust, mantle, outer core, and inner core. Activity in the outer core is believed responsible for producing the Earth's magnetic field.

The Earth was formed 4.6 billion years ago, and life probably developed within 1 billion years after that time. Geologists divide the Earth's history into four eras (the Precambrian, Paleozoic, Mesozoic, and Cenozoic eras), and we have a good understanding of Earth history and life history events during each.

■ Chapter Vocabulary

aberration of starlight	inner core	outer core	solar day
Bessel, Friedrich	ionosphere	ozone layer	stratosphere
Bradley, James	leap year	Paleozoic era	thermosphere
Cenozoic era	mantle	Precambrian era	tides
continent	mesosphere	precession	trigonometric parallax
continental shelf	Mesozoic era	Principle of Unifor-	tropical year
crust	nutation	mity	troposphere
Foucault, J.B.L.	ocean basin	sidereal day	wave
Foucault pendulum	ocean current	sidereal year	

■ Review Questions

1. Summarize the proof of Earth's orbital motion.
2. How do we use Earth's orbital motion to find the distances of stars?
3. How do we deal with the fact that Earth's year is not an integral number of days?
4. Distinguish between a solar day and a sidereal day.
5. Describe a Foucault pendulum.
6. Summarize the motions of the Earth other than orbit and rotation.

7. Summarize the basic features of the Earth's continents.
8. Summarize the motions of Earth's ocean water.
9. Why is ocean water "salty"?
10. Why are Earth's oceans important?
11. What gases are found in Earth's atmo-

sphere? What are the four layers into which Earth's atmosphere is divided?
12. Describe the four interior layers of the Earth.
13. Describe the probable cause of Earth's magnetic field.
14. Summarize the events of the four main eras of Earth history.

■ For Further Reading

Carr, Michael H. "Earth." In Carr, Michael H., editor. *The Geology of the Terrestrial Planets,* NASA, Washington, D.C., 1984.

Cloud, Preston. *Oasis in Space: Earth History from the Beginning.* Norton, New York, 1988.

Lutgens, Frederick K. and Edward J. Tarbuck. *Essentials of Geology.* Merrill, Columbus, OH, 1989.

Moody, Richard. *Prehistoric World.* Chartwell Books, Secaucus, NJ, 1980.

Ozima, Minoru. *The Earth: Its Birth and Growth.* Cambridge University Press, Cambridge, England, 1981.

Short, Nicholas M. et al. *Mission to Earth: Landsat Views the World.* NASA, Washington, D.C., 1984.

The Moon

CHAPTER OBJECTIVES

After studying this chapter, you should be able to

1. Summarize important aspects of the Moon's orbital and rotational motion.
2. Describe the various lunar phases and how they are caused.
3. Explain how the various types of solar and lunar eclipses occur.
4. Summarize the history of lunar spacecraft exploration.
5. Describe the important surface features of the Moon and how they were formed.
6. Explain why the Moon lacks an atmosphere.
7. Summarize what is known about the Moon's interior.
8. Summarize the basic events that occurred during each of the five eras of the Moon's history.

Earthshine. During the Moon's crescent phase, the portion not illuminated by the Sun can be faintly seen, because it is illuminated by light reflected from the Earth. (Photograph by Kerry Hurd, courtesy of Dennis Milon.)

The Earth has only one natural satellite, the Moon, which is sometimes called **Luna.** Most satellites are relatively small compared to the planet around which they orbit, averaging only a few percent of the planet's diameter. The Earth – Moon system is unusual in that the Moon's diameter is about one-quarter that of the Earth, making the two objects closer in size than any other planet – satellite pair except Pluto and Charon.

The Moon is the closest celestial object to the Earth, orbiting at an average distance of 384,404 kilometers (238,330 miles). Because of its proximity, it is the second brightest object in the sky, exceeded only by the Sun. Our natural satellite has inspired awe and interest in humans from earliest times. Its changing shape and position, as well as vague surface markings, which are often interpreted as depicting a face or an animal, set it apart from other heavenly objects. Its proximity to Earth made it a natural target for early space exploration. Its nearness and the fact that it is the only other object yet visited by humans make it the extraterrestrial object about which we know most. Because it is less active geologically than the Earth, study of lunar rocks and surface formations has given us a great deal of insight into the early history of the solar system, because even very old lunar rocks have remained relatively unchanged since their formation.

■ Appearance, Basic Properties, and Phases

The Moon's distance from the Earth was first determined by Earth-based triangulation, whereby the Earth's rotation permits observation of the Moon from two separate points. After allowing for the Moon's orbital motion during the observations, the observer can calculate its distance, based on the Moon's change of position among the background stars (Figure 15.1). Ptolemy was

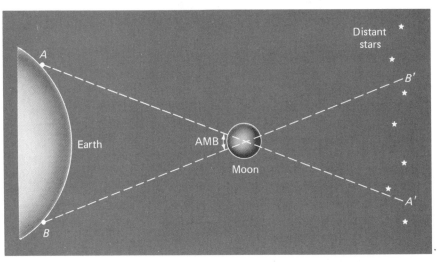

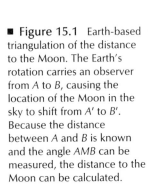

■ **Figure 15.1** Earth-based triangulation of the distance to the Moon. The Earth's rotation carries an observer from *A* to *B*, causing the location of the Moon in the sky to shift from *A'* to *B'*. Because the distance between *A* and *B* is known and the angle *AMB* can be measured, the distance to the Moon can be calculated.

apparently the first astronomer to make such a measurement, and he determined a value fairly close to the figure accepted today. Today, we can reflect radar from the Moon itself and laser light from reflectors left behind during the Apollo missions. Such measurements have revealed the distance with an accuracy of about a centimeter. The Moon's semi-major axis is 384,404 kilometers (238,330 miles). As a result of the Moon's eccentricity of 0.0549, its distance from Earth ranges between 356,334 kilometers (226,507 miles) and 406,610 kilometers (252,098 miles).

The Moon's diameter is 3476 kilometers (2160 miles), a figure that can easily be calculated from its angular diameter (about 31') and distance. The mass of the Moon, determined from Kepler's third law calculations, is 1 part in 81.3 times that of the Earth's, or 7.4×10^{22} kilograms (1.6×10^{23} pounds). The average density of the Moon is 3.33 grams per cubic centimeter, lower than any of the four terrestrial planets. This value is about the same as the estimated density of the Earth's mantle.

One of the most basic properties of the Moon, of course, is that it orbits the Earth once every 27.32 days. (Strictly speaking, it is not the Moon orbiting the Earth and the Earth orbiting the Sun, but the combined center of gravity of the Earth–Moon system orbiting the Sun. For our purposes, however, we can ignore this complication.) Because of the Moon's orbital motion, it slowly moves eastward (right to left, as seen in the Northern hemisphere) among the sky's background stars. This movement can be noticed by a careful naked-eye observer in only a few hours, and it is the cause of lunar occultations, which occur when the Moon covers a star while passing in front of it. Because the Moon is moving in such a way, its rising and setting times are later every day. The daily delay averages about 50 minutes, although the changing orientation of the Moon's orbital plane with respect to the horizon causes this delay to vary from less than half an hour to nearly one and one half hours.

Unlike most planetary satellites, the Moon does not orbit in its planet's equatorial plane. Instead, it orbits nearly in the ecliptic plane, with an orbital inclination of 5.15°. Various perturbations cause this value to vary between 4.95° and 5.33°. The fact that the Moon does not orbit in the Earth's equatorial plane may be a clue to its origin, indicating that it is a captured object. (We will discuss the origin of the Moon in detail in Chapter 29). The line connecting the nodes of the Moon's orbit (the two points where it crosses the ecliptic plane) is not fixed, but precesses around the sky with a period of about 18.6 years. This **regression of the nodes** is primarily the result of perturbations caused by the Sun, whose gravitational influence on the Moon is considerable.

The Moon's **phases**, or changes in apparent shape, are the most impressive result of the Moon's orbital motion and are visible to even a casual observer. Phases are caused by changes in the relative positions of the Moon with respect to the Sun as seen from the Earth. As shown in Figure 15.2, half of the Moon is illuminated at any time, whereas the remainder is dark. As the Moon orbits the Earth, we see varying amounts of the illuminated portion, causing the well-known phases, such as new moon, full moon, and first and last quarters. The Moon's orbital motion around the Earth causes the Moon to be visible at different times of the day and night during the various phases.

At **new moon,** the Moon is invisible, because it is aligned with the Sun and its unilluminated side faces Earth. During the first half of the phase cycle, the Moon appears to be getting larger, or **waxing.** After new moon, it gradually begins to show an illuminated crescent. One week after new moon comes **first quarter,** when one-quarter of the Moon can be seen (half of the side facing Earth). The first quarter moon is highest in the sky at sunset. During the next week, the Moon grows larger and is said to be **gibbous.** Finally, two weeks after new moon comes **full moon,** when all of the side facing Earth is illuminated. Because the Moon is now opposite the Sun, it rises at sunset and sets at sunrise. During the next two weeks, the Moon is said to be **wan-**

ing, as it passes through **last quarter** before reaching new moon again. The last quarter moon rises at midnight and is highest in the sky at sunrise, and the waning lunar phases are often seen after sunrise in the daytime sky.

If you observe the Moon with binoculars or a small telescope at any phase other than full moon, you will be able to see the most surface details near the **terminator,** which is the line where sunrise or sunset is occurring. This is because the shadows are longest in this particular area. If you observe the Moon when only a thin crescent is visible, you may notice a faint illumination on the otherwise dark portion. This phenomenon, called **earthshine,** is the result of sunlight reflected from the Earth illuminating the otherwise dark areas of the Moon. Just as the Moon goes through phase changes, an observer on the Moon would notice phase changes occurring on the Earth, and earthshine occurs because Earth is especially bright, being near "full Earth," while the Moon is near new.

The Moon's sidereal orbital period is 27.32 days (27 days, 7 hours, 43 minutes, and 11.5 seconds), but the phase cycle (or synodic period) requires 29.53 days (29 days, 12 hours, 44 minutes, and 2.8 seconds) to be completed. The syn-

■ **Figure 15.2** Lunar phases. (a) The Earth and Moon at eight positions as viewed from above the north pole. (b) The corresponding phases as they appear from Earth.

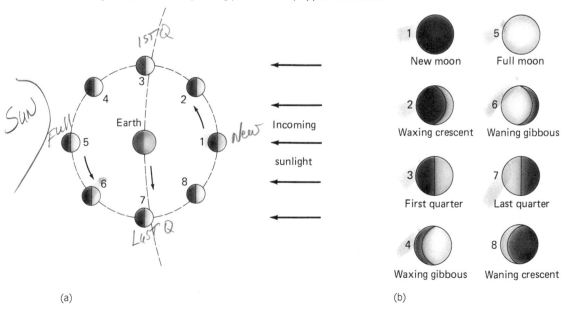

(a)

(b)

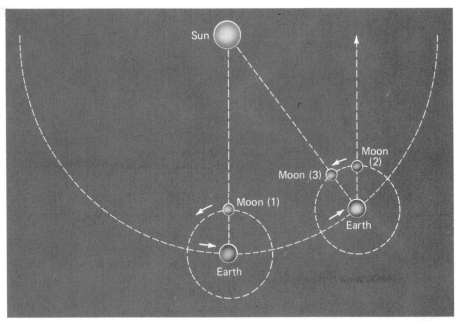

■ **Figure 15.3** Sidereal and synodic month. A sidereal month is the time required for the Moon to complete one orbit (from the point labeled *1* to the point labeled *2*). A synodic month is the time between successive new moons (from the point labeled *1* to the point labeled *3*).

odic period is longer than the sidereal period because the Earth's orbital motion around the Sun requires the Moon to travel a little "extra" between successive alignments with the Sun compared to completion of an orbit (Figure 15.3). The length of the month, of course, derives from this phase cycle, and many cultures throughout history have used lunar calendars based strictly on the Moon's phases. Lunar calendars, however, are limited in that the year does not contain an exact number of lunar months. In our calendar, we have 30- and 31-day months (except for February), which are longer than the Moon's synodic period. These months were devised so that we have a year containing exactly 12 months. The result of having 31-day months is that every few years, two full moons will occur during a month, usually on the 1st and 31st days. The second full moon in a month is called a "blue moon," and this is the source of the expression "once in a blue moon." (Some historians of astronomy believe that this expression originally derived from the very rare times when the Moon actually appears blue because of unusual terrestrial atmospheric conditions.)

No matter what lunar phase is occurring, the same side of the Moon always faces the Earth. Although one might guess that this means the Moon does not rotate, in actuality it is because the Moon is in perfect spin-orbit coupling, with its rotational and orbital periods both 27.32 days long (Figure 15.4). (The Moon's orbital and rotational motions are both prograde.) The Moon's rotational axis has an obliquity of 6.5° to the plane of its orbit, or 1.5° to the ecliptic plane. Even though people sometimes call the Moon's far side its "dark side," the name is inaccurate because it receives as much sunlight as the side facing us; any point on the Moon has 14.76 days of light, followed by 14.76 days of darkness. As a result of the Moon's spin-orbit coupling, astronomers had no knowledge of the appearance of the Moon's far side until the Soviet Luna 3 spacecraft flew behind the Moon in 1959.

As described in Chapter 7, the Earth's gravity is the cause of the Moon's 1:1 spin-orbit coupling, because tidal friction has greatly slowed its rotation from its original value. The Moon's gravity has also affected the Earth's rotation, as tidal friction has slowed the Earth's rotation over the

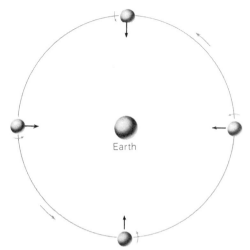

■ **Figure 15.4** Lunar spin-orbit coupling. As the Moon orbits the Earth, its rotation occurs at such a rate that one side of the Moon always faces the Earth. (From Abell, George, David Morrison, and Sidney Wolff. *Exploration of the Universe,* 5th ed. Saunders College Publishing, Philadelphia, 1987.)

course of geologic time, and the process continues today. This means that some of the Earth's rotational energy is being transferred to the Moon's orbital energy. How can we tell that the Earth's rotational rate has slowed? Evidence is obtained by examining daily and annual growth rings of fossil corals that lived during the Paleozoic era. These studies allow the number of days

in the year at that time to be counted. Because the absolute length of the year has not changed, such a determination allows the length of the day to be found. For example, there were over 400 days in the terrestrial year 400 million years ago, indicating a rotational period for the Earth of only about 22 hours. It is estimated that if the current trend continues, in about 50 billion years the Earth and Moon will both require a length of time equal to 47 current days to rotate, and the Moon's orbital period will also be that long. The Moon's distance from the Earth at that time will have increased to about 550,800 kilometers (340,000 miles). Theory indicates that the Sun will last only about 5 billion more years, so it is likely that no one will be around to witness this.

Although we can see only 50% of the Moon's surface at any time, a number of phenomena combine to cause the appearance of the Moon to change slightly with time as we occasionally "peek" around the edge of the Moon (Figure 15.5). As a result of this effect, called **libration,** terrestrial observers can see 59% of the Moon's total surface over a period of several years. The causes of libration include the variation in the Moon's orbital speed due to its eccentricity, the obliquity of the Moon's axis, small variations in the Moon's rotational period caused by its slight deviation from being perfectly spherical, and different vantage points for observers as the Earth rotates.

■ **Figure 15.5** The effects of lunar libration. Both photographs were taken at the same lunar phase but in different months. Libration caused the shift in the orientation of the Moon as seen from Earth. Note the change in location of features near the left edge of the Moon. (Lick Observatory Photographs.)

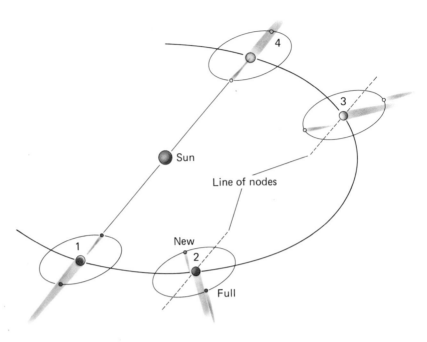

■ **Figure 15.6** The cause of eclipses. At points *1* and *4*, the line of nodes is aligned so that new and full phases occur when the Moon is on the ecliptic plane; therefore, eclipses occur. At points *2* and *3*, the Moon is north or south of the ecliptic at new and full phases, so no eclipses occur. (From Zeilik, Michael and Elske v.P. Smith. *Introductory Astronomy and Astrophysics,* 2nd ed. Saunders College Publishing, Philadelphia, 1987.)

■ Eclipses

One of the most spectacular astronomical events is an eclipse of the Sun or Moon. Eclipses occur whenever the Moon's shadow hits the Earth, or vice versa. Eclipses are fairly rare because of the 5° inclination of the Moon's orbit. Because of this inclination, the Moon's shadow generally misses the Earth at new moon, whereas the Earth's shadow generally misses the Moon at full moon. Twice each year, however, the nodes of the Moon's orbit are aligned so that new and full phases occur when the Moon is near the ecliptic (Figure 15.6). At these times when the Sun, Earth, and Moon are almost perfectly aligned, the Moon's shadow can touch the Earth and the Earth's shadow can touch the Moon, causing eclipses to occur. (The word ecliptic is derived from the word eclipse.) Eclipses do not occur on the same dates every year, because of the regression of the nodes mentioned earlier.

In order to understand eclipse phenomena, one must realize that an extended object like the Earth or Moon casts a two-part shadow when illuminated by an extended object like the Sun. The dark, inner shadow, called the **umbra**, is surrounded by an outer shadow, or **penumbra**, in which some sunlight is present (Figure 15.7).

One of the biggest coincidences of nature is that the Moon and the Sun have virtually the same apparent diameter in the sky. This allows the Moon to just cover the photosphere of the Sun, allowing the chromosphere and corona to be seen. When this happens, a total solar eclipse occurs (Figure 15.8). On the average, a given point on the Earth's surface will experience totality only once every 360 years. Because total solar eclipses are such spectacular sights, people travel far and wide to witness one. As the Earth rotates and the Moon orbits, the path of totality follows a long, narrow path on the Earth's surface, and totality lasts only a few minutes at a given point. (The maximum length of totality is 7 minutes and 31 seconds, but totality this long is rare.) Observers away from this track will experience a partial eclipse and will be shadowed only by the Moon's penumbral shadow, not the full, umbral one. Occasionally, when the Moon is near apogee (the most distant point in its orbit), a narrow ring of light will surround the dark Moon, yielding an annular eclipse. At least two, but no more than five, solar eclipses occur every year, of which zero to three may be total. Table 15.1 contains a list of upcoming solar eclipses visible from North America.

Lunar eclipses occur when the Moon passes into the Earth's shadow (Figure 15.9). Because of

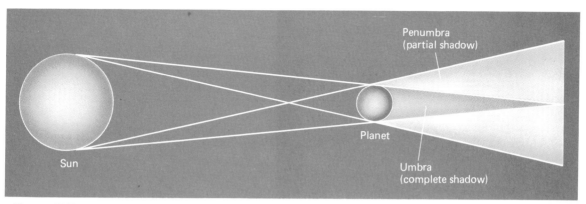

■ **Figure 15.7** Two-part shadows. An object such as the Earth or Moon casts a two-part shadow. The umbra *(inner shadow)* is a complete shadow that receives no sunlight, whereas the penumbra *(outer shadow)* is a partial shadow that receives some sunlight.

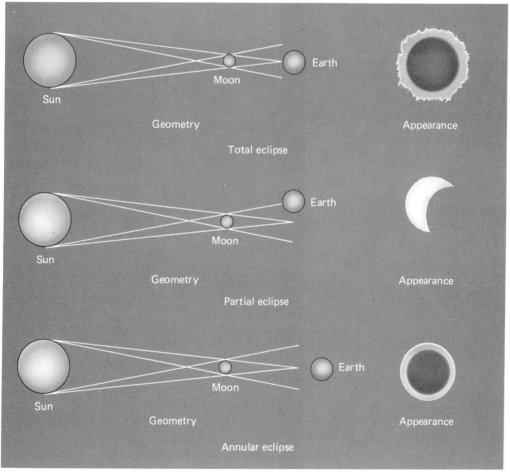

■ **Figure 15.8** Types of solar eclipses. See text for details.

214

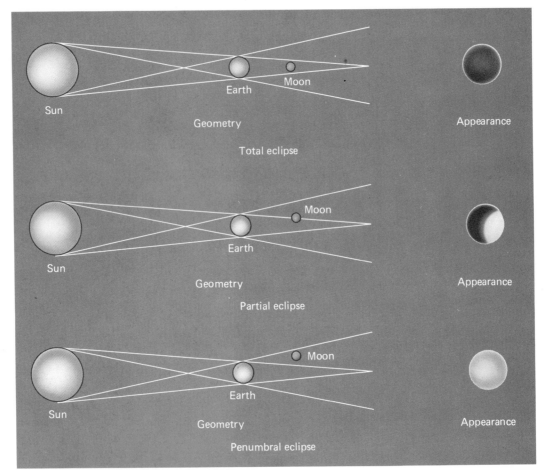

■ **Figure 15.9** Types of lunar eclipses. See text for details.

■ Table 15.1   Solar Eclipses Visible from North America During the 1990s

Date	Time of Maximum Eclipse	Type of Eclipse	Where Visible
July 11, 1991	2:04 P.M. EST	Total	CA, M
Jan 4, 1992	6:07 P.M. EST	Annular	w US
Dec 23, 1992	7:34 P.M. EST	Partial	A
May 21, 1993	9:09 A.M. EST	Partial	w NA
May 10, 1994	12:06 P.M. EST	Annular	US, e C
Oct 12, 1996	9:15 A.M. EST	Partial	n C
March 8, 1997	8:16 P.M. EST	Total*	A, nw C
Feb 26, 1998	12:26 P.M. EST	Total	CA
Aug 11, 1999	6:03 A.M. EST	Total*	n C
July 30, 2000	9:19 P.M. EST	Partial	A, w C
Dec 25, 2000	12:16 P.M. EST	Partial	s C, US, M

Abbreviations: A = Alaska, C = Canada, CA = Central America, M = Mexico, NA = North America, US = United States, e = eastern, n = northern, s = southern, w = western.
* Partial in North America.
Data calculated using the program *SunTracker*, by Charles Kluepfel, Zephyr Services, Pittsburgh, 1985.

the relative diameters of the Earth and Moon, the entire Moon can be covered by the Earth's umbra. This is called a total lunar eclipse. Lunar eclipses are easy to observe, because they are visible to everyone on Earth who can see the Moon. However, the totally eclipsed Moon does not disappear but turns an orange or coppery color, because it is illuminated by light refracted through the Earth's atmosphere. Sometimes, particularly when volcanic eruptions have placed considerable dust into the Earth's atmosphere, a totally eclipsed Moon will be so dark that it almost vanishes from view, but such instances are rare. If only part of the Moon enters the umbra, a partial umbral eclipse occurs, and if the Moon only enters the penumbra, a scarcely noticeable penumbral eclipse takes place. Counting penumbral eclipses, two to five lunar eclipses may occur in a given year; zero to three of these may be umbral. The maximum number of eclipses in a year, both solar and lunar, is seven. Table 15.2 contains a list of upcoming lunar eclipses visible from North America.

Just as total solar eclipses provide rare opportunities for the Sun's outer atmosphere to be seen and studied, lunar eclipses provide opportunities for conducting scientific research about the Moon. Prior to the space age, analysis of lunar light during an eclipse provided one of the few insights into the nature of the Moon's surface. The experimental procedure involved measuring the decrease in temperature of the lunar surface as direct sunlight was blocked. These experiments showed that the Moon, on average, cooled 150 K (270°F) in only 1 hour. This result indicated that the Moon's surface is made of materials that are incapable of storing large amounts of heat energy within them and, therefore, could not be solid rock, which retains heat and cools slowly. Later, the lunar landings revealed that the Moon's surface is covered with a layer of powdery regolith, and this powdery material is responsible for the surface's thermal properties.

■ Spacecraft Exploration

The first Earth satellite, the Soviet Sputnik 1, was launched October 4, 1957, marking the beginning of the space age. Once techniques had been

■ Table 15.2   Lunar Eclipses Visible from North America During the 1990s

Date	Time of Maximum Eclipse	Type of Eclipse
Jan 30, 1991	12:49 A.M. EST	Penumbral
June 26, 1991	10:11 P.M. EST	Penumbral
Oct 21, 1991	5:39 A.M. EST	Partial
June 14, 1992	11:54 P.M. EST	Partial
Dec 9, 1992	6:55 P.M. EST	Total
Nov 29, 1993	1:36 A.M. EST	Total
June 24, 1994	10:31 P.M. EST	Partial
Nov 18, 1994	1:48 A.M. EST	Penumbral
April 15, 1995	7:20 A.M. EST	Partial
April 3, 1996	7:13 P.M. EST	Total
Sept 26, 1996	9:50 P.M. EST	Total
Mar 23, 1997	11:38 P.M. EST	Partial
Mar 12, 1998	11:13 P.M. EST	Penumbral
Aug 7, 1998	9:21 P.M. EST	Penumbral
Sept 6, 1998	6:06 A.M. EST	Penumbral
Jan 31, 1999	11:04 A.M. EST	Penumbral
July 28, 1999	6:29 A.M. EST	Penumbral
Jan 20, 2000	11:36 P.M. EST	Total

Data calculated using the program *MoonTracker,* by Charles Kluepfel, Zephyr Services, Pittsburgh, 1985.

developed for launching vehicles into Earth orbit, attention turned to the Moon while the space age was still in its infancy. The United States launched several Pioneer spacecraft toward the Moon, beginning with Pioneer 1 in October, 1958, but none reached the lunar vicinity until Pioneer 4 passed within 60,430 kilometers (37,300 miles) of the Moon on March 4, 1959, before going into solar orbit. The first successful lunar mission, however, had already been accomplished by the Soviet Luna 1, which passed within 6040 kilometers (3730 miles) of the Moon in January, 1959. The Soviets, having taken an early lead in the "race to the Moon," were able to maintain it for nearly a decade.

On September 12, 1959, Luna 2 became the first probe to hit the Moon, after returning data indicating that the Moon lacked a magnetic field or radiation belt. On October 4, 1959, the Soviets launched Luna 3 into a greatly elongated Earth orbit that took it past the Moon at a time when much of the far side was illuminated by sunlight, and it photographed about 70% of the previously unseen far side. Its photographs were taken by film cameras, processed automatically, and scanned by on-board television cameras, which sent the images back to Earth. Although the images were of low quality, they provided the first look at the Moon's far side.

The United States' first program devoted exclusively to lunar exploration was Project **Ranger,** which consisted of nine flights from 1961 to 1965. The first six missions were failures, including several craft designed to hit the Moon that caused embarrassment to mission planners by missing it completely! Rangers 7, 8, and 9, however, were entirely successful, returning television pictures of the lunar surface before impacting into it. Their photographs were the first to reveal more details of the Moon's surface than those visible using large telescopes from Earth.

After their early lunar successes, both the Soviet Union and the United States moved to the next step in lunar exploration, soft landings. The first such attempts, conducted by the Soviets, were failures, and some researchers felt that the lunar surface might be covered by a dust layer thick enough to prevent such landings. However, Luna 9 successfully landed on February 3, 1966, followed by Luna 13 on December 24, 1966. The American **Surveyor** program attempted seven soft landings in 1966, 1967, and 1968; five of these were successful. Both the **Luna** and Surveyor landers returned photographs of the surface and performed analyses of surface samples using automated laboratories.

The early 1960s also saw the advent of manned spaceflight. As in other areas of space exploration at the time, the Soviet Union soon assumed what appeared to be an insurmountable lead in this endeavor. Nonetheless, in May, 1961, President John F. Kennedy established the goal of sending Americans to the Moon by the end of the decade. In part because the lunar landing would be a memorial to this popular president, who was assassinated in 1963, the United States spared no expense, and the first manned lunar landing was eventually accomplished on July 20, 1969.

Although the United States openly expressed its goal of manned lunar landings, Soviet goals were unannounced but believed to be the same. Both nations next concentrated on lunar spacecraft that would photograph the surface to help select suitable sites for manned landing at-

tempts. The Soviets sent four Zond spacecraft on looping trajectories around the Moon between 1965 and 1970. The last three of these spacecraft were designed to return their film to Earth, allowing much better image quality than would have been possible with televised images. Spacecraft were also placed into lunar orbit, starting with Luna 10, which was launched March 31, 1966. Three later Lunas also explored the Moon from orbit, as did five American **Lunar Orbiter** spacecraft, which photographed almost the entire lunar surface during 1966 and 1967.

The most exciting aspect of lunar exploration was, of course, the **Apollo** manned lunar landing program. Six successful lunar landings were accomplished between 1969 and 1972. (During two other missions, including the first successful manned lunar orbital mission, Apollo 8, in December, 1968, the spacecraft only orbited the Moon.) During each manned lunar landing mission, two astronauts descended to the lunar surface in the Lunar Module craft, whereas a third remained in orbit around the Moon in the Command/Service Module (Figure 15.10). Although most astronauts had been trained as test pilots, they were given enough geologic training so that they could do useful analysis of the lunar surface and choose the right samples to collect. (The final lunar landing mission included a geologist who had become a "scientist–astronaut.") A total of almost 400 kilograms (850 pounds) of lunar rocks was returned by the six lunar landing missions. Although the two astronauts who actually landed on the Moon overshadowed the one who remained in lunar orbit, a considerable amount of valuable information about the Moon was obtained from orbit, because much more of the Moon than the actual landing site could be photographed and analyzed from the orbiting Command/Service Module.

The first Apollo missions landed in smooth mare regions, where the chances for safe landings were considered greater, but later missions succeeded in landing in more rugged (and geologically diverse) terrain. (The locations of the Apollo landings are shown in Figure 15.11, and the astronauts and launch dates of these missions are listed in Table 15.3.) The first mission, Apollo 11, landed in Mare Tranquillitatis, and Apollo 12 landed in Oceanus Procellarum. After Apollo 13 was forced to abort its mission because of an on-

(a)

(b)

■ **Figure 15.10**   The vehicles used to explore the Moon during Project Apollo. (a) The Command/Service Module remained in lunar orbit. (b) The Lunar Module, shown here during a test in Earth orbit, landed on the Moon. (NASA Photographs.)

board explosion while on the way to the Moon, Apollo 14 landed in the Fra Mauro crater region, which was covered by ejecta from the impact that formed the Mare Imbrium basin. As a result, the age of the Imbrium feature could be determined.

Apollo 15 landed at the edge of Mare Imbrium, near Hadley Rille and the Apennine Mountains, and Apollo 16 landed in the highlands for the first time, in the Descartes region. Apollo 17 landed in the Taurus–Littrow region, near the edge of Mare

■ Table 15.3   Apollo Lunar Landing Missions

	Mission					
	11	12	14	15	16	17
Launch date	07–16–69	11–14–69	01–31–71	07–26–71	04–16–72	12–07–72
Astronauts						
On Moon	Armstrong	Conrad	Shepard	Scott	Young	Cernan
	Aldrin	Bean	Mitchell	Irwin	Duke	Schmitt
In orbit	Collins	Gordon	Roosa	Worden	Mattingley	Evans
Hours on Moon	21:36	31:31	33:30	66:54	71:14	74:59
Lunar walks						
Number	1	2	2	3	3	3
Distance (km)	0.25	2	3.3	27.9	27.0	35.0
Samples (kg)	21.7	34.4	42.9	76.8	94.7	110.5
Rock types	MB	MB	Br	Br, A	HB, Br	MB, Br, D
Oldest rocks*	3.72	3.37	3.96	4.09	3.92	4.48
Youngest rocks	3.48	3.15	3.85	3.28	3.84	3.77

Abbreviations: A = anorthite, Br = breccia, D = dunite, HB = highland basalts, MB = mare basalts.
* All rock ages in billions of years.

Data from Seeds, Michael A. *Foundations of Astronomy.* Wadsworth Publishing Co., Belmont, CA, 1984, and Baker, David. *The History of Manned Space Flight.* Crown Publishers, New York, Copyright 1981 by New Cavendish Books.

(a)

(b)

■ **Figure 15.11** Photographs and maps of the lunar surface. (a) The near side. Spacecraft landing sites and prominent surface features are labeled. (Lick Observatory Photographs. Maps from Abell, George, David Morrison, and Sidney Wolff. *Exploration of the Universe,* 5th ed. Saunders College Publishing, Philadelphia, 1987.) (b) The far side. (NASA Photograph.)

Serenitatis. Prior to the mission, many geologists had considered Taurus–Littrow to be a volcanic region. However, volcanic evidence was lacking, although the mission did succeed in recovering the oldest lunar rocks found.

Following the success of Apollo, the Soviets claimed that they had never intended a manned lunar landing, and they subsequently concentrated on unmanned lunar flights. (Spacecraft built for a manned lunar landing mission, but never used, were first shown by the Soviets to American visitors in late 1989.) In September, 1970, Luna 16 conducted the first automatic return of lunar samples to the Earth for study and analysis. Lunas 20 and 24 also returned lunar samples, in February, 1972 and August, 1976, respectively. Although the quantity of material that these craft returned was fairly small, they sampled areas not visited by Apollo. The most sophisticated Soviet lunar missions involved Lunokhod unmanned rover vehicles, which traversed considerable distances over the lunar surface after landing. Two such vehicles were placed on the lunar surface, Lunokhod 1 aboard Luna 17 in November, 1970 and Lunokhod 2 aboard Luna 21 in January, 1973. No lunar exploration has been accomplished since that time. Although several nations have proposed future lunar missions, no specific plans for any of them have yet been announced.

■ Large-Scale Surface Features

Even naked-eye observations of the Moon reveal that it does not have a uniform surface. The patterns of light and dark material on the lunar surface produce the "Man in the Moon" that is visible to those with vivid imaginations. The telescopic observations of Galileo began revealing the nature of the lunar surface. Galileo saw dark, smooth areas that he thought were bodies of water. These are now called **maria,** the plural of the Latin word mare, meaning "sea." (Even as late as the 1950s, some scientists believed that the rocks in maria were sedimentary, formed long ago when the Moon had oceans.) Today, of course, we know that the Moon currently lacks water and has probably never had any at its surface. Instead, as described in Chapter 10, maria are areas where lava flows have filled large impact basins. The remainder of the lunar surface, which is lighter in color and much more rugged, consists of areas called **highlands.** Highland regions have albedos of 9% to 12%, whereas the darker maria have albedos of only 5% to 8%.

As a result of the numerous lunar space missions, we know more about the surface of the Moon than any solar system object other than the Earth. Because the Moon is an airless body, its surface geology is totally different from that of our planet. One of the most significant differences is that very few changes are currently taking place, and the lunar surface we see is very old. Most of the features we will discuss were formed more than 1 billion years ago, making them much older than nearly all of the Earth's features.

Overall, about 17% of the Moon's surface is covered by maria (see Figure 15.11). Maria tend to have circular shapes, which provide an important clue to their origin: soon after the Moon formed, it experienced an era of intense bombardment by asteroid-sized bodies leftover from the accretion of the planets. These impacts produced huge, circular impact basins hundreds of kilometers in diameter and weakened the crust, allowing magma from the lunar interior to reach the surface and flood the basins. The lava fill was not instantaneous, as time was required for the magma to reach the surface. Over 1000 fissures where magma reached the surface have been identified in and around maria.

Spacecraft observations have revealed that both maria and highlands exist on the lunar far side. However, whereas the near side has a fairly high proportion of maria (about 30%), the far side is predominantly highlands (only 2% maria). The reason for the difference seems to be a variation in lunar crustal thickness between the two sides. The scarcity of far-side maria was the result of the far side having a thicker crust, which made it less likely for magma to reach the surface and flood impact basins than on the near side, whose crust is thinner. Although most maria are found in the lower portions of impact basins, some occur within highlands and on the floors of craters. It is estimated that mare rocks may be as thick as 4 kilometers (2.5 miles) in some places.

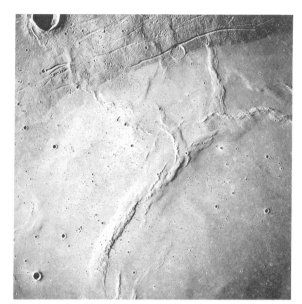

■ **Figure 15.12** The surface of Mare Imbrium. Note the ripples from lava flow, as well as the relatively small number of craters. (NASA Photograph, courtesy of the National Space Science Data Center and the Apollo 17 Principal Investigator, Frederick J. Doyle.)

Age dating of Apollo lunar samples indicates that the mare rocks, which are igneous rocks called **basalt**, formed when lava erupted during an interval from about 3.9 to 3.1 billion years ago. Ripples from the flowing lava are preserved in many maria, and relatively few craters are found in them, indicating that impact rates for the past 3 billion years have been fairly low (Figure 15.12). Basalt is a dark, fine-grained rock, and about 20 specific types of lunar basalts have been characterized based on composition. Most of the basalt types fall into one of three main categories (high-titanium, low-titanium, and high-aluminum), and indications are that these three rocks were produced consecutively as magma composition changed during the mare-formation interval (Figure 15.13).

The lunar highlands represent the original lunar crust, which formed when the Moon differentiated into layers. Being very old, it has been intensely bombarded and is heavily cratered. As a result, its geological history is difficult to interpret. The composition of the lunar crust apparently does not vary with depth, but the upper third of the crust differs from the rest in that it is highly fractured. The highland crust is 62 kilometers (38

miles) thick on the near side and 84 kilometers (52 miles) thick on the far side, for an average thickness of 71 kilometers (44 miles). Why is the crust of the near side thinner than that of the far side? One suggestion is that a huge impact, possibly the one responsible for forming the immense Oceanus Procellarum mare, removed a substantial amount of crustal rock.

The highland crust has an average density of 2.93 grams per cubic centimeter and is composed primarily of rocks containing the minerals plagioclase feldspar, low-calcium pyroxene, and olivine. Depending upon which of the three minerals is most abundant in the highland rocks, the rocks are called anorthosite (plagioclase feldspar most abundant), norite (pyroxene most abundant), and troctolite (olivine most abundant). **Anorthosite,** which is relatively rare on Earth, is the most common of the three rocks on the Moon, and its light color accounts for the high albedo of the highlands. Another material found in highland regions is KREEP breccia, named for its abundant potassium (chemical symbol K), rare-earth elements, and phosphorous. Most highland rocks occur as **breccias,** which are large rocks that contain irregular fragments of smaller rocks of one or more different varieties. The reason for the brecciation is the intense bombardment to which the highland areas have been subjected. Other highlands rocks are of a variety called **impact melt rocks,** which are igneous rocks formed from magma derived not from inside the Moon but from surface melted when impacts occurred. Examples of highland rocks are shown in Figure 15.14.

■ **Figure 15.13** A sample of mare basalt, collected during the Apollo 15 mission. (NASA Photograph.)

(a)

■ **Figure 15.14** Samples of highland rocks. (a) A rock containing anorthosite. (b) A breccia. (NASA Photographs.)

(b)

Occurring on both maria and highlands, but especially on the highlands, craters are one of the best known of all lunar surface features. The cratering process and the major features of craters were discussed in Chapter 10 and illustrated in Figures 10.11 and 10.15. Most highland areas are virtually saturated with craters, as old craters are cut by newer ones, and there are almost no flat, smooth areas (Figure 15.15). Although fewer craters exist in the maria, many large, conspicu-

■ **Figure 15.15** A region of the lunar highlands. Note the large number of craters, including many that are overlapping. (Lick Observatory Photograph.)

ous ones were formed after the mare lavas solidified (Figure 15.16). The rays associated with young lunar craters are composed of light-colored ejecta that splashed away in long arcs when the impact occurred. Rays eventually darken and match the rest of the lunar surface as a result of the activity of incoming solar wind particles and the mixing of the ray material with the underlying regolith because of continuous small impacts. For this reason, only craters younger than about 1 billion years have rays. One of the Moon's youngest craters, Tycho, has an impressive ray system. (Lunar craters are named for famous astronomers, other scientists, philosophers, and writers. Unfortunately for modern astronomers, the only unnamed craters left are small and insignificant!) On June 18, 1178, European monks saw a bright flash of light in the dark portion of the Moon, and some researchers today believe that they witnessed a meteorite impact. The young-appearing crater Giordano Bruno, which has one of the Moon's most impressive ray systems, is located approximately where the flash was seen, but there is no general agreement that it is that young.

Lunar impact structures larger than about 220 kilometers (135 miles) in diameter are called **basins** rather than craters. Lunar basins include some of the Moon's oldest features, and many formed prior to the period of mare flooding. Many basins are of the multiringed variety, similar to the Caloris basin on Mercury described in Chapter 12. Well-known examples are the Mare Imbrium basin on the near side, which has at least

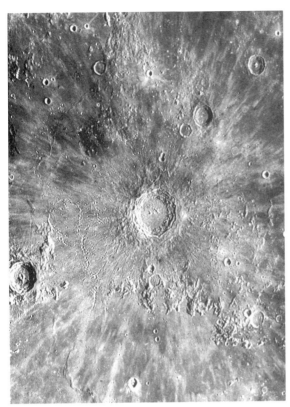

■ **Figure 15.16** The crater Copernicus, which is located in a mare region. Note its conspicuous rays. (Lick Observatory Photograph.)

three rings, and the Mare Orientale basin, which is located near the edge of the Moon's near side and thus poorly seen from Earth (see Figure 10.18). The Orientale basin is the youngest basin on the Moon and has at least four rings. It was only partially flooded by lava, so its structure is readily visible.

All of the mountain ranges on the Moon are parts of the rings of large impact basins, and none have been formed by plate tectonic activity as on Earth. In a surprising lack of originality, astronomers named these mountains for terrestrial ranges such as the Apennines and Pyrenees! Several isolated mountain peaks exist, some of which rise high above the otherwise smooth surface of a mare. Heights of these and other lunar features are calculated by triangulation from shadow lengths, because the Sun's elevation angle above the horizon as it illuminates each feature is known. Some peaks in Mare Imbrium are over 8 kilometers (5 miles) high (Figure 15.17). In addi-

tion to producing some spectacular mountains, the basin impacts also weakened the crust enough to allow magma emplacement and deposited thick and extensive blankets of ejecta over the older lunar highland surface.

A number of other surface features exist on the Moon. **Sinuous** (curved) **rilles** are apparently ancient lava channels or tubes. Most of them occur around the outer edges of maria and are probably relics of the actual basin-filling episode. The Apollo 15 spacecraft explored such a rille near the edge of Mare Imbrium (see Figure 10.10). Several **linear** (straight) **rilles** also exist on the Moon, with the best known being Straight Wall in Mare Nubium (Figure 15.18). Linear rilles are believed to be tectonic features, and Straight Wall is probably a 105-kilometer-long (65-mile-long), 300-meter-high (1000-foot-high) fault scarp. Alpine Valley is a 10-kilometer-wide (6-

■ **Figure 15.17** Photographs of Mare Imbrium showing some of the impressive mountain peaks that reach far above the lunar surface. (Lick Observatory Photograph.)

■ Figure 15.18 Lunar Orbiter IV photograph of Mare Nubium. Straight Wall, which is apparently a fault scarp, can be seen at the bottom of the picture, just left of center. The small white markings near Straight Wall were caused by processing defects in the spacecraft. (NASA Photograph, courtesy of the National Space Science Data Center and the Lunar Orbiter IV Principal Investigator, Leon J. Kosofsky.)

some of which have small depressions at the top (Figure 15.21). Cinder cones occur in the Marius Hills region, and small shield volcanoes occur in the Orientale basin region. Unfortunately, the ages of these features cannot yet be determined with certainty.

Some terrestrial observers have occasionally seen what appear to be clouds of gas partially obscuring lunar surface features. These **transient phenomena**, as they are called, seem to occur only at specific places, and it has been suggested that they are some form of venting of volcanic material. On November 3, 1958 and October 23, 1959, the Soviet astronomer Nikolai A. Kozyrev (1908–1983) obtained spectra of gas that apparently had emanated from the crater Alphonsus. Analysis indicated that the gas may have been the C_2 molecule. No evidence of lunar transient phenomena was obtained during the era of intensive exploration during the 1960s, and many researchers today tend to discount the earlier observations.

■ Figure 15.19 Alpine Valley. This 150-kilometer-long (90-mile-long) feature, which cuts through the Alps Mountains, is probably a graben, the bottom of which was flooded by mare lava. (NASA Photograph, courtesy of the National Space Science Data Center and the Lunar Orbiter V Principal Investigator, Leon J. Kosofsky.)

mile-wide), 150-kilometer-long (95-mile-long) graben (down-dropped, fault-bounded block) that cuts through the Alps Mountains (Figure 15.19). Other evidence for lunar faulting is the fact that some lunar craters have a hexagonal shape, probably as a result of joints and faults present in the area when meteorite impacts formed them (Figure 15.20).

One of the biggest unresolved questions about the Moon is whether volcanic activity still occurs there. Until the Apollo missions settled the question, some believed that lunar craters were formed by volcanic activity, and others believed that they were formed by the impact of meteorites. There appear to be some small-scale lunar volcanic features, particularly small **lunar domes**,

■ **Figure 15.20** The crater Alphonsus, at the center of the picture, has a somewhat hexagonal shape, which is probably a result of rock structures present prior to the time of impact. The dark markings inside the crater may be associated with transient phenomena that have been seen within it. The crater is 117 kilometers (73 miles) across. (NASA Photograph, courtesy of the National Space Science Data Center and the Apollo 16 Principal Investigator, Frederick J. Doyle.)

The Lunar Surface

The 12 men who walked upon the Moon's surface have seen a landscape unlike any that other human beings have ever experienced (Figure 15.22). The most noticeable difference between the Earth's surface and the Moon's is that the Moon's sky is pitch black even in daylight, the result of its having no atmosphere to scatter sunlight. The lunar surface is so bright, however, that the astronauts were able to see very few stars. Another striking feature of the lunar surface is its barrenness: it is completely devoid of life and very drab in color. Nearly all lunar rocks are gray or tan, with only a few exceptions to relieve the monotony. In addition, the surface is covered almost everywhere by a powdery coating of regolith. Footprints left in the lunar regolith by the Apollo astronauts will survive for centuries (Figure 15.23). This is because there are no surface processes to erase them, other than the occa-

sional impacts of small meteorites and the influx of solar wind particles, whose erosional effect is extremely small.

Atmosphere

The lack of a significant lunar atmosphere was demonstrated centuries ago when astronomers first observed occultations of stars by the Moon. When the Moon passes in front of a star, the star instantly vanishes from view. If the Moon had an atmosphere, it would cause a gradual dimming of the light. The reasons for the lack of a lunar atmosphere are its small mass and its high temperatures. The temperature of the lunar surface varies widely, from 370 K (207°F) at lunar noon to 120 K (−243°F) at lunar midnight. Although the length of the lunar day and night is partly responsible, the lunar temperature extremes are mainly due to the lack of an atmosphere to distribute thermal energy and moderate conditions. Because of the

■ **Figure 15.21** The Marius Hills region of the Moon. In addition to the prominent sinuous rille, this photograph shows several lunar domes that are possibly volcanic features. (NASA Photograph, courtesy of the National Space Science Data Center and the Lunar Orbiter V Principal Investigator, Leon J. Kosofsky.)

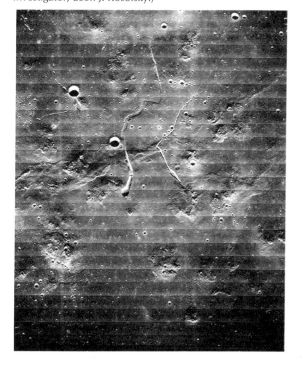

■ **Figure 15.22** A photograph of the lunar landscape, showing astronaut Eugene Cernan during the Apollo 17 mission. (NASA Photograph.)

high temperatures of the lunar day, even the heaviest gases are able to escape from it into space. (The Apollo manned lunar landings were conducted during daylight, shortly after sunrise, when temperatures were still relatively low. For this reason, artificial lighting was not needed, although air conditioning was required to keep the astronauts' spacesuits comfortable.)

■ **Figure 15.23** An astronaut's footprint in the lunar regolith, made during an Apollo mission. (NASA Photograph.)

During the Apollo missions, any possible lunar atmosphere was studied closely. The Moon was found to have an extremely tenuous atmosphere that is composed of components from the solar wind (hydrogen, helium, and neon), material outgassing from the spacecraft, and argon-40 from the decay of potassium-40 in lunar rocks. This atmosphere is almost equal to a vacuum by terrestrial standards, and it has essentially no effect on the Moon. The atmosphere is transient, with new material added to replace that which escapes into space.

■ Interior and Magnetic Field

Although the interior structure of the Moon is not yet fully understood, a great deal of progress has been made toward understanding it. The first evidence was obtained in the late 1960s from the five Lunar Orbiter spacecraft designed to photograph the lunar surface. Careful tracking of the orbits of these satellites was used to study variation in the Moon's gravity, which in turn gave information about the lunar interior. Gravity anomalies, or areas where the lunar gravity is slightly stronger or weaker than average, cause spacecraft to deviate from an ideal orbit. The presence of such an anomaly indicates that an area on the Moon lacks complete isostatic adjustment. The Lunar Orbiters found no major anomalies under the lunar highlands, indicating that they are isostatically compensated. However, deviations of the orbital paths occurred over many maria, indicating that mass concentrations, or **mascons**, were present under them. As an orbiter flew over these maria, the additional mass pulled it slightly closer to the Moon. At first, mascons were believed to be the buried remnants of impacting bodies, but it was later realized that these objects were totally destroyed when they impacted. Mascons are now considered to be the result of a lack of isostatic adjustment under the maria. When a mare impact occurred, it produced a basin whose lunar crust was thinner than average. At the time when the majority of these impacts happened, isostatic adjustments were still occurring on the Moon, and the mantle rose to compensate for the thinner crust under the large mare basins. Subsequently, however, the lunar crust and mantle stiffened enough to prevent further isostatic motion. Therefore, the mare lava flows that later filled the impact basins did not cause the crust to sink under their additional weight, producing the gravity anomalies responsible for the mascons.

The six Apollo missions left scientific equipment packages on the lunar surface. Their network of seismographs remained operational for almost a decade and sent back much information about "moonquakes" and the lunar interior. During the Apollo program, spent rocket stages and used spacecraft were intentionally crashed into the Moon to provide seismic disturbances of known energy. One natural impact on the far side was also recorded, and the seismic record from this event indicated that the moonquake waves traveled through a molten area within the Moon.

Moonquakes are all very weak and are comparable to the smallest terrestrial earthquakes. About 3000 were recorded annually, with a total annual energy release of 2×10^{13} ergs. (This is about the same amount of energy released by the explosion of 1 pound of dynamite.) By comparison, the total annual energy released by earthquakes is about 10^{24} ergs. Thirty-seven separate moonquake source sites were identified, mostly at depths of about 1000 kilometers (600 miles), although a few were less than 100 kilometers (60 miles) deep. Compared to the Earth, the Moon is a relatively inert body.

The structural details of the lunar interior are not fully known, and more than one model could fit the available data. One possibility is that a mantle of undetermined thickness that is composed of olivine lies below the crust. Deep in this mantle may be a zone of partial melting, and the Moon may have a metal core in its center, although its existence is far from certain. Based on measurements of the Moon's moment of inertia, it has been calculated that the Moon could have a metallic core with a radius of up to 450 kilometers (280 miles) or a somewhat larger core of iron oxide. One curious fact about the lunar interior is that the Moon's center of mass is offset from the center of the sphere. This offset, about 2 kilometers (1.2 miles) in the direction toward Earth, is probably a result of the Earth's gravitation and is related to the fact that the lunar crust is thinner on the near side. This gives the near side a greater average density, and the Earth has attracted this side during the development of the lunar spin-orbit coupling.

Although the Moon has no magnetic field today, some of its rocks contain remanent magnetism, indicating that a field was present when they formed. It is considered likely that a field existed until about 3 billion years ago, but the nature of the field, as well as why it disappeared, are unanswered questions. Part of the answer may be the fact that the Moon's core is now probably solid, which may not have always been the case.

■ History

Just as geologists have divided the Earth's history into four distinct eras, lunar researchers have divided the Moon's history into five. Each era witnessed different lunar geological processes and has associated with it specific units of rocks. Initially, lunar dating was relative, with events placed into sequence, but radiometric age dating of Apollo samples allowed an absolute age chronology to be established. Absolute age values for areas of the lunar surface have been instrumental in calibrating the crater counting system used for estimating ages of other solar system bodies with cratered surfaces.

The first era, the **Pre-Nectarian,** lasted from the origin of the Moon, about 4.6 billion years ago, until about 3.9 billion years ago. The first event during this era was the formation of the Moon as well as the rest of the solar system. Although we will discuss the origin of the Moon in detail in Chapter 29, most current theories suggest that the Moon's core was possibly captured by the Earth or broken from the Earth by impact, with the accretion process subsequently causing the Moon to grow to its present size. The next major Pre-Nectarian event was the differentiation of the Moon and the formation of its anorthosite crust, a process that probably required only a few hundred million years. It is considered likely that there was sufficient heat energy released from early impacts on the Moon that it was covered with a magma ocean. The mineral plagioclase crystallized in this magma, floated to the top, and formed the anorthosite crust, which after solidifying was marked by additional impacts. Olivine and pyroxene crystallized in the magma and sank, where they later provided the source material for the magma that flooded the mare basins. Approximately 30 impact structures large enough to be considered basins were formed during the Pre-Nectarian era.

The **Nectarian** era lasted from 3.9 billion years to 3.85 billion years ago. It is named for the impact that produced the Nectaris basin at the beginning of this time. This impact scattered ejecta that constitutes the Janssen formation. The Janssen formation and other impact-related strata are important because they can be identified over large portions of the Moon and used for determining the ages of other rock layers. About 10 other impact basins were formed during this time; those on the near side are filled with maria, whereas those on the far side generally are not.

The **Imbrian** era, which lasted from 3.85 billion years to 3.2 billion years ago, marked the end of the time of major impacts. This period began with the formation of the Imbrian basin, which covered much of the near side with an ejecta layer called the Fra Mauro formation. The formation of the Orientale basin was one of the last major impact events on the Moon. The Imbrian era also saw the beginning of the time when maria were formed as erupting lava filled many impact basins. This lava eruption began about 3.8 billion years ago, and the oldest mare basalts found during the Apollo missions are about 3.7 billion years old. Approximately two-thirds of the mare lava was extruded during the Imbrian era. During this time, the number of bodies hitting the Moon began to decrease.

The **Eratosthenian** era lasted from about 3.2 billion years to 1.0 billion years ago. During this time, volcanic activity continued, and the remainder of the mare rocks were formed. Some craters, but not large basins, continued to be formed during this time. The era is named for one of the more conspicuous crater examples. Craters formed during the Eratosthenian era lack light-colored rays but have well-preserved secondary craters.

The era of time encompassing the past billion years of lunar history is called the **Copernican** era, after an impressive crater formed during this time. Copernican craters can be identified because of their bright, conspicuous rays. The Moon has probably been volcanically dead throughout this time, so the only ongoing geological changes are those caused by impacts. This is certainly in great contrast to the Earth, where the majority of the geological features are much less than 1 billion years old. Although plate tectonic activity should continue to change the face of the Earth in the future, indications are that the Moon will probably not experience much internal geological change in the future.

■ Prospects for Future Exploration

Although lunar exploration has not occurred for over a decade, the proximity of the Moon will undoubtedly lead to renewed exploration in the future. Many people have proposed that a scientific outpost on the Moon would be a logical step for the manned space programs of the world, and it is reasonable to assume that such a lunar base might become a reality within several decades. In 1989, American space agencies began making preliminary plans for a possible 21st-century lunar base. Whether these plans become reality depends largely on congressional funding. There have also been proposals that many resources, including iron, aluminum, titanium, and other metals, could be mined on the Moon and processed there or in space. It will probably be well into the 21st century before such ambitious efforts are undertaken.

■ Chapter Summary

The Moon is the Earth's only natural satellite, and it is about one-quarter the diameter of our planet. The Moon orbits the Earth in just under one month, displaying its familiar phase cycle during that time. The Moon always presents the same side toward the Earth because of its 1:1 spin-orbit coupling. Periodic eclipses of the Sun and Moon produce some of the most spectacular events visible in the sky.

Many space missions have explored the Moon, and 12 people have walked on its surface. Because of this exploration, the Moon's surface geology is well understood. The lunar surface contains highland regions and basaltic maria; most of the maria occur on the near side. The Moon is heavily cratered as a result of numerous impact events. The Moon has no atmosphere, and although we are uncertain about the details of its internal structure, a solid metallic core may be present.

Researchers divide lunar history into five geological eras, during which specific events occurred.

■ Chapter Vocabulary

anorthosite	full moon	lunar dome	Ranger
Apollo	gibbous	Lunar Orbiter	regression of nodes
basalt	highlands	mare	sinuous rille
basins	Imbrian era	mascon	Surveyor
breccia	impact melt rocks	Nectarian era	terminator
Copernican era	last quarter	new moon	transient phenomena
earthshine	libration	penumbra	umbra
Eratosthenian era	linear rille	phases	waning
first quarter	Luna	Pre-Nectarian era	waxing

■ Review Questions

1. In what respect is the Earth–Moon system unusual?
2. Summarize the Moon's basic physical and orbital properties.
3. Describe the phases of the Moon and their cause.
4. Describe the Moon's spin-orbit coupling.
5. How has the Moon physically affected the Earth?
6. Describe what happens during a lunar eclipse and during a solar eclipse.
7. Why are eclipses infrequent?
8. Summarize the properties of lunar high- lands and lunar maria, emphasizing their differences. What types of rocks and minerals are found in each region?
9. Explain the difference between a crater and a multiringed basin.
10. List the major lunar surface features, and explain the origin of each.
11. Discuss the following properties of the lunar interior: gravity anomalies, moon-quakes, core, and magnetic field.
12. Summarize the major events of the five eras of lunar history.

■ For Further Reading

Cadogan, Peter. *The Moon: Our Sister Planet.* Cambridge University Press, New York, 1981.

Cortright, Edgar M., editor. *Apollo Expeditions to the Moon.* NASA, Washington, D.C., 1975.

French, Bevan M. *The Moon Book.* Penguin Books, New York, 1977.

Guest, J.E. and Ronald Greeley. *Geology on the Moon.* Wykeham Publications, London, 1977.

Mutch, Thomas A. *The Geology of the Moon.* Princeton University Press, Princeton, 1970.

Taylor, Stuart Ross. *Lunar Science: A Post-Apollo View.* Pergamon Press, New York, 1975.

Wilhelms, Don E. "Moon." In Carr, Michael H., editor. *The Geology of the Terrestrial Planets.* NASA, Washington, D.C., 1984.

Mars

Artist's conception of a Viking Orbiter spacecraft in orbit around Mars. (NASA/JPL Photograph.)

Mars, the most distant terrestrial planet from the Sun and the only one that has two satellites, was named for the Roman god of war because of its distinctive blood-red color when seen in the sky. The satellites of Mars are named Phobos and Deimos, which mean "panic" and "terror," appropriate companions for the war god.

The major orbital and physical properties of Mars are summarized in Table 16.1. The diameter of Mars is 6790 kilometers (4210 miles), about half that of the Earth but twice that of the Moon. Although Venus is physically similar to the Earth in bulk properties, Mars is Earth-like in many other ways. For example, its rotational period is just over 24.5 hours, and its 25.2° obliquity gives it seasons. It has a transparent atmosphere with occasional clouds and dust storms, polar ice caps, and seasonal changes in surface appearance. Even though Mars does not approach the Earth as closely as Venus can, it will be the obvious first choice for manned planetary exploration. It is likely, however, that the first such flights are many years away.

■ Appearance and Basic Properties

Unlike Mercury and Venus, Mars and the other **superior planets** can be seen high in the dark night sky all night long when at opposition. Mars is often called the "red planet," and its bright orange-red color allows it to be easily identified in the sky. (All of the other planets appear yellowish to the naked eye.) Mars has the longest synodic period of any planet (780 days), so its oppositions are just over two years apart. In the time between oppositions, Mars spends quite a while in the evening and then the morning sky, and its brightness varies greatly as its distance from Earth changes.

■ **Table 16.1   Summary of Mars' Basic Properties***

Semi-major axis: 1.52 astronomical units
Semi-major axis: 227.9 million kilometers
Semi-major axis: 141.6 million miles
Orbital eccentricity: 0.093
Orbital inclination: 1.85°
Orbital period: 687 days
Diameter: 6787 kilometers
Diameter: 4217 miles
Mass (Earth = 1): 0.107
Density: 3.94 grams per cubic centimeter
Sidereal rotational period: 24.62 hours
Axial obliquity: 25.2°
Number of satellites: 2

* For a complete table of planetary data, see Table A2.1 in
Appendix 2.

The orbit of Mars is much more eccentric than that of the Earth, so oppositions that occur when Mars is at perihelion are much more favorable for telescopic observers than those near aphelion (Figure 16.1). Despite its proximity to Earth, Mars has a rather small apparent size in the telescope because of its small diameter. For this reason, Earth-based observers have long had difficulty in properly interpreting its surface features.

By 1659, observers were able to determine Mars' rotational period by watching the movement of its surface markings. In addition, observers soon began to note that the Martian markings, although difficult to see clearly, exhibited changes with time (Figure 16.2). The polar ice caps grow and shrink according to season, and dark areas on the planet change shade and, to a lesser degree, areal extent. The latter observation led to speculation that vegetation was involved, as these areas are dark green, in contrast to the reddish-orange color of the remainder of the planet. However, we know today that Mars lacks vegetation, and as a result of spacecraft explorations, these light and dark albedo markings are now thought to be the result of movement of wind-borne sediment. The light-colored areas are places where the surface is covered by fine sediment, whereas darker areas are those where such material has been blown away, exposing the

■ **Figure 16.1** Oppositions of Mars from 1988 to 2003. As discussed in the text, the Earth–Mars distance can vary considerably between oppositions.

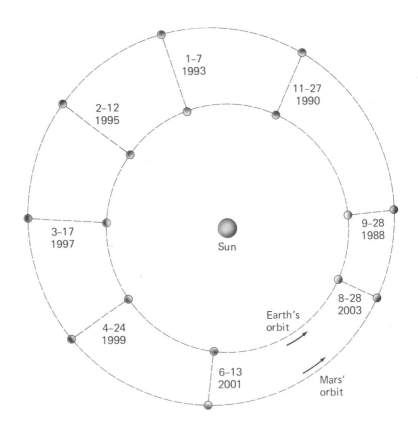

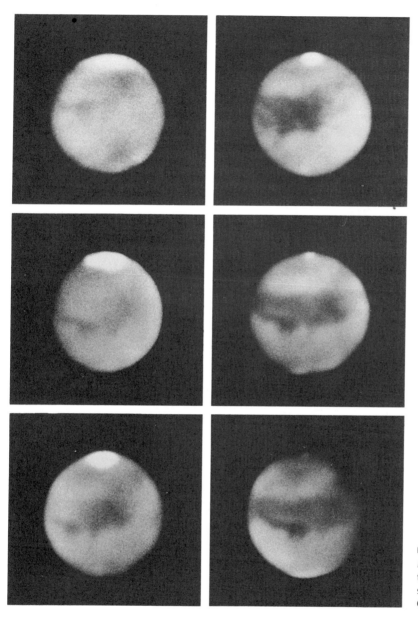

■ **Figure 16.2** Photographs of Mars, taken with a terrestrial telescope, showing the variation in surface features with time. (Lowell Observatory Photographs.)

underlying rock. In addition, we know that the dark areas of Mars are not as green as they appear, but that contrast with the surrounding reddish regions makes them appear greener than they actually are.

The most controversial aspect of Martian studies began in 1877, when an extremely favorable opposition occurred, during which Phobos and Deimos were discovered. Prior to this time, the Italian astronomer **Angelo Secchi** (1818–1878) had seen faint linear markings on Mars that he called canali, meaning "channels" or "canals." **Giovanni Schiaparelli** (1835–1910) mapped numerous canali in 1877, and they soon received a great deal of publicity. The word was generally translated into English as **canals,** a word that has the connotation of man-made features, and the general public soon had the idea that these Martian features had been artificially constructed!

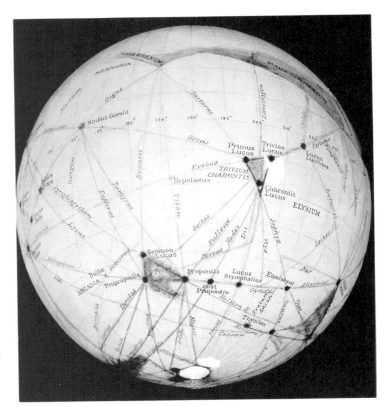

■ **Figure 16.3** A photograph of a globe of the intricate Martian canal system, as depicted by Percival Lowell. (Lowell Observatory Photograph.)

Before long, numerous astronomers were peering at highly magnified images of the tiny Martian disk, trying to see and map the canal system (Figure 16.3). Schiaparelli noted doubled canals in 1879, and the American astronomer **Percival Lowell** (1855–1916) built Flagstaff Observatory in Arizona mainly to take advantage of its clear air for Martian observations. Lowell took the canals much too seriously, as he began publishing books describing the Martian canals as irrigation ditches fashioned by a Martian civilization plagued by water shortages! Lowell saw large dark areas at intersections of canals and assumed these were cultivated areas!

Meanwhile, other competent planetary observers failed to see any canals at all, and some even reported seeing Martian craters! In 1909, for example, E. M. Antonindi noted that during a time of exceptionally good seeing, he saw Mars' surface covered with a maze of streaks and spots, not with a regular system of canals. He asserted that no one had ever seen a genuine Martian canal!

The mystery continued for many years and was complicated by the fact that observers who did see the canals generally agreed on their locations. Some photographs revealed hints of canals, but never with the clarity claimed by optical observers. (Even some contemporary photographs taken with small telescopes show vague linear markings.) Many astronomers proposed that the canals were optical illusions caused by the brain's unconsciously connecting diffuse markings that the eyes could not resolve. Experiments conducted by having nonscientists sketch simulated Mars globes from a distance supported this conclusion (Figure 16.4).

The matter was not settled until 1965, when Mariner 4 flew past the planet and saw craters instead of canals. It was then assumed that Mars was another Moon-like body, but later spacecraft observations proved that this assumption was not entirely correct, as we shall see. The whole canal controversy is instructive, because it points out one of the limitations of Earth-based optical as-

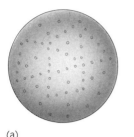 

(a)   (b)

■ **Figure 16.4** Canals as optical illusions. (a) A representation of small, vague markings on the Martian surface. (b) The human eye, unable to resolve the true markings, perceives instead a series of lines.

tronomy. The eye can indeed be fooled, and features at the limit of visibility can easily be misinterpreted.

■ Spacecraft Exploration

Although the Soviet Union has made significant advances in studies of Venus, its few Martian probes have generally not fared well. Two Soviet landers returned data, but only briefly. On the other hand, the United States has had more success with Mars than Venus and has done a superlative job of unlocking the secrets of Mars. American Martian exploration began on July 14, 1965, when the Mariner 4 spacecraft flew past the planet and returned over 20 photographs of its surface. In addition, it revealed that Mars lacks radiation belts and magnetic fields. Occultation studies of its radio signals allowed determination of the vertical structure of Mars' atmosphere. Mariners 6 and 7 flew past Mars on July 31 and August 5, 1969 and returned about 200 photographs, as well as spectroscopic data.

A great advance in Martian exploration occurred on November 13, 1971, when Mariner 9 went into orbit around the red planet, the first spacecraft to do so. Mariner 9 had the misfortune of arriving at a time when the entire planet was shrouded by a dust storm, but after the dust finally settled, many wonders were revealed. Mariner photographed volcanoes that dwarfed any found on Earth, stream erosional features indicat-

ing that subsurface water had been released by long-ago impacts, and a giant canyon next to which Earth's Grand Canyon would scarcely be noticed. Mariner 9 returned over 7000 images of the surface and also photographed the surfaces of Mars' satellites for the first time.

Even though none of the Mariner craft had found any evidence of canals, ruling out Lowell's fanciful ideas of a Martian civilization, many researchers felt that Martian life was possible. The next step in Martian exploration was the **Viking** program, which was designed to send two spacecraft into orbit and land two sophisticated remote laboratories on the surface to photograph it, analyze its rocks and soil, and search for life. Two Viking spacecraft were readied for the journey to Mars, each consisting of an orbiter with an attached lander.

Because the Martian air is too thin for parachutes alone to slow spacecraft for a soft landing, the landers were designed to use both parachutes and retrorockets. After separating from the orbiters, the landers were intended to enter the atmosphere, where a parachute would open at an altitude of 5800 meters (19,000 feet). The parachute would slow the craft until it reached 1220 meters (4000 feet); then, retrorockets would slow the final impact to only 9 kilometers per hour (5.5 miles per hour).

The Viking spacecraft were launched in 1975 and reached Martian orbit in the summer of 1976. It was initially planned that Viking 1 would land on July 4, in honor of America's bicentennial. However, orbiter photographs of the proposed landing site indicated roughness, prompting a search for another site. Earth-based radar was used to evaluate roughness of alternate landing areas, and a new site was chosen in the **Chryse Planitia** region at 23° north latitude. Viking 1 successfully landed on July 20, 1976, the seventh anniversary of the first manned lunar landing. Viking 2 was also sent to an alternate site for safety reasons. (Because both landing sites were chosen with regard to lander safety, they are probably not very representative of the entire Martian surface.) Viking 2 landed in the **Utopia Planitia** region, at 48° north latitude, on September 3, 1976.

The Viking missions sent vast quantities of information back to Earth. The orbiters had lifetimes of several years and obtained over 50,000 photographs of the Martian surface as well as of Phobos and Deimos. The landers also returned data for several years, because the surface conditions were not as detrimental to them as Venus' conditions are to spacecraft that land there. The Viking landers photographed the surface, analyzed its soil and rocks, searched for Martian life, and studied Martian weather conditions. (We will discuss the Viking lander results in detail later.) The Viking program ended in November, 1982, when contact with the Viking 1 lander was lost.

In July, 1988, the Soviet Union launched two Phobos spacecraft toward Mars. These craft were the first designed specifically to study the satellites of another planet rather than the planet itself. Contact was lost with both spacecraft, however — Phobos 1 about two months after launch and Phobos 2 shortly after it began orbiting Mars. Phobos 2 did operate long enough to obtain photographs of Phobos and Mars and gather some spectral data, but its failure to land on Phobos was a major disappointment.

■ Large-Scale Surface Features

A large number of geological processes have shaped the Martian surface, making its topography very diverse. (Major Martian surface features and topographic provinces are shown in Figure 16.5.) Although at first glance there is a slight resemblance to the Moon (both objects have heavily cratered areas and smooth plains), the Martian surface is much more complicated than that of our satellite. In contrast to the Earth, whose surface has a twofold distribution of elevation (ocean basin and continental surface), the Martian surface shows a threefold distribution of elevation: most of the planet is a mid-level terrain, but there are also lowland plains and mountains. The total relief on Mars is about 31 kilometers (19 miles), nearly 1.5 times that on Earth. Measurement of Martian elevation is based upon atmospheric pressure, with the level where the pressure is 0.00602 atmospheres used as the starting point. This value was chosen because it is the

lowest pressure at which liquid water can be stable.

Because of the wide diversity of Martian surface features, its craters are somewhat overshadowed by its more spectacular landforms. Even so, Mars has numerous impact features, as do nearly all of the solid bodies in the solar system. The craters of Mars are most conspicuous in topographic regions called heavily cratered terrain. Abundant cratering would be expected on Mars because of its proximity to the asteroid belt and the fact that its atmosphere is much less capable of stopping meteorites than that of the Earth. The fact that some areas of Mars have fewer craters than others is, of course, because subsequent geological activity has erased them.

In terms of cratering, Mars can be divided into two hemispheres. A hemisphere roughly corresponding to the southern hemisphere is covered with old, densely cratered terrain that superficially resembles that of the lunar highlands. (Martian craters, however, are more heavily eroded than those on the Moon.) Crater counting indicates that the age of this surface is about 3.9 billion years. The remainder of Mars (essentially the northern hemisphere) is covered with plains on which large craters are 10 to 100 times less abundant. The lack of craters in the northern hemisphere indicates that its surface has been modified by internal processes such as volcanism and external processes such as wind. This area has a lower average elevation than the southern hemisphere, and its surface has been shaped by processes acting over an extended period of time. Both Vikings landed in the plains of Mars' northern hemisphere.

Individual Martian craters are often different in form from lunar ones, in that crater ejecta on Mars is contained in discrete lobes that appear to have flowed outward from the site of the impact (Figure 16.6). This is believed to be the result of water or ice trapped beneath the surface and released by the heat of impact. A wide variety of sizes and forms of Martian craters exists, but recently formed craters are relatively rare. Erosion of Martian craters has occurred, primarily as the result of wind, and the effects of this wind erosion seem to be greater at higher latitudes.

MARS

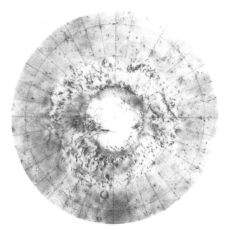

NORTH POLAR REGION

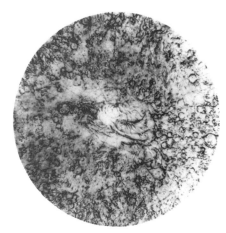

SOUTH POLAR REGION

OLYMPUS MONS

THARSIS BULGE
Viking 1 landing site

"Catastrophic Floods"

Viking 2 landing site

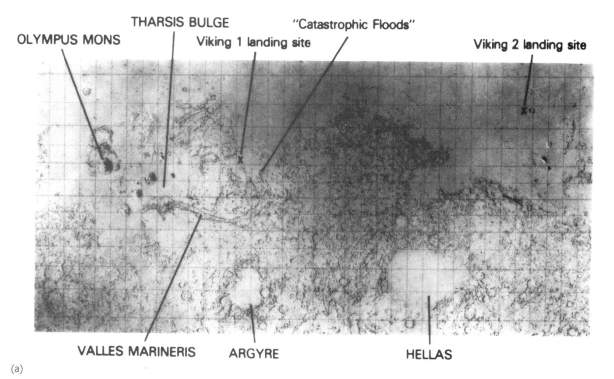

VALLES MARINERIS   ARGYRE   HELLAS

(a)

■ **Figure 16.5** Maps of Mars. (a) Shaded topographic maps. (NASA Photograph, courtesy of the National Space Science Data Center.) (b) Location and names of major landforms. (From Mutch, Thomas A. et al. *The Geology of Mars.* Copyright © 1976 by Princeton University Press, Princeton, NJ.) (c) Major topographical provinces of Mars. Polar units: **pi**-permanent ice, **ld**-layered deposits, **ep**-etched plains. Volcanic units: **v**-volcanic constructs, **pv**-volcanic plains, **pm**-moderately cratered plains, **pc**-cratered plains. Modified units: **hc**-chaotic hummocky terrain, **hf**-fretted hummocky terrain, **hk**-knobby hummocky terrain, **c**-channel deposits, **p**-undivided plains, **g**-grooved terrain. Ancient units: **cu**-undivided cratered terrain, **m**-mountainous terrain. (From Mutch, Thomas A. *The Geology of Mars.* Copyright © 1976 by Princeton University Press, Princeton, NJ.)

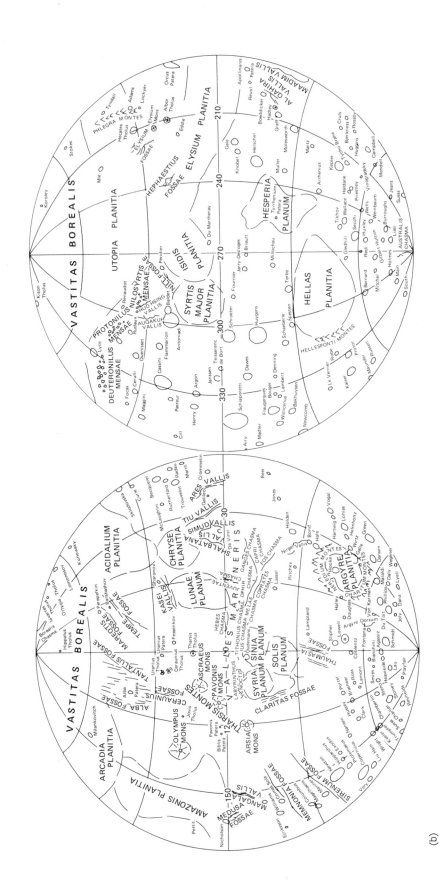

(b)

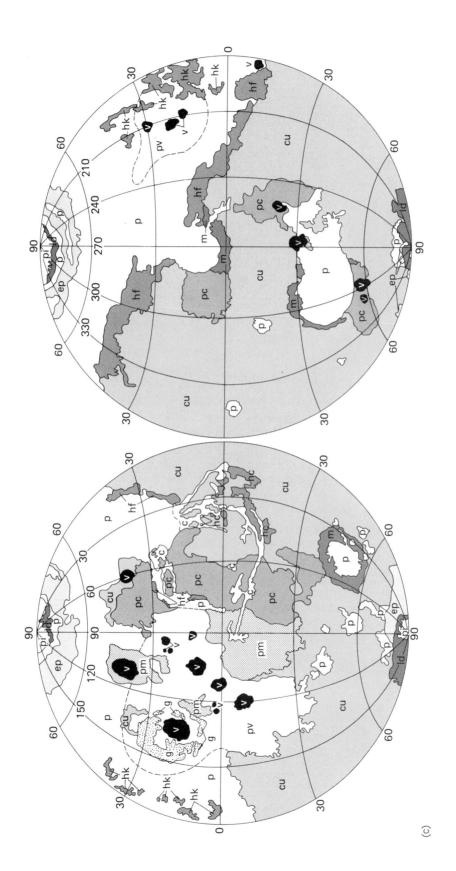

239

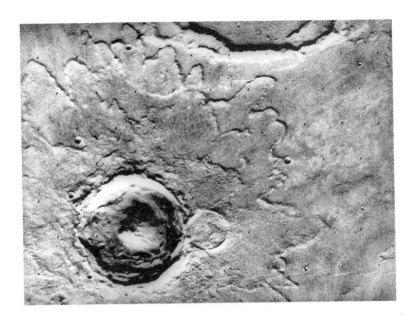

In addition to countless smaller craters, about 40 large impact basins have been identified on Mars, and they are among its oldest features. One of them, **Hellas**, is about 1600 kilometers (1000 miles) in diameter, making it the largest impact basin in the solar system. The floor of Hellas contains volcanic plains and wind-blown deposits.

Some of the most spectacular Martian features are the planet's numerous volcanoes. Several types of volcanic activity occur, including at least one not found on Earth, but all of the Martian volcanoes appear to be extinct. (Figure 16.7 shows relative sizes of some Martian and terrestrial volcanoes.) One of the most surprising discoveries made by Mariner 9 was that the Martian surface contains a number of huge shield volcanoes. These volcanoes are not uniformly distrib-

uted about Mars but exist in three provinces named **Hellas** (the oldest), **Elysium**, and **Tharsis** (the youngest). Tharsis, which is the largest such region, contains 12 volcanoes in a broad, 10-kilometer-high (6-mile-high) plateau about 4000 kilometers (2500 miles) across (Figure 16.8). Three volcanoes (**Arsia Mons, Pavonis Mons,** and **Ascraeus Mons**) are located in Tharsis, in virtually a straight line, and nearby is **Olympus Mons,** the largest known shield volcano in the solar system. (All Martian shield volcanoes contain the word Mons, the Latin word for "mountain," as part of their names.) Olympus Mons is 550 kilometers (340 miles) across, 25 kilometers (15 miles) high, and contains 50 to 100 times as much volcanic rock as Earth's largest volcano, Mauna Loa in Hawaii (Figure 16.9). The summit caldera itself, from which the magma last poured out, is about 70

■ **Figure 16.7** Sizes of Martian and terrestrial volcanoes drawn to scale with no vertical exaggeration.

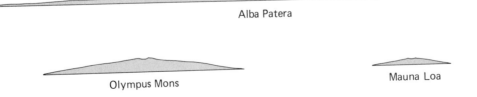

Alba Patera

Olympus Mons

Mauna Loa

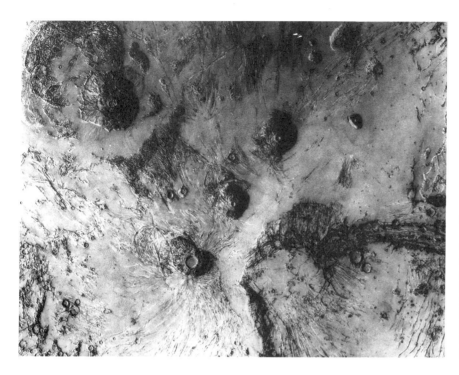

■ **Figure 16.8** A shaded relief map of the Tharsis volcanic province. Olympus Mons is located at upper left, and Arsia Mons, Pavonis Mons, and Ascraeus Mons lie along a diagonal line from lower left to upper right. (NASA Photograph, courtesy of the National Space Science Data Center.)

kilometers (45 miles) across! Like all shield volcanoes, the side of Olympus Mons has a gentle slope of only about 4°. For reasons that are not understood, Olympus Mons is surrounded by a steep cliff that is several kilometers high.

In addition to the shield volcanoes, Mars also contains some smaller, steeper dome volcanoes that are given the name **Tholus** (Figure 16.10). These may be the result of more viscous magma or the incorporation of pyroclastic material. A

■ **Figure 16.9** A photograph of Olympus Mons, the solar system's largest shield volcano, as viewed from orbit. (NASA/JPL Photograph.)

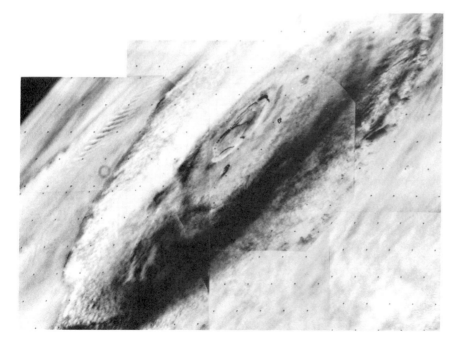

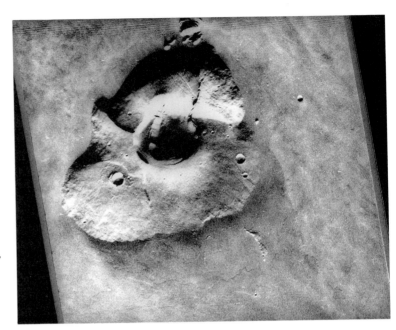

■ **Figure 16.10** Tharsis Tholus, a small, steep-sided Martian dome volcano. The volcano measures 170 by 110 kilometers (105 by 70 miles). (NASA/JPL Photograph, courtesy of the National Space Science Data Center.)

type of volcanic feature apparently unique to Mars is the **patera**, named for the Latin word meaning "saucer." Paterae are somewhat similar to shield volcanoes, but they are much larger and have extremely low relief. They appear to be old shield volcanoes that have either been eroded or collapsed internally. **Alba Patera** is about 1600 kilometers (1000 miles) across but less than 6 kilometers (4 miles) high (Figure 16.11). Its central caldera is 100 kilometers (60 miles) across, and a number of well preserved lava flows and ridges are visible on its sides. Alba Patera covers eight times the area of Olympus Mons, and it is the single largest volcanic feature known anywhere.

Volcanic activity seems to have become more restricted on Mars with time, and all of its major volcanoes are believed to be extinct. Ages of Martian volcanoes are estimated by counting the number of craters visible on their sides, because presumably these craters are all younger than the most recent eruption episodes. Olympus Mons appears to have become extinct most recently, with the end of eruptive activity estimated at between 100 and 300 million years ago. Some Viking photographs, however, have been interpreted as showing very recent magma from small volcanic vents. The dark rock pro-

duced would have been covered by lighter, wind-blown material if it had been very old, supporting the idea that it may have formed only recently.

Another Martian terrain type, canyonlands, is exemplified by the unusual feature **Valles Marineris** (Mariner Valley), a 4000-kilometer-long (2500-mile-long) system of canyons near Mars' equator (Figure 16.12). This extensive feature contains most of the canyonlands formations of Mars. Although Valles Marineris is often called a giant "Grand Canyon," its origin is certainly much more complex than the river-eroded Grand Canyon on Earth. First of all, a great deal of faulting has occurred in this and other similar areas of Mars, producing many low fault-bounded grabens. These grabens appear to have been formed when the Martian crust in this area was "stretched" as the Tharsis region bulged upward. In addition, mass wasting was also important in shaping this area, as was water released from either ground water or subsurface ice. The central part of Valles Marineris is about 2400 kilometers (1500 miles) long and consists of multiple, east–west oriented canyons, each up to 190 kilometers (120 miles) wide and 6 kilometers (4 miles) deep. The eastern end of Valles Marineris is

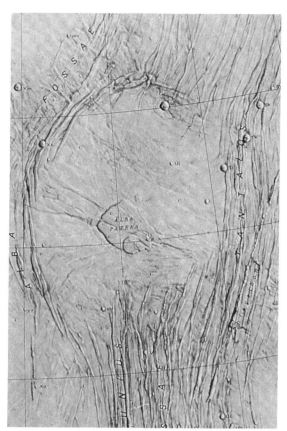

■ **Figure 16.11** A shaded relief map of the central portion of Alba Patera. Very little relief is visible; the volcano is marked largely by numerous faults. Note the caldera in the center. The area shown is about 1000 kilometers (600 miles) across. (U.S. Geological Survey.)

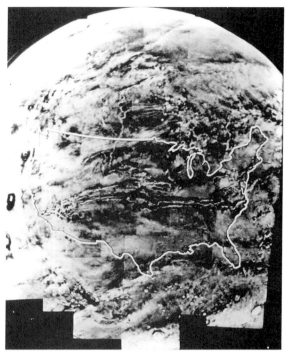

■ **Figure 16.12** A photographic montage of the area of Mars containing Valles Marineris. The outline of the United States has been superimposed for a size comparison. (NASA Photograph.)

a "chaotic terrain" of irregular, blocky hills, whereas the western end is a complex area of intersecting grabens. One of the biggest unsolved mysteries about Valles Marineris is the fate of the enormous amounts of material that were removed to make this impressive feature.

Although the exact role of surface running water in the formation of Valles Marineris is uncertain, Mars contains many smaller features that are definitely the result of stream erosion. These are stream channels and valleys that, although now dry, indicate that water once flowed over Mars' surface. Because Mars' surface pressure has probably always been too low to support liquid water indefinitely, suggestions are that ice trapped under the surface was melted and re-

leased by meteorite impacts. After flowing over the surface and performing geological work, the water evaporated. (It should be noted that these features are too small to be visible from Earth and are not associated with the old sightings of "canals.")

Three major types of Martian channel features have been identified (Figure 16.13). **Runoff channels** are small but increase in width downstream as tributaries enter them. They are typically tens of kilometers long and less than 1000 meters (3300 feet) wide, and they are commonly found in the heavily cratered section of Mars. **Outflow channels** are large features that start full grown from local sources and generally lack tributaries. The large volumes of water that flowed in them suggest that they may have formed when large bodies of flood waters broke through material that was surrounding them. These channels, which are not limited to one particular area of Mars, have also been suggested as being the result of glaciation or even features produced by

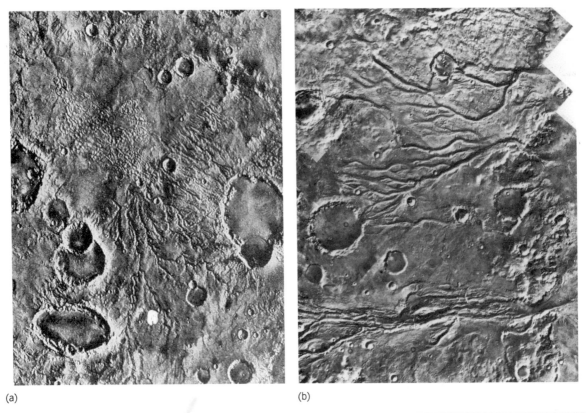

(a)

(b)

■ **Figure 16.13** Types of Martian channels.
(a) Runoff channels. (NASA/JPL Photograph.) (b)
Outflow channels. (NASA/JPL Photograph.) (c)
Fretted channels. (NASA/JPL Photograph,
courtesy of the National Space Science Data
Center.)

(c)

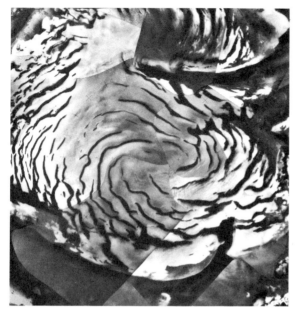

■ **Figure 16.14** The north polar cap of Mars, showing the residual cap, which does not disappear, even in summer. (NASA/JPL Photograph.)

surface lava flows. The final type, **fretted channels,** are wide, steep walled, and have smooth, flat floors. These channels, which intersect craters and other channels, are limited to certain latitude regions and are possibly runoff channels that were enlarged by mass wasting.

Although the Martian surface is dry today, ice or ground water may still exist at some point below its surface. Channel formation was probably more active in the past because volatile releasing impacts were more numerous then and because the subsurface water supply may now be somewhat depleted. The subsurface water supply is thought to have been originally produced by the condensation of steam brought to the Martian surface by magmas. Ronald Greeley has estimated that volcanic eruptions sufficient to produce the volcanic features visible today on Mars' surface would have released enough water to cover Mars to a depth of 46 meters (150 feet). Most of this water was released during the first 2 billion years of Mars' history, during which most of its volcanism occurred and most of the large outflow channels developed.

The Martian **polar caps,** which are visible telescopically from Earth, were long assumed to be water ice until they were examined spectroscopically from the Viking orbiters (Figure 16.14). The southern cap, which completely disappears during summer in that hemisphere, is actually a layer of **dry ice** (carbon dioxide ice) that is about 20 to 50 centimeters (8 to 20 inches) thick. The northern cap does not completely disappear, and the small residual cap remaining even in summer is thought to be a layer of water ice that is probably no more than a few tens of meters thick. (The remainder of the cap, which disappears during the summer, is dry ice whose thickness is similar to that of the southern cap.) In winter, each cap may extend more than 50° in latitude from the pole. The caps, particularly the northern permanent one, are not pure ice but also contain numerous layers of wind-blown dust.

At high latitudes (generally above 80°), the Martian surface is composed of ice-related deposits called **polar units** (Figure 16.15). The most important of these is called **layered terrain,** because layering is visible as dark and light bands when the material is cut and exposed by other

■ **Figure 16.15** Layered Martian polar units at 80° north latitude. (NASA/JPL Photograph.)

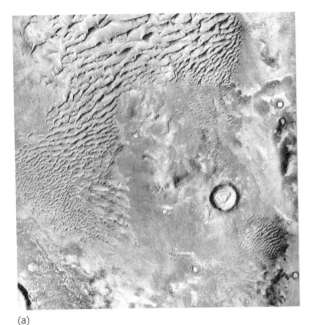

(a)

(b)

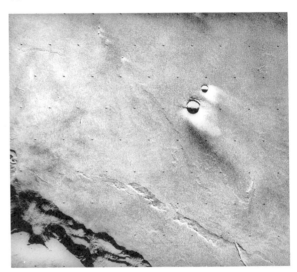

(c)

■ **Figure 16.16** Martian wind-related features. (a) An area containing numerous sand dunes. (b) A region with dark wind streaks associated with craters. (c) Two craters with bright wind streaks. (NASA/JPL Photographs, courtesy of the National Space Science Data Center.)

erosional processes. These bands are the result of horizontal layering, and the layers were probably formed by the accumulation of wind-blown material over long periods of time. The light and dark layers may be due to different types of materials alternately accumulating during periodic climate variations.

A final category of Martian landform includes those features formed by wind (Figure 16.16). Wind erosion appears to have been very important on Mars, because much of the surface is covered by loose material that is easily picked up and

transported. Unlike the Earth, where deserts are most common near the equator, Mars has more sand at high latitudes. The source of Martian sand is probably different from that of terrestrial sand, because water erosion produces most terrestrial sand. Mars has two major types of wind-erosional features, areas of deflation, where wind has removed material, and yardangs, which are streamlined, wind-eroded ridges. Yardangs are useful because their streamlining reveals wind direction. The major Martian wind-depositional features are sand dunes, which occur in many of the

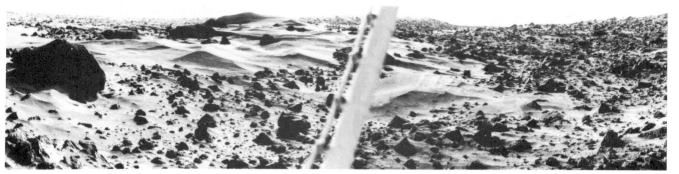

■ **Figure 16.17** A view of the Martian surface obtained by Viking 1. (NASA/JPL Photograph.)

same forms as on Earth, and **wind streaks,** which are deposits of material on the wind-protected sides of craters and other landforms. Wind streaks may be either dark or light, depending upon the type of material deposited.

■ The Martian Surface

The Viking landers were designed to photograph the Martian surface in color and analyze surface rocks and soil. The Viking 1 Chryse Planitia landing site resembled a lunar mare when seen from orbit, but from the lander, it had a gently rolling topography with rocks strewn everywhere (Fig-

ure 16.17). Within view of the spacecraft were boulders large enough to have toppled the spacecraft had it set down on any of them. The surface rocks at Chryse Planitia had a wide range of color, shape, and texture. Most were irregular in shape, had pitted surfaces, and appeared to have a basaltic origin. Fine-grained material occurred in wind-blown drifts, but some areas were bare. The overall color of the Martian surface was rust brown. Even the Martian sky had an orange tint because of the large amounts of suspended dust.

The Viking 2 landing site, Utopia Planitia, was a flat region, possibly part of an ejecta lobe from the crater Mie (Figure 16.18). The rocks at this landing site were generally very similar to one

■ **Figure 16.18** A view of the Martian surface obtained by Viking 2. (NASA/JPL Photograph, courtesy of the National Space Science Data Center and the Viking Lander Principal Investigator, Raymond E. Arvidson.)

another in both texture and size. The rocks were evenly distributed, and the scenery was remarkably uniform in all directions. Near the spacecraft were some small, interconnected, sand-filled troughs whose origin was uncertain. Viking 2 was far enough north that a thin layer of frost was deposited during the winter and dissipated when spring arrived.

Because the Viking landers were not designed to return samples to Earth, sophisticated analyses of Martian materials were conducted on board. Each lander had a 3-meter-long (10-foot-long) movable sample arm with a scoop at the end. The scoop was used to gather soil samples for the on-board laboratory, as well as to push small rocks aside so that fresh material could be obtained from underneath them. The composition of the soil was determined using X-ray fluorescence techniques. X-ray fluorescence works by bombarding materials with X-rays. When atoms are hit by these X-rays, they then emit X-rays of their own. A detector can analyze these secondary X-rays and determine which elements were present to produce them. Silicon, iron, magnesium, aluminum, calcium, and sulfur were found to be abundant, and both landing locations had remarkably similar compositions. These compositions are similar to those of igneous rocks, but exactly which minerals were present could not be determined. Because of the presence of water in the soil, it is now believed that hydrated iron-rich **clay minerals,** such as non-tronite, montmorillonite, and saponite, may be major components of Martian soil. Several magnets were placed on the lander to see how much of the soil was magnetic. Some material was held, and 1% to 7% of the soil was estimated to be magnetic. An additional compositional test was to search for organic materials in the soil. Organic compounds are those based on the element carbon, and the name derives from the fact that most organic materials on the Earth are produced by living organisms. Any Martian organic materials could have been produced by living organisms or brought to Mars by meteorites, some of which have been found to contain organic materials. Because organic-containing meteorites have undoubtedly struck Mars, some organic materials should be present on its surface. None were found, however, indicating that some process,

probably the influx of solar ultraviolet radiation, breaks down any organic materials present on the Martian surface.

■ The Search for Life

Because of the decades-long interest in Martian canals as well as the planet's many similarities to Earth, it is not surprising that a search for Martian life was of very high priority as the Viking missions were planned. Some of the more optimistic planners felt that detection of life would be easy, that vegetation would probably be visible in all directions, and that Martian "wildlife" might even occasionally wander across the fields of view of the television cameras! However, most believed that Martian life, if it existed at all, would be microscopic, so steps were taken to detect any such organisms that might be present in the soil. Great care was taken to sterilize the entire Viking spacecraft so that if any positive results were found they would not be due to "visiting" terrestrial microbes.

In order to search for life, one must know for what one is looking. In looking for extraterrestrial life, the phrase "life as we know it" is often used. Terrestrial life is based primarily on the atoms carbon (which has a very complex chemistry), oxygen, nitrogen, and hydrogen. Although it is possible that Martian life could have developed based on a totally different chemistry, it would be so different from life on Earth that it might be very difficult for us to recognize such Martian organisms as even being alive.

Life may be described as a series of processes occurring within an organism that is "alive." Organisms take in atmospheric gases as well as water and solid "food." (Food is not necessary for green plants, which produce their own by the process of **photosynthesis.**) Within the organism, chemical reactions occur, releasing energy that is used for movement, growth, and reproduction. Additionally, waste products (gaseous, liquid, and solid) are expelled. The basic unit of life on Earth is the cell. Some "simple" organisms are single-celled, but even a single such cell contains 1 trillion atoms! The human body contains 100 trillion cells; each of these must receive nourishment, just as the overall per-

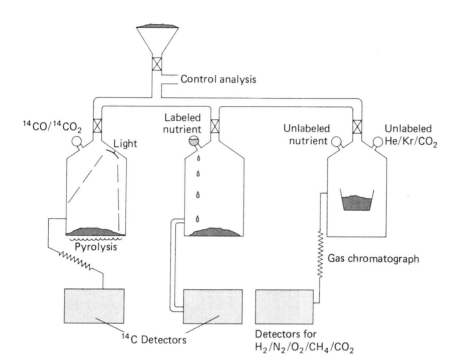

■ **Figure 16.19** Martian life-detection experiments. The pyrolitic-release experiment (*left*) contained a light source, a nutrient source, a heater, and a gas analyzer. The labeled-release experiment (*center*) contained a nutrient source and a gas analyzer. The gas-exchange experiment (*right*) contained a nutrient source and a gas analyzer. See text for details. (NASA diagram by Harold P. Klein. From Carr, Michael H. *The Surface of Mars.* Yale University Press, New Haven, CT, 1981.)

son must. The chemical process taking place within an organism and its individual cells is called **metabolism.**

The foregoing gives us the clues necessary to search for life. (The general aspects of searching for extraterrestrial life, as opposed to the specific search for life on Mars, will be described in Chapter 30.) Because most plants don't move, motion is not a clue to life. All organisms grow and reproduce, but the time scales of these processes might have been too long to witness during the Viking missions. Therefore, each Viking lander was designed to search for evidence of metabolism, which would prove the existence of Martian life. Three separate experiments were devised to search for signs of life in Martian samples that had been introduced into the appropriate sample chambers (Figure 16.19). These instruments could have detected terrestrial life no matter where they might have been placed on our planet — a tribute both to their sensitivity and to the ubiquity of life on Earth.

One experiment was **pyrolitic release,** designed to look for evidence of Martian photosynthesis. When plants photosynthesize, atmospheric carbon dioxide is incorporated into plant molecules. The pyrolitic-release apparatus con-

tained carbon dioxide especially formulated to contain a radioactive isotope of carbon. Such carbon dioxide is called "tagged" carbon dioxide because it can be easily traced during experiments, due to the presence of the radioactivity. Therefore, the carbon could be followed if it were taken from carbon dioxide in the air sample introduced into the test chamber and incorporated into plant material. The experiment was designed to expose a soil sample to tagged gas while shining light on it for five days. (Light is necessary for photosynthesis to occur.) Afterwards, the gas was pumped out and the sample heated in order to break down organic photosynthetic products and drive them out of the soil for detection. Any radioactive carbon found in this way would have indicated that photosynthesis had occurred. Seven of the nine pyrolitic release experimental runs gave positive results. However, positive results were obtained even if the light was off during the experiment or if the soil had been initially sterilized by baking. Researchers believe these results indicated that inorganic Martian molecules somehow are able to synthesize simple organic material. The exact method is uncertain, but the experimental results can be explained without life.

A second experiment, **labeled release,** was designed to "feed" any microbes present in the Martian soil with a nutrient medium composed of organic molecules in distilled water. As with pyrolitic release, these molecules contained carbon dioxide tagged with radioactive carbon. Because any microbes present would have consumed the nutrient and metabolized it, they would have emitted gases containing radioactive carbon. Once again, positive results were obtained. However, the Martian soil was determined to contain chemical compounds called **superoxides,** which would react with the formic acid contained in the nutrient medium to produce carbon dioxide. Terrestrial researchers were able to duplicate the labeled release results on Earth using only nonliving materials.

The third experiment, **gas exchange,** was similar, except that the nutrient medium was not labeled. Instead, an instrument for gas analysis known as a gas chromatograph was used to identify the particular gases emitted during the experiment. Carbon dioxide and oxygen were emitted in large quantities as soon as moisture was added, but there was no gradual release afterward, as would have been expected if life were present. This activity first indicated to Viking scientists the presence of superoxides.

Although cleverly designed, these life-detection experiments were apparently fooled by the unusual chemistry of the Martian regolith. Although the lack of organic compounds in the Martian soil is evidence that Mars is lifeless, it is now generally thought that Viking neither proved nor disproved the existence of life on Mars, but that samples will need to be returned to Earth for study before the matter can be settled. Most researchers feel that if life does not exist on Mars, the Earth is probably the only place in our solar system with life.

■ Atmosphere

A great deal is known about the Martian atmosphere as a result of the Viking explorations. The Viking landers served as the first long-term "weather stations" on another planet. The orbiters monitored clouds and dust storms, and oc-

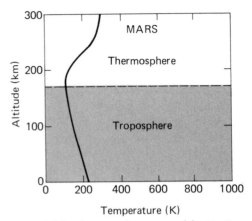

■ **Figure 16.20** The vertical structure of the Martian atmosphere. (From page 124 of Kivelson, Margaret G. and Gerald Schubart, "Atmospheres of the Terrestrial Planets," in Kivelson, Margaret G., editor, *The Solar System: Observations and Interpretations.* Copyright 1986. Adapted by permission of Prentice-Hall, Inc., Englewood Cliffs, NJ.)

cultation studies of their radio signals passing through the atmosphere allowed details of its vertical structure to be determined. (The vertical structure of Mars' atmosphere is shown in Figure 16.20.)

The Martian atmosphere is different from Earth's atmosphere in many respects. The surface pressure is only 8 millibars, compared to 1013 millibars on Earth, and the atmosphere is composed of 95.3% carbon dioxide, 2.7% nitrogen, and 1.6% argon. Because the total vertical relief on Mars is 31 kilometers (19 miles), there is a wider variation of air pressure with altitude on Mars than on Earth. Martian temperatures are much lower than terrestrial temperatures, as the Viking landers measured a range from about 190 K to 240 K ($-117°F$ to $-27°F$) during the Martian summer. The lowest temperature on Mars, 143 K ($-202°F$), occurs in winter at the south pole, whereas the highest temperature, 295 K ($72°F$), occurs in summer in southern hemisphere mid-latitudes. Mars' average surface temperature is 218 K ($-67°F$).

The amount of water vapor present in the Martian atmosphere is almost negligible, but the capacity of the air to hold moisture is very low. Therefore, even though the atmosphere is very dry, it is near saturation, and condensation often occurs, producing clouds. Martian clouds are visible from Earth, and several types were known even before the Mariner and Viking explorations. The **polar hood** forms over the poles in autumn as the polar caps are developing. Ground haze, which is essentially the same as morning fog on Earth, was found to be quite common by Viking. **Orographic clouds** form as a result of uplift of air over high regions and **wave clouds** form behind obstacles to wind flow, such as craters.

Mars, like Earth, has a global pattern of atmospheric circulation, but Mars' atmospheric circulation is simplified by the planet's lack of oceans, which store thermal energy on Earth. Conversely, the winds of Mars are more affected by topography and seasonal effects than those of the Earth. The Viking landers found summer winds to be 6 kilometers per hour (4 miles per hour) at night and 25 kilometers per hour (15 miles per hour) during the day. During Martian autumn, winds averaged 18 kilometers per hour (11 miles per hour), with gusts peaking at 36 kilometers per hour (22 miles per hour). During periods of high wind, dust storms occur on both local and global scales. Global dust storms tend to originate in the southern hemisphere near perihelion and expand to cover the remainder of the planet. Wind gusts during dust storms at the Viking landers reached almost 100 kilometers per hour (60 miles per hour), and changes in the sand as time passed could be seen at the landing sites.

The source of the Martian atmosphere is considered to have been gases that originated within the planet. The present atmosphere is probably just a small part of the total that originally outgassed, because the remainder would have escaped into space or been incorporated into the regolith or near-surface rocks. The various isotopes of argon present in the Martian atmosphere have implications for its origin and history. Mars has less argon-36 and argon-38 than Earth, which, in turn, has less of these isotopes than Venus. This is probably the result of differences in conditions at different locations within the solar system when the planets and their atmospheres were forming. Mars also has only about one-tenth as much argon-40 as Earth. Because argon-40 is produced by radioactive decay of potassium-40 in rocks, this means that Mars has had

less radioactivity and therefore has generated less internal heat.

A recent discovery about the Martian regolith may hold important implications for its atmospheric history. Spectroscopic evidence shows that the mineral scapolite may be an important constituent of the Martian soil. Scapolite can store carbonate ions inside of its crystal structure and may serve as a "storehouse" for much of the carbon dioxide that was formerly in Mars' atmosphere. Roger N. Clark has estimated that there may be as much as 1 atmosphere of carbon dioxide stored in this way, indicating that Mars' atmosphere may have been as dense as the Earth's atmosphere in the past.

■ Interior and Magnetic Field

Little evidence has been obtained about the Martian interior and magnetic field, primarily because of the lack of instrumentation. Only one American spacecraft, Mariner 4, carried a magnetometer, and although both Viking landers were equipped with seismographs, the instrument on Viking 1 failed and the one on Viking 2 detected more wind noise than "Marsquakes." The one working seismograph recorded a Marsquake of low intensity and probably local origin. The seismic data indicated that the crust at the Viking 2 site is about 16 kilometers (10 miles) thick.

Although there has definitely been faulting and volcanism on Mars, there is no evidence for Earth-style plate tectonics. There seems to be just one crustal plate at present, but the crust may have been more active in the past. It has been proposed that the difference between the low northern hemisphere crust and the higher southern hemisphere crust is the result of crustal drifting early in Mars' history, when the original northern crust drifted southward to underlie the original southern crust, forming an especially thick southern crust. Internal activity in Mars' past also was responsible for forming the Tharsis bulge, which is about 4000 kilometers (2500 miles) across and 10 kilometers (6 miles) high. Tharsis is not isostatically compensated, and it is the only appreciable gravity anomaly on Mars. The fact that this bulge is not compensated may be due to the thickness of Mars' lithosphere (crust and brittle upper mantle), which is estimated at 200 kilometers (125 miles). The Tharsis region is surrounded by numerous faults and fractures that were caused by crustal extensions as the bulge formed. Several complex models have been proposed to account for the Tharsis bulge, but there is no consensus yet regarding how it formed.

Mars appears to lack a magnetic field. One Soviet spacecraft measured a slight magnetic field, but that measurement has since been questioned. Even if a small field exists, it is much weaker than that of the Earth, although Mars rotates almost as fast as Earth. The absence of a Martian field is considered evidence that Mars probably lacks a molten core.

The structure of the Martian interior is believed to be similar to that of other terrestrial objects. Mars' uncompressed density is 3.71 grams per cubic centimeter, compared to the overall value of 3.93. Its moment-of-inertia coefficient of 0.365 indicates that there is central concentration of material and suggests that there is a core comprising 15% to 25% of the planet's total mass. As mentioned earlier, Mars' lithosphere is estimated to be fairly thick. The lower mantle may be somewhat similar to that of the Earth, except that there is undoubtedly much less convection. Two possible models for the core have been developed: one suggests that it is a molten mixture of iron and iron sulfide that extends to about half the radius of the planet, whereas the other proposes that a solid core of iron and nickel occupies only the inner third of the planet. The latter model would explain the lack of a Martian magnetic field. Which model is correct will not be known until further explorations are conducted.

■ History

The Martian surface is understood well enough that the relative ages of its rock units have been determined. These rock units, as well as Martian geologic history, have been divided into three main eras, **Noachian**, **Hesperian**, and **Amazonian**, each of which has been further subdivided to give eight time periods (and corresponding rock units). Ages of the various rock units are determined by the use of crater counts, but the lack of age-dated samples has caused Martian chronology to be much more uncertain than that of the Moon. Two major chronologies have been proposed, one of which has dates consistently earlier than the other. We will use the model with the earlier dates, which was proposed by G. Neukum and D. Wise.

The first period of Martian history was the Early Noachian, consisting of time before 3.92 billion years ago. This period corresponds roughly to Pre-Nectarian time on the Moon. Major Early Noachian events were the formation of the Martian lithosphere, the formation of many major impact basins, and the development of the extensive northern hemisphere lowland.

The Middle Noachian period lasted from 3.92 billion years to 3.85 billion years ago, and it corresponds to the Nectarian era on the Moon. This was the end of the period of heavy surface bombardment, and it was marked by the formation of smaller impact basins. Numerous small craters were formed, and the Tharsis region began to develop, as its earliest faults date from this time.

The Late Noachian period lasted from 3.85 billion years to 3.5 billion years ago and was marked by a decrease in cratering rates. Activity during this time included the formation of the early intercrater unit, which covers older highland material. There was volcanism in the highland areas and more faulting in the Tharsis region. This was the period of extensive runoff channel formation, and Valles Marineris also probably began to form at this time.

The early Hesperian period lasted from 3.5 billion years to 3.1 billion years ago. The cratering rate was low, but some large impact basins were still being formed. Faulting continued in the Tharsis region, and rifting continued in Valles Marineris. Volcanic activity formed numerous plains whose flow ridges are visible today and began to form Alba Patera.

The late Hesperian period lasted from 3.1 billion years to 1.8 billion years ago. This was a major period of outflow channel activity. The formation of the Tharsis bulge ended, and there was volcanism in a number of areas, including Elysium. Unconsolidated material was deposited at the south polar region, and layered deposits were formed in Valles Marineris.

The Early Amazonian period lasted from 1.8 billion years to 0.7 billion years ago. During this time, lava flows began to form the large shield volcanoes in the Tharsis region, including Olympus Mons, which probably attained most of its present size by the end of the period. The youngest impact basin, Lyot, formed during this time, and the amount of faulting, filling, and erosion occurring in Valles Marineris was greatly reduced.

The Middle Amazonian period lasted from 700 million years to 250 million years ago and was marked by continued volcanism in the Tharsis region, including Olympus Mons. Numerous lava flows occurred in the northern plains, and landslides occurred in Valles Marineris.

The Late Amazonian period began 250 million years ago. During this time, polar deposits have been reworked, as evidenced by the fact that their surfaces are unmarked by craters and are therefore very young. Volcanism in the Tharsis region ended at this time, with the last Martian volcanism probably occurring at Olympus Mons. Martian sand dunes were placed into their final form, and landslides occurred around the edge of Olympus Mons and in Valles Marineris.

■ Satellites

Mars' two small satellites are probably asteroids that have been gravitationally captured, although their orbits, which are nearly in Mars' equatorial plane, may indicate that they were accreted in their present locations. Although they were discovered by **Asaph Hall** (1829–1907) in 1877, Kepler had predicted them on the basis of numerology, because Earth had one satellite and Jupiter four! Jonathan Swift (1667–1745), aware of Kepler's prediction when he wrote the novel *Gulliver's Travels* in 1726, wrote that the Lilliputian astronomers had already discovered them. Swift described their orbits with uncanny accuracy, even though their actual discovery was 151 years away!

Phobos, the nearer moon, orbits only 9378 kilometers (5814 miles) from the center of Mars (Figure 16.21). Its period is 7 hours and 39 minutes, much less than a Martian day. It has an irreg-

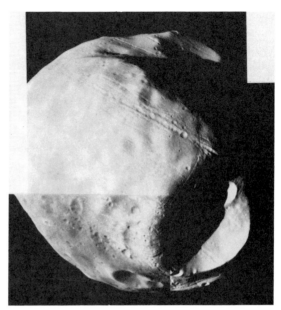

■ **Figure 16.21** Phobos. The large crater Stickney is at lower right. (NASA/JPL Photograph.)

ular, ellipsoidal shape, with dimensions of about 26 by 21 by 18 kilometers (16 by 13 by 11 miles). **Deimos** orbits at 23,459 kilometers (14,545 miles) and has a period of 30 hours and 18 minutes (Figure 16.22). It is similar in shape to Phobos but smaller, at 15 by 11 by 10 kilometers (9 by 7 by 6 miles). Neither satellite is massive enough to have had any tidal effects on Mars, but Mars has locked both satellites into 1:1 spin-orbit coupling. Phobos is gradually spiralling inward toward the

■ **Figure 16.22** Deimos. (NASA/JPL Photograph.)

planet and will impact in about 1 billion years, unless it breaks apart as a result of tidal forces while still in orbit and produces a ring around Mars.

The physical features of both satellites are similar. Both were photographed from relatively close proximity by Mariner 9 and the Viking orbiters. Their main characteristic is that they are heavily cratered. Phobos has three major craters, one of which, Stickney, is an impressive 10 kilometers (6 miles) across. Deimos' largest crater is only about 2.3 kilometers (1.4 miles) across. Some ejecta is present on each satellite, which is surprising in view of their low escape velocities. However, there are no secondary craters, because any fragments that could have caused them escaped during the impact process. Phobos' surface contains large grooves that are hundreds of meters wide and 15 to 30 meters (50 to 100 feet) deep. These grooves radiate outward from the large crater Stickney and appear to be a by-product of the impact that formed it.

Both satellites are covered by a thick layer of regolith. They are among the darkest bodies in the solar system, with albedos of 5% to 7%. Spectroscopic analysis indicates that they are similar in composition to a type of meteorite known as carbonaceous chondrites. Several asteroids also have this composition, supporting the theory that Phobos and Deimos may be captured asteroids.

■ Prospects for Future Exploration

Despite a relatively poor success rate with Martian spacecraft, the Soviet Union is planning a number of ambitious Mars missions in the coming years, including automated sample return missions and surface roving vehicles. The United States intends to launch a Mars Observer craft toward Mars in late 1992 to study Mars from orbit.

Just as human explorers will undoubtedly return to the Moon in coming decades, it is probably inevitable that human exploration of the red planet will eventually occur. Although such exploration promises rich scientific dividends, the voyage to Mars will be much longer than a trip to the Moon. It is not yet known with certainty what the long-term effects on the human body of such long periods in weightlessness would be. The Soviet Union is considering manned missions to Mars early in the next century, and there has been discussion of American missions or even joint Soviet–American efforts. Specific plans for manned flights will probably not be finalized for some time.

■ Chapter Summary

Mars, the red planet, has long been an interesting target for observers because of its changeability. The most controversial features of Martian observations were the canals, which were illusions caused by unresolved surface details. Many spacecraft have explored Mars, and the Viking landers returned considerable information from its surface.

Mars contains a variety of surface features, including craters, volcanoes, canyonlands, channels, polar caps, and wind-related features. The surface of Mars is covered with rock and sand and is probably composed of basaltic rocks and clay minerals produced by weathering of basalt.

Although evidence is still uncertain, it appears that there is probably no life on Mars. Its thin atmosphere is mostly carbon dioxide and nitrogen, and the planet is much colder than the Earth. Mars appears to lack a magnetic field, and the structure of its interior is not yet known with certainty. Martian geological history has been divided into three major eras, but an exact chronology is not yet possible because of the lack of age-dated samples. Mars has two small satellites that are probably captured asteroids.

■ Chapter Vocabulary

Alba Patera	gas-exchange	orographic clouds	Schiaparelli, Giovanni
Amazonian era	experiment	outflow channels	Secchi, Angelo
Antonindi, E.M.	Hall, Asaph	patera	Stickney
Arsia Mons	Hellas	Pavonis Mons	superior planet
Ascraeus Mons	Hesperian era	Phobos	superoxides
canal	labeled-release	photosynthesis	Tharsis
Chryse Planitia	experiment	polar caps	tholus
clay mineral	layered terrain	polar hood	Utopia Planitia
Deimos	Lowell, Percival	polar units	Valles Marineris
dry ice	metabolism	pyrolitic-release	Viking
Elysium	Noachian era	experiment	wave clouds
fretted channels	Olympus Mons	runoff channels	wind streaks

■ Review Questions

1. For what were Mars and its satellites named?
2. In what ways is Mars similar to Earth? In what ways is it different?
3. Describe the appearance of Mars in the sky and in a small telescope.
4. Describe the Martian surface features seen by early observers. Summarize the "canal" controversy and how it was resolved.
5. Summarize the highlights of space exploration of Mars.
6. Describe Martian craters and the way in which they differ from those on the Moon.
7. Describe Martian volcanic activity, and summarize similarities and differences from that on Earth.
8. Describe the appearance of Valles Marineris, and explain its possible origin.
9. What is the evidence that surface water erosion has occurred on Mars?
10. Describe the composition of the Martian polar caps and the periodic changes that occur in them. What sorts of material are deposited beneath them?
11. What is the evidence for Martian wind erosion?
12. Describe the landing sites of Vikings 1 and 2. How did the landers analyze these sites, and what did they discover?
13. How did the Viking landers search for Martian life? What is the current interpretation of the results obtained?
14. Summarize what the Viking landers learned about Martian weather.
15. What is known about the interior of Mars?
16. Summarize the geological history of Mars. When did its major features develop?
17. Describe the major properties of Phobos and Deimos.

■ For Further Reading

Arvidson, R. et al. "The Surface of Mars." In *Scientific American,* March, 1978.

Batson, R. et al., editors. *The Atlas of Mars.* NASA, Washington, D.C., 1979.

Carr, Michael H. *The Surface of Mars.* Yale University Press, New Haven, 1981.

Carr, Michael H. "Mars." In Carr, Michael H., editor. *The Geology of the Terrestrial Planets.* NASA, Washington, D.C., 1984.

Hartmann, William K. "What's New on Mars?" In *Sky and Telescope,* May, 1989.

Loevy, Conway B. "The Atmosphere of Mars." In *Scientific American,* July, 1977.

Veverka, Joe. "Phobos and Deimos." In *Scientific American,* February, 1977

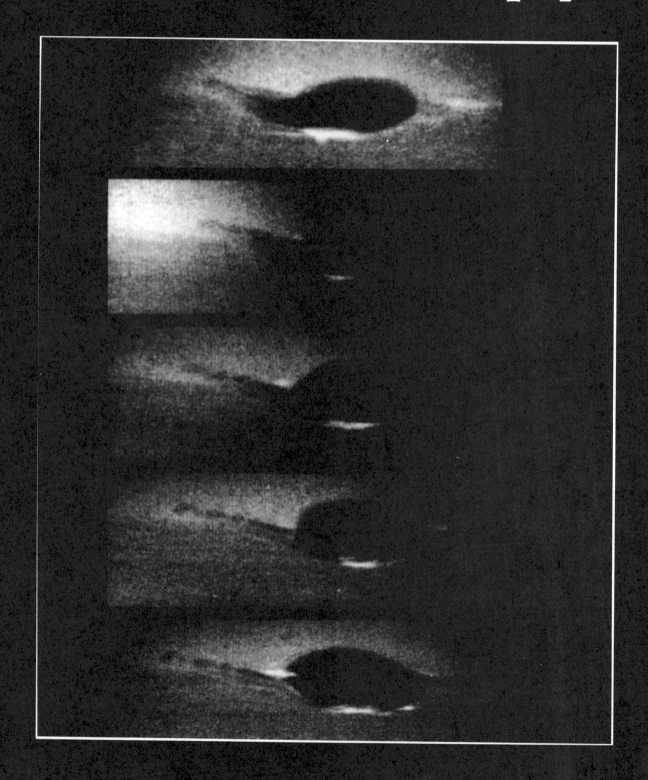

17

Overview of the Gas Giant Planets

160 14 174

CHAPTER OBJECTIVES

After studying this chapter, you should be able to

1. Summarize the basic physical properties of the gas giant planets.
2. Understand the differences between a typical gas giant and a typical terrestrial planet.
3. Understand the similarities and differences between a typical gas giant and the Sun.
4. Describe the effects of the gravity of the gas giants on other solar system objects.
5. Describe how the atmospheres and interiors of gas giants are studied.
6. List the components of the atmospheres and interiors of gas giants.
7. Summarize conditions within the atmospheres and interiors of gas giants.
8. Describe the magnetic fields of the gas giants and their effects.

Neptune's Great Dark Spot, photographed by Voyager 2 in 1989. Note the changes caused by atmospheric circulation during the time span during which these photographs were taken. (NASA/JPL Photograph.)

Beyond the orbit of Mars, the nature of the solar system changes dramatically. Between Mars and Jupiter lies the asteroid belt, where hundreds of thousands of small rocky and metallic bodies orbit the Sun. Beyond this lie the gas giant planets, along with their retinues of rings and satellites. The four gas giants, Jupiter, Saturn, Uranus, and Neptune, are quite different from the planets we have examined previously. Although we have relied extensively on geological concepts in describing the terrestrial planets, the fact that the gas giants lack solid surfaces gives them very little in common with the Earth with which we are so familiar. Fortunately, many secrets of the gas giants have been revealed during the space age, and we can now begin to understand these planets and how they compare with the rest of the solar system. We discuss general properties of the gas giants in this chapter; subsequent chapters will present details of the individual objects.

■ Basic Physical Properties

If we could place a typical gas giant next to a typical terrestrial planet, we would notice many contrasts (Figure 17.1). Terrestrial planets are small and dense, have solid surfaces, and may be surrounded by atmospheres. On the other hand, gas giants are large, of low density, and totally gaseous, so that no solid surfaces exist beneath the cloudy blankets that surround them. Gas giants range in size from Neptune, whose 49,524-kilometer (30,760-mile) diameter is 10 times that of Mercury, to Jupiter, whose 142,855-kilometer (88,730-mile) diameter is 11 times that of Earth. Even though these objects are very massive, their average densities are low, ranging from 1.75 grams per cubic centimeter for Neptune to 0.7 grams per cubic centimeter for Saturn. These densities are close to that of the Sun, and this fact is one of the primary lines of evidence that these planets, especially Jupiter and Saturn, which are

■ **Figure 17.1** Comparison of a typical terrestrial planet (Earth) with the gas giants Jupiter and Saturn. The three planets are shown to scale. (NASA Photograph. Composite prepared by Stephen Paul Meszaros.)

made mostly of hydrogen and helium, have compositions similar to that of the Sun. Although rocky or metallic material may be present near the centers of these planets, the extent of such material is necessarily small, or else the planets would have higher densities. Although these planets are composed largely of materials that are gaseous under the conditions found at the Earth's surface, conditions within the gas giants are quite different from those with which we are normally familiar. In fact, pressures within these planets are so high that the hydrogen and helium present there exist in a dense, liquefied state. Although we normally expect hydrogen and helium to liquefy only at low temperatures, they can also liquefy at high temperatures if the pressure is great enough, as is the case within the gas giants.

Unlike Jupiter and Saturn, which are largely composed of hydrogen and helium, Uranus and Neptune are believed to contain large quantities of water, methane, and ammonia near their centers, probably in a liquid or semisolid state due to the high pressure. This different composition probably reflects the fact that Uranus and Neptune were formed farther from the Sun, where water, ammonia, and methane ices were more abundant and the lightest elements may have been less abundant. As a result of the presence of these denser constituents, Uranus and Neptune have higher densities than Jupiter and Saturn.

When we look at a gas giant, what appears to be its outer surface is actually the outer layer of an atmosphere that descends deep into the planet. For this reason, gas giants lack the clear atmosphere–planetary surface demarcation present on a terrestrial planet like Earth. (Recall that the lack of a distinct surface is also a characteristic of the Sun, showing another resemblance between the Sun and the gas giants.) Although their atmospheres are primarily hydrogen and helium, the fact that the gas giants, particularly Jupiter, have colorful atmospheres shows that additional components must be present, because hydrogen and helium are colorless. Some of these components are dispersed throughout the atmosphere, whereas others are concentrated into bands of clouds. These atmospheric markings change with time, indicating that weather of some sort exists.

Observing atmospheric markings is one of the methods used to reveal the rotational rates of the gas giants. Although such huge objects might be expected to rotate very slowly, the gas giants are the solar system's fastest rotators, as Jupiter rotates in less than 10 hours and even the slowest-rotating gas giant, Neptune, spins faster than Earth. Because of their rapid rotation and the fact that gaseous bodies are not rigid, these planets have an oblate form. This polar flattening can easily be seen when viewing Jupiter and Saturn in even a small telescope. The reason that the gas giants are oblate and the Sun is not is probably due to the fact that the Sun rotates much more slowly, requiring about a month to spin on its axis, and that the Sun does not have the same internal structure as the gas giants. One feature that the rotation of the gas giants does have in common with the Sun is that all of these gaseous objects have different rotational rates at different latitudes. This means that observed rotational rates differ depending upon the latitude of the features whose motion is timed. For example, Jupiter's high latitudes require about 5 minutes longer to rotate than does its equatorial region. The "true" rotational rate of a gas giant is the rate at which the planet's dense core rotates. Because the core cannot be observed directly, its rotational rate must be measured by indirect methods. This is usually done by observing variation in radio emission produced by a planet's radiation belts or other phenomena associated with its magnetic field. Because magnetic fields are produced deep within the gas giants, the rotational rates of phenomena associated with them indicate how fast the core of the planet, not the atmospheric layer, is rotating.

The gas giants differ from the inner planets in other ways besides composition and structure. They each have numerous satellites and a ring system composed of numerous small bodies orbiting in belts or bands. Also, as predicted by Bode's law, the spacing between the gas giants is much greater than planetary separations within the inner solar system, and, as given by Kepler's third law, they require much longer to orbit the Sun.

■ Stars that Failed

Why are the gas giants so different from the inner planets, and why are they located in the outer solar system? The reason probably lies with the Sun. As our solar system was forming, the thermal radiation from the incipient Sun warmed the inner solar system so much that the majority of the light elements (hydrogen and helium) were driven off and volatile ices (water, ammonia, methane, and so forth) never condensed there. In the outer solar system, cooler conditions allowed hydrogen and helium to remain, and the gas giants formed by accumulating huge quantities of these light elements. In addition, volatile ices had condensed in the regions where Uranus and Neptune formed and were incorporated into them. Recall that the process by which the terrestrial planets reached their present size is considered to have been accretion, the combination of small rocky and metallic bodies. Although the cores of the gas giants may have formed by accretion, the planets then grew by the gravitational attraction of gas and ice molecules from the surrounding areas. In other words, material was probably added to Jupiter and the other gas giants bit by bit in the form of individual atoms and molecules, not by the influx of large bodies, which is the way in which the terrestrial planets were formed.

Although even mighty Jupiter has only about 0.1% of the mass of the Sun, we may argue that Jupiter is more like a star than a terrestrial planet. In terms of composition and density, this is certainly true. The gas giants are also star-like in that they radiate more energy into space than they receive from the Sun, although both quantities are relatively small. What is the source of this energy?

As described in Chapter 6, stars produce their energy by nuclear fusion, which can only occur where temperatures are in excess of 10 million K. The centers of newly forming stars (**protostars**) are initially heated to this temperature by the process of gravitational contraction, but they radiate some energy as a result of the contraction even before fusion begins. A protostar that is too small will not generate enough heat in its core for hydrogen to fuse into helium. This is what happened to Jupiter, which had a mass low enough that gravitational contraction could heat its core to only about 25,000 K. If Jupiter had been about 80 times more massive, fusion would have begun to occur in its core, and it would have become a star. (The least massive stars have only about 8% of the Sun's mass, whereas Jupiter has about 0.1% of its mass). Even though Jupiter failed to become a star, some thermal energy is produced in its core by gravitational pressure. This energy travels outward through its interior by convection, reaches the surface, and radiates into space. The same process occurs to a lesser degree in the other gas giants. Because of this behavior, the term **substar** is often applied to gas giant planets.

Suppose Jupiter had been massive enough to have become a star. In that case, the solar system would have been a binary (two-star) system. Many stars are not single like our Sun, but have one or more stellar companions. Some **binary stars** orbit each other very closely, whereas others are hundreds or even thousands of astro-

nomical units apart. Jupiter is far enough away from the Earth that had it become a star, it would have contributed less energy to our planet than the Sun, so it would not have greatly affected Earth's climate. A Jupiter massive enough to have become a star, however, would have had stronger gravity, which would have influenced the orbital spacings of all solar system objects.

■ Super Gravity and Grand Tours

Because of their large masses, the gas giants have strong gravitational fields, which (along with the cooler conditions in the outer solar system) allow them to retain hydrogen and helium in their atmospheres. More importantly, these planets' gravitational fields have a great deal of influence on other solar system objects. For example, the gas giants exert powerful tidal forces on their satellites, and nearly all satellites of gas giants have perfect 1:1 spin-orbit coupling. Furthermore, the gravity of the gas giants has allowed them to capture other objects (comets and asteroids), and many gas giant satellites are believed to be captured bodies. This is especially true of their outermost satellites, which can be identified as captured objects if they have retrograde or highly inclined orbits. (The four outermost satellites of Jupiter are examples.) In some cases, capture by gas giants may only be temporary. For example, calculations reveal that several comets have spent months or years as temporarily captured satellites of Jupiter before resuming their flights through the solar system. A small satellite of Jupiter "discovered" in 1975, but never seen again, may have been a temporarily captured object.

The influence of a gas giant's gravity, however, is not limited to its satellite system, as Jupiter's gravity is believed to have disrupted the accretion process in what is now the asteroid belt. As a result, the smaller objects there remained separate, forming the asteroids, rather than combining to form another terrestrial planet. Close approaches of some asteroids to Jupiter have resulted in their capture by that massive object or their deflection into new orbits.

As described in Chapter 3, the gravitational

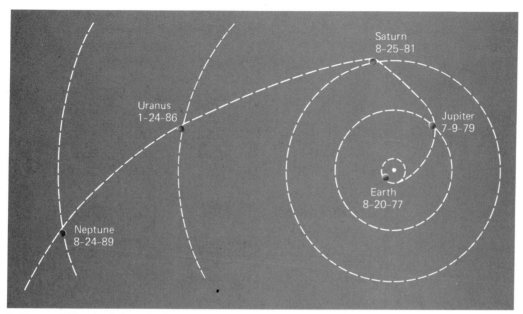

■ **Figure 17.2**   The flight path of Voyager 2.

effects of one body upon another object's orbit are called perturbations. The gas giants perturb the orbits of one another, and also of the inner planets. It was irregularities in Uranus' orbit that enabled the existence of Neptune to be deduced and its location determined. Comets traveling through the outer solar system are subject to perturbations and even capture. Some of the outer planet satellites may be captured comets, and many comets that escape capture have their orbits changed by the gas giants. In the 1930s, Comet Oterma barely escaped capture by Jupiter. Each gas giant has a "family" of comets, each of whose members has its aphelion near the orbit of the planet. For example, Halley's comet is a member of Neptune's family of comets.

As described in Chapter 5, one of the many interesting aspects of the gas giants is that their strong gravitational fields have been used to facilitate space travel. A spacecraft flying past a gas giant can obtain a gravitational "boost" from the planet's gravity to speed it to its next planetary destination. During the 1970s, a rare opportunity for planetary exploration presented itself. Because the gas giants have orbital periods ranging from 12 to 165 years, times when all four of them are approximately aligned on the same side of the

Sun are very infrequent. Such an alignment occurred in the late 1970s, making **Grand Tour** space missions possible. Pioneer 11 and Voyager 1 visited both Jupiter and Saturn, but Voyager 2, launched in 1977, had the most ambitious course of all. It flew by Jupiter in 1979, Saturn in 1981, and Uranus in 1986, gaining a gravitational boost at each planet. Voyager 2 passed Neptune in 1989, 12 years after launch; a direct flight from Earth to Neptune would have required 30 years. The trajectory of Voyager 2 is shown in Figure 17.2.

■ Atmospheres

When we observe the gas giants, we see their atmospheres, which are the easiest parts of the planets to study. This is particularly true of Jupiter and Saturn, whose atmospheric markings are easily visible from Earth. (Uranus and Neptune are more difficult to observe because of their greater distances from Earth.) No two of the gas giants look alike, and the reason for their atmospheric diversity is the temperature variation resulting from their different distances from the Sun. Changing temperatures affect the compositions

■ **Figure 17.3** Jupiter as it appears through a telescope. Note the distinctiveness of its various atmospheric markings, including the Great Red Spot. (Lunar and Planetary Laboratory, University of Arizona.)

of the atmospheres, because certain gaseous components may be removed if temperatures are low enough that they freeze. Jupiter's atmospheric markings are especially vivid and colorful, consisting of continuous bands that encircle the planet, as well as localized spots that are probably some sort of atmospheric storms (Figure 17.3). Jupiter's light-colored bands are called **zones**, and the dark-colored ones are called **belts**. The most notable of Jupiter's localized spot markings is the **Great Red Spot**, a long-lasting feature that is larger than the Earth. Voyager 2 revealed that Neptune's atmosphere also contains considerable weather activity, exceeded only by that of Jupiter's atmosphere. Neptune was found to have a **Great Dark Spot**, similar to Jupiter's Great Red Spot.

Information obtained during the Voyager missions has greatly increased our understanding of the atmospheres of the gas giants. For example, we have tracked the movements of gas in these atmospheres and can compare their global atmospheric circulation patterns to those of Earth. Even though Earth's atmosphere is obviously very different from that of a gas giant, the same physical laws govern the behavior of all atmospheres. For this reason, studying the meteorology of the gas giants helps us better understand the atmosphere of the Earth. Those who consider planetary exploration a waste of money should realize that information gained about other planets' at-

mospheres helps us improve weather forecasts on Earth.

One of the many important studies made of the gas giants' atmospheres is spectroscopic analysis of their gaseous components. Because the atmospheres of these planets are not separate entities like those of the terrestrial planets, they are probably representative of the planets' overall compositions. For this reason, the abundances of gases found in a gas giant's atmosphere is probably a good indicator of the planet's bulk composition.

The composition of the gas giants' atmospheres was determined only recently. The first information came in the 1930s, when methane and ammonia were identified spectroscopically in the atmospheres of Jupiter and Saturn. We now know that these are not very abundant components in the gas giants' atmospheres. However, the spectroscopic features of ammonia and methane are very intense, allowing them to be identified even if present in only small amounts. Molecular hydrogen (H_2) was first detected about 1960. Although it is the most abundant component in the atmospheres of the gas giants, its absorptions are very weak and difficult to detect. Helium cannot be detected from Earth, but it was identified by Voyager observations. On the whole, the gas giants' atmospheres have compositions approximately the same as that of the Sun: the Sun is 89% hydrogen and 11% helium; Jupiter is 90% hydrogen and 10% helium; and Saturn is 94% hydrogen and 6% helium. (These are abundances by number of atoms. Percentage abundances by mass would be different, because helium and hydrogen have different masses.) As mentioned previously, additional atmospheric components are present and are responsible for the intense atmospheric coloration, particularly of Jupiter. The exact identities of these substances are unknown, but sulfur compounds and exotic organic compounds have been suggested as likely candidates. Some of the coloration is produced in the clouds, as will be discussed shortly.

One of the best ways to study temperature, pressure, and other characteristics of the gas giants' atmospheres has been through the use of radio occultation as a spacecraft disappears from

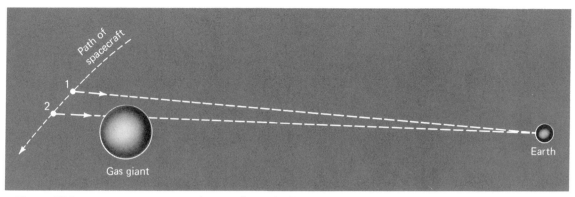

■ **Figure 17.4**   How a gas giant's atmosphere can be studied by observing the occultation of radio waves from a spacecraft as it transmits signals to Earth. As the spacecraft moves from point *1* to point *2*, its signals pass through successively deeper layers of the gas giant's atmosphere.

view behind a gas giant (Figure 17.4). This technique allows examination of the outermost layers of an atmosphere only; deeper inside a planet, its gas absorbs all radiation and is opaque to radio signals. Because of the great distances of these planets from the Sun, temperatures at the tops of their atmospheres are much cooler than Earth, even though the planets radiate more energy than they receive from the Sun. Jupiter is warmest at 170 K (−150°F) and Neptune is coolest at 59 K (−353°F). Unlike on Earth, the temperatures of the gas giants are independent of latitude, even though their equators (with the exception of highly oblique Uranus) receive more radiation from the Sun than do their poles. It is believed that this is the result of heat transport occurring deep within these planets, as well as the outflow of internal energy.

Radio occultation studies have shown that atmospheric temperature varies widely with altitude. Because gas giants lack solid surfaces, the starting point for altitude measurements on them is the point known as the **cloud top,** defined as the altitude where pressure is 100 millibars. This is also the altitude at which atmospheric temperature is lowest. Above this, temperature rises slightly for reasons that are not yet known, although the release of auroral energy is a likely cause. Temperature rises quickly below the cloud-top point because of the thermal energy coming from the planet's interior. Temperatures equivalent to normal room temperature occur

about 100 kilometers (60 miles) below the cloud tops on Jupiter and 300 kilometers (185 miles) deep on Saturn.

Voyager measurements revealed that three layers of clouds occur in the atmospheres of Jupiter and Saturn, and researchers believe that each layer has distinct characteristics: the top cloud layer is red and composed of **ammonia,** the middle layer is white and brown and made of **ammonium hydrosulfide,** and the bottom layer is blue and contains water in either solid or liquid form. It is not known whether this model is correct, as actual conditions may be more complicated as a result of vertical mixing of the various cloud components. Indications are that these three cloud layers could account for the vivid colors of gas giants, especially Jupiter. Blue clouds are deep, relatively warm, and visible only through holes in the upper clouds. Red markings are produced by the highest, coldest clouds, and white and brown markings are caused by clouds that are intermediate in both altitude and temperature. Neptune's cold temperature gives it a type of cloud found in no other gas giant atmosphere — high, white, wispy clouds of methane ice, similar in appearance to terrestrial cirrus clouds.

As described in Chapter 11, the Earth's atmosphere is divided into three circulation cells in each hemisphere. The fact that three cells are present instead of one is due to the Coriolis force caused by the Earth's rotation, which acts to break down into three distinct parts what would

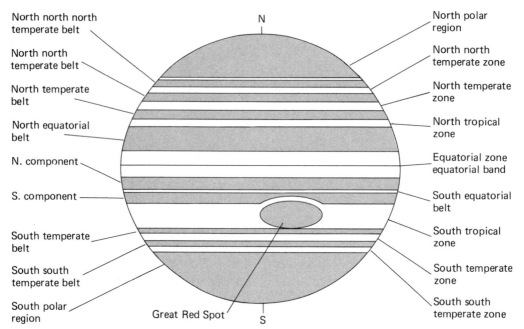

■ **Figure 17.5** Gas giant belts and zones. A view of the pattern of belts (*dark areas*) and zones (*light areas*) on Jupiter. (Redrawn after a NASA illustration.)

otherwise be a single cell. With their faster rotation and consequently greater Coriolis force, gas giants have even more complex circulation patterns on a scale much larger than the Earth's. (Unlike on Earth, the atmospheric circulation of the gas giants is driven mainly by internal heating, not by incident solar radiation.) Jupiter has five or six circulation cells (also called **zonal jets**) in each hemisphere. Saturn also has more circulation cells than the Earth, and its atmospheric circulation is even stronger than that of the larger Jupiter. The winds at Saturn's equator are about 1800 kilometers per hour (1100 miles per hour) in the direction of the planet's rotation! The boundaries of these circulation cells are marked by the atmospheric features that give the gas giants their characteristic striped appearance (Figure 17.5). Although these striped features are variable in color, width, and location, the wind velocities and the boundaries of the zonal jets appear to be fairly constant. Neptune has winds with speeds of about 1100 kilometers per hour (700 miles per hour) in the direction opposite to the planet's

rotation, the fastest such winds observed in any gas giant atmosphere.

The spots in gas giant atmospheres appear to be formed by eddies of stable atmosphere trapped between two zonal jets, as shown in Figure 17.6. The brilliant colors of many of these spots are probably the result of unusual atmospheric components trapped in these areas. As a result of the strong winds in relatively close proximity to them, distortions in the spots can occur and their appearance is variable. Furthermore, such spots tend to be temporary, rather than permanent, because the opposing winds will eventually shift slightly and cause the eddies (and related spots) to be sheared apart. Jupiter's mysterious Great Red Spot, which has somehow survived for over three centuries, is an intriguing exception. Time-lapse photographs of the Great Red Spot taken by the Voyagers show that it appears to be a huge storm somewhat similar to a terrestrial hurricane. Because of the huge size of the Great Red Spot, however, it is very difficult to understand what energy source would be sufficient to pro-

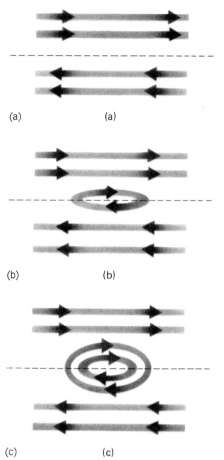

(a)   (a)

(b)   (b)

(c)   (c)

■ **Figure 17.6** Formation of atmospheric spots. A vortex develops between atmospheric areas of opposite wind direction. Compare with spots shown in the photograph of Jupiter in Figure 17.1.

duce such a large, long-lasting feature. However, computer simulations of Jupiter's atmosphere indicate that such a large feature might be permanent.

■ Interiors

One of the biggest mysteries about the gas giants is their internal structure. Although great progress has been made in understanding their interiors, many questions remain unanswered. Because seismic studies are impossible for these objects, other methods of study must be used.

It has long been known that the masses of the gas giants are centrally concentrated. This can be determined by study of the gas giants' moments of inertia, which can be measured by observation of the planets' responses to the tidal forces produced by their satellites. Moment-of-inertia coefficients for the gas giants range from 0.253 for Jupiter to an estimated 0.15 for Neptune. (Recall that the lower the value, the greater the degree of central concentration, and that the value for the Sun is 0.06.) Other clues to interior structure are these planets' low densities and Sun-like compositions. The rapid rotation rate and degree of oblateness have provided information about the strength of the internal components of the gas giants. The Voyager missions have gathered much information about the magnetic fields of the gas giants. Because these fields are internally generated, they indirectly reveal information about planetary interiors.

In some ways, studies of gas giant interiors bear more resemblance to the calculation of models of the solar interior than to techniques for investigating terrestrial planet interiors. Interiors are modeled using techniques similar to those described for the Sun in Chapter 6. For example, Jupiter was once thought to be entirely gaseous, like the Sun. A more modern model, proposed in 1938, suggested a dense inner core, an inner mantle of ice, and an outer mantle of hydrogen. The later discovery that heat is leaving Jupiter led to the realization that convection must be occurring within its interior, and the models were revised again. The interior models will be examined in more detail when we discuss individual gas giant planets in upcoming chapters.

The interiors of all four of the gas giant planets have many features in common. The major component of gas giant interiors is thought to be hydrogen, in the case of Jupiter and Saturn, and a mixture of ices (plus some hydrogen), in the case of Uranus and Neptune. Near the surface, below the outer atmospheric levels, hydrogen exists in the form of H_2 molecules, which contain two protons and two electrons that form a distinct unit. Deeper in the planets exists **metallic hydrogen,** an unusual form of hydrogen that can form only under very high pressures. In metals, electrons "drift" around and do not belong in specific atomic nuclei. The free electrons in metals allow them to be good conductors of electricity. Metallic hydrogen has not yet been produced in terrestrial laboratories, but the immense pressure inside Jupiter and Saturn probably allows it to exist there. (Uranus and Neptune are small enough that they have insufficient internal pressure to form metallic hydrogen.) All hydrogen inside gas giants is probably in a fluid state, with convection actively bringing thermal energy to the surface. The centers of the gas giants possibly contain rocky or nickel–iron cores, with some ices, of water, ammonia, and methane, surrounding them. A liquid or semisolid layer of water, ammonia, and methane probably constitutes the bulk of the interiors of Uranus and Neptune. Although we talk about "ices" inside these planets, these materials are not cold but are at temperatures of several thousand degrees. Such materials are called ices because they are solid (or semisolid) forms of materials that, at the Earth's surface, would be solid only if at low temperature. The term is still used even if it is high pressure, not low temperature, that causes them to be in the solid state.

■ Magnetic Fields

Prior to discussing the magnetic fields of the gas giants, let us review some of the basic features of Earth's field for comparison. The magnetic axis of Earth is tilted 11.4° to its rotational axis, and the field is oriented opposite to the rotational sense, although magnetic field reversals are known to occur periodically. (In other words, the Earth's north magnetic pole is now located close to our planet's south rotational pole.) The field, which is probably formed by currents within the Earth's conductive outer core, interacts with the solar wind, creating the Van Allen belt of charged particles.

Appropriately, the gas giants have intense and complicated magnetic fields. The magnetic fields of Jupiter and Saturn are probably produced by currents within the conductive metallic hydrogen in their interiors, whereas those of Uranus and Neptune are probably produced in the conducting layer of water, methane, and ammonia near each planet's core. Jupiter's field was discovered from Earth by analysis of radio emissions that come from the planet itself as well as from trapped particles in its radiation belts. Further study from the Pioneer and Voyager spacecraft revealed additional details of the field. The other three gas giants have fields too weak to be detected from Earth, but they were discovered by spacecraft. (It has been speculated that Saturn's extensive ring system may somehow interfere with its field and prevent its detection from the Earth.)

The fields of Jupiter, Saturn, Uranus, and Neptune are all oriented in the same sense as the planet's rotation. (In other words, the north magnetic poles of these worlds are close to their north rotational poles.) It is not yet known whether magnetic reversals occur in the fields of these planets. Saturn's magnetic axis is nearly perfectly aligned with its rotational axis, with an inclination of only 0.8°. Jupiter's magnetic axis is inclined 10°, Neptune's 50°, and Uranus' 58.6°. The overall strengths of the magnetic fields of the gas giants (total magnetic moments) are much greater than that of Earth, but because of the planets' larger diameters, the strength of the fields at the planets' surfaces is comparable to that of the Earth's field. Like that of the Earth, the magnetic fields of the gas giants interact with the solar wind to form magnetospheres and radiation belts, and these planets have satellites that are close enough to interact with the magnetosphere, producing "wakes" of stirred-up material. Considerable information was obtained by targeting flyby spacecraft near the magnetic shadows of the satellites of Jupiter, but interaction with these powerful fields was not without risks, because trapped particles in Jupiter's magnetic fields did considerable damage to the electronics of the Pioneer spacecraft. (Had these probes been manned, their crews would have been killed.)

Voyager photographs of the night surface of Jupiter revealed that auroras occur there, produced by a process similar to those on Earth (Figure 17.7). In both cases, the upper atmosphere is excited by energetic charged particles pouring in

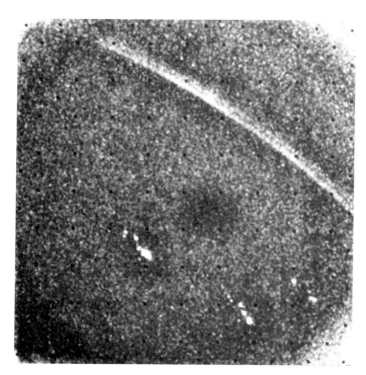

■ **Figure 17.7** Auroras in Jupiter's atmosphere. This photograph of Jupiter's night hemisphere, obtained by Voyager 1, shows auroras near the edge of the planet. The bright markings near the bottom of the photograph are lightning flashes. (NASA/JPL Photograph.)

near the poles, where the magnetic field does not provide protection. Auroras have also been detected in the atmosphere of Neptune.

■ Satellites

In 1610, Galileo discovered four satellites orbiting Jupiter, the first planetary satellites other than the Moon to be discovered. As telescopes improved, more and more satellites were discovered orbiting Jupiter and the other gas giants. In recent years, many new satellites have been discovered by the Pioneer and Voyager spacecraft, but even they may have missed objects because of the rapidity of their flights through the respective systems. All of the gas giant planets have satellites, ranging in number from 8 for Neptune to 17 for Saturn. The large number of satellites around the gas giants makes them resemble miniature versions of the solar system. (Neptune's satellite system is somewhat unusual, as its largest satellite, Triton, has a retrograde orbit.)

Although we will describe the satellite systems of the gas giants and the individual satellites in upcoming chapters, several features common to all gas giant satellites can be noted at this point. Gas giant satellites are all composed of rock and ice and have lower average densities than the four terrestrial planets. The surfaces of these satellites are exceedingly cold by terrestrial standards, ranging from about 105 K (−270°F) for those of Jupiter to about 40 K (−387°F) for those of Neptune. Some of the satellites of gas giant planets are substantial objects that are larger than the Moon, Mercury, and Pluto, whereas others are very small. The solar system's largest satellite is Ganymede, which orbits Jupiter. Even though Ganymede has a diameter of 5260 kilometers (3260 miles), it is still less than 4% of the diameter of Jupiter. No gas giant has satellites that are appreciable fractions of its size, as the Moon is to the Earth. Two satellites, Titan, which is Saturn's largest satellite, and Triton, which is Neptune's largest, are large enough to have atmospheres.

In addition to their satellite systems, the gas giants have **ring systems**, which are composed of countless small bodies orbiting in distinct belts. Although each component of a ring is technically an individual satellite, there are so many of them in a ring system that they are literally uncountable and cannot be considered individually. So much has been learned about planetary ring systems that we will devote the next chapter to a discussion of them.

■ Chapter Summary

Unlike the terrestrial planets, gas giants are large planets that are composed of light elements; consequently, these planets have low densities. They lack solid surfaces, and the only portions of the planets directly observable are their atmospheres. The gas giants are similar to the Sun in composition and emit more radiation than they receive, although they were not massive enough to become stars. Because of their powerful gravity, gas giants influence a wide range of solar system objects, from their satellites to more distant bodies.

Study of gas giant atmospheres reveals that they, and presumably the entire planets, are largely hydrogen and helium. Other atmospheric constituents produce colorful atmospheric markings. Atmospheric circulation produces a pattern of spots and other markings, which are particularly noticeable on Jupiter.

The interiors of Jupiter and Saturn contain both molecular and metallic hydrogen, along with helium and a solid core. Uranus and Neptune lack metallic hydrogen and contain more ices in their interiors. All gas giants have internally generated magnetic fields and associated features, and they also have numerous satellites and complex ring systems.

■ Chapter Vocabulary

ammonia	binary star system	Great Red Spot	ring system
ammonium hydro-sulfide	cloud top	metallic hydrogen	substar
	grand tour	molecular hydrogen	zonal jets
belt	Great Dark Spot	protostars	zone

■ Review Questions

1. Contrast the gas giants and the terrestrial planets.
2. How are the gas giants believed to have formed? How does their origin differ from that of the terrestrial planets?
3. What are the differences between a gas giant such as Jupiter and a full-fledged star?
4. How have the gas giants' strong gravitational fields affected other solar system bodies? How have they facilitated space exploration?
5. Describe the atmospheres of gas giants using comparisons to the Earth and the Sun.
6. How do we study the interiors of gas giants? What materials do we believe are contained within them?
7. Describe the magnetic fields of gas giants.
8. How is a typical gas giant satellite different from the planet around which it orbits?

■ For Further Reading

Hubbard, William. "Interiors of the Giant Planets." In Beatty, J. Kelly and Andrew Chaikin, editors. *The New Solar System,* Sky Publishing, Cambridge, MA, 1990.

Hunten, Donald M. "The Outer Planets." In *Scientific American,* September, 1975.

Ingersoll, Andrew P. "Jupiter and Saturn." In *Scientific American,* December, 1981.

Ingersoll, Andrew P. "Atmospheres of the Giant Planets." In Beatty, J. Kelly and Andrew Chaikin, editors. *The New Solar System.* Sky Publishing, Cambridge, MA, 1990.

Smith, Bradford A. "The Voyager Encounters." In Beatty, J. Kelly and Andrew Chaikin, editors. *The New Solar System.* Sky Publishing, Cambridge, MA, 1990.

18

Planetary Rings

CHAPTER OBJECTIVES

After studying this chapter, you should be able to

1. Describe the ring systems of each of the gas giants and how they were discovered.
2. Distinguish between forward and backward scattering of light from ring particles.
3. Understand the concept of the Roche limit.
4. Describe how ring particles orbit and how they are influenced by guardian satellites.
5. Explain how we determine the sizes and compositions of ring particles.
6. List the possible ways in which ring systems may have formed.

Saturn's rings, with the Earth shown at the same scale. (NASA Photograph. Composite prepared by Stephen Paul Meszaros.)

One of the most beautiful sights visible through any telescope is Saturn (Figure 18.1). Although the markings on the planet's disk are relatively subdued, the spectacular **ring system** surrounding it makes Saturn a favorite target of telescope users, and Saturn is often called the "ringed planet." Even though recent discoveries have shown that Saturn is not the only planet with rings, it remains the only one whose rings can be seen from Earth through a small telescope. In this chapter, we describe how the ring systems of the gas giants were discovered and summarize what is currently known about planetary rings.

■ The Discovery of Rings

Although Galileo made many important astronomical discoveries, his telescopes' optics were so poor that he never determined the true nature of Saturn. He described the planet as having "handles" and thought that it was perhaps accompanied by two large, nearby satellites. It was not until 1659 that Christiaan Huygens (1629–1695) was able to see that Saturn was circled by a ring system. In the three centuries since then, improved telescopes (and spacecraft visits) have given us increasingly clear views of Saturn's rings and enabled us to understand their physical nature. For most of that time, Saturn was assumed to be the only planet with rings, but this view changed during the 1970s.

On March 10, 1977, a group of researchers led by James Elliot observed an occultation of a faint star by the planet Uranus. They witnessed this event from a specially modified aircraft named the Kuiper Airborne Observatory, which contained astronomical instruments. The aircraft was flying over the southern Indian Ocean, as the occultation was only visible from that limited area of the Earth's surface. They hoped to obtain an improved value for the diameter of Uranus by measuring exactly how long the star was oc-

271

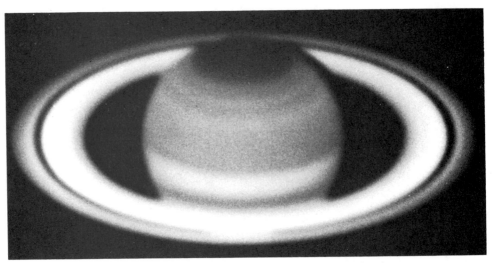

■ **Figure 18.1** Saturn as it appears through a telescope. Note the planet's most spectacular features, its rings. (Photograph courtesy of the Observatories of the Carnegie Institution of Washington.)

■ **Figure 18.2** Light intensity during the occultation that revealed Uranus' rings. The top tracing shows light drops prior to the occultation by the planet, the bottom tracing after the occultation. Note the symmetry of the two tracings. (Diagram by James L. Elliot. Reproduced, with permission, from the *Annual Review of Astronomy and Astrophysics*, volume 17. Copyright © 1979 by Annual Reviews, Inc.)

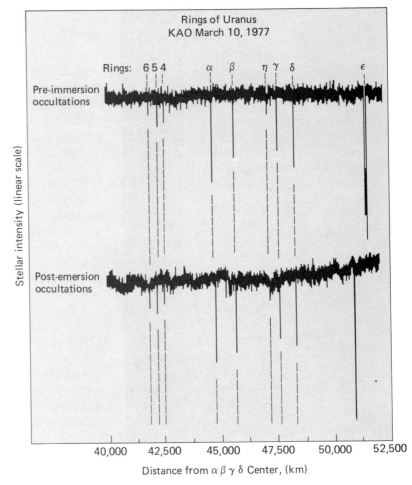

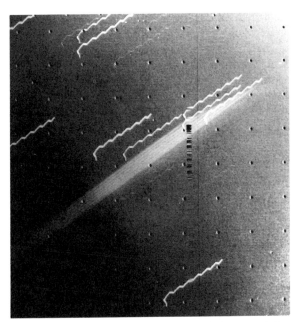

■ **Figure 18.3** The Voyager 1 photograph that revealed Jupiter's ring. Only the edge of the ring is shown. The elongated, wavy lines are images of stars that trailed because of the spacecraft's motion during this 11-minute exposure. (NASA/JPL Photograph, courtesy of the National Space Science Data Center and the Voyager Team Leader, Bradford A. Smith.)

culted. Surprisingly, the amount of light being received from the star dropped several times prior to the main occultation, as well as afterward (Figure 18.2). The light drops were found to be symmetrical on either side of the main occultation, leading to the conclusion that Uranus was a second planet with rings. These rings are narrow and faint, but they have since been photographed from Earth using a large telescope equipped with a mechanism to block out the light of the planet itself. They were also closely examined by Voyager 2 in January, 1986.

In 1973 and 1974, when Pioneers 10 and 11 passed closer to Jupiter than the orbits of any of its known satellites, they measured a pronounced dip in the number of charged particles in the planet's equatorial plane. A number of researchers believed that the data indicated that an unknown satellite or even a ring system might be present, close to the planet. Tobias Owen and Edward Danielson convinced the Voyager imaging team to have Voyager 1 photograph the suspicious area as it crossed the equatorial plane of Jupiter in 1979. When it did, the edge of a very faint ring was seen (Figure 18.3). Jupiter's ring cannot be seen from Earth, because it reflects very little sunlight back toward the Earth. (Particles in planetary rings can interact with light in two ways, as shown in Figure 18.4. **Backward scattering** reflects light back in nearly the same direction from which it came. **Forward scattering** sends light on in the general direction it was already headed. Because the gas giant planets are always farther from the Sun than the Earth is, we can only see backward-scattered light from their rings. Unlike Saturn's rings, which backward scatter light well, Jupiter's rings are very inefficient in backward scattering. They only appear bright to a spacecraft beyond them, which can observe the forward-scattered component of their light.)

In 1968, Edward F. Guinan observed an occultation of a star by Neptune. Unfortunately, his primary data were lost, and he did not initially analyze his less satisfactory back-up data. However, Guinan analyzed the data in 1982, because by that time it looked as though all gas giants might have rings. Guinan's data gave evidence that Neptune also has a ring system, but its existence was not verified, because observations of other occultations in the following years gave conflicting results. Results of some occultations

■ **Figure 18.4** Forward and backward light scattering. (a) Small particles tend to scatter light away from the source. (b) Larger particles tend to scatter light back toward the source.

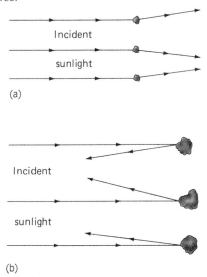

even seemed to suggest that Neptune's rings were discontinuous and did not completely circle the planet. In 1989, Voyager 2 verified that Neptune indeed has several narrow rings and also explained the apparent observations of discontinuous rings, as the outer ring has thick concentrations of material in some areas. Some occultation observers had seen only these portions of the rings and concluded that the rings were discontinuous.

In the remainder of this chapter, we examine some features that all ring systems have in common. Individual ring systems will be described in more detail in upcoming chapters.

■ The Roche Limit

Why do gas giants have ring systems? Why don't any terrestrial objects have them? Although the answers to these questions are not yet fully known, we do know why rings are located in the exact regions where they are found. Because all of the gas giants also have satellite systems, one might wonder why ring systems are also present. One important clue is that all of the ring systems are closer to the planets than are the major satellites.

Recall from our discussion of tides in Chapter 7 that there is a limit to how closely a satellite can orbit a planet. This is because tidal forces tend to "stretch" a satellite along the line connecting it to the planet. Naturally, the closer the objects, the stronger the tidal force and the greater the stretching (Figure 18.5). The **Roche limit**, described by **Edouard Roche** (1820 – 1883) in 1848, defines the closest distance one object can come to another without being destroyed by tidal forces. For a planet and satellite of identical densities, the Roche limit is 2.446 times the radius of the planet; a satellite coming any closer to the planet will be destroyed. If the objects have different densities, the Roche limit, L, is given by the equation

$$L = 2.446R\sqrt[3]{\frac{D_P}{D_S}},$$

where R is the radius of the planet, D_P is the density of the planet, and D_S is the density of the satellite. (The Roche limit neglects the internal strength of the materials in an object caused by atomic and molecular forces.)

The ring systems of the four gas giants lie within the Roche limits for their respective planets. Although the exact values are somewhat uncertain because the density of the ring material is unknown, the Roche limits for the gas giants are 174,640 kilometers (108,520 miles) for Jupiter, 146,760 kilometers (91,200 miles) for Saturn, 62,130 kilometers (38,610 miles) for Uranus, and 59,440 kilometers (36,850 miles) for Neptune.

■ **Figure 18.5**   The Roche limit. A satellite (far right) in orbit around a planet is slightly deformed by tides. If it moves closer to the planet, it is deformed even more and finally disintegrates when it passes within the Roche limit (*dotted line*), located 2.44 times the planet's radius from its center.

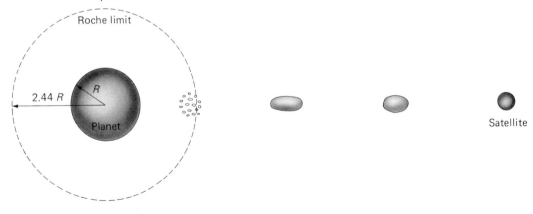

The lunar surface, as photographed during an Apollo mission. (NASA Photograph.)

The Moon during the partial eclipse of August 16, 1970. (Photograph by George East, courtesy of Dennis Milon.)

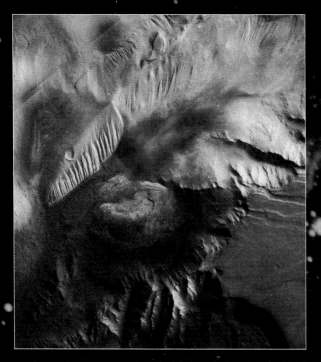

Viking Orbiter photograph of Martian canyonlands. (NASA/JPL Photograph.)

Viking Orbiter photograph of Olympus Mons, the largest shield volcano on Mars (and in the solar system). (NASA/JPL Photograph.)

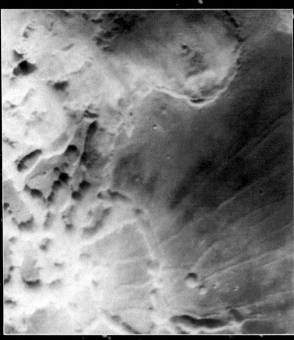

Viking Orbiter photograph of Mars showing the morning fog that sometimes is visible. (NASA/JPL Photograph.)

Viking 1 photograph of the Martian surface. (NASA/JPL Photograph.)

Photographs of the Earth and the four gas giant planets, shown at the same scale. (NASA/JPL Photograph.)

Voyager 1 photograph of Jupiter, as well as its satellites Io (*left*) and Europa (*right*). Note the colorful clouds of Jupiter. (NASA/JPL Photograph.)

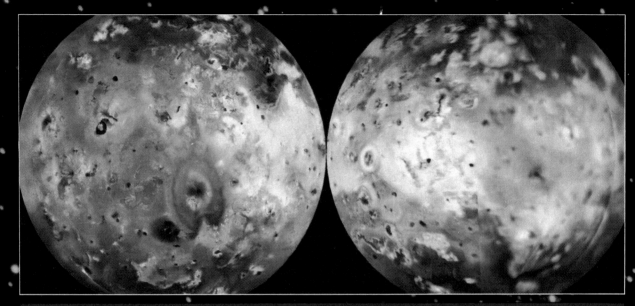

Photographs of both hemispheres of Io. (NASA/USGS Photograph.)

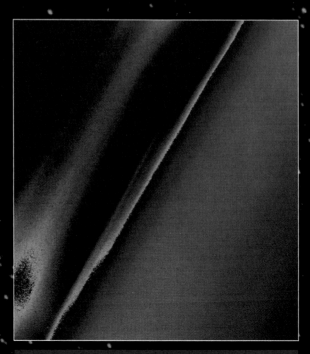

Color-enhanced photograph of Titan's upper atmosphere that shows its clouds and haze. (NASA/JPL Photograph.)

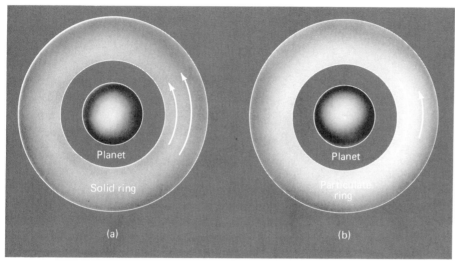

■ **Figure 18.6** Ring orbital motion. (a) If planetary rings were solid disks, the outer edge would be observed to move faster than the inner edge. (b) In actuality, the inner areas move faster, indicating that individual ring particles obey Kepler's laws.

For comparison, the Earth's Roche limit is 18,470 kilometers (11,470 miles). If the Moon ever ventured closer than this, it would break apart, and the Earth would have rings. (Artificial satellites can orbit closer to the Earth than this because the internal forces holding them together are much stronger than the tidal forces that such relatively small objects experience.)

The location of a gas giant's ring system inside the Roche limit has an implication regarding its origin: either a larger body wandered too close to the planet and was broken apart, or else small particles were prevented from accreting into a large satellite because they were located within the Roche limit. Their larger Roche limits may account for why only gas giants have rings. Because the terrestrial planets are so small, their Roche limits are physically much smaller than those of the gas giants. It is unlikely that debris (or a large body) could have ventured this close to them without impacting directly on the planets.

■ The Dynamics of Ring Systems

Someone seeing a photograph of Saturn for the first time might assume that the ring system is a solid disk rotating as a single unit. Unfortunately,

the dynamics of planetary rings are not that simple. In the 1800s, two results, one theoretical, the other observational, proved that Saturn's rings could not be solid disks. James C. Maxwell (1831–1879) calculated that such solid rings would be torn apart. James Keeler (1857–1900) examined the spectrum of Saturn's rings in 1895, and his investigations of Doppler shifts of light from the rings revealed that the regions closer to Saturn were orbiting the planet faster than the outer ones, something that would be impossible in a solid disk. The modern interpretation of planetary rings is that they are composed of numerous small, orbiting bodies, each of which obeys Kepler's laws (Figure 18.6).

Early telescopic observations of Saturn's rings revealed gaps or divisions in the ring structure. These features, now called the Cassini division, Encke division, Keeler division, and so forth, are named for the observers who discovered them. Occultation studies of Uranus' rings have also indicated that several divisions are present in them. This shows that although ring systems may have a wide lateral extent, the radial distribution of material within them is not uniform.

Prior to the space age, the origins of the gaps in Saturn's rings were believed to be well understood, and they were thought to be caused by **orbital resonances** with the satellites orbiting the

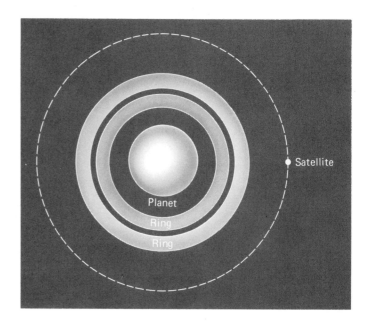

■ **Figure 18.7**   Ring resonance gaps. A satellite orbiting outside of the rings will disrupt the orbit of ring particles at a resonance point, clearing a gap in the ring.

planet. For example, a ring particle located in the **Cassini division** in Saturn's rings would have an orbital period exactly half that of the satellite Mimas. This 1:2 orbital resonance would cause frequent alignments of the ring particles with Mimas, which in turn would disrupt the orbits of the ring particles, causing them to move away from the resonance point (Figure 18.7).

The Voyager images of Saturn revealed that our knowledge of the structure of its rings was incomplete. Instead of containing just the few divisions visible from Earth, Saturn's rings were shown to contain myriad divisions (Figure 18.8). Saturn's rings are composed of literally thousands of **ringlets**, and the structures of both its and Uranus' ring systems are so complicated that res-

■ **Figure 18.8**   A Voyager 1 photograph of Saturn's rings showing the numerous ringlets that compose them. (NASA/JPL Photograph, courtesy of the National Space Science Data Center and the Voyager Team Leader, Bradford A. Smith.)

■ **Figure 18.9** Ringlets in the Cassini division in Saturn's rings. This photograph, taken from the unlit side of the rings, shows the Cassini division as a relatively bright region because it contains particles that scatter light forward. Intricate details of ringlets in the Cassini division can be seen. (NASA/JPL Photograph.)

onances alone cannot account for them. As an additional complication, even the major gaps in Saturn's rings turned out not to be empty but to contain numerous ringlets whose combined light is too faint to be visible from Earth (Figure 18.9).

A surprising discovery made during Voyager flybys was that the edges of planetary rings and ringlets are generally sharp and distinct, even though gravitational interactions would be ex-

pected to cause diffuse boundaries. Voyager found that several of Saturn's sharply bounded rings are "guarded" by satellites that orbit nearby. These **guardian,** or shepherd, **satellites** exert perturbing forces on the orbiting ring particles. These forces confine the particles to a zone with limited width (Figure 18.10). (The guardian satellites can exist inside the Roche limit because they are small enough to experience only mini-

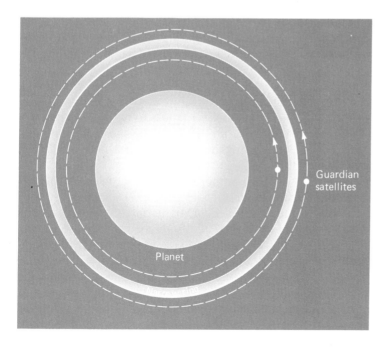

■ **Figure 18.10** Guardian satellites. Satellites orbiting just inside of and outside of a planetary ring will keep the edges of that ring very sharp.

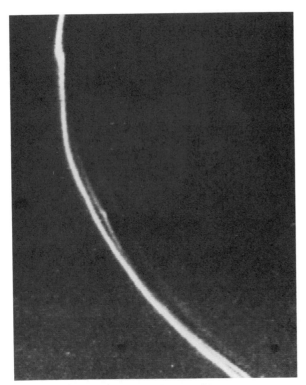

■ **Figure 18.11**   The braided F-ring of Saturn, as photographed by Voyager 1. (NASA/JPL Photograph.)

mal tidal forces.) The fact that Uranus' rings are composed of several narrow ringlets led researchers to expect to find many guardian satellites during the Voyager 2 flyby, but only one pair of guardians was found. Obviously, either there are indeed guardians present in all cases, with some of them too small to see, or else another, currently undiscovered process is involved. It is somewhat hard to imagine that all of the thousands of ringlets in Saturn's rings could have guardians associated with them.

Voyagers 1 and 2 discovered other surprising phenomena, especially in Saturn's rings. Saturn's outer "F-ring" was shown by Voyager 1 to be composed of several thin, braided strands, a situation that apparently results from gravitational interactions among F-ring particles and the rings' two guardian satellites (Figure 18.11). A problem lacking an explanation is why some of Saturn's ringlets and Uranus' major rings are elliptical instead of circular. Guardian satellites with eccentric orbits may be responsible. Voyager also

photographed dark, **radial spokes,** which are temporary features in Saturn's rings (Figure 18.12). Previously seen by telescope users on nights of exceptionally fine seeing, these seem to follow Keplerian orbits and are thus torn apart after a short time, because the inside part of the spoke orbits Saturn faster than the outside. Although their exact origin is unknown, they are possibly clouds of fine dust particles levitated above the rings by electrostatic forces.

One of the most surprising discoveries made by Voyager 2 about Neptune's ring system was that its outer ring is not uniform but contains three concentrations of material, named "sausages" by Voyager scientists. The sausages are limited to a single 35° section of the ring. Such concentrations would be expected to disperse along the ring, but it is believed that a guardian satellite, not seen by Voyager, may be keeping the material in this unusual configuration.

■ Characterization of Ring Particles

Modern Earth-based remote-sensing techniques and Voyager results have helped answer several long-standing questions about planetary rings: How thick are the overall rings? How big are the ring particles? What is their composition?

Saturn's rings are thick enough to cast shadows on the planet, and Saturn's shadow can be seen distinctly on them. On the other hand, the rings are thin enough that they are invisible from Earth when viewed edge on. This observation limits their thickness to no more than a few kilometers. Results from Voyager 2 indicate that Saturn's rings are only about 200 meters (600 feet) thick.

The diameters of the particles in Saturn's rings have been investigated by seeing how well the rings reflect radar. (Radar reflections have not yet been received from any other ring systems.) They are very good radar reflectors, which indicates that the ring particles are probably of a diameter comparable to the wavelengths of the radar used, which range from centimeters to tens of meters. This would indicate that the particles range from the size of ping-pong balls to perhaps

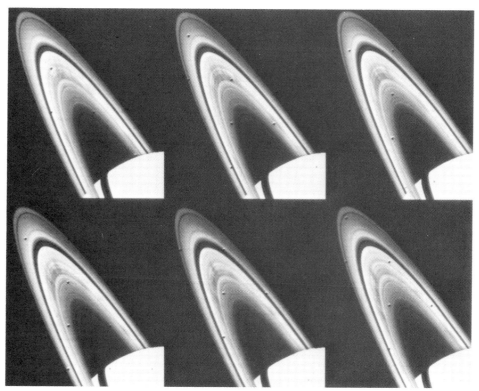

■ **Figure 18.12**   Radial spokes in Saturn's rings, as photographed at 15-minute intervals. As described in the text, these dark features appear to be dust suspended above the main portion of the rings. (NASA/JPL Photograph.)

the size of houses. Although objects of these sizes would have been too small to have been photographed individually by Voyager, the fact that kilometer-sized ring particles were not seen tends to confirm these radar findings. In addition, particles of these sizes would be expected to backward scatter light. Because the rings of Jupiter, Uranus, and Neptune primarily forward scatter sunlight, they are probably composed of much smaller particles.

Spectroscopic studies of Saturn's rings, although inconclusive, indicate a composition of either water ice or ammonia ice. Current thinking is that the ring particles are either fragments of solid ice or else rocky fragments covered with ice. Some silicates or carbon may also be present. The particles in Uranus' rings have much lower albedos, which may indicate a composition different from that of the particles in Saturn's rings. Observational evidence suggests that Uranus' rings are composed of rocky, rather than icy, particles. This is unexpected, because the area where Uranus formed was cold enough that ices would have condensed there.

■ Formation and Stability of Ring Systems

The modern theory of planetary formation proposes that as each planet formed it was surrounded by a cloud of gas and dust, particles of which eventually accreted into its satellites. Based on this theory, three possible scenarios can be developed that describe the formation of a ring system. Planetary rings may be a remnant of material left inside the Roche limit. Because of its proximity to the planet, this material could not accrete but remained in its initial form. Alternately, a large body may have passed within the Roche limit of the planet and been fragmented as a result of tidal stress. Such a body might have come from outside the planet's satellite system, or it may have been a satellite that somehow spiraled in too close to the planet. A third possibility is that numerous small bodies remained outside the Roche limit at the end of the accretion period without ever having formed larger satellites. Collisions involving these bodies may have caused numerous smaller particles to travel inside the Roche limit and become trapped there, forming the rings.

Why have rings become so flat and, in most cases, circular? Suppose the particles inside the Roche limit originally had a wide range of inclinations and eccentricities. Theoretical models of their orbits indicate that numerous collisions among these bodies would have occurred. One result of these collisions would have been the "cancellation" of the objects' inclinations and eccentricities, leading to what are described as **circular, coplanar, collisionless orbits.** This is a good description of what we see in ring systems today. (It is likely that ring particles were not originally oriented with as wide a range of inclinations as eccentricities, as the materials accreting around the planets were distributed in a disk-shaped fashion.)

Because of the large number of ring particles and their close spacings, rings are very active places, and it is possible that individual ring particles do not have infinite lifetimes. Although collisions are probably rare, several processes could lead to loss of ring particles. First, icy particles would be expected to disappear eventually by evaporation. Most calculations, however, indicate that the evaporation rate would be very slow for larger icy particles, such as those in Saturn's rings. Another hazard for ring particles is that collisions or near misses might eject particles away from the ring system. The fact that ring systems exist shows that the processes that tend to destroy ring particles are relatively slow and therefore unimportant, or else are countered by processes that introduce new particles into the ring system. For example, Jupiter's ring is so close to the planet that material is probably lost from the rings and spirals inward to Jupiter's outer atmosphere. The ring is probably replenished by small particles that were knocked from Jupiter's innermost satellites by impacts of small objects hitting them from space. Therefore, we can assume that the rings of Jupiter and the other gas giants should have lifetimes as long as those of planets and satellites.

■ Chapter Summary

Although Saturn's rings were discovered in the 1600s, it was not until recent decades that it was learned that Jupiter, Uranus, and Neptune also have rings. The invisibility from Earth of rings other than those of Saturn results from the forward, rather than backward, scattering of light by the particles in them.

Rings exist inside the Roche limits of various planets, the closest an object can approach without being disrupted by tidal forces. Individual ring particles orbit according to Kepler's laws and are distributed in numerous ringlets, apparently as a result of guardian satellites and orbital resonances.

Ring systems are relatively thin, and Saturn's rings are composed of either solid ice or ice-covered rock particles that range from the size of a ping-pong ball to the size of a house. The rings of Jupiter, Uranus, and Neptune are composed of smaller, darker particles.

Ring systems probably developed from particles that failed to accrete into planetary satellites, or else from satellites that ventured too close to their planets and were broken apart.

■ Chapter Vocabulary

backward scattering
Cassini division
circular, coplanar,
 collisionless
 orbits

forward scattering
guardian satellites
orbital resonances

radial spokes
ring system
ringlets

Roche, Edouard
Roche limit

■ Review Questions

1. Summarize the discovery of the ring systems of each of the gas giants.
2. Explain the two ways in which ring particles can interact with light.
3. What is the Roche limit? How does it explain the presence of planetary ring systems?
4. How do orbital resonances cause gaps in ring systems? Why do Voyager results indicate that additional phenomena are responsible for ring structure?
5. Describe the thickness of Saturn's ring system, and the size and composition of the particles in it.
6. What is the modern theory of ring formation?
7. Why are ring systems so flat?
8. What evidence is there that individual ring particles do not have infinite lifetimes? How does this affect ring systems?

■ For Further Reading

Burns, Joseph A. "Planetary Rings." In Beatty, J. Kelly and Andrew Chaikin, editors. *The New Solar System,* Sky Publishing, Cambridge, MA, 1990.

Elliot, James and Richard Kerr. *Rings.* MIT Press, Cambridge, MA, 1984.

Pollack, James B. and Jeffrey N. Cuzzi. "Rings in the Solar System." In *Scientific American,* November, 1981.

Jupiter

Voyager 1 photograph of Jupiter. Note that the satellite Io can be seen, as well as its shadow on Jupiter's disk. (NASA/JPL Photograph.)

Jupiter is named for the king of the gods in Roman mythology. The name is certainly appropriate, as Jupiter is the largest planet in our solar system and is over twice as massive as the other eight planets combined. Despite its huge bulk, it is the fastest-rotating planet, with a rotational period of just under 10 hours. It is the closest gas giant planet to Earth and the one about which we know most. It has been visited by four spacecraft and is the only gas giant for which the United States has plans for future space exploration. Jupiter's major orbital and physical properties are summarized in Table 19.1.

■ Appearance and Basic Properties

Jupiter is an imposing sight in the night sky, as it is normally the second brightest planet after Venus. (Mars can be slightly brighter than Jupiter, but only at very close oppositions.) Unlike Venus, though, Jupiter's pure yellow light at times can be seen all night long and high in the sky, making it especially impressive. Because Jupiter takes 12 years to orbit the Sun, it spends 1 year in each of the dozen zodiacal constellations. It is this property that is believed to be the main reason that Jupiter was named for the king of the gods, as the zodiac was looked upon with great importance in ancient times.

Jupiter is one of the easiest planets to observe through a telescope, and it presents a fine view to observers (Figure 17.3). With the exception of Venus when it is closest to the Earth in crescent phase, Jupiter's apparent diameter is greater than that of any other planet. When viewed through a small telescope, three things catch an observer's attention. The first is Jupiter's oblateness, which causes the planet's disk to have an elliptical shape. Jupiter's equatorial diameter is 9254 kilometers (5750 miles) greater than its polar, giving it an oblateness of 0.065.

■ Table 19.1 Summary of Jupiter's Basic Properties*

Semi-major axis: 5.20 astronomical units
Semi-major axis: 778.3 million kilometers
Semi-major axis: 483.6 million miles
Orbital eccentricity: 0.049
Orbital inclination: 1.30°
Orbital period: 11.86 years
Diameter: 142,796 kilometers
Diameter: 88,733 miles
Mass (Earth = 1): 317.8
Density: 1.33 grams per cubic centimeter
Sidereal rotational period: 9.92 hours
Axial obliquity: 3.12°
Number of satellites: 16

* For a complete table of planetary data, see Table A2.1 in
  Appendix 2.

The second feature visible to observers of Jupiter is the planet's striped appearance, which is caused by light areas called zones and dark areas called belts. As described in Chapter 17, these atmospheric markings are the result of Jupiter's atmospheric circulation, with the different colors due to different cloud layers and possibly also to variation in atmospheric composition. Fine details in these markings can be followed as the planet rotates, and their motion is used to determine the planet's rotational rate. Jupiter, like the other gas giants, is not a solid body, so areas of different latitudes on the planet are seen to rotate at different rates. Equatorial regions rotate in about 9 hours, 50 minutes, and 30 seconds, whereas ones at high latitudes require about 9 hours, 55 minutes, and 40 seconds. The atmospheric markings of Jupiter are constantly changing as a result of disturbances analogous to weather on Earth (Figure 19.1). A number of spots and storms are usually visible, and one of these, the Great Red Spot, has been observed since the 1600s, although its color, location, and rate of rotation around the planet are all variable.

The third thing visible in a small telescope is the changing pattern of Jupiter's four Galilean satellites, Io, Europa, Ganymede, and Callisto. At least a couple of them are usually visible, and all four are often visible simultaneously (Figure 19.2). The positions of these moons change every night, and eclipses and transits involving them and Jupiter can also be witnessed. Unlike most planetary satellites, the Galileans are bright and easy to see, and they can easily be spotted in binoculars. In fact, they are bright enough that they would be visible to the unaided eye if they were not overpowered by Jupiter's brilliance.

■ Figure 19.1 Composite photographs of Jupiter's atmosphere, obtained by Voyager 1 in March, 1979 (top) and Voyager 2 in July, 1979 (bottom). Note the changes that the atmospheric markings underwent between the two photographs. (NASA/JPL Photograph.)

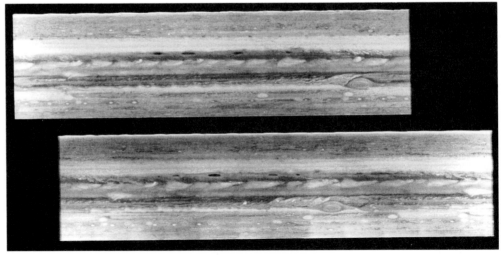

■ **Figure 19.2** A view of Jupiter and its four Galilean satellites as they appear when observed with a small telescope. In this photograph, two of the satellites are to the lower left of the planet, and two are at upper right. The object directly to the right of Jupiter is a star. (Lick Observatory Photograph.)

■ Spacecraft Exploration

Four spacecraft flew past Jupiter between 1973 and 1979. The first two of these were Pioneers 10 and 11, in 1973 and 1974. As their names imply, the Pioneers were less advanced craft that paved the way for the later Voyagers. **Pioneer 10** was launched March 2, 1972 and flew within 130,000 kilometers (80,600 miles) of Jupiter's cloud tops on December 3, 1973. **Pioneer 11** was launched April 5, 1973 and passed even closer, within 34,000 kilometers (21,080 miles), on December 2, 1974. The Pioneers returned the first close-up photographs of Jupiter and measured its atmospheric composition, magnetic field, and radiation belt structure with great precision. Pioneer 11 was "boosted" toward Saturn by Jupiter's gravity, the first time this technique was used in the outer solar system. Most importantly, the Pioneers proved that spacecraft could survive Jupiter's powerful radiation belt and helped the designers of the Voyagers improve those space probes and plan their missions.

The Voyagers were much more advanced than the Pioneers and sent back magnificent photographs of the planet and its moons. Because the Pioneers passed very close to the planet, their electronics were somewhat damaged by the charged particles trapped in Jupiter's radiation belts. For this reason, the Voyagers did not pass as close to the planet. Even so, their improved imaging systems returned more detailed pictures than did the Pioneers. **Voyager 1** was launched September 5, 1977 and flew past Jupiter's cloud tops at a distance of 277,570 kilometers (172,100 miles) on March 5, 1979. **Voyager 2** was launched August 20, 1977 and flew by at a distance of 650,600 kilometers (403,400 miles) on July 9, 1979. (Although Voyager 2 was launched first, Voyager 1 traveled a more direct trajectory and arrived first.) Both Voyagers gained gravitational boosts from Jupiter and continued toward Saturn. Numerous discoveries were made by the Voyagers: they provided extensive new information about the atmosphere, magnetic field, and radiation belts of Jupiter, and also examined interactions among these regions. Lightning and auroras were found in Jupiter's atmosphere, and significant changes were noted during the four months separating the two encounters, as well as since the Pioneer flybys. The Galilean satellites were closely examined by the Voyagers, and three new satellites, as well as the ring system, were discovered.

The next step in Jupiter explorations is the Galileo orbiter/atmospheric probe. Originally scheduled for launch aboard a Space Shuttle in the early 1980s, Galileo was delayed several times, most notably as a result of the *Challenger* disaster, before it was finally launched from the Space Shuttle *Atlantis* on October 18, 1989. Because, for safety reasons, the Shuttle was not able to carry a rocket large enough to boost Galileo directly to Jupiter, a complex mission using three gravity-assist encounters is being used. Galileo will pass Venus in February, 1990; Earth in December, 1990; and Earth again in December, 1991; each time it will get a "slingshot" acceleration. Galileo will then travel toward a December 1995 encounter with Jupiter. At that time, Galileo will go into orbit around Jupiter and observe it for a period of at least two years, during which time close encounters with the Galilean satellites will occur. An atmospheric probe will be sent into the clouds of Jupiter and will transmit information about conditions it encounters, until it is crushed by the enormous atmospheric pressure. As a bonus, Galileo will encounter two asteroids on the way to Jupiter — Gaspra in October, 1991 and Ida in August, 1993.

■ Atmosphere

As described in Chapter 17, Jupiter is of Sun-like composition, being made primarily of hydrogen and helium. Both of these elements have been spectroscopically identified in the planet's upper atmosphere, and water, methane, and ammonia are also present in minute quantities. Although all of these gases are colorless, Jupiter's atmosphere is marked by vivid features of virtually every hue, making it the most colorful gas giant. As indicated in Chapter 17, Jupiter's many belt and zone atmospheric markings are probably related to circulation cell boundaries and to the planet's cloud layers. There are more circulation cell boundaries in Jupiter's atmosphere than in Earth's because Jupiter's more rapid rotation produces a greater Coriolis effect. The colors of the cloud markings are probably in part due to temperature variations, but compositional effects, perhaps from sulfur or organic compounds, may also be partly

responsible. Even though Jupiter's atmosphere is very colorful, it must be remembered that a great deal of "color enhancement" was used when producing the Voyager images commonly seen. This was done to increase color contrast and make faint details easier to see. An observer on board the Voyagers would have seen much more subtle color variation than that shown in photographs. Jupiter's major belt and zone markings are generally permanent, but changes do occur, as in 1989, when one of the planet's dark equatorial belts faded from view.

Because Jupiter has no real surface, an artificial boundary surface has been chosen — the level where the atmospheric pressure is 100 millibars, which also is where the temperature is lowest. At this point, temperature is about 110 K ($-260°$F). About 40 kilometers (25 miles) below this point, the pressure increases tenfold to 1000 millibars ($= 1$ bar), and the temperature is 170 K ($-150°$F). About 135 kilometers (85 miles) below the 100 millibar point, pressure is 10 bars and temperature is 350 K ($170°$F). Between the 100 millibar point and about 100 kilometers (60 miles) below it are three cloud layers. In order of increasing depth, they are ammonia clouds, which are reddish; ammonium hydrosulfide clouds, which are whitish or brownish; and water clouds, which are bluish (Figure 19.3). Above the 100 millibar point, temperature rises slowly and pressure decreases.

Wind plays an important role in forming Jupiter's striking atmospheric features. Measurements show that winds vary by latitude and that velocities range from over 160 kilometers per hour (100 miles per hour) in the westward direction (opposite the rotational direction) to almost 500 kilometers per hour (300 miles per hour) in the eastward direction. The most rapid westward winds occur in dark belts at 20° north and south latitude, whereas the highest eastward velocities were measured near the equator and in a bright zone between 20° and 30° north latitude (Figure 19.4). Vertical air movement also occurs, as light zone markings are the site of rising warm gas and dark belts are places where cooler gas sinks.

The best-known feature in Jupiter's atmosphere is the **Great Red Spot**, which was first observed by Robert Hooke (1635 – 1703) in 1664.

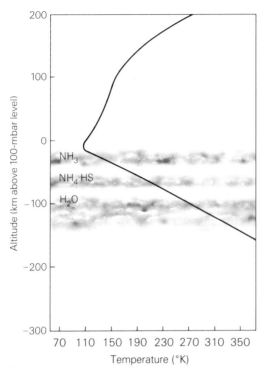

■ **Figure 19.3** Vertical structure of Jupiter's atmosphere. Altitude in kilometers relative to 100-millibar level is plotted against temperature in degrees Kelvin. The three dark bands are the cloud layers described in the text. (NASA diagram, from Abell, George, David Morrison, and Sidney Wolff. *Exploration of the Universe*, 5th edition. Saunders College Publishing, Philadelphia, 1987.)

Although records for the next two centuries are incomplete, it is known that the Great Red Spot became very conspicuous in the 1800s and has been present continuously since at least 1878. The Great Red Spot covers 10° of latitude on Jupiter and is bigger than the Earth (Figure 19.5). Three white ovals lie just to the south of the spot but wander in longitude and are not always adjacent to it. Time-lapse photography from Voyager clearly showed counterclockwise wind circulation around the Great Red Spot at speeds over 320 kilometers per hour (200 miles per hour). (The Great Red Spot is a high-pressure system that is located in Jupiter's southern hemisphere, which accounts for the direction of circulation.) Approximately six days are required for circulation around the Great Red Spot, which is somewhat similar to a terrestrial hurricane. However,

hurricanes last only a few days, after which their energy source is gone, so the major problem with the Great Red Spot is explaining its extreme longevity. There are several possible sources for the energy that has enabled the Great Red Spot to survive for so long: it may be obtained from condensation occurring in gases below it, from the circulation cells in the surrounding atmosphere, or perhaps from the absorption of smaller eddies and vortices.

■ Interior and Magnetic Field

On the whole, Jupiter is very similar to the Sun in both density and composition. Both objects are made primarily of hydrogen and helium. Based on numbers of atoms (not mass), Jupiter is 90% hydrogen and 10% helium, whereas the Sun is 89% hydrogen and 11% helium. (Note that for both objects, these compositions are based upon surface observations. If the interiors of these ob-

■ **Figure 19.4** A graph of wind velocities on Jupiter as a function of latitude. Positive velocities are those in the direction of planetary rotation, and negative velocities are those in the opposite direction. (NASA diagram, from Abell, George, David Morrison, and Sidney Wolff. *Exploration of the Universe*, 5th edition. Saunders College Publishing, Philadelphia, 1987.)

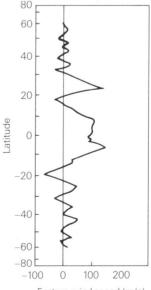

■ **Figure 19.5** A Voyager photograph of Jupiter's Great Red Spot (*upper right*) with a superimposed image, to scale, of the Earth. (NASA Photograph. Composite prepared by Stephen Paul Meszaros.)

jects differ significantly from what we believe them to be, the actual overall compositions would be different.) Because they are primarily composed of these light elements, Jupiter's overall density is 1.33 grams per cubic centimeter, whereas the Sun's is 1.4.

Just as the internal structure of the Sun has been deduced by the use of theoretical models that cannot be tested by direct observation, similar techniques have given us an idea of what the inside of Jupiter is like (Figure 19.6). Under the outer atmospheric layer lies an area of molecular hydrogen that is fluid and in convection. Below this region is a layer in which hydrogen is in the metallic form as a result of Jupiter's immense internal pressure. The boundary between the molecular and metallic hydrogen layers probably occurs about one-quarter of the way from the surface of Jupiter into the center. The temperature at the boundary is about 10,000 K, the density is about 1.2 grams per cubic centimeter, and the pressure is about 3 million atmospheres. A core, whose composition and exact size are uncertain, lies at the center of the planet. The boundary between the metallic hydrogen layer and the core probably occurs about 80% of the way into the center. At this point, the temperature is probably 20,000 K, the density jumps from

about 4 to 15 grams per cubic centimeter over the transition, and the pressure is about 40 million atmospheres. Jupiter's core probably contains silicates and nickel – iron metal, and the center of Jupiter is thought to be at a temperature of 25,000 K, with a density of 20 grams per cubic centimeter and a pressure of 80 million atmospheres. Although these last values are much more extreme than conditions in the center of the Earth, they are mild compared to those in the center of the Sun! As noted in Chapter 17, Jupiter's core was warmed enough by the release of gravitational potential energy when the planet formed that considerable internal heat was produced. Even today, enough energy radiates outward from the planet that Jupiter emits more energy than it receives from the Sun. However, its core was heated to less than 1% of the temperature that would have been required for it to have become a star.

Jupiter has the strongest magnetic field of any planet in the solar system. The field is probably generated by currents in Jupiter's metallic hydrogen region. Jupiter's total magnetic moment is about 20,000 times stronger than Earth's, but its surface field strength is only 4.3 times Earth's, because the surface is so far from the internal source of the field. As a result of its powerful magnetic field, Jupiter has an extensive system of belts of trapped, charged particles, similar to Earth's Van Allen belts. Jupiter's magnetic field, like the Earth's, allows charged particles to enter

■ **Figure 19.6** Jupiter's interior layers. The numbers represent the fraction of Jupiter's total radius. See text for detailed descriptions.

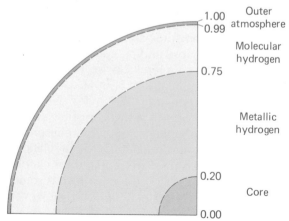

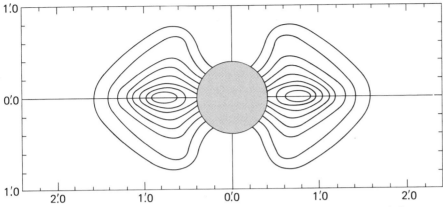

■ **Figure 19.7** Intensity contours of decimetric radiation from Jupiter's radiation belts. The areas of maximum intensity to the left and right of the planet mark the centers of the radiation belts. (Reprinted with permission from Berge, Glenn L., "An Interferometric Study of Jupiter's Decimetric Radio Emission," *The Astrophysical Journal,* volume 146, page 767, 1966.)

the planet's atmosphere only at the poles, where they interact with Jupiter's ionosphere to form auroras (see Figure 17.7).

Because neither magnetic fields nor radiation belts can be studied directly using visible light, our knowledge of these areas prior to the flyby missions came only from radio telescopic observations. Jupiter is one of the most powerful radio sources in the sky, and intense bursts of radio emission from the planet were first detected in 1955. Study of radio emission at **decametric** (approximately 10-meter) wavelengths revealed a periodicity of 9 hours, 55 minutes, which is the planet's "true" (internal) rotational period. Although the periodicity seems to indicate emission from a region of Jupiter's ionosphere associated with the magnetic field, additional factors complicate the simple model. The most important observation is that Jupiter's closest large satellite, Io, somehow seems to stimulate the emission of radio bursts when at certain points in its orbit. Apparently, Io disturbs Jupiter's radiation belts as it passes through them. Jupiter also emits radio signals at **decimetric** (approximately 0.1-meter) wavelengths, apparently from fast-moving electrons trapped in Jupiter's radiation belts by its magnetic field. Detailed mapping of this radiation reveals the extent of Jupiter's radiation belts (Figure 19.7).

Jupiter's magnetosphere extends outward to about 65 times the planet's radius. The magnetosphere, which extensively interacts with the solar wind, is so large that if visible to the unaided eye, it would appear to be twice the size of the full moon (Figure 19.8)! Some aspects of Jupiter's magnetic field activity are apparently unique to that world. A cloud of sodium atoms is associated with the satellite Io, forming a transient atmosphere around it as well as a **torus** (doughnut-shaped ring) of material spread out along its orbit. These atoms are knocked loose from Io by the high-energy impacts of particles in the radiation belt.

■ Rings

As described in Chapter 18, one of the many interesting discoveries made by Voyager 1 was that Jupiter has a ring system. Because rings were suspected on the basis of a drop in particles counted by the Pioneers as they traveled through Jupiter's equatorial plane, Voyager's camera was aimed at a point in the plane and a long exposure of the region was taken (see Figure 18.3). The photograph revealed the outer edge of a faint, narrow ring. Additional photographs were taken that revealed the full extent of the rings (Figure 19.9). Jupiter's rings are the faintest of all those in the solar system and are much too faint to be seen from the Earth. They also appear to have the least complex structure of any ring system yet known.

The ring system has three components. The **main ring** lies between 1.71 and 1.81 times the

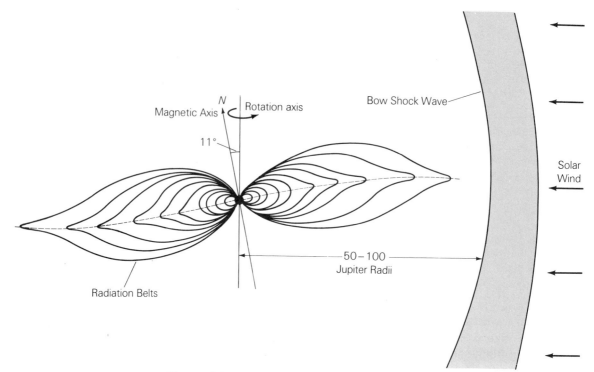

■ **Figure 19.8**   Map of Jupiter's magnetosphere. (From Zeilik, Michael and Elske v.P. Smith, *Introductory Astronomy and Astrophysics,* 2nd edition. Saunders College Publishing, Philadelphia, 1987.)

radius of Jupiter away from the planet's center, or between 50,600 and 58,000 kilometers (31,400 and 36,000 miles) above the cloud tops (Figure 19.10). This is a very narrow ring, only about 7130 kilometers (4420 miles) wide. The outer edge of the main ring is just inside the orbit of the satellite Metis, which may act as a guardian and cause it to have a fairly sharp outer boundary. The main ring is probably less than 30 kilometers (20 miles) thick. The **halo** extends from the inner edge of the main ring to a point just 20,000 kilometers (12,000 miles) above Jupiter's cloud tops, or 1.28 times Jupiter's radius. The halo is not confined to Jupiter's equatorial plane but extends about 3000 kilometers (1800 miles) above and below it at its maximum extent. The halo appears to contain the same total quantity of ring particles as the main ring, but because they are so dispersed, the ring is very inconspicuous. The final ring component, the **gossamer ring**, extends outward from the main ring to a distance nearly 140,000 kilometers (87,000 miles) above the cloud tops, or almost three times Jupiter's radius. The gossamer ring is the faintest component of the rings.

The fact that Jupiter's rings are readily visible only when seen in forward-scattered light indicates that most of the particles in them are approximately 10^{-6} meters in diameter. Such small particles would have short lifetimes in the rings (on the order of decades to centuries) and would have to be replenished constantly. Presumably, these small particles are generated when meteorites impact onto satellites outside the ring system. Some of these fine particles then spiral in and eventually reach the ring area. The ring particles have low albedos and are presumably composed of silicates.

■ Satellites

Jupiter's large retinue of satellites contains 16 objects, 3 of them discovered by the Voyagers. These satellites are given names of mythological characters associated with Jupiter. (Table 19.2 lists their major properties.) The most important satellites are Io, Europa, Ganymede, and Callisto, known collectively as the **Galileans** because they

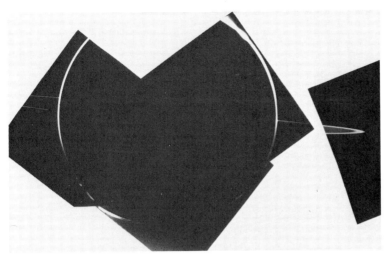

(a)

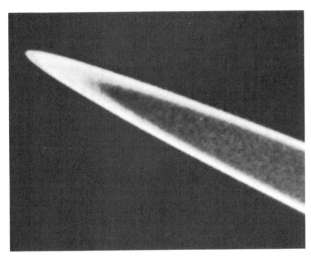

(b)

■ **Figure 19.9** Photographs of Jupiter's rings obtained using forward-scattered light. (a) Crescent Jupiter with ring. (b) The outer edge of the ring. (NASA/JPL Photographs.)

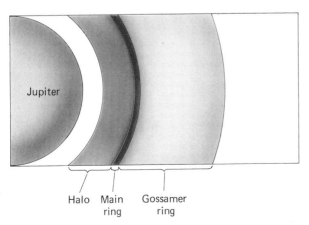

Jupiter

Halo   Main ring   Gossamer ring

■ **Figure 19.10** Map of Jupiter's ring system, as seen from above the planet's north pole.

■ Table 19.2   Summary of the Basic Properties of Jupiter's Satellites*

| Name | Semi-Major Axis | | Orbital Period | Diameter | | Density |
	10^3 km	10^3 mi	(days)†	km	mi	(g/cm^3)
Metis	128.0	79.5	0.2948	? × 40 × 40	? × 25 × 25	?
Adrastea	129.0	80.2	0.2983	25 × 20 × 15	16 × 13 × 9	?
Amalthea	181.3	112.7	0.4981	270 × 164 × 150	168 × 102 × 93	?
Thebe	221.9	137.9	0.6745	? × 110 × 90	? × 68 × 56	?
Io	421.6	262.0	1.769	3630	2256	3.57
Europa	670.9	416.9	3.551	3138	1950	2.97
Ganymede	1070	665	7.155	5262	3270	1.94
Callisto	1883	1170	16.689	4800	2983	1.86
Leda	11,094	6894	238.72	16	10	?
Himalia	11,480	7134	250.57	180	112	?
Lysithea	11,720	7283	259.22	40	25	?
Elara	11,737	7293	269.65	80	50	?
Anake	21,200	13,174	−631	30	19	?
Carme	22,600	14,044	−692	44	27	?
Pasiphae	23,500	14,603	−735	70	43	?
Sinope	23,700	14,727	−758	40	25	?

* For a complete table of satellite data, see Table A3.1 in Appendix 3.
† Negative period indicates retrograde orbit.

were discovered by Galileo on January 7, 1610. The Galileans are all somewhat comparable in diameter to the Moon and are located between 422,000 and 1,880,000 kilometers (262,000 and 1,166,000 miles) from Jupiter's center. They were extensively studied by Voyagers 1 and 2 and will be discussed in the next chapter.

Between the Galileans and the outer edge of the rings are four small satellites. Amalthea is an elliptical, reddish-gray, cratered object about 270 by 150 kilometers (170 by 90 miles) in size (Figure 19.11). It is oriented so that it orbits with its long axis always pointed toward Jupiter. Adrastea, Metis, and Thebe were discovered by the Voyagers, but no surface details were visible on photographs of these tiny worlds, and very little is known about them. Adrastea, Metis, and Amalthea all orbit within the gossamer ring.

Beyond the Galileans are two sets of satellites, each with distinct types of orbits. Between 11.1 and 11.7 million kilometers (6.9 and 7.3 million miles) from Jupiter are Leda, Himalia, Lysithea, and Elara. These objects all have prograde orbits with high inclinations and are dark, rocky objects. The largest, Himalia, is 170 kilometers (105 miles) in diameter. None of them have been observed by spacecraft. Between 20.7 and 23.7

million kilometers (12.8 and 14.7 million miles) distant are Anake, Carme, Pasiphae, and Sinope. These outermost satellites have high-inclination retrograde orbits, indicating that they are presumably captured comets or asteroids. They are small, dark objects about which little is known, as the Voyagers did not photograph them.

■ Figure 19.11   Voyager 1 photograph of Amalthea. (NASA/JPL Photograph.)

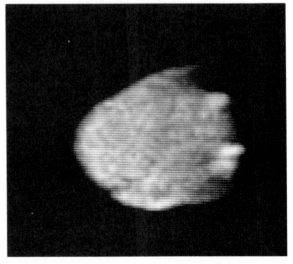

■ Chapter Summary

Jupiter, the solar system's most massive planet, is an impressive sight both to the unaided eye and through a small telescope. It has so far been observed by four spacecraft, and the Galileo probe should eventually add to our knowledge of the planet and its satellites.

Jupiter's cloud-covered atmosphere presents a dynamic view to observers. Although primarily hydrogen and helium, the atmosphere contains other gases that provide color. Many atmospheric features, such as the immense Great Red Spot, result from winds and global circulation.

Jupiter's interior contains layers of molecular hydrogen, metallic hydrogen, and a silicate–metal core. Jupiter's magnetic field, presumably generated in the metallic hydrogen region, generates radio emission that can be detected from the Earth.

Jupiter's ring system, invisible from the Earth, is relatively simple and composed of dark, fine-grained silicate particles. Although Jupiter has 16 satellites, little is known about those other than the four large Galileans.

■ Chapter Vocabulary

decametric radiation
decimetric radiation
diffuse disk
Galilean satellites

gossamer ring
Great Red Spot
halo

main ring
Pioneer 10
Pioneer 11

torus
Voyager 1
Voyager 2

■ Review Questions

1. For what is Jupiter named, and why is the name appropriate?
2. What features of Jupiter are apparent to someone observing it through a small telescope?
3. Describe the missions of Pioneers 10 and 11 and Voyagers 1 and 2 and what they told us about Jupiter.
4. What should the Galileo spacecraft accomplish?
5. Describe the composition, vertical structure, and markings of Jupiter's atmosphere.
6. Describe the interior structure of Jupiter.
7. Explain how Jupiter's magnetic field and its effects can be studied using radio astronomy.
8. Describe the main characteristics of Jupiter's ring system.
9. Describe the four groups of satellites of Jupiter.

■ For Further Reading

Morrison, David and Jane Samz. *Voyage to Jupiter.* NASA, Washington, D.C., 1980.

Wolfe, John H. "Jupiter." In *Scientific American,* September, 1975.

20

Jupiter's Galilean Satellites

CHAPTER OBJECTIVES

After studying this chapter, you should be able to

1. Understand how the discovery of the Galileans influenced astronomy.
2. Explain why there are compositional changes from Io to Callisto.
3. List the basic characteristics of each Galilean, and describe the geology of its surface.
4. Explain why volcanism occurs on Io.
5. Describe a palimpsest crater.

Montage of Voyager 1 photographs of Jupiter and its four Galilean satellites. (NASA/JPL Photograph.)

One of Galileo's first telescopic discoveries was that Jupiter is accompanied by four bright companions. As he observed these companions in January, 1610, he became convinced that they were small objects in orbit around Jupiter (Figure 20.1). These four objects are known collectively as the Galilean satellites, in honor of their discoverer. Each has been named for a mythological figure associated with Jupiter, the king of the gods. In order of increasing distance, they are Io, Europa, Ganymede, and Callisto. They are also designated by the Roman numerals I, II, III, and IV.

The discovery of the Galilean satellites greatly influenced astronomy and other sciences. It was strong evidence against the geocentric system of the solar system, because the motion of these satellites proved that the Earth was not the center of all orbital motion. In addition, observations of these objects proved that Kepler's laws are valid for all orbital motion, not just for that around the Sun. After Newton refined Kepler's third law to account for masses, he was able to calculate Jupiter's mass, as well as that of the other planets known to have satellites at that time, the Earth and Saturn.

The Galilean satellites played an important role in Roemer's 1675 measurement of the speed of light. Because of Jupiter's varying distance from Earth, the time at which eclipses and transits of the Galileans are witnessed on Earth lags behind the actual time of occurrence by different amounts. Celestial mechanics allowed the determination of the times when the events actually occurred, and subsequently the speed of light could be calculated.

As the mathematical capabilities of celestial mechanics specialists improved during the 1700s, they were able to determine the masses of the Galileans. This was done by measuring the perturbations that these bodies produced in one another's orbits. Calculations of their masses in this way, as well as measurements of their diameters, allowed their densities to be determined.

295

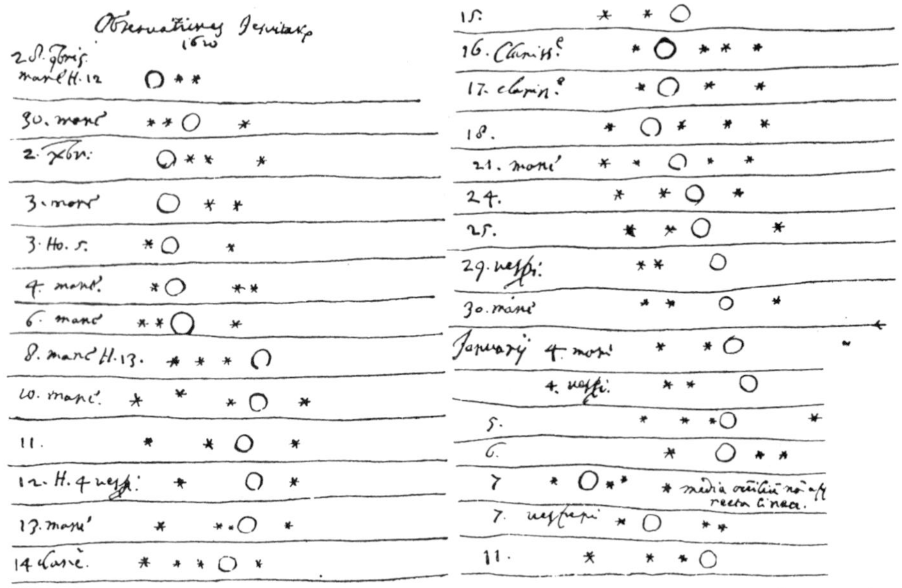

■ **Figure 20.1**   Galileo's sketches recording his observations of Jupiter's satellites. (Yerkes Observatory Photograph.)

■ A Miniature Solar System

Relatively little was known about the Galileans prior to the Voyager missions. Because of differences in albedo and density, the inner Galileans were considered rocky bodies and the outer ones icy. Io's sodium atmosphere and associated sodium torus, described in Chapter 19, were discovered by Earth-based observations. The Voyager missions changed our view of the Galileans from four points of light to four distinctive, complex worlds (Figure 20.2). As a result of the two flybys, we now know what their surfaces are like and can reconstruct some of their geological history.

The Galileans are all relatively large, ranging in diameter from 3126 kilometers (1942 miles) for Europa to 5276 kilometers (3279 miles) for Ganymede, which is the solar system's largest satellite. Large as they are, however, they are insignificant when compared with enormous Jupiter. One of the most noticeable features of the four satellites is that their densities depend upon their distance from Jupiter, as the density values decrease with increasing distance from the planet. Io and Europa have densities similar to that of the Moon, whereas Ganymede and Callisto are much lighter. This is believed to be the result of the heat

that was radiating from Jupiter even as the satellites were forming. Although the Galileans formed in the outer solar system, where temperatures were generally cool enough for volatile materials such as ices to exist, the heat of Jupiter

■ **Figure 20.2**   Composite photograph, to scale, of Jupiter's four Galilean satellites: Io (top left), Europa (top right), Ganymede (bottom left), Callisto (bottom right). (NASA/JPL Photograph.)

drove volatile materials away from the inner satellites Io and Europa, causing them to be dense and rocky. Ganymede and Callisto, being farther from Jupiter, remained cool enough that they retained their volatile materials; therefore, they are less dense and contain large quantities of ices. This pattern is similar to that observed in the solar system as a whole, where dense, rocky planets occur near the Sun and low-density gas giants and icy planets occur at greater distances. Such patterns of compositional differences provide important evidence of how solid materials condensed from the solar nebula, as will be discussed in Chapter 29.

All of the Galileans exhibit perfect spin–orbit coupling with Jupiter. In addition, the first three Galileans interact gravitationally with one another, producing strong tidal forces. Because individual Galileans are being "tugged" both by Jupiter and their satellite neighbors, internal heating may be occurring in them, especially in the case of Io, the one closest to Jupiter. Io is small enough that any internal heat generated when the satellite formed should have dissipated long ago, but the tidal forces that it experiences generate the heat responsible for Io's current volcanic activity. Such heating probably also occurs inside Europa.

■ Io

Io, located about 422,000 kilometers (262,000 miles) from Jupiter's center, has a diameter of 3632 kilometers (2257 miles) and a density of 3.53 grams per cubic centimeter. (Note that these values are very similar to those for Earth's Moon, suggesting some internal similarities even though their surfaces are different.) Io rotates and orbits in 1 day, 18 hours, and 28 minutes. Prior to the Voyager missions, Io was known to have some peculiar properties. It exhibits a phenomenon known as **posteclipse brightening:** when Io first emerges from Jupiter's shadow, it is initially much brighter than average but then fades to its normal brightness. In addition, a torus of sodium gas is spread over Io's orbit, which is located within the magnetosphere of Jupiter. As mentioned in Chapter 19, the generation of Jupiter's bursts of radio emission also seem to be related to Io.

The first Voyager photographs of Io showed a remarkable world: there were no impact craters visible, indicating that the surface is very young (Figure 20.3). This was the first time that spacecraft exploration had revealed a solid world completely lacking impact craters. The surface is also very colorful, with red, yellow, white, and orange markings, and on the whole, Io is even redder

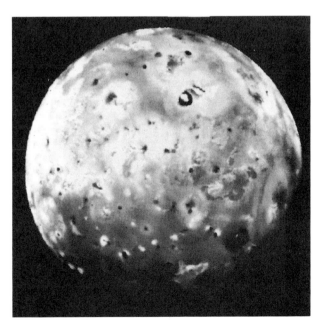

■ **Figure 20.3**   Voyager 1 photograph of Io. (NASA/JPL Photograph.)

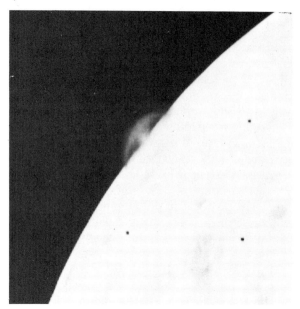

■ **Figure 20.4** Voyager 1 photograph of a volcanic plume on Io. (NASA/JPL Photograph.)

than Mars. These colors are believed to be due to elemental sulfur or sulfur compounds, which can change color when phase changes occur as temperature varies. The unusual chemistry of Io's surface provides the explanation for posteclipse brightening: during an eclipse, Io cools sufficiently that elemental sulfur changes from yellow to white. As soon as the eclipse ends, the higher-albedo white surface appears brighter until the sulfur warms and changes back to its darker, yellow form.

As Voyager 1 flew past Io, the numerous photographs returned to Earth failed to explain why its surface is so young. As the spacecraft sped on, it took a picture looking back at Io to fix its position against the stars for navigational purposes. The computer program analyzing the image noted that Io's image was not circular, and examination revealed a plume of material streaming away from one limb (Figure 20.4). It was soon realized that volcanoes were erupting, and the rest of the time before approaching the next target was spent looking back at Io for additional study.

Io is the only solar system object other than the Earth and Neptune's satellite Triton with active volcanoes, and in relation to its size, Io is even

more volcanically active than our planet. Nine volcanic plumes were seen by Voyager 1, and eight of them (along with two new ones) were also erupting when Voyager 2 passed by four months later (Figure 20.5). These volcanoes spray erupting material as high as 300 kilometers (185 miles) above Io's surface and deposit it as far as 600 kilometers (370 miles) from the volcanic vents. Most volcanic features have little relief, indicating that the magma issuing from Io is of very low viscosity.

The mechanism for Io's volcanic activity is not completely understood, although it is known to be far different from that occurring on Earth. The magma is either pure molten sulfur or a molten rock with a high sulfur content. It is possible that sulfur dioxide gas may drive the volcanoes by forcing material out from within Io, in which case they would be more similar to terrestrial geysers

■ **Figure 20.5** Voyager 2 photograph of Io that shows two volcanic plumes, illuminated by sunlight, rising above the satellite's limb. (NASA/JPL Photograph.)

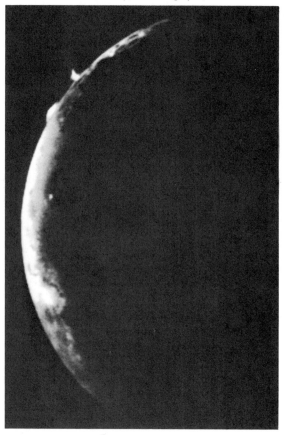

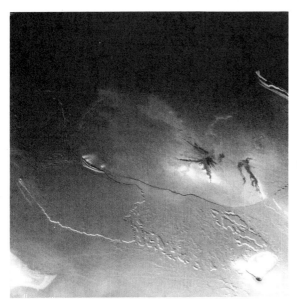

■ **Figure 20.6** Voyager 1 photograph of plains units on Io. (NASA/JPL Photograph, courtesy of the National Space Science Data Center and the Voyager Team Leader, Bradford A. Smith.)

piter, Europa is close enough that its gravity exerts forces on Io that produce heating by internal friction. Coincidentally, only weeks before the Voyager 1 encounter, some researchers had proposed that Io might have volcanic activity because of such heating. Radioactive decay also contributes to heating Io, whose total heat flow has been estimated to be 30 times that of Earth's!

About one-third of Io's surface was photographed at 5-kilometer (3-mile) resolution, and three types of surface geological units have been identified. **Plains units** are the most widespread and have a variety of colors, although they tend to be darker in the polar regions (Figure 20.6). The plains are composed of material that has fallen from volcanic plumes as well as material that has flowed out from the volcanoes, and the plains are therefore smooth. Some plains are cut by scarps and grabens, and it is believed that this faulting may be the result of collapse caused by the withdrawal of magma from beneath the surface. A second surface unit, **vent material**, includes calderas, fissure vents, and their associated lava flows (Figure 20.7). Some of the caldera vents are more than a kilometer deep, and some of Io's lava flows can be traced hundreds of kilometers from their sources. The third unit is **mountain material**, consisting of blocks of rock over 100 kilometers (60 miles) across that can rise up to 9 kilometers (5

than to volcanoes. Some of the gas expelled during eruptions remains close to Io for a while, producing a tenuous atmosphere. The heat for magma production is generated internally by tidal interactions among Jupiter, Io, and Europa. Although Io has 1:1 spin–orbit coupling with Ju-

■ **Figure 20.7** Voyager 1 photograph of vent material on Io, showing a caldera and the flows associated with it. (NASA/JPL Photograph.)

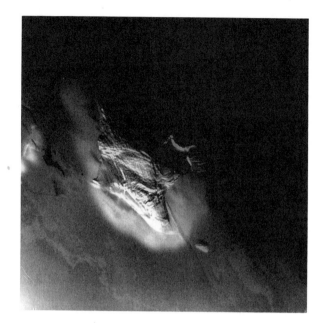

■ **Figure 20.8** Voyager 1 photograph of mountainous terrain on Io. This feature is about 200 kilometers (125 miles) across. (NASA/JPL Photograph, courtesy of the National Space Science Data Center and the Voyager Team Leader, Bradford A. Smith.)

■ **Figure 20.9** Cross-sectional views showing the interior structures of the Galilean satellites. (From Greeley, Ronald. *Planetary Landscapes*. Allen and Unwin, London, 1985. Reproduced by kind permission of Unwin Hyman, Ltd.)

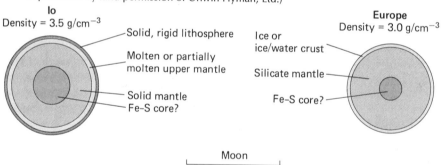

Io
Density = 3.5 g/cm^{-3}

Solid, rigid lithosphere

Molten or partially molten upper mantle

Solid mantle
Fe–S core?

Europe
Density = 3.0 g/cm^{-3}

Ice or ice/water crust

Silicate mantle

Fe–S core?

Moon

Mercury

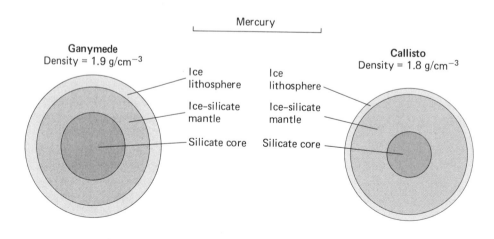

Ganymede
Density = 1.9 g/cm^{-3}

Ice lithosphere

Ice-silicate mantle

Silicate core

Callisto
Density = 1.8 g/cm^{-3}

Ice lithosphere

Ice-silicate mantle

Silicate core

miles) above the surrounding plains (Figure 20.8). These mountains are of uncertain origin, as neither volcanic nor tectonic activity seems to have formed them, and they are probably not composed of sulfur, as it would not support the steep slopes found in them.

The interior of Io is probably rich in silicates. (Cross-sectional diagrams of the Galilean satellites are shown in Figure 20.9.) The satellite has a low content of volatile materials, partly because it formed close to warm Jupiter but also because volatile materials have been sprayed out into space by the active volcanism. Io's core is probably composed of iron and sulfur. The mantle of Io is probably a mixture of silicates and some sulfur, and the upper mantle is at least partially molten. A rigid lithosphere surrounds the mantle. Indications are that Io "turns itself inside out" as a result of its volcanic activity, with interior material burying surface features. For this reason, impact features are eradicated more rapidly on Io than on most other planets and satellites.

■ Europa

Europa is the smallest and least understood Galilean, partly because it received the poorest photographic coverage during the Voyager missions. Located 671,000 kilometers (417,000 miles) from Jupiter, it is 3126 kilometers (1942 miles) in diameter and has a density of 3.17 grams per cubic centimeter. It rotates and orbits once every 3 days, 13 hours, and 14 minutes. Europa's surface has very low relief and exhibits a lack of craters, especially larger ones (Figure 20.10). This indicates a young surface, but the surface's exact age and amount of ongoing geological activity are unknown.

Europa has the highest albedo of any Galilean satellite, and its surface is presumed to be mostly water ice. It has very little relief, and its topographic features are caused more by color differences than elevation differences. Two types of surface units have been identified. One is **mottled terrain,** which is brown or gray in color. Brown mottled terrain contains small hills, whereas gray mottled terrain is smoother. The second surface unit is **plains,** which are generally

■ **Figure 20.10** Voyager 2 photograph of Europa. Note the lack of craters and the abundance of linear markings, which are difficult to see from the distance this photograph was taken. (NASA/JPL Photograph, courtesy of the National Space Science Data Center and the Voyager Team Leader, Bradford A. Smith.)

smooth but may be cut by numerous streaks and lineations and marked by spots. Some craters have been identified on Europa, but their total number is less than one dozen.

The most conspicuous surface features on Europa are numerous streaks, which appear to be cracks that formed in an icy crust and were subsequently filled in (Figure 20.11). (These features closely resemble what earlier observers believed to be the "canals" on Mars!) Because Europa's density is comparatively high, any water ice contained within it must be limited to a thin, near-surface crust. Presumably, the cracks formed in the following way (Figure 20.12): Europa's surface at one time was a smooth, icy crust, with liquid water underneath. This water was kept in a liquid state by heat from internal radioactive decay, but most of the liquid eventually froze as Europa cooled. When that occurred, expansion of the freezing material caused cracks in the upper sur-

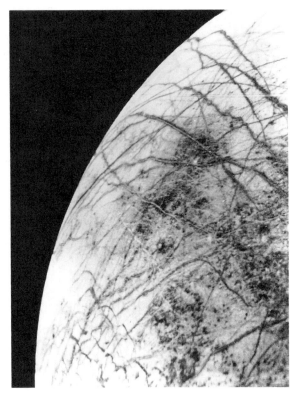

■ **Figure 20.11** Voyager 2 photograph of Europa showing the satellite's numerous surface cracks. (NASA/JPL Photograph.)

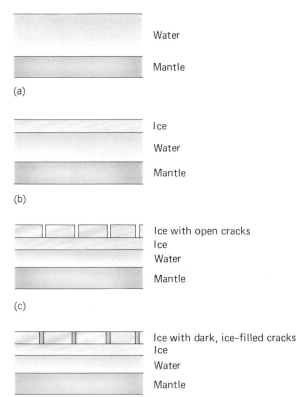

■ **Figure 20.12** Formation of Europa's surface cracks. (a) The surface of Europa's original water ocean froze, producing a thin layer of ice (b). A lower layer of the ocean then froze (c), and its expansion cracked the upper layer. (d) The cracks in the upper layer were filled with water that seeped up from below. This water subsequently froze, producing dark, linear markings.

face layer, with water from below seeping into and freezing in the cracks as they formed.

The interior of Europa is probably somewhat similar to that of Io. In order to account for its lower density, Europa's iron–sulfur core must be proportionally smaller than Io's, and its silicate mantle larger. The outermost 100 kilometers (60 miles) of Europa is soft ice or even liquid water covered with a thin layer of ice. (Internal heating by tidal friction may occur inside Europa as it does in Io, and this would enable some water to remain in liquid form.) It is possible that volcanic activity of a sort may occur on Europa if there is liquid water remaining. The water would be under pressure, and if any reached the surface through a crack, it would spray into space. It has also been suggested that the watery layer in Europa may be the only other place in the solar system besides Earth where life could have developed, but there is absolutely no evidence yet to support the idea that Europan life may exist.

■ Ganymede

Ganymede, with a diameter of 5276 kilometers (3279 miles), is the largest planetary satellite in the solar system. It orbits 1,070,000 kilometers (665,000 miles) from Jupiter, and both its orbital and rotational periods are 7 days, 3 hours, and 43 minutes. Ganymede is relatively light, with a density of only 1.99 grams per cubic centimeter.

At first glance, Ganymede looks very Moon-like, as it has dark regions on a brighter background (Figure 20.13). However, there is a major difference, because the dark regions on Ganymede are old and rough, whereas the lighter regions are young and smooth, unlike the Moon,

■ **Figure 20.13** Voyager 2 photograph of Ganymede. Note the distinct albedo differences between various regions on the satellite. (NASA/JPL Photograph, courtesy of the National Space Science Data Center and the Voyager Team Leader, Bradford A. Smith.)

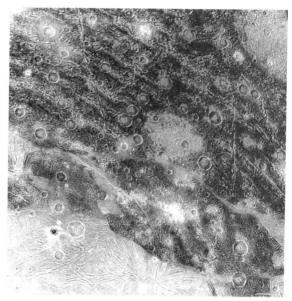

■ **Figure 20.14** Voyager 2 photograph of dark, cratered terrain on Ganymede. This material may be the original crust of Ganymede. The area shown is about 1700 kilometers (1050 miles) wide. (NASA/JPL Photograph, courtesy of the National Space Science Data Center and the Voyager Team Leader, Bradford A. Smith.)

■ **Figure 20.15** Voyager 2 photograph of light, grooved terrain on Ganymede. These grooves appear to be the result of faulting. The area shown is about 1500 kilometers (930 miles) wide. (NASA/JPL Photograph, courtesy of the National Space Science Data Center and the Voyager Team Leader, Bradford A. Smith.)

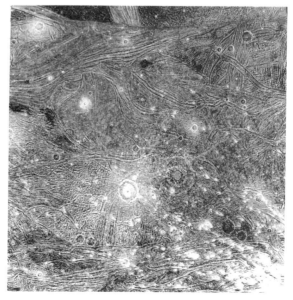

whose dark areas are smooth and light areas are rough. These two regions are the main surface units on Ganymede. **Dark, cratered terrain** is believed to be the original crust of the satellite (Figure 20.14). The largest area of this material is named **Galileo Regio,** after Galileo. The other surface unit, which is the most abundant, is **light, grooved terrain,** which covers 60% of the surface area photographed (Figure 20.15). As the name implies, it contains grooves that are parallel sets of ridges and troughs with a total relief as high as 700 meters (2000 feet). These grooves cover much of the surface and form rather intricate patterns. Grooved terrain is probably the result of tensional faulting or the release of water or slush from beneath the surface. The release of light-colored material in this way would explain the fact that grooved terrain areas seem to have grown at the expense of the dark, cratered terrain.

Ganymede (and also Callisto) are more normal solid bodies than the inner two Galileans, because their surfaces are heavily cratered. However, the large impact features on these icy worlds are much different from those on the Moon and Mars. The largest craters and impact

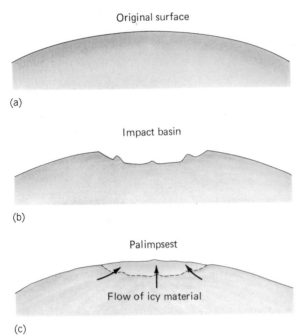

Original surface

(a)

Impact basin

(b)

Palimpsest

Flow of icy material

(c)

■ **Figure 20.16** Formation of palimpsests. The original surface (a) experiences an impact that forms a large basin (b). Flow of icy material fills the basin but preserves the outline of the earlier structure, forming a palimpsest (c).

basins on these satellites have very little relief, a result of isostatic adjustment of their icy surfaces. They look like craters when viewed from above but have almost no vertical relief (Figures 20.16 and 20.17). These large "phantom craters" are called **palimpsests**, a term originally applied to reused ancient writing materials on which older writing was still visible underneath newer writing. Palimpsest craters range in diameter from about 50 to 400 kilometers (30 to 250 miles). Craters smaller than these palimpsests have more normal cross-sectional profiles, and some have central peaks. Both bright and dark rays of ejecta material occur around Ganymede's craters. Rays tend to be bright for craters formed on grooved terrain and dark for craters formed in the dark, cratered terrain.

Because of Ganymede's low density, it is estimated to be approximately half water ice. Ganymede is believed to have a silicate core extending out to about half the planet's radius. Surrounding this is a mantle composed of both ice and silicates and a crust that is probably a thick layer of water ice.

■ Callisto

Callisto, the most distant Galilean satellite, is 1.88 million kilometers (1.17 million miles) from Jupiter and requires 16 days, 16 hours, and 32 minutes to orbit and rotate. It is 4829 kilometers (2988 miles) in diameter and has a density of only 1.76 grams per cubic centimeter.

Callisto's surface is the darkest of any Galilean, and the dark color is believed to indicate that the surface is covered by a layer of **carbonaceous** (carbon-containing) **material.** Callisto's surface is dominated by a large number of impact craters but has very little relief (Figure 20.18). Callisto has the highest crater density of any Galilean satellite's surface, and therefore probably has the oldest surface. Callisto's craters are nearly continuous, and there are no plains units where craters formed during the early period of heavy bombardment have been obliterated by later internal processes. Callisto is the only cratered body in the solar system that lacks such plains, leading to the conclusion that it has never experienced basaltic volcanism or volcanism involving other forms of material.

Callisto has craters of many sizes, and, as a

■ **Figure 20.17** Voyager 1 photograph of a barely discernible palimpsest basin on Ganymede, which is marked by an *arrow*. (NASA/JPL Photograph, courtesy of the National Space Science Data Center and the Voyager Team Leader, Bradford A. Smith.)

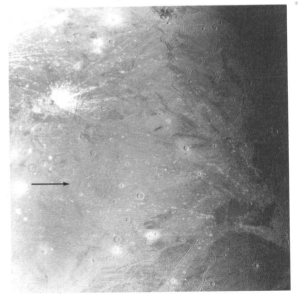

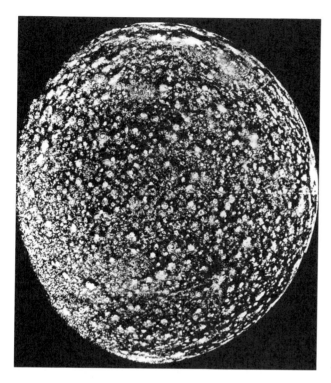

■ **Figure 20.18** Mosaic of Voyager 2 photographs of Callisto. Note the extremely large number of craters covering the surface. (NASA/JPL Photograph.)

rule, the larger the diameter of the crater, the lower its relief. Callisto's large craters are of the same palimpsest variety found on Ganymede, and its largest surface feature is a multiringed basin named **Valhalla** (Figure 20.19). It is comparable in size to the Orientale and Caloris structures on the Moon and Mercury, respectively, but it lacks relief and is essentially just a light-colored albedo feature. Valhalla's central bright zone is 600 kilometers (370 miles) across, and it is surrounded by concentric rings extending as far as 2000 kilometers (1200 miles) from its center. Although Valhalla is the largest impact basin on Callisto, at least seven others have been identified.

Because of its low density and low relief, Callisto is believed to contain a great deal of ice. Its interior is probably very similar to that of Ganymede, except that its silicate core is proportionately smaller and its icy crust is probably thinner, accounting for the lower amount of relief.

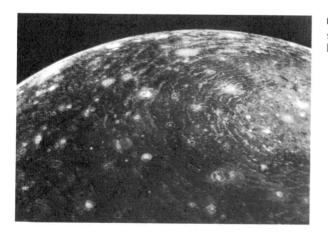

■ **Figure 20.19** Mosaic of Voyager 2 photographs showing the Valhalla impact basin on Callisto. Note the lack of relief in this large feature. (NASA/JPL Photograph.)

Prospects for Future Exploration

When the Galileo spacecraft eventually orbits Jupiter, it will have numerous encounters with the Galilean satellites and will photograph their surfaces with even greater resolution than the remarkable images returned by Voyagers 1 and 2. If Galileo is successful, the 1990s should bring additional advances in our understanding of the geology of these fascinating objects.

Chapter Summary

The Galilean satellites, Io, Europa, Ganymede, and Callisto, were discovered by Galileo in 1610. The first satellites other than Earth's Moon to be discovered, they greatly influenced astronomy. Much of what we know about these bodies was learned during the Voyager flybys.

The densities of the Galileans decrease from Io to Callisto, probably as a result of heating from Jupiter as they were forming. All of them have 1 : 1 spin–orbit coupling with Jupiter, and at least two of them, Io and Europa, probably generate internal heat by tidal friction.

Io has a colorful, relatively smooth surface that is relatively young as a result of resurfacing by volcanic activity. Io's volcanoes appear to erupt either sulfur or sulfur-rich magma.

Europa has a smooth surface that appears to be water ice, possibly covering an ocean of liquid water. Its only conspicuous surface markings are lines resulting from cracks in its icy crust.

Ganymede and Callisto are both icy and have older, cratered surfaces. Both of these satellites have craters of palimpsest form, with relief eliminated by isostatic rebound. Ganymede's surface has old, dark, rough areas and younger, lighter, smooth areas, whereas Callisto's surface is entirely old and heavily cratered.

Chapter Vocabulary

Callisto	Galileo Regio	mottled terrain	posteclipse
carbonaceous	Ganymede	mountain material	brightening
material	Io	palimpsest	Valhalla
dark, cratered terrain	light, grooved terrain	plains units	vent material
Europa			

Review Questions

1. What further scientific discoveries resulted from the discovery of the Galilean satellites?
2. What changes occur in the Galileans as one moves from Io out to Callisto, and why do they occur?
3. Describe Io's posteclipse brightening and its volcanic activity.
4. Summarize Io's major surface features.
5. What process is believed to be responsible for the streaks on the icy surface of Europa?
6. Summarize Europa's major surface features.
7. Describe the characteristics and origin of palimpsest craters.
8. Summarize Ganymede's major surface features.
9. Summarize Callisto's major surface features.

■ For Further Reading

Johnson, Torrence V. "The Galilean Satellites." In Beatty, J. Kelly and Andrew Chaikin, editors. *The New Solar System.* Sky Publishing, Cambridge, MA, 1990.

Johnson, Torrence V. and Laurence A. Soderblom. "Io." In *Scientific American,* May, 1981.

Morrison, David. "Four New Worlds." In *Astronomy,* September, 1980.

Soderblom, Laurence A. "The Galilean Moons of Jupiter." In *Scientific American,* January, 1980.

21

Saturn

CHAPTER OBJECTIVES

After studying this chapter, you should be able to

1. Describe Saturn's appearance to the unaided eye and through a small telescope.
2. Summarize the history of spacecraft exploration of Saturn.
3. Describe the basic features of Saturn's atmosphere and interior and how they differ from Jupiter's.
4. List and describe the components of Saturn's ring system.

Voyager 1 photograph of Saturn showing it in a crescent phase, which is never visible from the Earth. (NASA/JPL Photograph.)

Saturn, the second largest planet, was named for the father of the Roman god Jupiter. It is the most distant planet normally visible to the naked eye and was believed to mark the outer boundary of the solar system until Uranus was discovered in 1781. It is often called the "ringed planet," and the name is still appropriate, even though it is now known that other planets have rings. This is because Saturn's remarkable ring system is the only one that can be viewed easily from Earth. Saturn has the lowest density of any planet and is the only one less dense than water Incredible as it seems, Saturn would float if a large enough container of water could be found! Saturn's major orbital and physical properties are summarized in Table 21.1.

■ Appearance and Basic Properties

Saturn requires nearly 30 years to orbit the Sun, so it moves rather slowly through the zodiacal constellations. It appears as a yellow object in the sky, somewhat like Jupiter, but it is considerably fainter because of its greater distance from Earth. Saturn's appearance to the naked eye is rather undistinguished and gives no clue to the magnificent splendor awaiting the telescope user (see Figure 18.1).

Almost from the beginning, telescopic observers have paid more attention to the rings of Saturn than anything else about the planet. Galileo, the first observer to look at the planet, was unable to understand what he saw and assumed Saturn had two close companions that made the planet look as though it had "handles." As described in Chapter 18, improved telescope optics allowed the true nature of the rings to be understood by the mid-1600s. Later, sharp-eyed observers noticed gaps in the rings, and the separate ring components were given individual names.

■ **Table 21.1   Summary of Saturn's Basic Properties***

Semi-major axis: 9.56 astronomical units
Semi-major axis: 1429.4 million kilometers
Semi-major axis: 888.2 million miles
Orbital eccentricity: 0.056
Orbital inclination: 2.49°
Orbital period: 29.46 years
Diameter: 120,000 kilometers
Diameter: 74,568 miles
Mass (Earth = 1): 95.2
Density: 0.70 grams per cubic centimeter
Sidereal rotational period: 10.5 hours
Axial obliquity: 26.73°
Number of satellites: 17

* For a complete table of planetary data, see Table A2.1 in
  Appendix 2.

Because Saturn's axis has an obliquity of 26.7°, the orientation of its rings changes during its 29.5-year orbital period, and their appearance from Earth varies (Figure 21.1). Every 15 years, the rings seem to disappear as they are seen edge-on. New Saturnian satellites have been discovered during times of ring disappearances, because close moons normally lost in the glare of the rings are then visible. Midway between these disappearances, the rings are fully opened toward Earth and present the best view to telescope observers.

Saturn's disk receives far less attention than its rings, not just because the rings are so beautiful but because the disk of Saturn is relatively bland. Saturn's surface markings are rather muted compared with Jupiter's, and there are not many colorful spots and storms. Saturn's disk, like Jupiter's, shows flattening due to the rapid, 10.5-hour rotation, and Saturn is the most oblate planet in the solar system. Saturn has a number of satellites visible through small telescopes, but only the largest, Titan, is as conspicuous as Jupiter's Galilean moons. Titan is noteworthy because it is the only satellite in the solar system known to have a substantial atmosphere. (Neptune's satellite Triton has an atmosphere, but it is very tenuous.)

■ Spacecraft Exploration

Saturn has been visited by three flyby spacecraft, all of which first passed by Jupiter and obtained gravitational boosts during those encounters. **Pioneer 11**, which flew past Jupiter on December 2, 1974, flew past Saturn on September 1, 1979. Although some mission planners had wanted to target the craft between the rings and the planet, a more conservative trajectory was selected. Pioneer crossed the ring plane at a distance well beyond the outside edge of the rings, and then swung under the rings for its closest approach to

■ **Figure 21.1**   The changing appearance of Saturn's rings as seen from Earth. Note that the rings cannot be seen when oriented edge-on. (Lowell Observatory Photographs.)

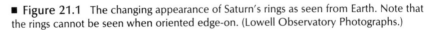

the planet, some 21,000 kilometers (13,000 miles) above the planet's cloud tops. At the time of this encounter, Pioneer 11 was traveling 114,000 kilometers per hour (71,000 miles per hour). Pioneer survived the encounter with no ill effects, even though it almost collided with a satellite, one of two that it discovered! Pioneer also discovered a new ring section, as well as Saturn's magnetic field and magnetosphere, and it verified that Saturn has an internal heat source. For the first time, Saturn's rings were observed from the unlit side. Photographs made using transmitted light are much different from previous views of the rings made using reflected light (Figure 21.2). In transmitted-light views, Saturn's rings have bright areas where they are thin and contain relatively few particles to block the light. Areas of the rings that contain many particles, which look bright when viewed from Earth because they reflect light so well, are dark in transmitted light because they block most of the light. Areas without any ring material also appear dark.

Voyager 1, which encountered Jupiter on March 5, 1979, passed through the Saturn system in November, 1980. A major objective was photography of the satellite Titan, but its atmosphere was found to be opaque, like Venus' atmosphere, and its surface was therefore invisible. Voyager 1 passed only 4000 kilometers (2500 miles) above the clouds of Titan on November 11, 1980. The next day, November 12, Voyager 1 made its closest approach to Saturn, at 124,000 kilometers (77,000 miles), and encountered many other satellites. Voyager 1 also obtained high-resolution photographs of Saturn's rings, revealing for the first time the extreme complexity of their structure. In addition, Voyager observed a star as it was occulted by the rings, and this technique allowed the detection of even finer details of ring structure than those that could be photographed directly. Because of the nature of its trajectory through the Saturn system, Voyager 1 could not be sent to any other planets.

Voyager 2, which had visited Jupiter on July 9, 1979, encountered Saturn on August 25, 1981. It passed only 32,000 kilometers (20,000 miles) beyond the outer edge of the rings, or a distance of 113,000 kilometers (70,000 miles) from Saturn's cloud tops. The first pictures returned after

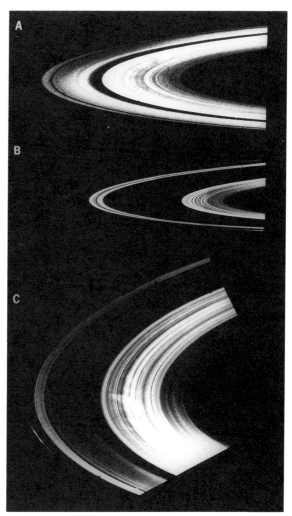

■ **Figure 21.2** A comparison of Saturn's rings when seen in reflected light (a) and transmitted light (b and c). Note that ring components that are faint in reflected light are bright in transmitted light. (NASA/JPL Photographs, courtesy of the National Space Science Data Center and the Voyager Team Leader, Bradford A. Smith.)

the spacecraft's close passage through the ring plane were blank, and fears mounted that collisions with ring particles had damaged the imaging system. Later analysis revealed that the lubricant had failed on the heavily used instrument aiming system, and the camera platform had not been aimed at its targets. Subsequently, pictures were able to be obtained by changing the entire spacecraft's orientation so that its camera platform would be pointed in the right direction. After the

■ **Figure 21.3** Voyager 1 photograph of Saturn and its rings. (NASA/JPL Photograph.)

spacecraft had left Saturn, engineers unjammed the camera platform by carefully moving it to redistribute the lubricant. These repairs allowed the platform to function during subsequent encounters with Uranus and Neptune. The August, 1981 encounter gave the spacecraft the proper boost for a journey to Uranus and Neptune. Both of the Voyager encounters sent back remarkably detailed photographs to Earth and greatly increased our knowledge of Saturn and its satellites.

The next mission to Saturn will be Cassini, an American spacecraft currently scheduled for launch in April, 1996. Cassini will be somewhat similar to the Galileo craft in that it will split into two parts upon reaching Saturn in 2002. The orbiter will study Saturn and its satellites for an extended period of time, and one of its many goals will be mapping the surface of Titan using radar to penetrate its opaque atmosphere. The second part of Cassini, the Titan Atmosphere Probe, will enter Titan's atmosphere and transmit data before striking the satellite's surface.

■ Atmosphere

Saturn's atmosphere is similar to Jupiter's in many ways. Saturn, like Jupiter, has belts, zones, and various spots in its atmosphere, but Saturn's atmospheric markings lack the color contrast of those on Jupiter and are much less apparent (Figure 21.3). It is believed that Saturn's atmospheric markings are muted in this way because its cloud layers are deeper in its atmosphere than the clouds of Jupiter and therefore more difficult to see. Saturn also lacks any atmospheric markings as long-lived as the Great Red Spot of Jupiter. Another difference between the planets is that Saturn has a bit less helium in its atmosphere than Jupiter (6% versus 10%), indicating that some past process has depleted the helium in Saturn's atmosphere. One aspect in which Saturn's atmosphere is not inferior to Jupiter's is in wind velocity, which ranges from about 80 kilometers per hour (50 miles per hour) in the westward direction (opposite to the rotational direction) at 40° latitude to nearly 1800 kilometers per hour (1100 miles per hour) in the eastward direction at the equator (Figure 21.4).

As with all gas giants, the arbitrary starting point for making atmospheric measurements on Saturn is the 100-millibar level. This is the point of minimum temperature, abut 80 K (−315°F). As on Jupiter, both temperature and pressure rise below this point, although at a slower rate on Saturn (Figure 21.5). About 100 kilometers (60 miles) below this point, pressure has risen to 1 bar, and the temperature is 130 K (−225°F). About 300 kilometers (180 miles) below the 100-millibar point, the pressure is 10 bars and the temperature is 300 K (80°F). Saturn's three cloud layers, composed of ices of ammonia, ammonium hydrosulfide, and water, in order from top to bottom, occur between about 100 and 300

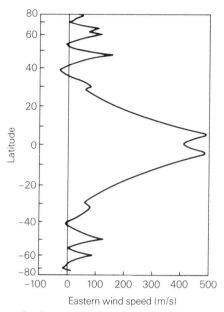

■ **Figure 21.4** A graph of wind velocities on Saturn as a function of latitude. Positive velocities are those in the direction of planetary rotation, negative velocities are those in the opposite direction. (NASA diagram from Abell, George, David Morrison, and Sidney Wolff. *Exploration of the Universe,* 5th ed. Saunders College Publishing, Philadelphia, 1987.)

kilometers (60 and 180 miles) below the 100-millibar point. As mentioned earlier, the fact that these clouds are lower in altitude on Saturn than on Jupiter probably explains why Saturn is not as colorful as Jupiter. In addition, the three cloud layers are farther apart on Saturn because that planet's weaker gravity allows its atmosphere to have a greater vertical range than Jupiter's.

■ Interior and Magnetic Field

Saturn, like Jupiter, is somewhat Sun-like, as it is 94% hydrogen and 6% helium (by number of atoms), compared to the Sun, which is 89% hydrogen and 11% helium, and Jupiter, which is 90% hydrogen and 10% helium. Saturn is by far the least dense of these three objects, with a density of 0.70 grams per cubic centimeter. The interior of Saturn is believed to be somewhat similar to that of Jupiter, except for differences caused by its smaller mass (Figure 21.6). One result of the lower mass is that it causes less internal pressure, which causes less compression, allowing Saturn's

overall density to be so low. Also, the proportion of Saturn's interior that contains metallic hydrogen is probably much smaller than that of Jupiter, because pressures high enough to form metallic hydrogen would only exist very close to the center of Saturn.

Below Saturn's outer atmosphere lies its molecular hydrogen layer, which probably extends about halfway down to the center of the planet. The molecular hydrogen layer is depleted in helium and probably enriched somewhat in water and other heavier materials. At the base of this layer, the temperature is about 9000 K, the density is about 1.2 grams per cubic centimeter, and the pressure is about 3 million atmospheres. Below this is the metallic hydrogen layer, which extends down to a point about 75% of the way to the center of Saturn. The upper part of the metallic hydrogen region is thought to be nonhomogeneous and enriched in helium; helium "rain-

■ **Figure 21.5** Vertical structure of Saturn's atmosphere. Altitude in kilometers relative to 100-millibar level is plotted against temperature in degrees Kelvin. The three dark bands are the cloud layers described in the text. (NASA diagram from Abell, George, David Morrison, and Sidney Wolff. *Exploration of the Universe,* 5th ed. Saunders College Publishing, Philadelphia, 1987.)

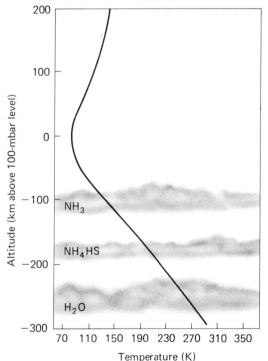

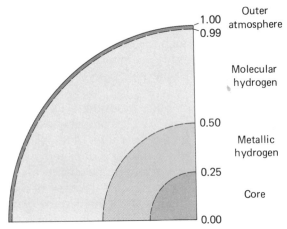

1.00 — Outer atmosphere
0.99

Molecular hydrogen

0.50

Metallic hydrogen

0.25

Core

0.00

■ **Figure 21.6** Saturn's interior layers. The numbers represent the fraction of Saturn's total radius. See text for detailed descriptions.

drops'' form there and sink to lower levels. Beneath the metallic hydrogen layer is the planet's silicate–iron core. At the metallic hydrogen-core boundary, the temperature is about 12,000 K, the density jumps from about 2 to 7 grams per cubic centimeter across the transition, and the pressure is about 8 million atmospheres. Saturn's center is thought to have a temperature of 20,000 K, a density of 15 grams per cubic centimeter, and a pressure of 50 million atmospheres.

Although Saturn's central conditions are less extreme than Jupiter's, Saturn is a better overall heat source, emitting 2.8 times as much energy as the solar energy it receives, in contrast to the value of 2.0 times for Jupiter. It is possible that Saturn is such an efficient heat source because energy is produced by the separation of helium in

Saturn's atmosphere and the subsequent ''raining'' of the gas down to deeper levels. This process would convert potential energy to heat energy and also account for the helium deficiency in the atmosphere.

Saturn's magnetic field, unlike Jupiter's, exhibits no phenomena detectable from Earth, and it was discovered by spacecraft observation. Saturn's total magnetic moment is almost 600 times stronger than Earth's, but its surface field strength is only two-thirds that of the Earth. Saturn's field is unique in the solar system in that there is almost no inclination between the rotational axis and the dipole magnetic axis. The reason why Saturn's spin and magnetic axes are virtually the same is not yet known. Pioneer 11 determined that Saturn's magnetosphere extends outward to about 20 times the radius of the planet. A large torus of neutral hydrogen circles Saturn, extending from the orbit of the satellite Rhea outward to a point just beyond the orbit of Titan. The source of much of this hydrogen is believed to be the photodissociation of methane from Titan's atmosphere.

■ Rings

Saturn's rings are a huge aggregate of small objects orbiting the planet in its equatorial plane. As described in Chapter 18, the particles in the ring are believed to be either icy fragments or rocks coated with ices. The ring particles are very bright, with albedos of about 80%, and are believed to range in diameter from centimeters to tens of meters. Although the rings are probably

■ Table 21.2 Components of Saturn's Rings

Designation	Extent (Saturn Radii)	Extent (kilometers)	Extent (miles)
D ring	1.11–1.21	67,000–73,200	41,600–45,500
C ring	1.21–1.53	73,200–92,200	45,500–57,300
B ring	1.53–1.95	92,200–117,500	57,300–73,000
A ring	2.01–2.26	121,000–136,200	75,200–84,600
F ring	2.33	140,600	87,400
G ring	2.8	170,000	105,600
E ring	3.5–5.0	211,000–302,000	131,000–188,000

Data from Morrison, David. *Voyages to Saturn.* NASA, Washington, D.C., 1982.

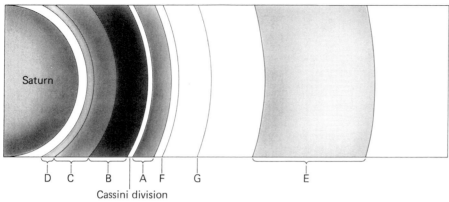

■ **Figure 21.7** Map of Saturn's ring system, as seen from above the planet's north pole.

no more than a few hundred meters thick, their lateral extent is enormous, from 1.11 to about 5 times the radius of the planet away from Saturn's center (74,000 to 303,000 kilometers, or 42,000 to 188,000 miles). The rings are designated by letters of the alphabet, and the A, B, C, and D rings are visible from Earth. Other rings (E, F, and G) have been discovered by spacecraft. The components of Saturn's ring system are listed in Table 21.2 and shown in Figure 21.7.

The outermost ring visible from Earth, the **A ring**, is the second brightest ring and is 15,000 kilometers (9400 miles) wide. It has several divisions near its outer edge, the largest of which, the **Keeler division**, is about 270 kilometers (170 miles) wide. (The Keeler division is also known as the Encke division. Encke first saw the feature in 1838, but his observation was disputed until Keeler verified it in the late 1880s.) Voyager data showed that the outer edge of the A ring is kept sharp by a guardian satellite, Atlas.

Saturn's most conspicuous ring is the **B ring**, which is 25,600 kilometers (15,600 miles) wide and brighter than all of the other rings combined. It is located just inside the A ring and is separated from it by the 3500-kilometer-wide (2200-mile-wide) **Cassini division**, discovered by **Giovanni D. Cassini** (1625–1712) in 1675. This division, which appears empty from Earth, was shown by Voyager 1 to contain numerous ringlets (see Figure 18.9). As described in Chapter 18, the division is probably associated with an orbital resonance from the satellite Mimas. The B ring itself is opaque and casts its shadow on Saturn, leading to

the belief that it contains most of the mass in the ring system. The Voyagers revealed that it contains the most extensive small-scale structure of the ring system and somewhat resembles a grooved phonograph record (Figure 21.8). They

■ **Figure 21.8** Detailed structure in Saturn's B ring, as photographed by Voyager 2. The photograph shows an area 6000 kilometers (3700 miles) wide and reveals details only 10 kilometers (6 miles) wide. (NASA/JPL Photograph.)

also provided evidence of radial variation in composition and size of the ring particles within the B ring.

The **C ring**, or crepe ring, is the innermost ring visible in telescopes, but it is very faint and difficult to see. Its inner boundary is sharp, but its outer edge is simply a transition to the B ring, a phenomenon that is not well understood. The crepe ring is 19,400 kilometers (12,000 miles) wide and has numerous structural features, including two prominent gaps. Investigations have revealed that the C ring is deficient in small particles compared to the A and B rings.

Inside the C ring is the extremely faint **D ring**, discovered photographically in 1969 by the French astronomer Pierre Guerin. This ring cannot be seen visually, but it occasionally causes a slight darkening of Saturn's equatorial clouds. It is 6000 kilometers (3800 miles) wide and composed of a series of thin ringlets. Despite the scarcity of material in it, the D ring has the same intricate structure as the denser outer rings. (The designation "D ring" has also been applied to rings seen beyond the A ring by terrestrial observers. Many observers have glimpsed such a ring, and one was photographed by W. A. Feibelman in 1966. The Voyager probes revealed several ring components, now called E, F, and G, beyond the A ring, and they probably account for these observations. Since the Voyager missions, most terrestrial observers have used the designation "D' ring" for sightings of these outer rings.)

Beyond the A ring lies the narrow **F ring**, which was discovered by Pioneer 11 and extensively studied by the Voyagers. It is only a few hundred kilometers wide and bounded by two guardian satellites, Prometheus and Pandora. It appears to be made of several discontinuous strands. Voyager 1 saw these in several braided kinks, which was totally unexpected (see Figure 18.11). The braided appearance of these rings was less noticeable, but still present, when they were photographed by Voyager 2.

The outermost rings, G and E, lie far beyond Saturn's Roche limit and the main part of the ring system. Discovered by spacecraft, they are rather tenuous and lack fine-scale structure. The **G ring**, like the F ring, is relatively narrow, whereas the **E ring**, which is 91,000 kilometers (57,000 miles) wide and diffuse, is believed to be related somehow to the satellite Enceladus. There have been suggestions that some form of volcanic activity has occurred on Enceladus, and if some of the erupted material escaped from the satellite, it could have remained in orbit around Saturn as a component of this ring.

■ Satellites

At present, Saturn holds the record for the most satellites of any planet, with 17. Prior to the space age, nine satellites of Saturn were known. A tenth satellite, discovered in 1966, was revealed by later observations to be actually two satellites in nearly identical orbits! These **co-orbital satellites**, as they are called, are the only case yet known in the solar system of two objects in the same orbit other than those at Lagrangian points. Spacecraft observations and telescopic views during the 1980 edge-on ring orientation revealed six other satellites, three ring guardians and three Lagrangians. Because good images were obtained of most of these objects, we know a good bit about their geology, and we will discuss the satellites in detail in the next chapter.

■ Chapter Summary

Saturn is the solar system's second largest planet, and its ring system certainly makes it the most beautiful planet when observed through a telescope. It is the most distant planet visible to the unaided eye. Three spacecraft have flown past Saturn and provided us with much information about it.

Saturn's atmosphere is similar to Jupiter's, except that Saturn's atmospheric markings are much less conspicuous, probably as a result of its cloud layers being deeper and thus partly hidden from view. Saturn's interior is also similar to Jupiter's but is not as compressed. As a result, the metallic hydrogen region is smaller, and the planet has a low enough density that it is the only solar system object known to be less dense than water.

Saturn's ring system contains several components visible from Earth and others dis-covered only by firsthand spacecraft observations. Saturn's satellite system contains 17 satellites, the most of any planet.

■ Chapter Vocabulary

A ring	Cassini division	F ring	Pioneer 11
B ring	co-orbital satellites	G ring	Voyager 1
C ring	D ring	Keeler division	Voyager 2
Cassini, Giovanni D.	E ring		

■ Review Questions

1. List some of Saturn's unique characteristics.
2. Describe Saturn's appearance through a small telescope. Summarize the ways in which it differs from Jupiter's appearance.
3. Discuss the accomplishments of the three flyby missions to Saturn.
4. Describe the atmosphere of Saturn.
5. Describe the interior of Saturn.
6. Describe the rings of Saturn that are visible from Earth.
7. What rings were discovered by the Voyager spacecraft?
8. Explain why several of Saturn's satellites are considered to have unusual orbits.

■ For Further Reading

Dooper, Henry S. F. *Imaging Saturn — The Voyager Flights to Saturn.* Holt, Rinehart, and Winston, New York, 1982.

Morrison, David. *Voyages to Saturn.* NASA, Washington, D.C., 1982.

Saturn's Satellites

Photographs of Saturn's major satellites, reproduced to the same scale as a silhouette of Titan, its largest satellite. (NASA/JPL Photograph. Composite prepared by Stephen Paul Meszaros.)

Saturn's 17 satellites include a wide variety of objects, from Titan, a large object with a substantial, haze-filled atmosphere, to Enceladus, which has the highest albedo of any solar system object, to several tiny objects invisible from Earth. These satellites, along with their major properties, are listed in Table 22.1. Saturn's satellite system is very regular, as all except the outermost two (Iapetus and Phoebe) travel in low-eccentricity, low-inclination orbits. Phoebe also has a retrograde orbit and may be a captured object. All of the satellites except Phoebe (and possibly Hyperion) exhibit 1:1 spin-orbit coupling. Because of this, one hemisphere of a satellite leads as its orbits, while the other side trails. These two hemispheres, the **leading** and **trailing**, are frequently mentioned when discussing satellite geology. This is because the surfaces of each hemisphere are quite different on many satellites, due mainly to the fact that the leading hemisphere is more likely to collide with debris as the satellite orbits.

Saturn's satellite system differs from Jupiter's in that there is no pattern based upon density or composition. Icy and rocky objects are found throughout the satellite system, but there are no satellites as dense as Io and Europa. This indicates that while Saturn's satellites were forming, Saturn was not generating enough heat to drive away volatile materials in the inner part of its satellite system. This is not surprising, as Saturn is only about one-third as massive as Jupiter and experienced less internal heating. Water ice has been detected in the spectra of all of the satellites.

■ Guardian and Co-orbital Satellites

Saturn has three satellites that serve as guardian, or shepherd, satellites for parts of its ring system. Although these three objects are obviously insufficient to explain the complex structure of the

319

■ Table 22.1   Summary of the Basic Properties of Saturn's Satellites*

Name	Semi-Major Axis 10³ km	Semi-Major Axis 10³ mi	Orbital Period (days)†	Diameter km	Diameter mi	Density (g/cm³)
Atlas	137.6	85.5	0.602	38 × ? × 28	24 × ? × 17	?
Prometheus	139.4	86.6	0.613	140 × 100 × 74	87 × 62 × 46	?
Pandora	141.7	88.1	0.629	110 × 86 × 66	68 × 53 × 41	?
Epimetheus	151.42	94.09	0.694	140 × 118 × 100	87 × 73 × 62	?
Janus	151.47	94.12	0.695	220 × 190 × 160	137 × 118 × 99	?
Mimas	185.5	115.3	0.942	394	245	1.17
Enceladus	238.0	147.9	1.370	502	312	1.24
Tethys	294.7	183.1	1.888	1048	651	1.26
Telesto	294.7	183.1	1.888	? × 24 × 22	? × 15 × 14	?
Calypso	294.7	183.1	1.888	30 × 26 × 16	19 × 16 × 10	?
Dione	377.4	234.5	2.737	1118	695	1.44
Helene	377.4	234.5	2.737	36 × ? × 30	22 × ? × 19	?
Rhea	527.0	327.5	4.518	1528	949	1.33
Titan	1221.9	759.3	15.95	5150	3200	1.88
Hyperion	1481.1	920.4	21.28	350 × 240 × 200	217 × 149 × 124	?
Iapetus	3561.3	2213.0	79.33	1436	892	1.21
Phoebe	12,952	8048	−550.5	230 × 220 × 210	143 × 137 × 130	?

* For a complete table of satellite data, see Table A3.1 in Appendix 3.
† Negative period indicates retrograde orbit.

rings, they do account for some of its properties, namely, the sharpness of the A ring's outer edge and the narrowness of the F ring. Figure 22.1 is a composite photograph of the three ring guardians, as well as Saturn's other small satellites. The Voyagers did not obtain extremely good images of any of Saturn's small satellites. However, enough detail was seen to show that they have rough surfaces, are irregular in shape, and are heavily cratered. The objects are all probably made of ice.

As mentioned in Chapter 21, the outer edge of the A ring is extremely sharp and well defined. This is because it is shepherded by the satellite Atlas, which was first seen by Voyager 1. Atlas orbits only about 1100 kilometers (700 miles) beyond the A ring's outer edge and is an irregular object only 30 kilometers (19 miles) across.

Saturn's F ring is extremely narrow and sharply bounded. The reason for this was learned by the Voyager spacecraft, which photographed two guardian satellites, one just inside the F ring, the other just outside. (One of these had been detected indirectly by Pioneer 11 but not photographed.) The guardian located to the inside of the rings, Prometheus, and the one located to the outside, Pandora, have orbits separated by only 2400 kilometers (1500 miles). Both of these satellites are irregular in shape, with Prometheus, whose largest dimension is about 140 kilometers (90 miles), being the larger of the two.

In 1966, when Saturn presented an edge-on view of its rings, a new inner satellite, subsequently named Janus, was discovered from Earth. Later study indicated that discrepancies existed in the original observations. In 1980, after Pioneer 11's flyby, new Earth-based observations, and Voyager 1's close-up observations of the satellites, the true situation was learned: Janus was actually two satellites in virtually the same orbit! Although discarding the name Janus to avoid confusion was initially considered, it was decided to retain this name for one object and call the other one Epimetheus. These two objects orbit some 151,600 kilometers (94,000 miles) from Saturn in orbits only 50 kilometers (30 miles) apart, with periods of 16.7 hours. However, because the dimensions of the two irregular objects are 220 by 200 by 160 kilometers (140 by 125 by 100 miles) for Janus and 140 by 120 by 100 kilometers

■ **Figure 22.1** Composite photograph of Saturn's small satellites, to scale. Left to right: Atlas; Pandora *(top)* and Prometheus *(bottom)*; Janus *(top)* and Epimetheus *(bottom)*; Calypso *(top)* and Telesto *(bottom)*; Helene. (NASA/JPL Photograph.)

(90 by 75 by 60 miles) for Epimetheus, this is not enough separation to prevent a collision! What happens is that the inner co-orbital satellite catches up to the outer one every four years, and the two objects exchange places (Figure 22.2)! This phenomenon was unexpected and is probably unique in the solar system.

■ Mimas

Mimas, the major satellite closest to Saturn, is only 390 kilometers (240 miles) in diameter and has a density of 1.2 grams per cubic centimeter.

Mimas has numerous craters, most of which are deep and bowl-shaped (Figure 22.3). Many of its large craters have central peaks, and ejecta deposits and bright rays are generally lacking. Mimas has some large grooves on its surface that are probably related to the forces involved with crater formation, as the satellite is too small to have any internally generated tectonic activity.

The most notable feature on Mimas is the giant crater **Herschel,** which is on the satellite's leading hemisphere (Figure 22.4). Herschel, named for Mimas' discoverer, is 130 kilometers (80 miles) across, fully one-third of Mimas' diame-

■ **Figure 22.2** How co-orbital satellites exchange orbits. Recall that an object in an inner orbit always moves faster than one in an outer orbit. The white satellite approaches the slower-moving shaded one *(1)*. Gravitational forces decelerate the shaded and accelerate the white *(2)*, and the shaded drops into a lower-energy (but higher-velocity) orbit *(3)*. Meanwhile, the white satellite moves upward into a higher-energy (but lower-velocity) orbit.

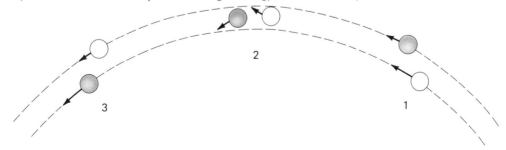

■ **Figure 22.3** Voyager 1 photograph of Mimas that shows the heavily cratered nature of its surface. (NASA/JPL Photograph.)

ter! This is about the largest impact that would have been possible on Mimas, for anything much larger would have shattered the satellite. Herschel is nearly 10 kilometers (6 miles) deep and has a central peak almost 6 kilometers (4 miles) high. This tremendous relief is still present because there has not been enough isostatic adjustment to reduce it.

■ Enceladus

Enceladus is 500 kilometers (310 miles) in diameter and has a density of 1.2 grams per cubic centimeter. It is noteworthy in having the highest albedo of any object in the solar system, virtually 100%. In fact, Enceladus reflects more light than does newly fallen snow on the Earth! This high albedo, as well as its low density, indicates that Enceladus is probably pure ice, with very little rocky material, especially at the surface.

Enceladus is also remarkable for the diversity of its surface features, which indicate a complex geological history (Figure 22.5). Its surface has been divided into four principal types of terrain. **Cratered terrain** has abundant craters about 15 kilometers (10 miles) in diameter. Some cratered terrain has shallow craters, apparently the result of isostatic adjustment, whereas other regions have deeper craters, indicating a more rigid surface. **Cratered plains** have smaller, bowl-shaped craters, whereas **smooth plains** are relatively crater-free. Some of the smooth plains contain grooved terrain. A final terrain type consists of **ridged plains**, which contain kilometer-high ridges, approximately parallel, with smooth areas between them. Most of the ridged plains lie on Enceladus' trailing hemisphere and seem to have replaced the older cratered terrain.

One fact indicated by Enceladus' surface features is that some mechanism has apparently been responsible for resurfacing at least part of the satellite. Like Mimas, Enceladus is too small for internal radioactive heating and no tectonic activity would be expected. However, it is believed that volcanic activity has occurred on Enceladus, although there is no proof that it has indeed happened. The heat source for this vol-

■ **Figure 22.4** Voyager 1 photograph of Mimas that shows the huge impact crater Herschel. (NASA/JPL Photograph.)

■ **Figure 22.5** Voyager 2 photograph of Enceladus. Note the wide variety of surface features. (NASA/JPL Photograph.)

canism is Enceladus' orbital resonance with the satellite Dione, which may generate enough tidal heating to keep Enceladus' interior at least partly molten. (Enceladus' orbital period is half that of Dione's, and the interactions have caused Enceladus' orbit to be somewhat eccentric.) Because of its composition and density, we know that volcanic activity on this icy world would not have involved molten silicates, but water, either in a liquid or slushy, semifrozen state. The thin E ring of Saturn is believed to be associated with Enceladus and may consist of material spewn forth from previous volcanic activity on its surface.

■ Tethys, Calypso, and Telesto

Tethys is 1050 kilometers (650 miles) across and has a density of 1.2 grams per cubic centimeter. On the average, it has a high albedo, but its lead-

ing hemisphere is a bit darker than the trailing one. Larger than Mimas and Enceladus, Tethys has much more evidence of tectonic activity. Tethys' surface can roughly be divided into **hilly cratered terrain**, which is rugged and heavily cratered, and **plains terrain**, which is flatter and has fewer craters. Some faint lines and albedo variations are visible in the latter terrain.

Tethys has two very impressive surface features. **Odysseus** is an impact crater 400 kilometers (240 miles) across (Figure 22.6). Unlike the similarly huge Herschel on Mimas, Odysseus is very shallow, and its floor matches the curvature of Tethys. This indicates that isostatic adjustment of the surface of the crater floor has occurred, making this feature similar to the palimpsest craters on Ganymede and Callisto. A second spectacular feature of Tethys is **Ithaca Chasma**, a branching valley system extending three-quarters of the way around the moon (Figure 22.7). It is 100 kilometers (60 miles) wide and up to 4 kilometers (2.5 miles) deep. Two mechanisms have been proposed to explain this huge feature. It may have been formed as a result of the forces generated by the impact that formed Odysseus, and the location of Ithaca Chasma supports this theory. Another suggestion is that Ithaca Chasma formed by cracking as Tethys froze and expanded, but this model fails to explain the localized nature of this terrain.

■ **Figure 22.6** Voyager 2 photograph of Tethys that shows the large impact crater Odysseus. (NASA/JPL Photograph.)

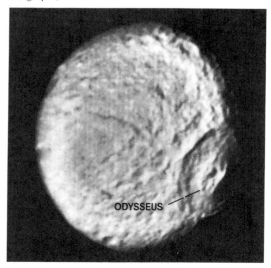

ODYSSEUS

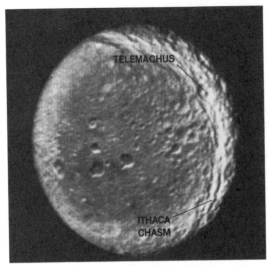

■ **Figure 22.7** Voyager 2 photograph of Tethys that shows Ithaca Chasma. (NASA/JPL Photograph.)

Tethys has Lagrangian satellites at both the leading and trailing positions. They were discovered from Earth when Saturn's rings were edge-on in 1980, and these observations were verified by the Voyager spacecraft. Calypso orbits 60° ahead of Tethys, whereas Telesto orbits 60° behind. These irregularly shaped objects, each 34 kilometers (21 miles) in its longest dimension, appear to be nothing more than chunks of icy material, and little is known about their surfaces.

■ **Figure 22.8** Voyager 1 photograph of Dione that shows the bright streaks covering the dark, trailing hemisphere. (NASA/JPL Photograph.)

■ Dione and Helene

Dione is 1120 kilometers (700 miles) across and has a density of 1.4 grams per cubic centimeter. This density, second only to Titan's among Saturn's satellites, indicates a high rock content in addition to ice. The portions of Dione imaged by the Voyagers have been divided into three geological units. **Heavily cratered terrain**, the most common, is rough and contains numerous large craters. In addition, there are **cratered plains**, which record fewer impacts, and **smooth plains**, which contain almost no craters.

Dione has a wide range of surface brightnesses. Its leading side is bright, whereas its trailing side is dark but covered by a series of bright, wispy markings (Figure 22.8). A bright feature, possibly impact-related, is located in the center of the pattern that these markings form, but the wispy marks do not appear to be ejecta, because they follow irregular paths. Both hemispheres are heavily cratered; the leading hemisphere is especially battered (Figure 22.9). One crater on Dione is surrounded by bright rays, the only instance of

■ **Figure 22.9** Voyager photograph of Dione that shows the numerous craters on its surface. Note the trough feature near the terminator at the top of the image. (NASA/JPL Photograph, courtesy of the National Space Science Data Center and the Voyager Team Leader, Bradford A. Smith.)

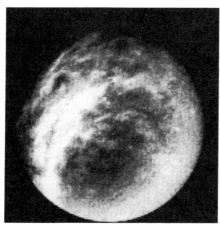

■ **Figure 22.10** Voyager 1 photograph of Rhea. Note the contrast between the leading (bright) hemisphere and the trailing (dark) hemisphere. (NASA/JPL Photograph.)

rays yet detected on a satellite of Saturn. There are indications that the cratering took place in two distinct episodes, one before and one after the formation of Dione's smooth plains. Dione's surface also contains tectonic features, such as ridges and fault scarps up to 100 kilometers (60 miles) long and troughs that can reach similar lengths. These fault-related features probably resulted from changes in Dione's size produced by expansion and contraction of its interior due to changes in internal temperature.

Sharing Dione's orbit is a satellite located at the leading Lagrangian position, 60° ahead of Dione. The object, named Helene, was discovered telescopically in 1980, when Saturn's rings were edge-on, and verified by Voyager 1. Helene is irregular in shape, with dimensions of 36 by 32 by 30 kilometers (22 by 20 by 19 miles). One large crater was visible on its surface during the flyby missions, but little else is known about this presumably icy body.

■ Rhea

Rhea is Saturn's second-largest satellite; it is 1530 kilometers (950 miles) in diameter and has a density of 1.3 grams per cubic centimeter. Rhea has a "split personality" in that one hemisphere is much brighter than the other (Figure 22.10). The leading hemisphere is bright and featureless,

whereas the trailing hemisphere is dark except for bright wisps similar to those found on Dione. When closely examined during the Voyager flyby missions, Rhea was found to be a heavily cratered body that somewhat resembled the highlands of the Moon (Figure 22.11). Although Rhea is an icy object like Ganymede and Callisto, its craters retain topographic relief and are not of the palimpsest variety. Although most of the surface is old and cratered, some sections are less heavily cratered, indicating that some form of surface regeneration has occurred. This regeneration was probably the result of volcanic activity involving liquid or water "magma" that froze to produce a smooth surface. After this new surface was formed, the number of impacting particles was low, causing subsequent craters to be much less numerous.

Rhea contains at least two multiringed basins. One is 450 kilometers (280 miles) across and fairly well-preserved (Figure 22.12). A second is more than twice as large but highly degraded, although as many as seven rings may be traced. Some tectonic features, such as troughs and scarps, also occur.

■ Titan

Titan is the largest satellite of Saturn and one of only two planetary satellites that have atmospheres (Neptune's satellite Triton is the other). It

■ **Figure 22.11** Photograph showing the heavily cratered hemisphere of Rhea. (NASA/JPL Photograph.)

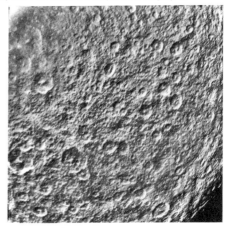

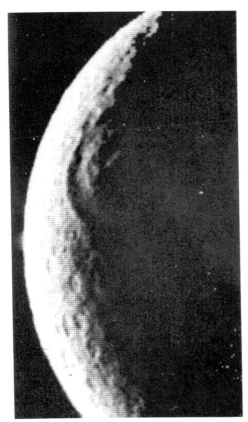

■ **Figure 22.12** Voyager 1 photograph of a large, multiringed basin on Rhea. (NASA/JPL Photograph.)

cal reactions caused by heating as meteorites entered the original atmosphere disrupted the ammonia molecules, producing the nitrogen and the various hydrocarbons.

Titan's atmosphere is not transparent, but it contains reddish haze that prevents observation of its surface (Figure 22.13). This haze is contained in two layers, a dense one approximately 300 kilometers (185 miles) above the surface and a thin one about 510 kilometers (315 miles) high (Figure 22.14). Titan's atmosphere is spread over a much greater vertical range than Earth's because of the satellite's lower gravity. Surface conditions on Titan are unusual, because methane rain and snow probably occur, and a smog composed of various hydrocarbons is also present. Titan might even be covered with an ocean of liquid ethane and methane, because its surface temperature of 92 K ($-294°$F) would allow these substances to exist in all three phases, similar to ice, liquid water, and water vapor on Earth.

The surface geology of Titan and exactly what processes have shaped its surface are unknown at this time. As on Venus, clouds prevent observation of Titan's surface except with radar. In July, 1989, terrestrial radar signals were reflected from Titan for the first time. Most of the reflected signals were weak, but on one day, a much stronger echo was received. A possible explanation for the stronger signal is that it was from a rocky or icy continent, with the weak signals from surrounding seas of ethane or methane. Radar mapping by Cassini, if that spacecraft is successful, should reveal many details of Titan's surface geology. On the whole, Titan is probably similar to Callisto and Ganymede in bulk properties, because it formed fairly far from Saturn. Therefore, large craters on Titan (if they even exist) are probably of the palimpsest form as a result of isostatic adjustment. Titan's interior is probably made of equal parts of rock and ice, with the rock concentrated at the core.

Titan is a likely target for future spacecraft exploration for two reasons. One is that radar images of its surface should answer many questions about what geological processes have occurred there. In addition, study of its chemical processes involving hydrocarbons might yield insight into the processes that led to the formation of organic molecules (and life) on Earth.

is 5120 kilometers (3180 miles) across and has a density of 1.88 grams per cubic centimeter. Titan was the first Saturnian satellite discovered (by Huygens in 1655), and prior to the space age, it was considered the largest satellite in the solar system. Even though Jupiter's moon Ganymede is now known to be larger, Titan is unique because of its atmosphere. Methane was first detected spectroscopically in 1944 and was long considered to be the major component, but Voyager studies revealed that the atmosphere is actually about 95% nitrogen and 5% methane, at a pressure of about 1.6 times that of Earth's atmosphere. Titan's atmosphere contains many **hydrocarbons** (compounds of hydrogen and carbon), such as acetylene, ethane, and ethylene, and some argon is also believed to be present. It has been suggested that Titan's atmosphere was originally ammonia and methane, but that chemi-

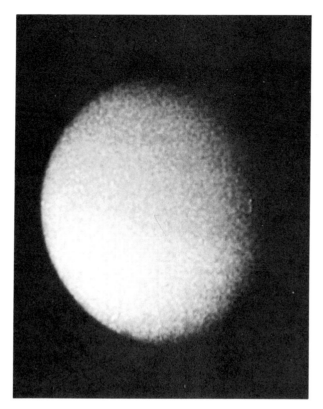

■ **Figure 22.13** Voyager photograph of Titan. Because of the opaque atmosphere, no surface details are visible. (NASA/JPL Photograph.)

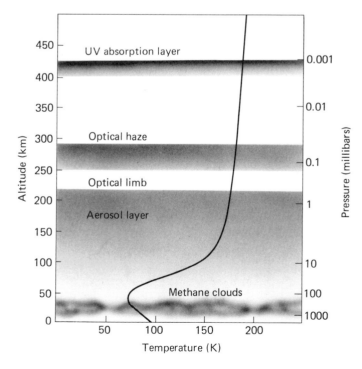

■ **Figure 22.14** Vertical structure of Titan's atmosphere. Height in kilometers is plotted against temperature in degrees Kelvin. Cloud and haze levels are shown. (Gary E. Hunt and Patrick Moore.)

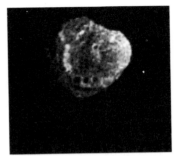
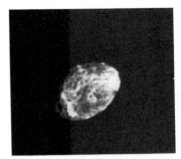
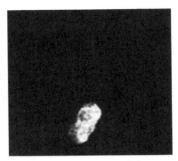

■ **Figure 22.15** Voyager 2 photographs of Hyperion. Note the irregular shape of the satellite. (NASA/JPL Photograph.)

■ Hyperion

Hyperion is one of Saturn's most unusual satellites, as it is shaped somewhat like a hamburger or hockey puck, with dimensions of 410 by 260 by 220 kilometers (255 by 160 by 135 miles). Unfortunately, its density is unknown, so its composition is uncertain, although it is probably icy. Hyperion is very heavily cratered, indicating that it is a very old object whose surface has not been reworked (Figure 22.15). Some of Hyperion's craters are 120 kilometers (75 miles) across, with relief as high as 10 kilometers (6 miles). The impacts that formed these craters may have been responsible for Hyperion's irregular shape, and it is possibly a fragment of an object that originally was much larger. One mystery involving Hyperion is its orbital orientation. An irregularly shaped object in orbit around Saturn would be expected to be aligned with its long axis pointed toward the planet; however, Hyperion is not aligned in this fashion, for reasons that are not yet understood.

■ Iapetus

Iapetus, which is 1440 kilometers (895 miles) across and has a density of 1.2 grams per cubic centimeter, has been a mysterious object ever since its discovery by Cassini in 1671. The mystery is that Iapetus is much brighter when on one side of Saturn than on the other. Because Iapetus has 1:1 spin-orbit coupling, we see one hemisphere of Iapetus when it is on one side of Saturn and the other hemisphere when on the other side, indicating that the satellite's surface does not have a uniform albedo. It has been deter-

mined that the leading hemisphere of Iapetus has an albedo of 5% and the trailing hemisphere, an albedo of about 50%. (Surprisingly, the boundaries of these two regions are rather well defined [Figure 22.16].) Most researchers believe that the leading hemisphere is dark because it has a covering of some extremely dark substance on top of brighter material.

Prior to the Voyager missions, most theories suggested that the dark material on Iapetus was derived from another satellite. One suggestion was that Iapetus had been partially coated by dark material that was somehow knocked from the outermost satellite, Phoebe. Spectroscopic comparisons of the two satellites, however, revealed sufficient compositional differences to eliminate this theory. Most post-Voyager models suggest that the dark material was somehow erupted from within Iapetus. This theory is supported by the fact that the dark material seems common in low areas such as crater floors.

Iapetus was not photographed at extremely high resolution, but a number of craters were seen, some as large as 120 kilometers (75 miles). They tend to have sharp rims and contain central peaks. A larger feature, about 200 kilometers (120 miles) across, was also seen, and it is possibly a multiringed basin.

■ Phoebe

Phoebe, the most distant Saturnian satellite, is only 220 kilometers (135 miles) in diameter. It was photographed by Voyager 2 at fairly low resolution, but its mass, and therefore its density, were not measured (Figure 22.17). Phoebe is the dark-

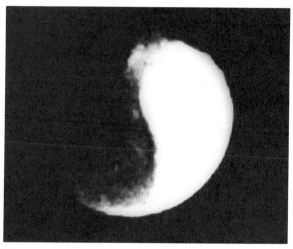

(a)

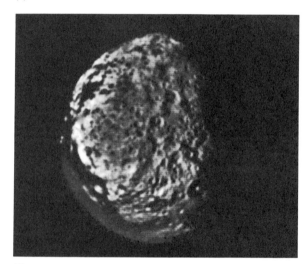

(b)

■ **Figure 22.16** Voyager photographs of Iapetus.
(a) Full-disk view showing the contrast between the light
and dark regions. (NASA/JPL Photograph, courtesy of the
National Space Science Data Center and the Voyager Team
Leader, Bradford A. Smith.) (b) Photograph illustrating the
rough nature of Iapetus' surface (NASA/JPL Photograph.)

■ **Figure 22.17** Voyager 2 photograph of
Phoebe. (NASA/JPL Photograph.)

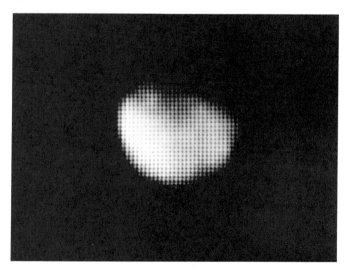

est satellite of Saturn, and its very low albedo, about 5%, indicates that its surface may be similar to those of carbonaceous (carbon-containing) asteroids. This, along with its retrograde orbit, indicates that Phoebe may be a captured object.

Phoebe's surface exhibits albedo markings, with light areas that may indicate impact features.

The fact that Phoebe is spherical was unexpected, because it is so small that it would have lacked sufficient gravitational force to pull itself into that shape.

■ Chapter Summary

Saturn has 17 satellites, about half of which are large enough that Voyager photographs have allowed interpretation of their surface geology. Three of its satellites are ring guardians; two, called co-orbital, share what is essentially a single orbit; and three other satellites occupy Lagrangian positions in orbits of major satellites.

Unlike Jupiter's major satellites, Saturn's satellites have no pattern of density or compositional distribution. The innermost satellites, Mimas, Enceladus, Tethys, Dione, and Rhea, have low densities and are largely made of ices.

The larger of these bodies appear to have experienced internally generated tectonic activity, possibly a result of "volcanism" involving semisolid ice "magma."

Saturn's largest satellite, Titan, has a hazy, opaque atmosphere of nitrogen and methane. Hyperion is small and irregularly shaped, whereas Iapetus is noteworthy because of the considerable differences in albedo between its leading and trailing sides. Phoebe, the outermost satellite, has an unusual orbit and may be a captured asteroid.

■ Chapter Vocabulary

cratered plains	Herschel	leading hemisphere	ridged plains
cratered terrain	hilly, cratered terrain	Mimas	smooth plains
Dione	hydrocarbons	Odysseus	Tethys
Enceladus	Hyperion	Phoebe	Titan
heavily cratered terrain	Iapetus	plains terrain	trailing hemisphere
	Ithaca Chasma	Rhea	

■ Review Questions

1. Where are Saturn's guardian satellites located? What do they do?
2. Describe the behavior of Saturn's co-orbital satellites.
3. Describe Mimas, and explain why Herschel is so unusual.
4. What characteristics of Enceladus are unusual? What is its relation to the E ring?
5. What are the impressive surface features of Tethys?
6. Summarize the surface features of Dione.
7. Summarize the surface features of Rhea.
8. Describe the atmosphere of Titan. Why is Titan a likely target for future space exploration?
9. What is unusual about Hyperion?
10. What is a likely explanation for the differences in the albedos of Iapetus' two hemispheres?
11. Why is Phoebe probably a captured object?

■ For Further Reading

Owen, Tobias. "Titan." In *Scientific American,* February, 1982.

Soderblom, Laurence and Torrence V. Johnson. "The Moons of Saturn." In *Scientific American,* January, 1982.

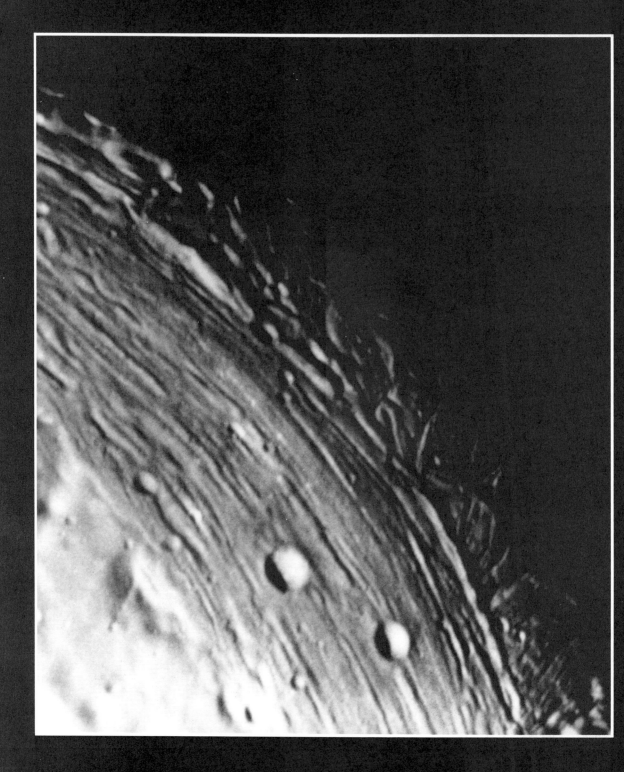

Uranus

The surface of Uranus' satellite Miranda, which has some of the most interesting surface features of any object in the solar system. (NASA/JPL Photograph.)

Prior to the year 1781, when the planet Uranus was discovered, Saturn was considered to mark the outer boundary of the solar system. In the two centuries since then, the solar system has "grown" considerably with the discovery of three planets (Uranus, Neptune, and Pluto) as well as numerous smaller bodies. Uranus and its virtual twin Neptune, the outermost gas giants, are approximately half-scale versions of Jupiter and Saturn. Because of their sizes, Uranus and Neptune have properties that are somewhat different from those of the larger gas giants. This is especially evident when interior models of the smaller gas giants are calculated. In this chapter, we discuss the third gas giant, Uranus, whose major orbital and physical properties are summarized in Table 23.1.

■ Discovery

Uranus, the first planet discovered telescopically, was found in 1781 by **William Herschel** (1738–1822), a German living in England. Herschel made his living as a musician, but he was an avid amateur astronomer and one of the greatest observers of the 18th century, discovering numerous star clusters and nebulae. On March 13, 1781, Herschel observed a new "star" that was not located on any of his charts. The new object moved too slowly to be a comet, but enough motion had occurred by the end of the year that the object was deduced to be a planet.

Herschel wanted to name the object "The Georgian Planet" in honor of his patron, King George III of England, and there was also sentiment for naming the object for Herschel himself. Eventually, however, the object was named for a mythological figure in the traditional fashion. Uranus, the father of Saturn, was an appropriate designation considering the planet's position as the next planet beyond Saturn. Incidentally, this name was first proposed by Johann Bode, the astronomer for whom Bode's law was named.

■ Table 23.1   Summary of Uranus' Basic Properties*

Semi-major axis: 19.22 astronomical units
Semi-major axis: 2.875 billion kilometers
Semi-major axis: 1.787 billion miles
Orbital eccentricity: 0.046
Orbital inclination: 0.773°
Orbital period: 84.0 years
Diameter: 50,800 kilometers
Diameter: 31,567 miles
Mass (Earth = 1): 14.50
Density: 1.30 grams per cubic centimeter
Sidereal rotational period: 17.2 hours (retrograde)
Axial obliquity: 82.14°
Number of satellites: 15

* For a complete table of planetary data, see Table A2.1 in Appendix 2.

Uranus is just bright enough to be visible to the naked eye, and it is surprising that it escaped detection for so long. Even a sharp-eyed ancient Greek astronomer could have spotted it. Investigations after Herschel's discovery showed that Uranus had been charted over 20 times between 1690 and 1781 by people who failed to detect its motion while making star maps. The earlier observations were not wasted but gave additional positional information that was useful in calculating Uranus' orbital elements after Herschel discovered it in 1781.

Appearance and Basic Properties

Because of its vast distance, Uranus presents only a small disk when seen through the telescope. Even with a small telescope, however, the disk can be seen clearly enough to show that the object is not a star. The disk appears somewhat greenish in color, and faint albedo markings are visible (Figure 23.1). (Voyager 2 confirmed that Uranus' atmospheric markings are much less noticeable than either Jupiter's or Saturn's, even when viewed from close proximity.)

Uranus is noteworthy because it has the lowest orbital inclination of any planet other than the Earth, whose inclination is defined as zero. It also has the highest axial obliquity, 82.1°, and rotates

retrograde, the only planet other than Venus and Pluto to do so. (Uranus' obliquity is sometimes given as 97.9° with prograde rotation, which describes the equivalent situation.) It is possible that Uranus' high obliquity may be the result of a glancing impact that disturbed Uranus as it was forming. The difficulty with such a scenario is that it cannot explain why Uranus' rings and satellites orbit in the planet's tilted equatorial plane.

Because of its high obliquity, Uranus and its rings and satellites present an unusual orientation for observation. Currently, the south pole of the planet is facing the sun, so the rings actually encircle the planet's disk. (The rings, as we will see later, are too faint to be observed from Earth unless special techniques are used to block the planet's light.) The satellites can also be observed throughout their entire orbits. Because the polar region now faces the Sun (and the Earth) and the orbital period is 84 years, it will be continuously illuminated for decades. Uranus therefore has seasons that last about 21 years each.

Spacecraft Exploration

Uranus has been visited by only one spacecraft, **Voyager 2**, the only spacecraft to have explored all four gas giants. Voyager 2 sped by the planet at a velocity of 16 kilometers per second (10 miles per second) on January 24, 1986, nine and one-half years after its launch and after previously encountering Jupiter in 1979 and Saturn in 1981. Voyager came within 81,600 kilometers (50,700 miles) of the tops of the planet's clouds and transmitted numerous photographs of the planet, its

■ Figure 23.1   Photograph of Uranus taken from Earth. Note that very little atmospheric detail is visible. (Lunar and Planetary Laboratory, University of Arizona.)

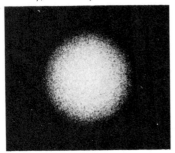

rings, and satellites. In addition to studying previously known objects, Voyager discovered ten new satellites and two new rings. Because of Uranus' high obliquity and the fact that its rings and satellites lie in its equatorial plane, Voyager did not make a leisurely cruise through the system as it did at Jupiter and Saturn. Voyager's path more resembled that of an arrow passing through a bull's-eye target, making its encounter sequence very rushed. In addition, only the southern hemispheres of the planet and its satellites were seen, as their northern hemispheres were in darkness.

It is fortunate that the Voyager 2 encounter went well and much data were sent back to Earth, because no further Uranian missions are planned. The majority of what we now know about the planet, its rings, and its moons was learned during this one brief encounter. In fact, so much information was gathered during this mission that it will be many years before researchers are able to interpret all of it.

■ Atmosphere

Contrary to expectations, Voyager 2 was not able to see much more atmospheric detail than is visible from Earth. Uranus was found to be sheathed in a murky haze, and even with computer enhancement, only a few details were visible in the pictures (Figure 23.2). One finding was that the polar region facing the Sun is slightly brighter and redder than the clouds nearer the equator.

Earth-based spectroscopy had discovered hydrogen and methane in the Uranian atmosphere prior to the Voyager encounter. Uranus, in fact, was the first planet in whose atmosphere hydrogen was detected from Earth. Methane absorbs red light, and this process accounts for Uranus' greenish cast. The International Ultraviolet Explorer (IUE) spacecraft observing Uranus from Earth orbit found that acetylene is also present in the atmosphere, where it is probably produced by the breakdown of methane molecules. Voyager results indicate that Uranus' atmosphere is 12 to 15% helium, a slightly greater abundance than that of the Sun (11%), Jupiter (10%), and Saturn (6%).

The uppermost region of Uranus' atmosphere was found to be very hot, as temperatures range from 750 K (890°F) in the sunlit region to 1000 K (1340°F) in the dark. The higher temperature of the night side may be explained by the fact that there are so few molecules in the upper atmosphere that cooling proceeds very slowly. As in Jupiter and Saturn, temperature drops with depth, reaching a minimum of 50 K (−370°F) at

■ **Figure 23.2** Voyager 2 photographs of Uranus' polar region. The photograph at left reveals very little detail, whereas the one at right has been contrast-enhanced in order to reveal additional details. (NASA/JPL Photographs.)

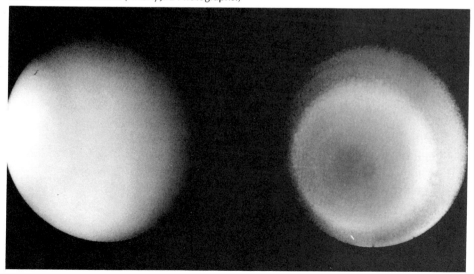

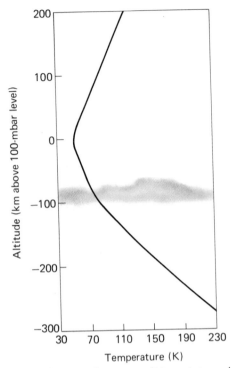

■ **Figure 23.3** Vertical structure of Uranus' atmosphere. Altitude in kilometers relative to the 100-millibar level is plotted against temperature in degrees Kelvin. The location of the cloud deck is shown.

the level where the pressure is 100 millibars, the "cloud top," which is the starting point for measurements of atmospheric elevation (Figure 23.3). Temperature and pressure both rise below the 100-millibar point, reaching about 75 K (−330°F) and 1 bar at a point 90 kilometers (55 miles) below it. A thick cloud deck, probably composed of methane ice, occurs slightly below this at a level where the pressure is about 1.6 bars. Very little variation in temperature with latitude was found at this depth, despite the unusual pattern of solar heating.

Uranus' smog-like atmospheric haze is believed to consist of trace amounts of hydrocarbons slightly deeper in the atmosphere than the 100-millibar level. Only a few clouds were seen with certainty, and the motion of two of them indicated atmospheric rotation of 16 and 16.9 hours. The different rotational periods are due to atmospheric currents such as those found on the other gas giants. (As we will see later, Voyager indicated that the interior of Uranus rotates more slowly than the atmosphere, with a period of 17.24 hours. Given the pattern of solar heating on Uranus, the reason why the atmosphere rotates faster than this has not yet been explained.)

■ Interior and Magnetic Field

The overall density of Uranus is about 1.30 grams per cubic centimeter, which is slightly lower than that of Jupiter (1.33 grams per cubic centimeter). However, Uranus' density figure is the result of the presence of denser elements within it, unlike Jupiter, whose high density is the result of internal compression. Whereas Jupiter and Saturn are believed to be largely made of hydrogen and helium, it is estimated that only about 10% of Uranus' total mass is composed of these elements. Because of its smaller mass and internal pressure, it is very likely that none of the hydrogen present inside Uranus is in the metallic state.

Most models of Uranus' internal structure suggest that three layers are present (Figure 23.4). Beneath the atmosphere is a layer of molecular hydrogen that extends about 30% of the way to the center of the planet. In addition to molecular hydrogen, helium and methane are probably also major components. The base of this layer has a density of about 0.4 grams per cubic centimeter, a temperature of about 2500 K, and a pressure of about 0.2 million atmospheres.

■ **Figure 23.4** Uranus' interior layers. The numbers represent the fraction of Uranus' total radius. See text for detailed descriptions.

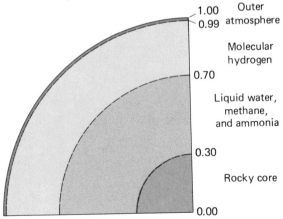

Below this region is a layer composed primarily of water, methane, and ammonia. Extending down to about 70% of the way to Uranus' center, this layer is probably a liquid that is churning in slow, convective motion. Because Uranus has a magnetic field, a conducting layer must be present somewhere within it, and such a liquid mantle would suffice if conducting ions of the three main constituents were present. Experiments have indicated that these materials would break into conducting ions at the conditions believed to be present in this layer. The density of this layer rises from 1.3 to 4 grams per cubic centimeter from top to bottom, the temperature at its base is 7000 K, and the pressure there is 6 million atmospheres.

The center of Uranus probably contains a rocky core. The density of the core is about 13 grams per cubic centimeter, the central pressure is 17 million atmospheres, and the central temperature is over 7200 K.

Voyager 2 revealed that Uranus, like Jupiter and Saturn, has a magnetic field. Voyager entered the magnetosphere 467,000 kilometers (290,000 miles) from Uranus' center and found that the overall magnetic moment is about 50 times stronger than the Earth's, although the surface field is slightly less than Earth's because of the larger diameter of Uranus. Like the other gas giants and unlike Earth, the polarity of Uranus' field is aligned in the rotational direction. However, the magnetic axis is tilted 59° from Uranus' rotational axis. Even more surprisingly, the magnetic "center" of the planet is offset from the physical center nearly 8000 kilometers (5000 miles) toward the north (currently unlit) pole. Neither of these peculiarities has yet been adequately explained. Bursts of energy from charged particles trapped by the magnetic field lines were used to time the true physical rotation period of Uranus, which was found to be 17.24 hours. Voyager found that many protons and electrons are trapped inside the magnetosphere, and a weak aurora was observed on the planet's night side.

■ Rings

As mentioned in Chapter 18, Uranus' rings were discovered during an occultation of a star by the

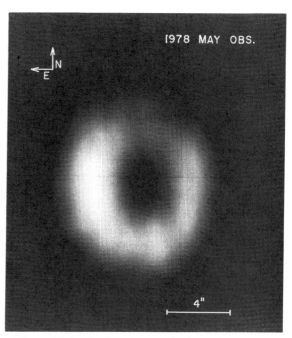

■ **Figure 23.5** The first photograph of Uranus' rings ever taken from Earth, by Matthews, Neugebauer, and Nicholson. This image was made from two scans at infrared wavelengths. The light of Uranus was removed from the scans, producing an image of the rings only. (Photograph courtesy of Keith Matthews. From Matthews, Keith et al. "Maps of the Rings of Uranus at a Wavelength of 2.2 Microns." In *Icarus*, vol. 52, p. 126. Copyright 1982 by Academic Press, Inc.)

planet in 1977. Between their discovery and the 1986 Voyager flyby, the rings were observed during numerous additional occultations, and a considerable amount of information was learned about them. The rings are normally too faint to be directly visible from Earth, but images of them have been obtained by telescopes using special devices to block the much brighter light from Uranus itself (Figure 23.5).

A total of nine rings had been identified prior to the Voyager 2 encounter. They are known as 6, 5, 4, alpha, beta, eta, gamma, delta, and epsilon, in order outward from the planet. (These are only provisional designations pending assignment of permanent names by the International Astronomical Union. A diagram of the rings is shown in Figure 23.6.) The rings range from 1.6 to 1.95 radii from Uranus' center, well inside its Roche limit. **Epsilon,** the outermost ring, is the widest and most conspicuous of the nine.

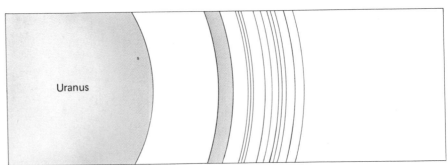

■ **Figure 23.6**   Map of Uranus' ring system, as seen from above the planet's north pole. From left to right, the rings are designated 1986U2R, 6, 5, 4, alpha, beta, eta, gamma, delta, 1986U1R, and epsilon. All except 1986U2R are so narrow that they are indicated on this map as lines.

As Voyager approached Uranus, it was able to photograph the nine rings clearly (Figure 23.7). A tenth ring (designated 1986U1R) was discovered between the two outermost rings, delta and epsilon. A faint, broad band of material inside ring 6 was also discovered and designated 1986U2R.

■ **Figure 23.7**   Voyager 2 photograph of Uranus' rings made using reflected light. The nine previously known ring components, as well as the one first seen in this photograph, are shown. The most conspicuous ring at upper left is the epsilon ring. (NASA/JPL Photograph.)

This ring is much wider than the others (about 2500 kilometers, or 1600 miles), and its inner edge is only about 37,000 kilometers (23,000 miles) from the planet's center, or 10,900 kilometers (6700 miles) above its cloud tops.

After Voyager passed Uranus, it looked back and photographed the forward-scattered light from the rings (Figure 23.8). Viewed in this way, the main rings were still visible, but about 100 bands of material that were previously invisible were revealed. These new features were actually brighter when viewed in forward-scattered light than were the previously known rings, indicating that they are composed of small, dust-sized particles that are missing from the main rings. Some partial arc rings were also found.

Voyager also studied details of the rings' structures as fine as 8 meters (25 feet) in width by viewing the light of bright stars as they were occulted by the rings. One result was that three new rings, each less than 150 meters (500 feet) wide, were found outside the epsilon ring. In addition, the gamma ring was found to be only 600 meters (2000 feet) wide. The gamma and delta rings each showed a single structure in one place and three strands elsewhere, similar to the kinks and wiggles in Saturn's F ring. Some of the rings were found to have dense, narrow cores surrounded by diffuse, broader components.

One major difference between Uranus' rings and the rings of Saturn is that the Uranian rings are very narrow. This was known prior to Voyager, because occultation studies had indicated that the first nine rings discovered were only 8 to 16 kilometers (5 to 10 miles) wide. Voyager results

■ **Figure 23.8** Voyager 2 photograph of Uranus' rings made using transmitted light. Note how much more detail is visible here than in Figure 23.7. (NASA/JPL Photograph.)

particles in the main rings are charcoal black, with albedos of only about 5%. Because Uranus' satellites are largely composed of ices, the absence of bright, icy material from its ring system is difficult to explain.

■ Satellites

Prior to Voyager 2, five satellites had been detected orbiting Uranus (Figure 23.9). In contrast to normal astronomical tradition, they were named not for mythological figures but for characters from the writings of William Shakespeare and Alexander Pope: Miranda, Ariel, Umbriel, Titania, and Oberon. Voyager not only photographed these five worlds and found them to be extremely interesting geologically but also discovered ten new ones, all of which are closer to Uranus than Miranda, the innermost of the "big five." All 15 satellites orbit in Uranus' equatorial plane, and because of the planet's obliquity, Voyager was able to photograph only their southern hemispheres.

Because the Voyager flyby was just a few days prior to the explosion of the Space Shuttle

■ **Figure 23.9** Photograph of Uranus and its five large satellites, obtained from Earth using a large telescope. (Photograph courtesy of William Liller, National Optical Astronomy Observatories.)

verified these figures, with the exception of those for the epsilon ring, which was found to be somewhat wider. Six of the nine pre-Voyager rings are somewhat eccentric in shape. For example, the distance of the epsilon ring from Uranus varies by about 800 kilometers (500 miles). The long axes of the eccentric rings are not all aligned in the same direction, and these eccentric rings also have variable widths, being narrow where closest to Uranus and widest where farthest away. For example, the width of the epsilon ring ranges from 20 to 96 kilometers (12 to 60 miles). Prior to Voyager, guardian satellites were thought to be responsible for the narrowness of the rings. One of the biggest surprises in the Voyager images was that only two guardians were found, one on either side of the epsilon ring. Because the other rings are even narrower than epsilon, they must have undiscovered guardians around them or else some unknown process is keeping their boundaries so sharp.

Radio occultations indicated that ring particles in the nine main rings are relatively large, ranging from softball-sized to car-sized. These rings seem to lack smaller, dust-sized particles. Particles in the ring bands that are visible only in forward-scattered light are thought to be much smaller, only about 2×10^{-5} meters in size. The

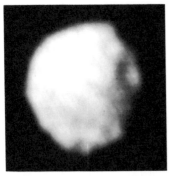

■ **Figure 23.10** Voyager 2 photograph of Puck. (NASA/ JPL Photograph.)

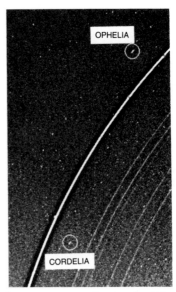

■ **Figure 23.11** Voyager 2 photograph of Cordelia and Ophelia, Uranus' innermost known satellites and guardians of its epsilon ring. (NASA/JPL Photograph.)

Challenger, there was originally sentiment for naming seven of the new satellites for the astronauts killed in that tragedy. However, the new satellites were also named for characters from Shakespeare's and Pope's works. All of them are small and dark. Except for the outermost one, **Puck,** which is 170 kilometers (105 miles) in diameter, all of the new satellites are between 40 and 80 kilometers (25 and 50 miles) across, and their albedos are all less than 10%. The only one of the new satellites on which any surface details were visible, Puck, appeared to be nearly spherical, and its surface was marked by several large impacts (Figure 23.10). Surprisingly, no albedo markings were seen. The two innermost satellites, Cordelia and Ophelia, are the guardian satellites of the epsilon ring (Figure 23.11).

Miranda is the closest of the five main satellites to Uranus at 130,000 kilometers (80,600 miles), and it proved to have the most complicated geology of any of them. (A composite photograph of the five main satellites is shown in Fig-

ure 23.12.) Miranda is 485 kilometers (300 miles) across and has a density of 1.26 grams per cubic centimeter and an albedo of 34%. Much of Miranda's visible surface is covered by densely cratered plains produced during the early impact stage (Figure 23.13). In addition, three unusual areas called **coronae** are superimposed on the plains regions (Figure 23.14). These features are between 200 and 300 kilometers (125 and 185 miles) across and are much younger than the plains. Each corona is oval or trapezoidal in shape and consists of bands of ridges, grooves, and other linear features marked by albedo variations. The appearance of coronae somewhat resembles

■ **Figure 23.12** Composite photograph, to scale, of Uranus' five major satellites. Note the differences in their albedos. Left to right: Miranda, Ariel, Umbriel, Titania, and Oberon. (NASA/JPL Photograph.)

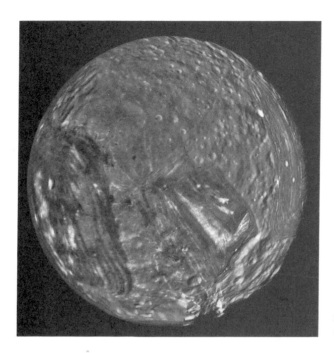

■ **Figure 23.13**   Voyager 2 photograph of Miranda. (NASA/JPL Photograph.)

■ **Figure 23.14**   Voyager 2 photograph of a corona region on the surface of Miranda. (NASA/JPL Photograph.)

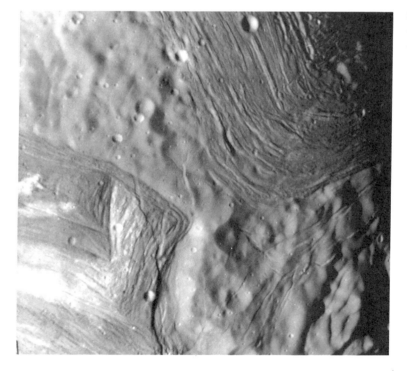

■ **Figure 23.15** Voyager 2 photograph of Ariel. Note the complex system of fault valleys. (NASA/JPL Photograph.)

that of racetracks on the Earth, and nothing like them has been seen on any other solar system body.

Although the origin of the coronae is not yet fully understood, it has been suggested that Miranda was once literally broken into pieces by large impacts. These pieces then reassembled as a result of their mutual gravitation, but some areas

previously in Miranda's core ended up at the surface. This dense material sank to the center again, producing disturbances that formed the coronae. It has also been suggested that Miranda was never disrupted, but that the coronae were produced by differentiation of its originally uniform material.

Uranus' second major satellite, **Ariel**, is located 191,000 kilometers (119,000 miles) from Uranus and is 1160 kilometers (720 miles) in diameter. Ariel is the brightest Uranian satellite, with an albedo of 40%, and is also the densest, at 1.65 grams per cubic centimeter. Ariel shows evidence of ancient impacts, and many of its larger craters are of palimpsest form (Figure 23.15). The most conspicuous feature on its surface, however, is a huge, complex system of fault valleys, some as deep as 16 kilometers (10 miles). Just as terrestrial grabens can be the site of igneous activity, so were the fault valleys on Ariel. However, the "magma" on Ariel was not molten rock, but probably a warm mixture of ice and rock that behaved somewhat like a flowing terrestrial glacier. Some of this icy material was probably responsible for burying part of Ariel's cratered terrain.

Umbriel, the third large satellite, orbits 266,000 kilometers (165,000 miles) from Uranus and is 1190 kilometers (740 miles) in diameter. It is the darkest major Uranian satellite (albedo 19%) and has a density of 1.44 grams per cubic

■ **Figure 23.16** Voyager 2 photograph of Umbriel. (NASA/JPL Photograph.)

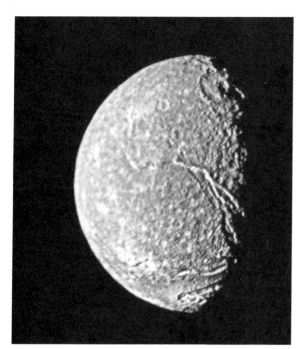

■ **Figure 23.17** Voyager 2 photograph of Titania. (NASA/JPL Photograph.)

centimeter. Umbriel is the most heavily cratered satellite of Uranus and appears to have the oldest surface (Figure 23.16). It somewhat resembles the lunar highlands, but no crater rays are present, a fact that may indicate that Umbriel's dark surface layer extends deep enough so that no light ejecta was produced during impacts. Umbriel has only two bright regions on its surface, one of them a bright ring covering the floor of one impact crater, the second another crater's bright central peak. The satellite's dark covering, which may consist of dark hydrocarbons caused by decomposition of methane at the surface or else of carbonaceous material brought by impacting objects, is probably thin or absent in these bright areas.

The fourth major satellite, **Titania,** is 436,000 kilometers (271,000 miles) from Uranus and, with a diameter of 1610 kilometers (1000 miles), is the largest Uranian satellite. Titania's albedo is 28%, and its density is 1.59 grams per cubic centimeter. Titania is heavily cratered, but there is evidence that many of its oldest craters have been erased by some internal process, a process that was

more effective on some portions of the surface than on others (Figure 23.17). This removal of the early craters was probably accomplished by the extrusion of icy "magma" from within Titania while some internal heat was still present. When Titania cooled, its crust and interior froze and expanded, producing a series of fault valleys on its surface. Other features visible on Titania include some young, bright-rayed craters.

Oberon, Uranus' outermost satellite, orbits 583,000 kilometers (362,000 miles) from Uranus and is 1550 kilometers (965 miles) in diameter. Its albedo and density are 24% and 1.50 grams per cubic centimeter, respectively. Oberon's surface is heavily cratered, indicating that there have been no processes active there to erase the marks left during the early period of bombardment (Figure 23.18). A large mountain that was seen at the edge of Oberon may be the central peak of a crater that was not otherwise visible. Some of Oberon's craters have bright rays emanating from them, and patches of very dark material are found on the floors of some craters, indicating that "dirty water" may have risen through cracks in

■ **Figure 23.18** Voyager 2 photograph of Oberon. (NASA/JPL Photograph.)

the crust to fill the crater floors. Some features resembling faults were seen on Oberon, but it shows less evidence of tectonic activity than the other large Uranian satellites.

These five major satellites are as diverse a group of objects as found in any satellite system in the solar system. Although geologists will be studying Voyager 2 images of these worlds for many years, it is unfortunate that additional exploration of these objects is probably many decades away.

■ Chapter Summary

Uranus, the third gas giant, was discovered accidentally by Herschel in 1781. It is smaller than Jupiter and Saturn and has different physical properties. Uranus' most unusual characteristic is its high obliquity, whose origin is uncertain but may be impact-related. Much of what we know about Uranus was learned during the Voyager 2 flyby in 1986.

Uranus' atmosphere contains very few markings, primarily because of the presence of a hydrocarbon haze. Uranus' interior consists of a molecular hydrogen layer; a liquid conducting ionic region; and a rocky core. Uranus' ring system, discovered in 1977, consists of ten major bands, all of which are narrow and composed of dark, rocky material.

Of Uranus' 15 satellites, 5 are large enough that Voyager photographs gave insight into their surface geology. All of them are low-density, icy bodies marked by impacts, and they possibly have experienced "ice–magma" volcanism. Miranda's strange surface markings suggest that it may have been disrupted by a major impact and subsequently re-accreted.

■ Chapter Vocabulary

Ariel	Herschel, William	Puck	Umbriel
coronae	Miranda	Titania	Voyager 2
epsilon ring	Oberon		

■ Review Questions

1. In what respects is Uranus noteworthy?
2. Describe the discovery of Uranus.
3. In what ways does Uranus' high obliquity cause it to present unusual orientations for observation?
4. Describe the encounter of Voyager 2 with Uranus. In what ways was this encounter different from those with Jupiter and Saturn?
5. Describe the atmosphere of Uranus.
6. Describe the interior of Uranus. In what way is its interior different from those of Jupiter and Saturn?
7. In what way is Uranus' magnetic field unusual?
8. Describe Uranus' ring system. In what ways is it different from Saturn's?
9. Summarize the properties of Uranus' five major satellites.
10. Describe the unusual surface features of Miranda.

■ For Further Reading

Cuzzi, Jeffrey N. and Larry W. Esposito. "The Rings of Uranus." In *Scientific American,* July, 1987.

Hunt, Garry and Patrick Moore. *Atlas of Uranus.* Cambridge University Press, Cambridge, England, 1989.

Ingersoll, Andrew P. "Uranus." In *Scientific American,* January, 1987.

Johnson, Torrence V. et al. "The Moons of Uranus." In *Scientific American,* April, 1987.

Laeser, Richard P. et al. "Engineering Voyager 2's Encounter with Uranus." In *Scientific American,* November, 1986.

24

Neptune

CHAPTER OBJECTIVES

After studying this chapter, you should be able to

1. Understand how Neptune's existence was predicted and how the planet was discovered.
2. Describe the appearance of Neptune from the Earth.
3. Summarize the major discoveries made by Voyager 2.
4. Describe the properties of Neptune's atmosphere and interior.
5. Describe the properties of Neptune's ring system and satellites.

Voyager 2 photograph of Neptune and its large satellite Triton. Both objects appear in the crescent phase. (NASA/JPL Photograph.)

N eptune is the most distant, smallest, and densest gas giant planet. Although there are some differences, Neptune is physically similar to Uranus. (Neptune's major orbital and physical properties are summarized in Table 24.1.) Because of its extreme distance from the Earth, Neptune is much harder to observe than Uranus, and less was known about it than the other gas giants until Voyager 2 encountered it in 1989. Neptune has eight satellites, of which only two were known prior to the Voyager 2 encounter. Neptune's largest satellite, Triton, has a complex surface geology, and volcanic activity was witnessed by Voyager. Neptune also has a ring system whose properties were not well understood until Voyager 2 returned firsthand pictures.

■ Discovery

Neptune has the distinction of being the first planet discovered after its existence was predicted. Following the discovery of Uranus in 1781, Uranus' orbital elements were determined, enabling its location in the sky to be predicted in advance. However, the planet did not move entirely as expected, and by as early as 1790, it was obvious that something was unusual about Uranus' path. By the 1830s, it was the consensus of astronomers that Uranus' deviations from its expected positions were caused by perturbations from a distant planet that had not yet been discovered.

By the early 19th century, celestial mechanics had advanced to the point where the position of a perturbing object could be calculated based upon the deviations of an object (like Uranus) from its expected path. In the mid-1840s, **John Adams** (1819 – 1892) in England and **Urbain Leverrier** (1811 – 1877) in France independently used these methods to calculate where the perturbing object would be found. Adams presented his results to the British Astronomer Royal,

■ Table 24.1   Summary of Neptune's Basic
Properties*

Semi-major axis: 30.11 astronomical units
Semi-major axis: 4.504 billion kilometers
Semi-major axis: 2.799 billion miles
Orbital eccentricity: 0.009
Orbital inclination: 1.77°
Orbital period: 164.8 years
Diameter: 48,600 kilometers
Diameter: 30,200 miles
Mass (Earth = 1): 17.2
Density: 1.76 grams per cubic centimeter
Sidereal rotational period: 16.05 hours
Axial obliquity: 29.56°
Number of satellites: 8

* For a complete table of planetary data, see Table A2.1 in
Appendix 2.

George Airy (1801–1892), who carried out only a half-hearted search and failed to discover the object. Airy and James Challis (1803–1882), who assisted him, were hampered by a lack of good star charts of the area where the new object was predicted to be. They lost time plotting charts of the stars in that part of the sky, so that they could detect an object that moved between observations. Challis eventually found Neptune, but not until another astronomer had found it first.

Meanwhile, Leverrier had failed to convince the French Academy to look for his object. (Both he and Adams were relatively young at this time, and older astronomers were not entirely convinced of their skill.) Leverrier then sent his prediction to **Johann Galle** (1812–1910), a German, who fortunately possessed accurate star charts of the predicted region of the new planet. On September 23, 1846, Galle found the planet in about half an hour's time. It was within 1° (twice the diameter of the full Moon) of its predicted location.

In keeping with the tradition of naming planets for major mythological figures, the new solar system member was named Neptune for the god of the seas because it appeared to be sea-green in color when viewed through the telescope. Although Leverrier's prediction was the basis for the actual discovery, Adams had obtained the same result, and both men are now credited equally with the discovery. They met for

the first time after the discovery and became life-long friends.

As a curious footnote to this story, historians of astronomy have recently realized that while observing Jupiter during the winter of 1612 and 1613, Galileo actually saw Neptune and sketched it as being within the same telescopic field of view as Jupiter. Amazingly, he noted Neptune's motion, but apparently did not attach enough significance to it to investigate the matter further. In addition, Joseph J. Lalande (1732–1807) observed Neptune on May 8 and 10, 1795 while compiling a star catalog. On May 10, he noted that one of the "stars" measured two nights before had shifted, but he concluded that his earlier measurement had been erroneous!

The discovery of Neptune struck a blow to Bode's law, as the planet is much closer to the Sun than the law would predict the eighth planet to be (30.1, instead of 38.8, astronomical units). Even though both Adams and Leverrier had based their predictions on the assumption that Neptune would be located at the distance given by Bode's law, the sky positions they obtained were still accurate enough to allow its discovery.

■ Appearance and Basic Properties

With an average brightness of about eighth magnitude, Neptune is much too faint to be visible to the unaided eye. Even its telescopic appearance is disappointing. Unlike Uranus, which shows a disk even when observed with a small telescope, Neptune appears star-like except in very large telescopes. Therefore, its identity cannot be ascertained unless a precise sky map of the area in question is available, or if motion can be verified by looking at it over a period of several days. A few faint markings are visible on its disk in larger telescopes, as is the disk's overall greenish cast (Figure 24.1).

Neptune has several noteworthy properties. Although Neptune's rotational rate is comparable to that of the other gas giants, its orbital period is the longest, at 165 years. It orbits the Sun so slowly that it has not yet completed one orbit since its discovery! It has the lowest eccentricity

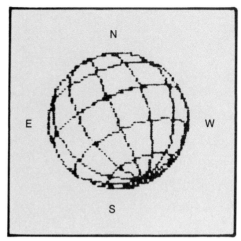
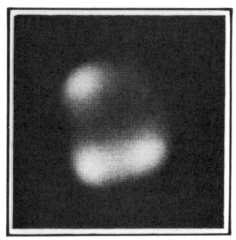

■ **Figure 24.1** Infrared image of Neptune taken from Earth. The diagram at left shows the orientation of the photograph. Note the cloud markings in Neptune's atmosphere. (R. J. Terrile and B. A. Smith, Las Campanas Observatory, Carnegie Institution, and NASA/JPL.)

of any gas giant (0.009), and its orbital inclination is 1.77°. Unlike Uranus, Neptune has a moderate obliquity of 29°, giving it seasons that last over 40 years each!

■ Spacecraft Exploration

Neptune is the most distant planet yet visited by spacecraft. **Voyager 2** encountered the planet on August 24, 1989, just over 12 years after its August 20, 1977 launch from Earth. This made Voyager 2 the first spacecraft to visit all four gas giants, having encountered Jupiter in 1979, Saturn in 1981, and Uranus in 1986. The spacecraft was aimed for a point 29,500 kilometers (18,300 miles) from the planet's center, or about 4900 kilometers (3000 miles) above its cloud tops. Voyager skimmed over the planet's north pole and then encountered the large satellite Triton from a distance of about 40,000 kilometers (25,000 miles). The only other previously known satellite, Nereid, was observed from a distance of 4.7 million kilometers (2.9 million miles). Six additional satellites were also discovered. Because terrestrial observations prior to the flyby indicated the existence of a ring system, the trajectory was planned so that it would be avoided. The spacecraft not only avoided collisions with ring particles but returned numerous photographs of the rings, allowing

their dimensions and other properties to be determined.

Because Neptune is so distant, there was less illumination for photography than at Uranus, requiring very long time exposures, during which the spacecraft's motion would have caused blurring of the image. This problem was solved by having the spacecraft turn during long-exposure photographs so that its motion was counteracted, a technique first used at Uranus. Another problem caused by Neptune's extreme distance was that the radio signal received from Voyager 2 was very weak. Prior to the flyby, improvements were made to the radio telescopes on Earth that were used to receive the faint signals from Voyager, and there was no difficulty in receiving the data it transmitted.

■ Atmosphere

Neptune and Uranus can be considered twins in the same sense that Earth and Venus are; although they are not identical, many resemblances can be noted. One example of this similarity is in the atmospheres of Uranus and Neptune. Both planets have atmospheres that are similar in composition and vertical structure. Hydrogen and methane had been detected spectroscopically in Neptune's atmosphere prior to the

Voyager flyby, and methane, which absorbs red light, is responsible for the planet's blue color. Helium is probably present in Neptune's atmosphere, but it has not yet been detected spectroscopically. Ammonia and water could not exist as gases in Neptune's atmosphere, but they could be present as ices in the planet's clouds.

One of the mysteries of Neptune's atmosphere is why its cloud-top temperature is about the same as Uranus', even though Neptune's greater distance from the Sun allows it to receive far less incident radiation. Voyager 2 indicated that the average cloud-top temperatures on Neptune are 59 K ($-353°$F), whereas those on Uranus are 50 K ($-370°$F). The apparent explanation for this phenomenon is that some mechanism within Neptune is releasing considerably more thermal energy than is being generated inside Uranus. Voyager 2 failed to reveal why the two planets differ in this manner.

Prior to the Voyager encounter, estimates for Neptune's rotation rate based on cloud-marking movements ranged from 11 to 22 hours, with the uncertainty due in large part to the difficulty of observing such markings from the Earth. Cloud features observed by Voyager 2 indicated that the atmosphere rotated in 18.4 hours far south of the equator and in 15.8 hours near the equator. (Because of Voyager 2's trajectory, the majority of its observations were of features in Neptune's southern hemisphere.) Voyager 2's observations of Neptune's internal radio emissions showed that the interior of the planet rotates in just over 16 hours, faster than most researchers had anticipated. This means that Neptune's atmosphere is rotating slower than its interior, except in areas south of 53° south latitude. As a result, most of Neptune's winds blow in the east-to-west direction, opposite to the internal rotation. Neptune has the fastest east-to-west winds of any planet, 1170 kilometers per hour (730 miles per hour). In areas near the south pole, atmospheric winds, moving west to east in the same direction as the planet rotates, have speeds of 70 kilometers per hour (45 miles per hour).

Voyager 2 discovered several atmospheric markings, primarily in Neptune's southern hemisphere. These markings revealed that Neptune's atmospheric circulation is exceeded in complexity only by that of Jupiter, which has a somewhat

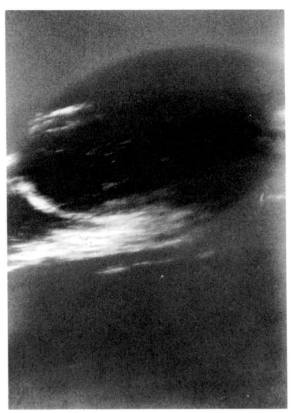

■ **Figure 24.2** Neptune's Great Dark Spot. Note the white clouds comprising the "bright companion" to the dark spot. (NASA/JPL Photograph.)

similar cloud pattern. The most conspicuous feature is the **Great Dark Spot**, located at 20° south latitude (Figure 24.2). This feature, similar in location and circulation pattern to Jupiter's Great Red Spot, measures 12,000 by 8000 kilometers (7500 by 5000 miles) in size and has a rotational period of 18.3 hours. South of the Great Dark Spot, at 33° south latitude, is a white feature called the bright companion. This feature seems to be 50 kilometers (30 miles) higher in the atmosphere than the Great Dark Spot. (Neptune's white clouds are believed to be located this high above the remainder of the methane cloud deck because winds around dark features, such as the Great Dark Spot, cause atmospheric gases to rise high enough that the temperature drops below the freezing point of methane. In this respect, the white clouds are similar to the high cirrus clouds in the Earth's atmosphere. As shown in Figure 24.3, shadows cast by Neptune's high clouds are

■ **Figure 24.3** High, thin clouds in Neptune's atmosphere. Note the shadows cast on the lower cloud deck. (NASA/JPL Photograph.)

■ **Figure 24.4** Cloud features in Neptune's southern hemisphere. At left center are the Great Dark Spot and its "bright companion." Farther south is the "scooter," and near the bottom of the picture is the southern dark spot, D2. (NASA/JPL Photograph.)

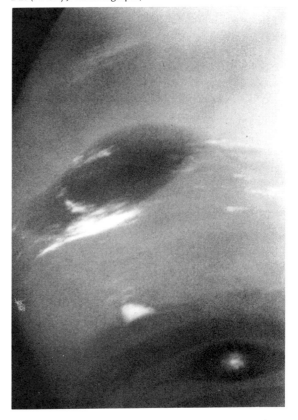

visible on the lower cloud deck.) Further south, at 42° south latitude, is an even brighter white cloud, which was termed the scooter because of its rapid motion. Another dark spot, named D2, was located at 55° south latitude (Figure 24.4). Rotation this far south is more rapid than at the latitude of the Great Dark Spot, and as a result, D2 overtakes the Great Dark Spot every five days (Figure 24.5). At 70° south latitude lies the south polar feature, a band of clouds whose configuration constantly changes. In addition to these specific cloud features, a high, thin haze was detected over the entire atmosphere, and its effect on the vertical structure of temperature is uncertain. This haze had been observed from Earth and seen to thin and re-form in a cycle lasting from days to weeks.

■ Interior and Magnetic Field

Neptune's bulk composition and internal structure are probably very similar to those of Uranus (Figure 24.6). Neptune is denser, indicating that it probably has a more massive core and higher central pressure and density. Neptune's outer layer of molecular hydrogen and methane probably extends about one-quarter of the way to the planet's center. Below this, a convective (and possibly ionic) layer of water, methane, and am-

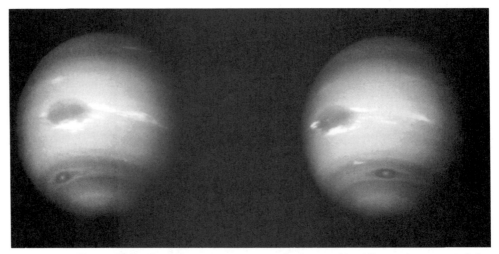

■ **Figure 24.5** Changes in Neptune's atmosphere caused by differential rotation. At left, the Great Dark Spot and D2 are nearly aligned. At right, D2 has moved relative to the Great Dark Spot. The two photographs were taken 17.6 hours apart. (NASA/JPL Photograph.)

monia extends about 70% of the way to the center. Neptune's core is probably composed of silicates and iron, and the planet's central pressure may be about 22 million atmospheres.

Neptune's magnetic field was not discovered until the firsthand observations by Voyager 2. Recall from Chapter 23 that Voyager 2 found Uranus' magnetic field orientation to be very unusual, and that researchers assumed that Uranus' high obliquity was responsible, or else that the planet's field direction was in the process of re-

■ **Figure 24.6** Neptune's interior layers. The numbers represent the fraction of Neptune's total radius. See text for detailed descriptions.

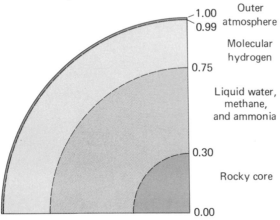

versing. Surprisingly, Neptune's field axis was found to be oriented at an angle of 47° to Neptune's rotational axis and offset 14,000 kilometers (8700 miles) from the center of the planet. Because of the different axial obliquities of Uranus and Neptune, another explanation for their field orientations must be found, especially as it would be improbable that both planets' fields are simultaneously in the process of reversing. One possible explanation is that the fields of the two outermost gas giants may be generated much closer to their surfaces than for any other planets. This would explain the extreme variability of Neptune's surface field strength, which ranges from 0.06 to 1.2 gauss, the widest range of any planet. Auroras were seen in Neptune's atmosphere, and although the planet has a trapped radiation belt, it is weaker than that of Jupiter and even of the Earth.

■ Rings

As mentioned in Chapter 18, occultation data obtained by Edward Guinan in 1968, but only analyzed much later, indicated that Neptune has a ring system consisting of two rings, one about 2900 kilometers (1800 miles), the other 6800

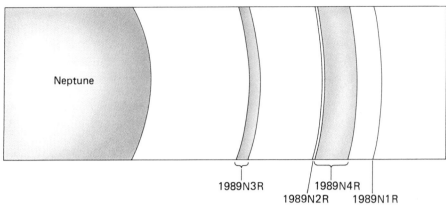

■ **Figure 24.7** Map of Neptune's ring system, as seen from above the planet's north pole.

1989N3R 1989N4R 1989N1R
1989N2R

kilometers (4200 miles) above the clouds. Between 1982, when Guinan announced these results, and 1989, when Voyager 2 flew past Neptune, over 40 Neptune occultations were carefully observed but gave contradictory results. Fewer than ten occultations indicated the presence of rings, whereas the rest recorded no light decreases produced by ring features. One pre-Voyager explanation of these observations was that Neptune's rings were discontinuous features, or **ring arcs.**

Voyager 2 verified that Neptune has rings and revealed them to be complex, unique features. Four major ring components were seen (Figures 24.7 and 24.8). The outermost ring, termed 1989N1R, lies 62,900 kilometers (39,000 miles) from the center of the planet and is probably less than 50 kilometers (30 miles) wide. This ring somewhat resembles Saturn's unusual F ring in that it is nonuniform. In fact, 1989N1R contains three concentrations of material in one 33° section of the ring. These "sausages," as some Voyager scientists termed them, were the first parts of the rings photographed by the incoming spacecraft and resembled the ring arcs (Figure 24.9). It is considered likely that these clumps of material in the outer ring explain the terrestrial observations that suggested the presence of ring arcs. Explaining the clumps is difficult, however, as no satellites were found that could be responsible for shepherding the clumps of material.

The narrowest ring component, 1989N2R, is located 53,200 kilometers (33,000 miles) from

the planet and is about 15 kilometers (9 miles) wide. Unlike the outermost ring, 1989N2R lacks concentrations of material. Extending outward from 1989N2R is a faint, wide, diffuse ring component termed 1989N4R and also called the "plateau." It extends from 53,200 to 59,000 kilometers (33,000 to 36,700 miles) from Neptune. The innermost ring component, 1989N3R, is centered 41,900 kilometers (26,000 miles) from Neptune and appears to be nearly 1700 kilometers (1050 miles) wide.

■ **Figure 24.8** Photograph of Neptune's rings. Taken when Voyager 2 was beyond Neptune, this image shows the forward-scattered light from the rings. The three major ring components are visible, but the "plateau" feature is invisible in this image. (NASA/JPL Photograph.)

■ **Figure 24.9** Photograph of Neptune's outermost ring component, 1989N1R, showing the three "sausages" that were detected from Earth during occultations and considered ring arcs. (NASA/JPL Photograph.)

■ **Figure 24.10** Photograph of Neptune and its satellites Triton (*arrow,* near planet) and Nereid (*arrow,* upper right). (Lick Observatory Photograph.)

Neptune's rings are relatively unsubstantial, and it has been estimated that if all of the ring material could be collected into a sphere, it would be no greater than 5 kilometers (3 miles) across. Indications are that the rings are losing small, dusty particles, and some mechanism must be replenishing this material. One possible source of new ring material is collisions of comets with Neptune's small satellites.

■ Satellites

Two satellites of Neptune were known prior to the Voyager 2 flyby: Triton, discovered in 1846, and Nereid, discovered in 1949 (Figure 24.10). Data from an occultation in 1981 seemed to indicate the presence of one or two additional satellites, but this result was not verified until the Voyager encounter. Voyager 2 discovered six new satellites that have not yet received names but are known by temporary designations. All of the new satellites are closer to Neptune than Triton and Nereid are, and all of them are small, low-albedo, heavily cratered objects ranging in diameter from 50 to 400 kilometers (30 to 250 miles). The outermost new satellite, 1989N1, is Neptune's second largest satellite and is distinctly nonspherical in shape (Figure 24.11). The next new satellite in order toward Neptune, 1989N2,

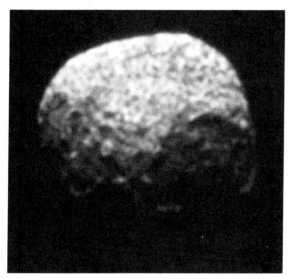

■ **Figure 24.11**   Neptune's second largest satellite, designated 1989N1. Note its rough surface and irregular shape. (NASA/JPL Photograph.)

was probably responsible for the 1981 occultation results (Figure 24.12). Satellite 1989N4 orbits just inside the outermost ring, 1989N1R, and may act to shepherd the ring particles, but it cannot be responsible for the clumpiness of the ring. In simi-

lar fashion, satellite 1989N3 orbits just inside ring 1989N2R. The other two new satellites, 1989N5 and 1989N6, orbit between the inner two rings, 1989N2R and 1989N3R (Figure 24.13).

Triton is Neptune's largest satellite, with a diameter of 2705 kilometers (1680 miles). Prior to the Voyager flyby, very little was known about its physical properties, but its unusual orbit set it apart from other planetary satellites. Whereas Triton's orbit is nearly circular, Triton moves in the retrograde direction, the only large satellite in the solar system to do so. Triton orbits 354,000 kilometers (220,000 miles) from Neptune, and dynamical studies have indicated that its orbit is decaying, causing it to slowly approach Neptune. Early calculations predicted that Triton will survive in orbit only another 10 to 100 million years, but recent results indicate that the decay is much slower and the satellite will survive for billions of years.

Data obtained during the Voyager encounter revealed Triton's physical properties to be no less unusual than its orbital properties (Figure 24.14). Triton's density was found to be 2.02 grams per cubic centimeter, indicating a composition of ice and rock common among outer solar system satellites. With a surface temperature of 37 K

■ **Figure 24.12**   Neptune's satellite 1989N2. Like 1989N1, it appears to have a rough surface and is irregular in shape. (NASA/JPL Photograph.)

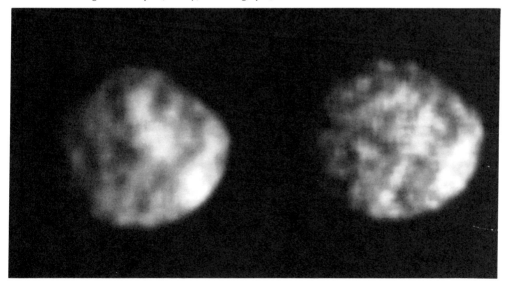

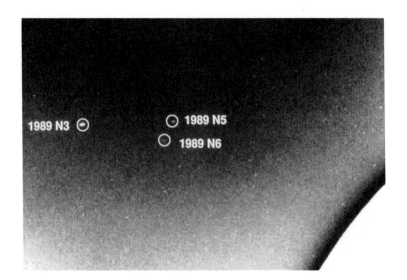

■ **Figure 24.13** Three of Neptune's small satellites discovered by Voyager 2. (NASA/JPL Photograph.)

(−392°F), Triton is the coldest solar system object yet observed by spacecraft. Pre-Voyager spectroscopic evidence indicated that Triton might have an atmosphere of methane and nitrogen, and Voyager verified that Triton, like Titan, is a satellite with an atmosphere of mostly nitrogen with some methane. However, Triton's atmosphere is very tenuous, with a pressure only 10^{-5} that of Earth's atmosphere. Unlike Titan's atmosphere, Triton's is transparent, except for a thin haze composed of very fine particles located between 5 and 10 kilometers (3 and 6 miles) above its surface (Figure 24.15). For reasons that are not yet understood, the temperature in Triton's atmo-

■ **Figure 24.14** Mosaic of Triton's Neptune-facing hemisphere from several combined Voyager 2 images. Note the lack of impact features and the variety of surface features discussed in the text. (NASA/JPL Photograph.)

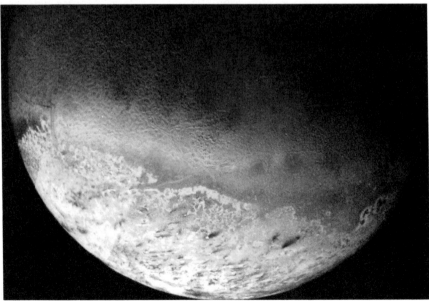

■ **Figure 24.15** Voyager 2 photograph of Triton that shows the layer of thin haze in its atmosphere. (NASA/ JPL Photograph.)

sphere rises to 100 K (−280°F) 600 kilometers (325 miles) above the satellite's surface.

The surface of Triton, composed of rock and ices of water, nitrogen, and methane, has a high albedo (60%–90%) and is white with a pinkish cast in some areas. It is brightest at the poles, where polar caps of condensed atmospheric nitrogen lie (Figure 24.16). (The south polar cap appears to be dissipating, as the southern hemisphere is currently experiencing summer. Dark

■ **Figure 24.16** Triton's polar ice cap. Note dark streaks due to volcanic activity and other albedo variations caused by dissipation of the ice cap. (NASA/JPL Photograph.)

■ **Figure 24.17** Basin on the surface of Triton that has been flooded by the extrusion of icy "lava." (NASA/JPL Photograph.)

regions within the cap indicate areas that have apparently just lost their cover of nitrogen frost.) Voyager photographs revealed a world that lacks craters from the heavy bombardment era early in the solar system's history, indicating that internal activity has refigured Triton's surface. Much of the resurfacing was done by the extrusion of "ice magma," as on several other outer planet satellites. Voyager photographed 30-kilometer-wide

(20-mile-wide) fault valleys and larger basins that had been flooded in this fashion (Figure 24.17). It was initially assumed that Triton's internal activity had ceased long ago and that internal heating may have been generated by tidal forces when Triton was initially captured by Neptune, a scenario that is suggested by its retrograde orbit. However, several photographs indicate that volcanic activity is ongoing on Triton, making it the

■ **Figure 24.18** Voyager 2 photographs of a volcanic eruption of a dark plume of material into the atmosphere of Triton. The two photographs are identical, but *arrows* have been added to the bottom image to show the extent of the dark plume, as well as the location of its shadow. (NASA/JPL Photograph.)

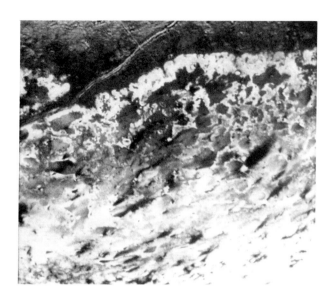

■ **Figure 24.19**   Region of Triton's surface near its south pole. The numerous dark streaks are deposits of material that has erupted from within the satellite. (NASA/JPL Photograph.)

only place other than the Earth and Io with active volcanoes. Geyser-like vents shoot dark particles 40 kilometers (25 miles) high above the surface, where winds carry them 70 kilometers (45 miles) downwind, depositing long, dark streaks on the surface (Figures 24.18 and 24.19). Because Triton is not currently heated by tidal interactions and any heat source from radioactivity would have dissipated, most researchers believe that some other process must be responsible for heat generation. It is considered likely that pools of liquid nitrogen may be trapped under pressure beneath Triton's surface. When these break through cracks in the surface, geyser-like volcanic activity would result.

Neptune's outermost satellite, **Nereid**, is much smaller, only 340 kilometers (210 miles) in diameter (Figure 24.20). Nereid is noteworthy because it has the most eccentric orbit of any planet or satellite in the solar system, 0.749. As Nereid orbits, its distance from Neptune varies from 1.38 to 9.71 million kilometers (0.86 to 6.03 million miles). Voyager 2 observed Nereid from a very great distance, but it determined that Nereid's albedo is about 14% and that it is irregular in shape. Because irregular objects may vary in brightness as they rotate, the irregularity of Nereid explains earlier, pre-Voyager observations of periodic brightness variations that suggest that Nereid's rotational period is probably between 8 and 24 hours.

Why are the orbits of Triton and Nereid so unusual? Although several possibilities have been proposed, the answer to this question remains unknown. One possibility is that both Triton and Nereid are captured objects. Another possibility involves Pluto, the final planet, and suggests that Pluto and its satellite were once also satellites of Neptune, but that another object disrupted Neptune's original satellite system. We will discuss Pluto, as well as this theory, in more detail in the next chapter.

■ **Figure 24.20**   Neptune's outermost satellite, Nereid. This Voyager 2 photograph shows little surface detail but reveals that the satellite is irregularly shaped. (NASA/JPL Photograph.)

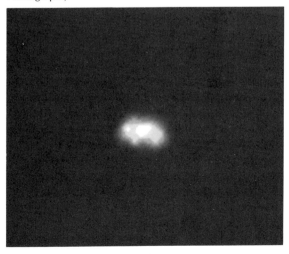

■ Chapter Summary

Neptune, the most distant gas giant, was discovered in 1846 after its existence had been predicted. Neptune had revealed itself by causing perturbations to Uranus. Neptune is so distant that it shows little detail, even in large telescopes, and most of what we know about it was revealed by Voyager 2 in 1989.

Neptune's atmosphere is similar to Uranus' but contains more markings. Because Neptune's interior rotates faster than most of the atmosphere, the planet has high winds in the westward direction. Voyager 2 revealed the presence of several dark and light atmospheric markings, the most notable of which has been called the Great Dark Spot. The interior of Neptune is believed to be similar to that of Uranus. Their magnetic fields resemble one another greatly despite the planets' differences in obliquity.

Neptune has a ring system that consists of three narrow bands and a wide, diffuse "plateau" region. The outermost ring contains three sausage-like concentrations of ring particles that had been detected from Earth as ring arcs. Voyager 2 discovered six new satellites and revealed considerable information about previously known Triton, Neptune's largest satellite. Icy Triton has a thin nitrogen–methane atmosphere and a relatively young surface that is marked by ongoing volcanic activity. Nereid, the outermost satellite, is relatively small and has an extremely eccentric orbit.

■ Chapter Vocabulary

Adams, John
Galle, Johann
Great Dark Spot
Leverrier, Urbain
Nereid
ring arcs
Triton
Voyager 2

■ Review Questions

1. Describe the events leading to the discovery of Neptune.
2. What is the appearance of Neptune in a telescope?
3. Summarize the results of the Voyager 2 encounter with Neptune.
4. What are the major features of Neptune's atmosphere? In what ways are its winds unusual?
5. Why is the similarity between the magnetic fields of Uranus and Neptune surprising?
6. Describe the basic features of Neptune's rings.
7. Summarize the features of Triton's atmosphere and surface.
8. In what ways are the orbits of Triton and Nereid unusual?

■ For Further Reading

Anonymous. "Voyager's Last Picture Show." In *Sky and Telescope*, November, 1989.
Anonymous. "Neptune and Triton: Worlds Apart." In *Sky and Telescope*, February, 1990.
Beatty, J. Kelly. "Getting to Know Neptune." In *Sky and Telescope*, February, 1990.
Grosser, Morton. *The Discovery of Neptune.* Harvard University Press, Cambridge, MA, 1962.

Kaufmann, William J. "Voyager at Neptune — A Preliminary Report." In *Mercury,* November/December 1989.

Kinoshita, June. "Neptune." In *Scientific American,* November, 1989.

Moore, Patrick. "The Discovery of Neptune." In *Mercury,* July/August 1989.

Moore, Patrick. *The Planet Neptune.* Wiley, New York, 1989.

25

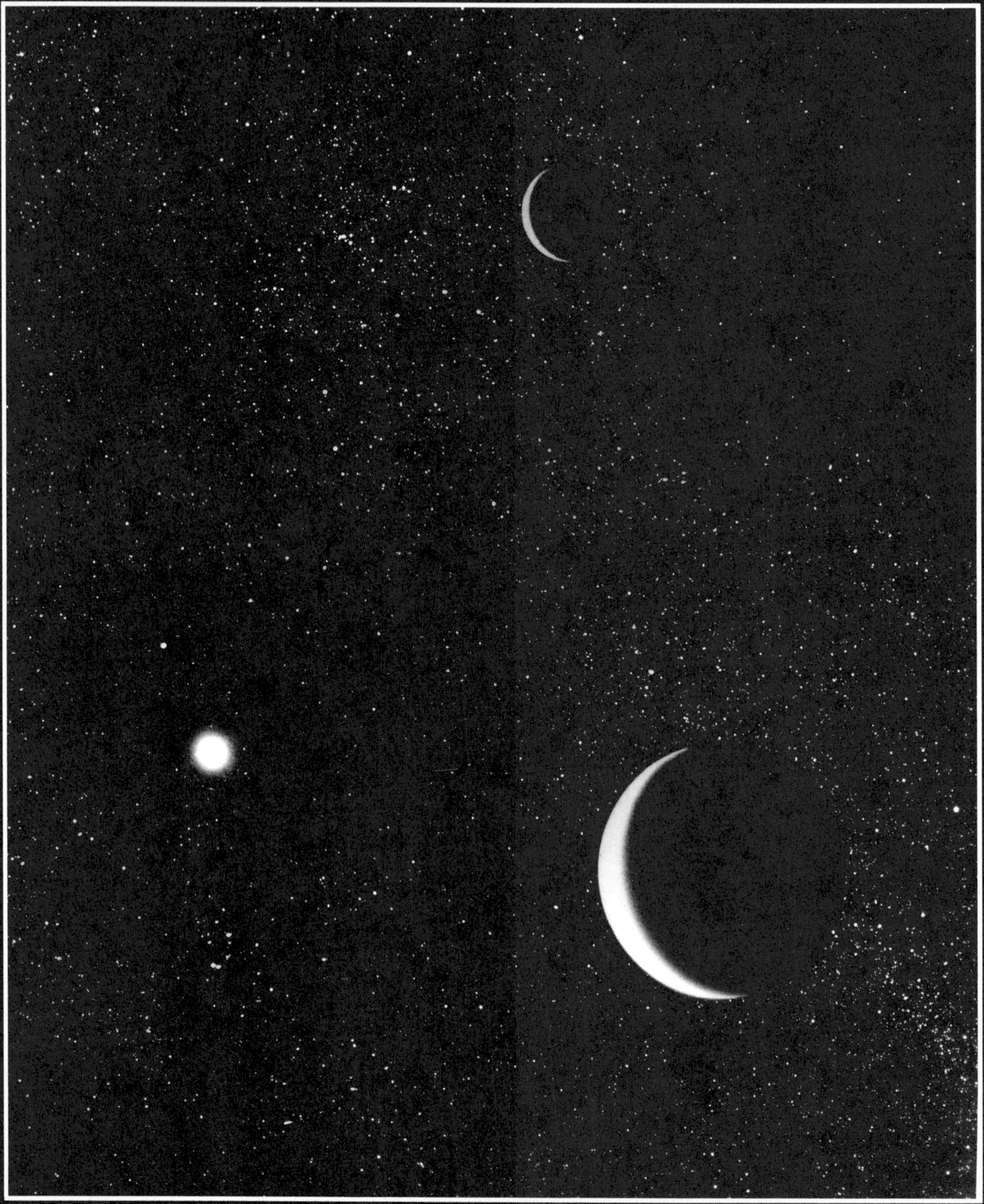

Pluto

CHAPTER OBJECTIVES

After studying this chapter, you should be able to

1. Describe the events involved in the discovery of Pluto and its satellite Charon.
2. Summarize the basic orbital and physical properties of Pluto and Charon.
3. Describe what researchers believe that the atmosphere, surface, and interior of Pluto are like.
4. Explain the relationship between Neptune and Pluto and why they may have been together in the past.
5. Summarize what steps have been conducted to search for additional planets and what their results have been.

Artist's conception of Pluto and its satellite, Charon. Since no missions to the outermost planet are planned, it will be a long time before actual views of this system will be obtained. (NASA Photograph, courtesy of the National Space Science Data Center.)

eyond Neptune lies one more planet, Pluto. Although undiscovered planets may exist, faraway Pluto currently lays claim to being the one whose semimajor axis is largest. Pluto is so far from the Sun that from Pluto, the Sun would not present a visible disk, although it would be 1600 times brighter than the full Moon. Although much has been learned about this distant object, its small size and extreme distance make it very difficult to study, and the planet unfortunately remains somewhat shrouded in mystery. Unfortunately, there are no current plans to send a spacecraft to its vicinity, so additional information about it will have to come from telescopic observations.

Pluto's major orbital and physical properties are summarized in Table 25.1. Pluto is the smallest planet in terms of diameter, mass, and volume. At least seven planetary satellites—the Moon, Io, Europa, Ganymede, Callisto, Titan, and Triton—are larger than it is. Pluto's small size has led some astronomers to suggest that it shouldn't even be considered a full-fledged planet, but an asteroid. (Some asteroids, like Pluto, may have satellites, and some asteroids exist in the outer solar system.) The fact that Pluto has a tenuous atmosphere, however, sets it apart from all asteroids, and it will probably continue to be considered a planet.

Pluto might actually be considered a double planet because its satellite Charon is about half its diameter. Pluto is a unique planet in that it is neither a rocky and metallic terrestrial object nor a gaseous body like its nearest planetary neighbors. Rather, it is largely an ice-covered world, similar in many ways to the satellites of the gas giants.

■ Discovery of Pluto and Charon

After the discovery of Neptune in 1846, the new planet's perturbing influence was taken into ac-

■ Table 25.1   Summary of Pluto's Basic Properties*

Semi-major axis: 39.44 astronomical units
Semi-major axis: 5.90 billion kilometers
Semi-major axis: 3.67 billion miles
Orbital eccentricity: 0.250
Orbital inclination: 17.17°
Orbital period: 247.7 years
Diameter: 2245 kilometers
Diameter: 1395 miles
Mass (Earth = 1): 0.0025
Density: 1.84–2.14 grams per cubic centimeter
Sidereal rotational period: 6.39 days (retrograde)
Axial obliquity: 62° (?)
Number of satellites: 1

* For a complete table of planetary data, see Table A2.1 in
   Appendix 2.

count when calculating the position of Uranus. Surprisingly, this failed to account for all of Uranus' motion, and even Neptune behaved as though something might be perturbing its orbit. Therefore, astronomers began to calculate where a "trans-Neptunian" planet might be found. Predictions were made, but there was no quick discovery, as had been the case with Neptune. Unlike Adams and Leverrier, the astronomers who predicted another new planet generally were not in agreement. Most of the predictions were of a single gas giant planet, but one researcher even suggested that two bodies were present.

Percival Lowell (1855–1916), the noted observer of Mars, published his prediction for the location of the object he called "Planet X" in 1905. A search was conducted from 1905 to 1907, but nothing was found. A second search in 1911 was also fruitless, although it was later learned that Pluto was photographed then but not recognized. Lowell revised his prediction in 1915 and began another search, but he died the next year. All of these searches had involved looking for a planet that would show a disk. In 1929, **Clyde Tombaugh**, a new observer at what by then was called Lowell Observatory, began to search using a new technique, examining photographs with the aid of a device known as a **blink comparator**. The same area of the sky (near the opposition point, where a planet would be beginning retrograde motion) would be photographed several days apart, after which the two photographic plates were placed in a device that "blinked" them so that the person checking the plates saw first one, then the other, in rapid succession. This technique is very effective for locating objects that have moved. On February 18, 1930, Tombaugh identified the images of the new planet, which had been taken on January 21 and 29 (Figure 25.1). The results were announced on March 13, which was Lowell's birthday and also the anniversary of the discovery of Uranus.

The planet was named for the god of the dark, cold underworld, with Lowell's initials incorporated into the first two letters of the planet's name. The name seemed particularly appropriate, because the god Pluto could make himself invisible, which is what the planet itself seemingly had done for so long. As more has been learned about Pluto over the years, it has been found to

■ **Figure 25.1**   The photographs on which Clyde Tombaugh discovered Pluto. They were taken on January 23, 1930 *(left)* and January 29, 1930 *(right)*. (Lowell Observatory Photographs.)

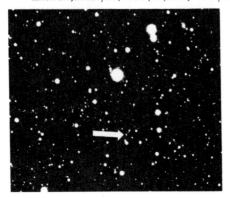 

be far too low in mass to account for the unexplained perturbations. This means that its discovery was due partly to luck but more to the determination of Lowell and especially Tombaugh, who didn't give up the search for "Planet X" until it was found. Pluto was, however, found within 6° of the point predicted by Lowell, and whether this represents luck or Lowell's computational skill remains hotly debated.

If Pluto is too low in mass to have caused the perturbations, how can we explain the irregularities of Uranus and Neptune? Some modern researchers claim that the older observations were inaccurate and that the supposed perturbations were erroneous. Other modern researchers believe that the perturbations were produced by another planet whose orbit is greatly inclined to the ecliptic. This planet would now be far above the plane of the solar system, causing its perturbations of Uranus and Neptune to have effectively ceased. Centuries from now, it may once again be aligned with the other outer planets as it crosses the ecliptic, and its renewed perturbing effects may allow it to be found at that time.

Another exciting discovery occurred in 1978, when **James Christy** noticed a "bump" on several photographic images of Pluto. These had been taken with a large telescope, but the graininess of the photographic process left the photographs very unclear. Christy came to the conclusion that these images revealed a satellite of Pluto, one too close to produce a separate image (Figure 25.2). This object was named **Charon** for a mythological figure associated with Pluto. From 1984 to 1990, Charon's orbital plane was oriented so that it both eclipsed Pluto and was eclipsed by it. These events led to more accurate figures for the diameters of both objects, as will be described later.

■ Appearance and Basic Properties

Pluto was a disappointment to its discoverers because it failed to reveal a disk. Even today, photographs taken through large telescopes fail to show

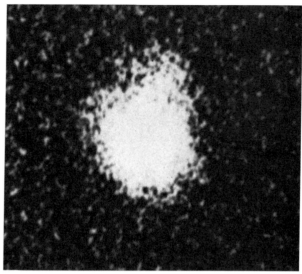

■ **Figure 25.2** The photograph on which James Christy discovered Charon. The two objects cannot be resolved, but Charon causes a "bump" at the upper right of Pluto's image. (U.S. Naval Observatory Photograph by James Christy.)

the disk clearly. At maximum brightness, Pluto is only about 14th magnitude, and the planet usually cannot be seen in a telescope with an aperture of less than 30 to 36 centimeters (12 to 14 inches). Because of the planet's faintness, extremely good star charts are required to identify Pluto, or else its motion must be noticed between observations to confirm the identity of the planet. Pluto's surface features are obviously invisible from Earth, and most of what we know about Pluto's surface has been learned from spectroscopy and occultation measurements.

One of Pluto's many unusual features is its strange orbit, the most eccentric and highly inclined of any planet. Its distance from the Sun varies between 4.42 billion kilometers (2.75 billion miles) at perihelion to 7.37 billion kilometers (4.58 billion miles) at aphelion. Pluto's orbit is so eccentric that for two decades of its nearly 250-year orbit, it is actually closer to the Sun than Neptune, as is currently the case between 1979 and 1999. The fact that these planets' orbits cross might suggest that they will someday collide. However, Pluto's high orbital inclination precludes this, and their closest approach, 18 astronomical units, occurs when Pluto is at aphelion.

There is a 3:2 orbital resonance between Neptune and Pluto, as Neptune completes three 165-year orbits as Pluto completes two 247.7-year orbits. The significance of this resonance with respect to the history of the two planets is not yet fully known.

Researchers Gerald J. Sussman and Jack Wisdom have performed computer calculations in which motions and perturbations of the outer planets were followed for a period of 845 million years. This research indicated that many of Pluto's orbital elements vary over long periods of time, but that close encounters between Pluto and Neptune will not occur. Because the orbital element variation was not perfectly regular, Pluto is thought to exhibit chaotic orbital motion, which means that Pluto could once have had much lower eccentricity and inclination but evolved to its present orbit.

Even such a basic quantity as diameter has been hard to ascertain for Pluto. The planet was originally thought to be nearly as large as Mars, but such estimates have shrunk over the years. In 1965, Pluto's failure to cover a star during what was expected to be an occultation placed an upper limit on the diameter of 6800 kilometers (4200 miles). Later, methane was found in the spectrum of Pluto, and Pluto was assumed to be covered with methane ice. Knowing the albedo of methane, researchers then estimated how large Pluto would have to be to shine at its observed brightness. That figure was about 2900 kilometers (1860 miles). During the 1980s, a number of observations, including visual studies of Pluto–Charon eclipses, infrared measurements by the Infrared Astronomy Satellite (IRAS), and observations made when Pluto occulted a star on June 9, 1988 allowed refined values of Pluto's and Charon's diameters to be determined. The currently accepted values are 2245 kilometers (1395 miles) for Pluto and 1119 kilometers (695 miles) for Charon. These values make Charon the largest satellite, relative to its primary, in the solar system.

Because of variations in its brightness, Pluto's rotational period has long been known to be 6.39 days. Once Charon was discovered, its orbital and rotational periods were also found to be 6.39 days, making this an example of a totally spin-orbit coupled system. Only 19,640 kilometers (12,200 miles) separate the two objects. Pluto's rotation and Charon's rotation and orbital motion are all in the retrograde direction.

Perhaps the most significant result of Charon's discovery was that it allowed mass determination using Kepler's third law. The combined mass of both objects has been found to be only 0.0025 that of Earth. Assuming a Pluto–Charon mass ratio of 10:1, the masses would be 0.0023 for Pluto and 0.0002 for Charon. These low mass values illustrate the mystery of how Lowell was able to predict even approximately Pluto's location, because its mass would have to be at least 0.1 that of Earth to influence Uranus and Neptune.

Although the individual densities of Pluto and Charon are not precisely known, William B. McKinnon and Steve Mueller have determined that the average density of the combined Pluto–Charon system is 1.99 grams per cubic centimeter. These researchers feel that Pluto's density lies between 1.84 and 2.14 grams per cubic centimeter; although Charon's density is not as well known, it is probably between 1 and 3 grams per cubic centimeter.

■ Surface

Because it will undoubtedly be many, many years before a spacecraft is sent to Pluto, our only knowledge of its surface comes from Earth-based remote sensing. Pluto–Charon mutual eclipses, which occurred during the 1980s, allowed each object's contribution to the total spectrum to be identified. This spectroscopy indicated that Pluto's surface is probably covered with methane ice. Curiously, the spectrum of Charon revealed no evidence of methane or ammonia, but it indicated that water ice is present on the satellite's surface. Careful observation during partial eclipses of Pluto by Charon has allowed rough mapping of color and albedo features on Pluto's surface. Different observational results have led to more than one model for Pluto's surface. Figure 25.3 illustrates one such model. Although Pluto's surface is probably cratered, the number of its craters, as well as whether any sort of internal activity occurs and renews its surface, is unknown.

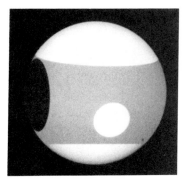

■ **Figure 25.3** Model of albedo markings on the surface of Pluto prepared by Marc Buie while at the University of Hawaii. This model was created by reproducing the rotational light curve of the Pluto–Charon system. It successfully matches light curves seen during Pluto–Charon occultations. (Diagram courtesy of Marc Buie.)

■ Atmosphere

Spectroscopic evidence obtained in 1980 first suggested that Pluto has a methane atmosphere with a pressure of about 0.1 atmospheres. During the next several years, as Pluto approached its 1989 perihelion, scientists theorized that Pluto's atmosphere was temporary, existing only near perihelion when the planet was warmest. Because of the planet's small mass, it could not maintain a permanent atmosphere, as methane gas would eventually escape. However, its temperature is low enough that escape of gas would not be immediate, permitting a short-term atmosphere to exist.

Proof that Pluto has an atmosphere came on June 9, 1988, when the first occultation of a star by Pluto was witnessed. The event, which lasted about 100 seconds, was observed by the Kuiper Airborne Observatory as it flew over the southern Pacific Ocean. The light from the star dimmed gradually as Pluto passed in front of it, clearly indicating passage of the light through atmospheric gases.

The presence of an atmosphere explains one of the mysteries of Pluto's surface, namely, why it is so bright. Solar radiation and cosmic rays should have converted methane on Pluto to a surface layer of dark hydrocarbons, which are not observed. Because Pluto's atmosphere is probably temporary, the "snowing" of methane as Pluto recedes from the Sun and cools is likely to renew its surface with a bright coating of new methane ice. This hypothesis is supported by the observation that Pluto has darkened in recent decades as it has approached perihelion, presumably as a result of methane evaporation revealing underlying dark material.

■ Interior

Because of a lack of firsthand observations, very little can be said about the interiors of Pluto and Charon. The moments of inertia of these objects are unknown, and the only evidence about their internal structure is the improved density value for Pluto announced by McKinnon and Mueller in 1988. Based on the value of 1.84 to 2.14 grams per cubic centimeter, McKinnon and Mueller predicted that Pluto is 68% to 80% rock, a much higher ratio than for the icy satellites of the gas giants. They have also proposed a model for Pluto's interior (Figure 25.4). Under an outer layer of methane ice that is a few kilometers thick lies a mantle of water ice that extends to a depth 20% to 25% of the way to the planet's center. The planet's core is probably composed largely of silicate minerals.

One possible reason why Pluto's density is so high is that Charon (or the particles from which it accreted) were derived from Pluto's icy mantle when an impact knocked them loose. The loss of this icy material left Pluto enriched in high-den-

■ **Figure 25.4** Pluto's interior layers. The numbers represent the fraction of Pluto's total radius. See text for detailed descriptions.

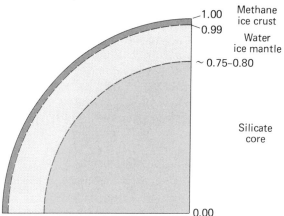

sity materials. The validity of this model will probably not be known until Charon's density is determined to a greater accuracy than that presently known.

■ Origin

One of the biggest mysteries about Pluto is why its unusual orbit crosses Neptune's. Although Pluto's orbit is sufficiently inclined that there appears to be no chance of a future collision with Neptune, it has been suggested that there is a relationship between the two planets. The unusual orbits of Neptune's satellites themselves can be used as arguments that something extraordinary occurred. One computer model has indicated that a body moving past Neptune could have disrupted the orbits of Triton and Nereid and thrown Pluto and Charon, both former satellites of Neptune, out of the system. Although such a scenario is possible, there is no way at present to tell for certain whether it happened.

■ Beyond Pluto

Are there planets beyond Pluto? Tombaugh spent the next 13 years after he discovered Pluto searching near the ecliptic for other planets, but he found nothing brighter than 16th magnitude. Since then, several specific predictions for the locations of new planets have been published, but none have proved to be correct. If other planets have high inclination orbits, they could lie in areas of the sky far from the ecliptic and would be extremely hard to track down. Even a large planet beyond Pluto would appear faint and lack a visible disk.

The paths of the Pioneer and Voyager spacecraft, now leaving the solar system after exploring the gas giants, are being closely watched to look for gravitational perturbations from unknown objects. None have been detected so far, but the paths of these crafts lie close to the ecliptic, and they would be unable to detect objects far from the solar system's plane.

Although we know of no planets beyond Pluto, there are many other solar system objects beyond it. As we will see in the next chapter, the Oort cloud of comets is believed to lie about 50,000 astronomical units from the Sun. Closer than that lies the outermost edge of the Sun's magnetosphere and the solar wind, which the Pioneer and Voyager spacecraft will probably reach before too many years.

■ Chapter Summary

Pluto, the solar system's smallest planet and the only one with an icy surface, was discovered by Tombaugh in 1930. A satellite, Charon, was discovered by Christy in 1978. Pluto has the most unusual orbit of any planet and spends part of its orbit closer to the Sun than Neptune.

Because Pluto is ice-covered, some of its methane surface material evaporates when the planet is near perihelion, giving it a temporary atmosphere. Although details of Pluto's interior are uncertain, its density is about 2 grams per cubic centimeter and it is probably 68% to 80% rock.

Although the reason for Pluto's unusual orbit is unknown, some scenarios suggest that Pluto and Charon may have been satellites of Neptune at one time until the system was disrupted by another body. Although there have been predictions of planets beyond Pluto, there is not yet any definite evidence that additional planets exist.

■ Chapter Vocabulary

blink comparator	Christy, James	Lowell, Percival	Tombaugh, Clyde
Charon			

■ Review Questions

1. In what ways is Pluto unusual?
2. Describe the events leading to the discovery of Pluto.
3. Why is Pluto so hard to observe telescopically?
4. Summarize what is known about Pluto's basic physical properties.
5. Of what is Pluto's surface probably composed? Why are Pluto's surface and atmosphere related?
6. Describe the basic properties of Charon.
7. What is Pluto's density? What does its value tell us about the planet's interior?
8. Describe how the Pluto and Neptune systems may be related.
9. Explain what exists beyond the orbit of Pluto.

■ For Further Reading

Tombaugh, Clyde W. and Patrick Moore. *Out of the Darkness: The Planet Pluto.* Stackpole Books, Harrisburg, PA, 1980.

Whyte, A.J. *The Planet Pluto.* Pergamon Press, Toronto, 1980.

26

Saturn, as photographed by Voyager 2. Three satellites are visible at the bottom, and the shadow of one of the satellites is visible on the planet. (NASA/ JPL Photograph.)

Voyager 2 photograph of Uranus showing the planet in a crescent phase. (NASA/JPL Photograph.)

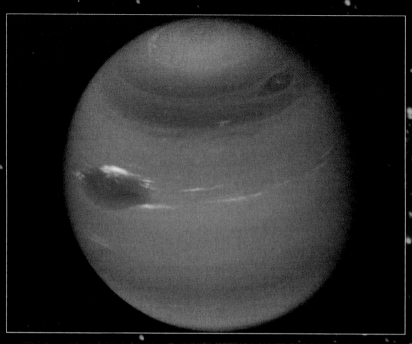

Voyager 2 photograph of Neptune. The Great Dark Spot is visible near the left edge of the planet. (NASA/JPL Photograph.)

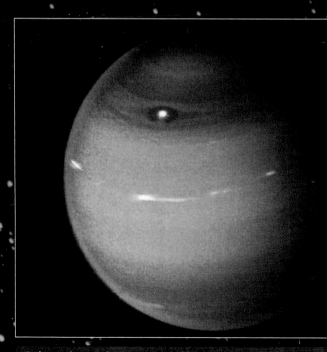

Color-enhanced Voyager 2 photograph of Neptune that shows its atmospheric haze, which appears red in this image. (NASA/JPL Photograph.)

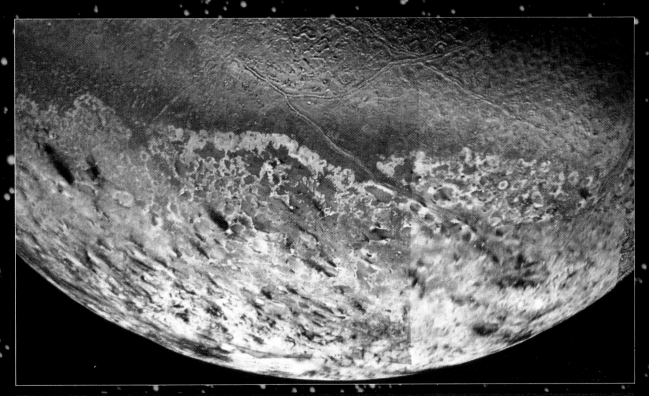

Mosaic of Voyager 2 photographs of Triton. Triton's polar cap is visible at bottom. (NASA/JPL Photograph.)

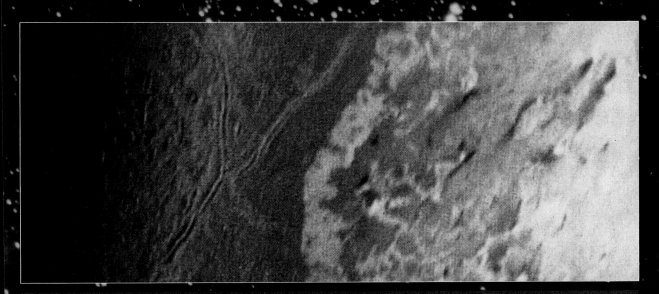

Voyager 2 photograph of Triton that shows its various terrain types. (NASA/JPL Photograph.)

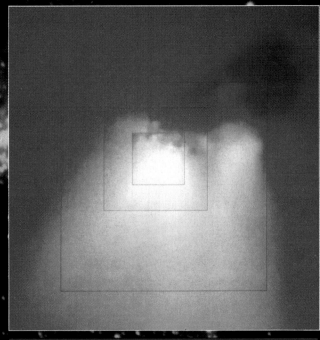

Color image of the nucleus of Halley's comet, obtained by the Halley Multicolor Camera on board the European Space Agency's Giotto spacecraft. This is a composite of four images, each of which was taken at a different distance from the nucleus and thus has different resolution. The three boxes on the photograph mark the boundaries of the four images. (Courtesy of H. U. Keller, Max-Planck-Institut fur Aeronomie, Lindau/ Harz, West Germany. Copyright 1986 by Max-Planck-Institut fur Aeronomie.)

Comet West, as photographed on March 8, 1976. (Photograph by George East, courtesy of Dennis Milon.)

Photograph of a polished surface of the Allende carbonaceous chondrite. Note the white inclusions, which contain mineral grains that crystallized early in the solar system's history. (Photograph by the author.)

Photograph of a polished surface of the Steinbach stony–iron meteorite. Note the Widmanstatten pattern visible in the metallic areas of the surface. (Photograph by the author.)

Comets

CHAPTER OBJECTIVES

After studying this chapter, you should be able to

1. Understand how it was learned that comets are Sun-orbiting members of the solar system.
2. Summarize the highlights of spacecraft exploration of comets.
3. List and describe the physical parts of a comet.
4. Describe the orbits of comets and distinguish between long- and short-period comets.
5. Understand why Oort concluded that numerous comets surround our solar system.

The spectacular Comet West, photographed on March 7, 1976. (Lick Observatory Photograph.)

Comets, which are small, icy objects that orbit the Sun, generally in very eccentric orbits, remain one of the more mysterious groups of objects in our solar system, even though astronomers have already learned many of their secrets. A large part of the mystery of comets is their unpredictability, because new ones are constantly being discovered and old ones do not always behave as expected. The main reason that comets are so unpredictable is that the large eccentricity of most comet orbits prevents them from being observed constantly. Only when a comet is closest to the Sun can it be observed, and these often-infrequent visits sometimes bring surprising changes in comet characteristics.

The word comet comes from the Greek word kometes, which means "long-haired." A comet appears fuzzy and indistinct when seen in the sky and usually has a tail of material streaming behind it (Figure 26.1). Comet tails somewhat resemble long hair blowing in the breeze, and this accounts for the origin of the name. Although noting a resemblance to hair may be a colorful way to describe a comet, it does little to explain the actual nature of what is being seen in the sky: light reflected from material streaming away from the icy mass of comet material.

Whereas modern science seeks to understand comets, their unpredictability made them evil and unwelcome in ancient times. Although their understanding of the solar system was imperfect, ancient people understood planetary motion and knew in advance where the Sun, Moon, and planets would be located in the sky. This was not the case with the comets, which appeared without warning, were visible in the sky for weeks or months, and then vanished as suddenly as they had appeared. Comets were considered portents of doom, and the sighting of a new comet was certainly not a welcome occurrence!

■ **Figure 26.1** The appearance of a comet in the sky. This photograph of Comet Bennett, taken April 10, 1970 using a 35-millimeter camera and normal lens, illustrates how a moderately bright comet appears to the unaided eye. Note the conspicuous tail. (Photograph by the author.)

■ The Discovery of Comets

It was only a few centuries ago that the true nature of comets began to be understood. At first, many people had considered comets to be some sort of atmospheric phenomenon (like clouds or auroras), or else some astronomical phenomenon occurring between the Earth and the Moon. In 1577, **Tycho Brahe** (1546–1601) noted two facts about comets that were the first clues to their true nature. Tycho first realized that the motion of a comet through the sky is much too slow for an object closer to the Earth than the Moon is. He also noted that comets did not exhibit parallax when viewed from different points on the Earth. In other words, all terrestrial observers see a comet at the same point with respect to the background stars. Because even the Moon shows parallax, Tycho concluded that comets are independent of the Earth–Moon system.

In 1705, **Edmond Halley** (1656–1742), the British Astronomer Royal, made an important advancement toward the understanding of comets. Halley had encouraged Newton to publish his theories of motion and gravity, and he believed that comets, like planets, had orbits that could be explained using the laws of Kepler and Newton. Halley studied the past appearances of bright comets and searched for ones that might have been periodic. He proposed that the bright comets seen in 1531, 1607, and 1682 were all the same object. Noting the 75 or 76 years between these appearances, Halley speculated that the comet would return in 1758. Although Halley died before he could see whether he was correct, the comet was seen by an amateur astronomer on Christmas night of 1758 and was named **Halley's comet** in his honor. The same comet also returned in 1835, 1910, and 1986, and will next return in 2061. Halley's work proved that comets, like planets, are objects that orbit the Sun. However, comets typically have orbits that are much more eccentric and highly inclined than those of any planets.

After the time of Halley, comet hunting became a very fashionable pursuit among astronomers, many of whom scanned the skies in an attempt to be the first to see an approaching comet. After a new comet was found, its orbital elements could be calculated and future positions determined. Although this was a difficult task in the 1700s, anyone with a personal computer can do it today. One of the favorite activities of many amateur astronomers today is spending long hours at the telescope searching for new comets, because anyone fortunate enough to find a comet has the honor of having it named for himself. (Comets, following the tradition begun with Halley's comet, are the only solar system objects named for their discoverers.) Almost all comet discoveries are made telescopically, because there are so many amateurs looking for comets that one hardly ever remains undetected until it gets bright enough to be seen with the naked eye. Most comets are now discovered by amateurs because professionals are too busy observing known objects to spend time specifically searching for comets. However, professionals occasionally discover comets accidentally while photographing other objects.

On the average, between 15 and 30 comets are discovered each year. About half of these are comets that have never been seen before, whereas the rest are periodic comets that have been seen previously but are being "recovered," that is, seen for the first time on their current approach to the Sun. Halley's comet, for example, is called P/Halley (the "P" signifies "periodic") and is not renamed each time someone first sees it during each appearance.

Someone looking for new comets tries to find objects with "fuzzy" appearances as he or she looks through the telescope. The difficulty in doing this is that the sky is filled with fuzzy-looking star clusters, nebulae, and galaxies. When a suspicious object is found, the observer checks detailed star atlases to see if it is something other than a comet. If the object is uncharted, it is watched for movement over a few hours. If motion is seen, indicating that the object is a comet, the discoverer notifies one of the observatories that serves as a clearinghouse for such discoveries, and the discovery is then announced. As many as three co-discoverers may have their names attached to the comet. For example, Comet Ikeya–Seki was discovered independently by two people in 1965. Comet IRAS–Araki–Alcock was discovered by the IRAS spacecraft in 1983, as well as by two earth-based observers.

In addition to its name, a comet receives two other designations. One is a provisional designation based upon the order of its discovery in a given year: the first comet discovered in 1982 was 1982a, the second was 1982b, and so on. For example, Halley's comet, which was recovered October 16, 1982, was the ninth comet found that year and was designated 1982i. A second designation is the comet's permanent one, which is based upon the order in which comets pass perihelion in a given year: the first comet that reached perihelion in 1982 was 1982I, the second 1982II, and so on. Because a comet might not pass perihelion until long after its discovery, the years of discovery and perihelion need not be the same, as was the case with P/Halley, which reached perihelion on February 9, 1986 and was designated 1986III.

■ **Figure 26.2** Comet Finsler, photographed in 1937. Note the coma and the tail, which is not yet well developed. (Lick Observatory Photograph.)

■ Physical Properties and Spacecraft Exploration

Because comets are such small, icy bodies, they are undoubtedly the least substantial members of the solar system. However, their unusual orbits cause marked changes in their physical properties, and a comet's appearance changes dramatically depending upon its distance from the Sun. When a comet is far from the Sun, it is a cold, dark, icy mass at most a few tens of kilometers across and totally invisible from the Earth. As the comet's orbit brings it closer to the Sun, the Sun's heat begins to vaporize some of the comet's ices and release grains of dust trapped within these ices. The material produced in this way forms a cloud, called a coma, that surrounds the comet's small, solid nucleus (Figure 26.2). The coma serves as an efficient reflector for sunlight and also generates light by emission, and the comet becomes visible from Earth. Finally, the solar wind and the radiation pressure of the Sun's light push a stream of gas and dust away from the comet, forming a feature called a tail (see Figure 26.1). Comet tails generally begin to develop when the comet crosses the orbit of Mars.

During the 1980s, spacecraft missions to two comets, Giacobini–Zinner and Halley, provided much knowledge about the physical nature of comets. These spacecraft verified that many of our earlier theories about comets were correct and also provided new information upon which additional theories will be based.

The first comet mission was the International Comet Explorer (ICE), which flew through the tail of Comet Giacobini–Zinner on September 11, 1985. International Comet Explorer was originally launched into Earth orbit by the United States in 1978 as the International Sun–Earth Explorer 3 (ISEE-3), and it investigated conditions in the solar wind and the Earth's magnetosphere. In the early 1980s, mission controllers realized that the ISEE-3 could be sent toward Comet Giacobini–Zinner by using gravitational boosts from encounters with the Moon. These maneuvers were successfully accomplished, and the renamed spacecraft passed through Comet Giacobini–Zinner's tail 7860 kilometers (4880 miles) behind its nucleus. International Comet Explorer studied the electric and magnetic fields associated with the comet's tail but was not equipped to return images.

The long-awaited 1986 return of Halley's comet provided an opportunity for investigation of history's most famous comet, and a veritable armada of spacecraft was launched toward it between late 1984 and the middle of 1985. These craft all encountered Halley's comet during March, 1986 (Figure 26.3). The Soviet Union sent two spacecraft, Vega 1 and 2, toward Halley to pass through its coma and photograph its nucleus. Recall from Chapter 13 that the Vegas passed Venus on their way to the comet and sent atmospheric and lander craft toward that planet. Vega 1 passed through Halley's coma 8892 kilometers (5523 miles) from the nucleus on March 6, and Vega 2 passed at a distance of 8034 kilometers (4990 miles) on March 9. Both obtained good photographs of the nucleus. The European Space Agency's Giotto spacecraft passed only 541 kilometers (336 miles) from the nucleus on March 14 and obtained the best photographs of the nucleus. Meanwhile, two Japanese spacecraft explored the comet from greater distances, flying past it on the sunward side. Suisei, designed to study the hydrogen halo surrounding the comet, flew by the comet at a distance of 151,000 kilometers (94,000 miles) on March 8, whereas Sakigake, designed to investigate solar wind conditions "upwind" of the comet, encountered

■ Figure 26.3 Paths of the spacecraft that explored Halley's comet. The distance scale, which is in kilometers, is not linear but logarithmic. It measures the distance from the comet's nucleus. (European Space Agency Photograph, courtesy of *ESA Bulletin* and R. Reinhard.)

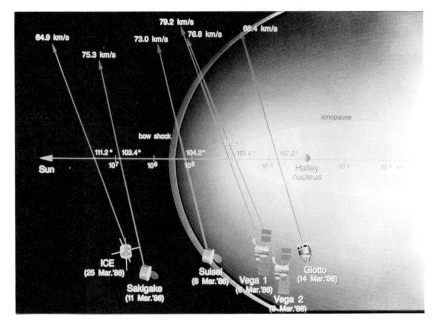

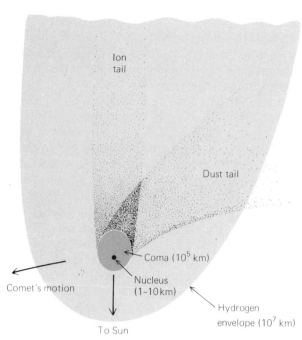

Ion
tail

Dust tail

Coma (10^5 km)

Nucleus
(1–10 km)

Comet's motion

Hydrogen
envelope (10^7 km)

To Sun

■ **Figure 26.4** The parts of a comet. The nucleus and coma are collectively known as the head and are surrounded by the hydrogen halo. Extending away from the comet are its straight, type I (ion) tail and the curved, type II (dust) tail. The comet's direction of motion and the direction toward the Sun are indicated. (From Abell, George, David Morrison, and Sidney Wolff. *Exploration of the Universe,* 5th ed. Saunders College Publishing, Philadelphia, 1987.)

Halley at a distance of 7.1 million kilometers (4.4 million miles) on March 11. On March 28, ICE encountered Halley's comet at a point 32 million kilometers (20 million miles) ahead of it, measuring electric and magnetic field conditions.

Let us examine the parts of a comet in more detail (Figure 26.4). The only substantial part of a comet is its solid **nucleus.** Prior to the mid-20th century, most astronomers believed that comet nuclei were aggregates of sand and rock. In the 1950s, however, **Fred Whipple** proposed that comet nuclei were instead composed primarily of various ices, with rocky and dusty material mixed in. Whipple's idea, usually called the **dirty snowball model**, gained rapid acceptance and was verified when spacecraft observed Halley's comet in 1986.

Prior to these spacecraft observations, comet nuclei had escaped direct observation be-

cause they are totally shielded by comas. However, spectroscopic studies of comas showed the presence of water vapor, methane, and ammonia, so these substances, in solid form, were considered likely prospects for nuclear material. Complex compounds such as **hydrates** (minerals containing water in their crystal structure) or **clathrates** (substances in which water is physically trapped in voids) were also suggested as constituents of comet nuclei. The Giotto spacecraft detected the polymer polyoxymethylene in Halley's coma. **Polymers** are chains of linked molecules, and polyoxymethylene is composed of linked formaldehyde molecules. This polymer is dark in color and could account for the fact that comet nuclei have low albedos.

Observations by the Giotto and Vega spacecraft revealed that the nucleus of Halley's comet was elongated, with dimensions of about 8 by 11 kilometers (5 by 7 miles). The surface was very dark, with an albedo of only 5% (see Figure 26.5). The nucleus was seen to contain some bright spots, which mark the locations of jets of vapor streaming from the nucleus. Most of the streaming gas was water vapor, and production reached 27,000 kilograms (30 tons) per second at some points (Figure 26.6)! (Such gas jets from comet nuclei act like rocket engines and can cause the comet's orbital elements to change slightly.) In addition to gas, dust particles also stream away from the nucleus. The overall density of Halley's

■ **Figure 26.5** The nucleus of Halley's comet as photographed by the Vega 2 spacecraft. (Courtesy of the U.S.S.R. Academy of Sciences.)

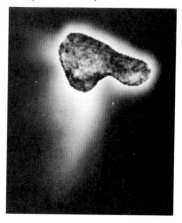

■ **Figure 26.6** The nucleus of Halley's comet as photographed by the Halley Multicolor Camera on board the European Space Agency's Giotto spacecraft. Note the bright jets of material streaming away from the nucleus. (Courtesy of H.U. Keller, Max-Planck-Institut fur Aeronomie, Lindau/Harz, West Germany. Copyright 1986 by Max-Planck-Institut fur Aeronomie.)

nucleus is estimated to be only one-tenth to one-fourth that of water ice. The nucleus rotates with a period of 7.4 days, and its axis precesses every 2.2 days, although some researchers feel that these figures might be reversed, and others have obtained different values.

Because the nucleus is the source of material for the coma and tail, it grows smaller on each trip through the inner solar system. Eventually, perhaps after hundreds of passes, the volatile materials are used up and the comet "dies." Several comets have been observed to split into two or more parts, presumably as the result of the breakup of the nucleus. Some comets have been disappointing in that they have not achieved the brightness predicted for them. Many researchers feel that these are comets making their first trip into the inner solar system and that their surfaces are covered by a hard crust of some sort.

Surrounding the nucleus of a comet and derived from it is the **coma**, which forms a cloud of gas as large as hundreds or thousands of kilometers across. A coma generally begins to appear once a comet approaches to within about 3 astro-nomical units of the Sun and reaches its maximum diameter between 1.5 and 2 astronomical units from the Sun. Closer than that, solar ultraviolet radiation begins to dissociate the molecules, making the coma less spectacular. Spectroscopic examination of comas reveals that most of the molecules present are neutral, rather than ionized.

The nucleus and coma together are called the **head** of the comet. Surrounding the coma is a large, invisible hydrogen cloud, up to hundreds of thousands of kilometers across. This hydrogen **halo** is believed to contain remnants of gas molecules from the comet that were dissociated into their various elements by sunlight. The immense diameter of the halo may be appreciated using the following analogy: If the nucleus of Halley's comet were the size of a peanut, the coma would be 270 meters (890 feet) across and the halo would be 33 kilometers (20 miles) across.

The most spectacular portion of a visible comet is its **tail**, which may stretch up to 30° across the sky and extend for millions of kilometers through space. The tail is formed by material pushed away from the coma by the Sun's radiation pressure and the solar wind. For this reason, the tail always points away from the Sun and may actually precede the comet in its journey through space, depending upon the comet's position in its orbit (Figure 26.7). Tails are composed of two types of material, and a comet may have one or both types (Figure 26.8). A **type I tail** contains ionized gases and appears bluish in color, producing light mainly by the emission process. This tail is composed mainly of CO^+, N_2^+, and CO_2^+ and points almost directly away from the Sun, because the solar wind acts on its charged particles. The **type II tail** is composed of dust, presumably particles of very fine silicates about 10^{-6} meters in size that are derived from the comet's nucleus. Type II tails shine by reflecting sunlight and are therefore generally yellowish in color. Type II tails are curved because radiation pressure pushes the solid grains into higher Keplerian orbits, with lower velocities and longer periods. In these new orbits, they lag behind the head of the comet. This is in contrast to the ions in the type I tails, which are low enough in mass that gravitational forces on them are negligible compared to the solar wind.

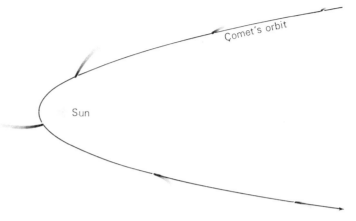

■ **Figure 26.7** Orientation of a comet's tail. Note that the tail develops fully only when the comet is close to the Sun, and that it always points away from the Sun. (From Abell, George, David Morrison, and Sidney Wolff. *Exploration of the Universe,* 5th ed. Saunders College Publishing, Philadelphia, 1987.)

Several unusual phenomena involving comet tails occur. One is a disconnection event, during which the type I tail breaks free from the comet, followed by the formation of another one (Figure 26.9). Spacecraft observations have revealed that these events occur when the comet crosses a boundary over which the magnetic polarity of the solar wind reverses. Another unusual tail feature is an anti-tail, which appears to be pointing in the opposite direction of the normal tail (Figure 26.10). This is generally a result of our point of view from the Earth and the alignment of the comet's tails.

Perhaps the most interesting fact about comet tails is that the Earth has passed through them on several occasions, the most noteworthy passage being through the tail of Halley's comet in 1910. Announcements by astronomers that comet tails contain poisonous gases such as cyanide caused panic in many people, but tails are so tenuous that they have no effect on the Earth's surface. In fact, the only effect of Earth's passage

■ **Figure 26.8** Comet Mrkos, photographed in 1957. Note that both tails are clearly visible: the straight tail of ionized gas and the curved tail of dust. (Palomar Observatory Photographs.)

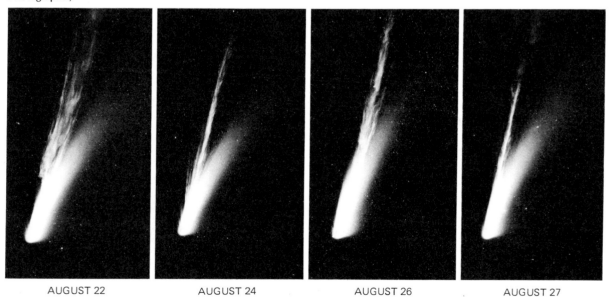

| AUGUST 22 | AUGUST 24 | AUGUST 26 | AUGUST 27 |

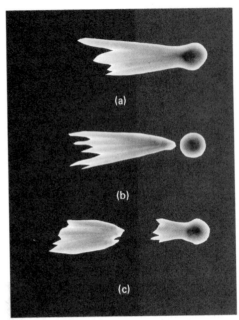

■ **Figure 26.9** Appearance of a comet's ion tail during a disconnection event. (a) Normal appearance of comet and tail. (b) Tail disconnects. (c) As the old tail dissipates, a new one begins to form.

through Halley's tail in 1910 was that many entrepreneurs made fortunes selling "comet pills" to the gullible for protection!

If a comet nucleus hit the Earth, the results could be far more serious, as even an icy object could do significant damage. Although there is no conclusive record of such a cometary impact upon Earth, it is believed that such an incident occurred on June 30, 1908, when a tremendous explosion rocked the Tunguska area of Siberia. Although it was assumed that a giant meteorite had hit the area, later expeditions found no crater, which such an impact should have caused. Despite the lack of a crater, there was definite evidence of great energy release, as trees were flattened over a large area (Figure 26.11). One modern explanation for this event is that an undiscovered comet entered the Earth's atmosphere but vaporized before it could hit the surface. The damage during the **Tunguska event** was caused by the atmospheric shock wave that resulted, even though the object did not survive long enough to strike the surface. Calculations suggest that the Tunguska object was 90 meters (300 feet) across and weighed 10^8 kilograms (10^5

tons). (It is not unreasonable that a comet could fail to be detected in such a way. On August 30, 1979, the Solwind/P78-1 satellite, which studied the Sun from Earth orbit, photographed a comet actually hitting the Sun. The satellite subsequently noted five more such impacts, and other spacecraft have made similar discoveries. None of these comets had been seen by Earth-based observers.)

As comets break down by heating and are dispersed by the tail-producing solar wind and radiation pressure, vast amounts of debris are emplaced in the solar system. Large amounts of this dusty material have accumulated in the ecliptic plane, and sunlight reflecting from it appears as a faint light above the horizon before sunrise or after sunset, a phenomenon called the **zodiacal light** (Figure 26.12). Orbits of comets have concentrations of material strewn along them, and the passage of the Earth through this material results in meteor showers, as will be described in Chapter 28.

■ **Figure 26.10** Comet Arend–Roland, photographed in 1957. Both the normal tail, pointing upward, and the anti-tail, pointing downward, are visible. (Palomar Observatory Photograph.)

■ **Figure 26.11** Damage to the Tunguska region of Siberia, which occurred on June 30, 1908. The cause of this devastation was probably the entry of a comet nucleus into the atmosphere. (Smithsonian Institution Photograph, courtesy of Brian Mason.)

■ Orbits

Comets generally have the most unusual orbits of any solar system objects. As we shall see, comets seem to originate far beyond the orbit of Pluto, and because many approach the Sun closer than Mercury, they typically have high-eccentricity orbits. The eccentricity of Halley's comet, for example, is 0.967, and its perihelion and aphelion distances from the Sun are 0.59 and 35 astronomical units, respectively. Comet orbits are not limited to the ecliptic plane, and many have high inclinations. (Halley's comet has an inclination of 162°. Recall that inclinations greater than 90° in-

■ **Figure 26.12** The zodiacal light, as revealed by a time-exposure photograph. This photograph, taken with a wide-field camera, shows nearly the entire sky. The zodiacal light extends upward from the horizon at lower left. The three lines across the photograph were produced by the camera, whereas the dark object at lower right is an observatory dome. (Yerkes Observatory Photograph.)

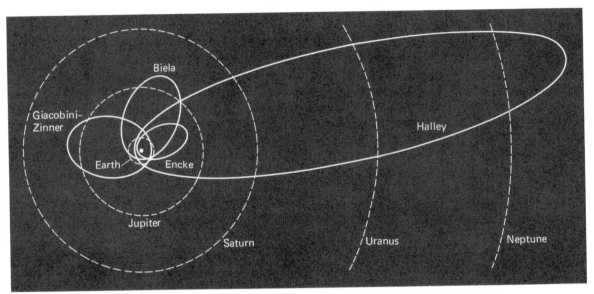

■ Figure 26.13   Orbits of several short-period comets.

dicate retrograde orbital motion.) The orbits of several comets are shown in Figure 26.13.

When a comet is discovered, it must be observed for as long as possible to determine its precise orbital elements. Unfortunately, comets generally can be observed only when they are relatively close to the Sun, a time during which a comet typically covers only a small section of its orbit. Accordingly, there is much uncertainty in the orbital elements of many comets, especially those with large semi-major axes. Another complication is that comets, being relatively small and lightweight, are susceptible to perturbations resulting from interactions with planets, especially the gas giants. As mentioned earlier, even the jets of gas streaming from comet nuclei can serve to cause orbital changes. Halley's comet reached perihelion in 1986 four days later than predicted because of this effect.

Comets are divided into two major groups depending upon their orbital periods. The first group, **long-period comets**, contains those with orbital periods longer than two centuries. Some of these comets have periods as long as 10^5 or 10^6 years, with eccentric orbits carrying them as far as 50,000 astronomical units from the Sun. Because we see only a small portion of their orbits when they are near perihelion, it is almost impossible to tell whether their orbits are elliptical or parabolic. Although many orbits appear parabolic, this would mean that these objects are not gravitationally bound to the solar system, which is unlikely. Table 26.1 lists the orbital elements of several notable long-period comets. A well-known example of a long-period comet was Kohoutek, discovered in 1973. It was discovered while still quite far from perihelion, and astronomers speculated that it would become bright enough to be seen even during the daytime. However, the comet was much fainter than expected, leading researchers to conclude that it was making its first trip into the inner solar system.

The second group of comets, **short-period comets**, contains those with periods shorter than two centuries. These comets are believed to have once been long-period comets whose orbits were changed by perturbations from one of the gas giant planets. Each of the gas giants has a "family" of comets whose aphelia lie in its general vicinity. For example, Halley's comet has its aphelion just outside Neptune's orbit, indicating that it was probably perturbed by that planet. Some outer planet satellites, especially Jupiter's retrograde ones, are probably captured comet

■ Table 26.1   Representative Long-Period Comets

Name	Semi-Major Axis (AU)	Period (yr)	Eccentricity	Inclination (°)	Perihelion Distance (AU)	Only Observed Perihelion
Ikeya – Seki	92	880	0.999	141.85	0.008	1965
Kohoutek	?	?	1.000	14.31	0.142	1973
West	?	?	0.999	43.07	0.196	1976
Tago – Sato – Kosaka	?	?	0.999	75.82	0.472	1969
Bennett	?	?	0.996	90.04	0.538	1970
Thiele	70	583	0.982	56.39	1.213	1906
Humason	204	2900	0.990	153.28	2.13	1962

Data from Marsden, Brian G. *Catalogue of Cometary Orbits*, 3rd ed. Smithsonian Astrophysical Observatory, Cambridge, MA, 1979.

nuclei. As discussed in Chapter 17, Comet Oterma barely escaped temporary capture by Jupiter in the 1930s, and a satellite of Jupiter that was "discovered" in 1975, but never seen again, was probably a temporarily captured comet.

Because a comet moving in a prograde orbit would have a greater probability of encountering a planet and having its orbit perturbed by the planet, the majority of short-period comets have prograde orbits. Table 26.2 lists the orbital elements of several notable short-period comets. Periods of these comets range from a low of 3.3 years for P/Encke to 155 years for P/Herschel–Rigollet. Halley's comet, whose earliest known visit was recorded by the Chinese in 240 B.C., is certainly the best-known comet. Although its 1910 visit was bright and remarkable, orbital configurations were such that the comet was never

very close to Earth or bright during 1985 and 1986, and most people, expecting a magnificent sight, were disappointed (Figures 26.14 and 26.15). Unfortunately, few of them will get a second chance to see the comet, as it will not return until 2061!

Some long-period comets belong to a family known as **Sun-grazing comets**, whose perihelion distance is only 0.005 to 0.007 astronomical units. These comets often break apart as a result of the gravitational strain of passing so close to the Sun. Spacecraft observations show that many of the comets that do this escape detection by Earth-based observers. As mentioned earlier, some of them have been seen to fall into the Sun. The smallest perihelion distance of any short-period comet is that of P/Machholz, which is 0.13 astronomical units.

■ Table 26.2   Representative Short-Period Comets

Name	Semi-Major Axis (AU)	Period (yr)	Eccentricity	Inclination (°)	Perihelion Distance (AU)	First Observed Perihelion
Encke	2.21	3.31	0.846	11.94	0.34	1786
Giacobini – Zinner	3.48	6.52	0.715	31.71	0.99	1900
Biela	3.52	6.62	0.756	12.55	0.86	1772
Tuttle	5.74	13.8	0.822	54.37	1.02	1790
Tempel – Tuttle	10.27	32.9	0.904	162.71	0.98	1366
Halley	17.78	76.1	0.967	162.21	0.59	240 B.C.
Herschel – Rigollet	28.84	155	0.974	64.20	0.75	1788

Data from Marsden, Brian G. *Catalogue of Cometary Orbits*, 3rd ed. Smithsonian Astrophysical Observatory, Cambridge, MA, 1979.

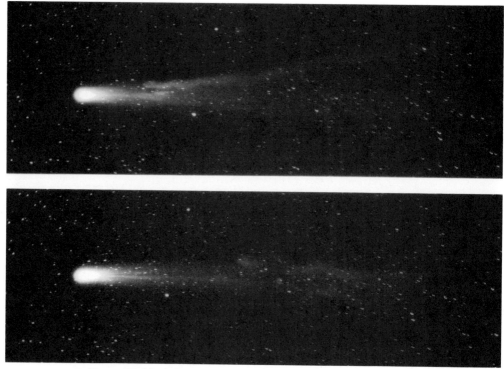

■ **Figure 26.14** Halley's comet, as photographed on June 6 and 7, 1910. (Lick Observatory Photographs.)

■ **Figure 26.15** Halley's comet, as photographed on February 22, 1986 from the United Kingdom Schmidt telescope in Australia. (Copyright © Royal Observatory, Edinburgh. Used by permission.)

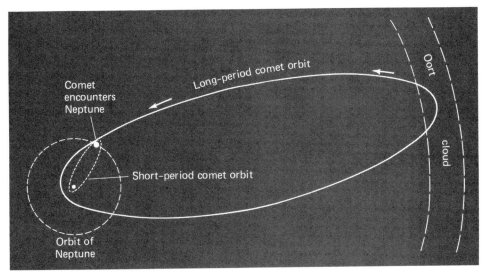

■ **Figure 26.16**   Change of comet orbit. A long-period comet encounters a gas giant planet such as Neptune. This causes its orbit to change, making it a short-period comet.

■ The Oort Cloud

In the early 1950s, **Jan Oort** investigated the problem of comet origins. He made a careful statistical analysis of the orbits of long-period comets and concluded that they comprise a vast swarm of comets orbiting the Sun at a distance ranging from 20,000 to 100,000 astronomical units, with the greatest concentration at about 50,000 astronomical units. He estimated that this swarm, now called the **Oort cloud**, consists of about 200 million comets, with a total combined mass only about one-tenth that of the Earth. Oort noted that the orbital inclinations of long-period comets are totally random. Therefore, he concluded that unlike the inner solar system, where orbits are generally restricted to the ecliptic plane, the Oort cloud is probably spherically symmetrical, totally surrounding the solar system. This distribution is probably a result either of the way in which comets were emplaced into the Oort cloud or of gravitational interactions from passing stars.

Presumably, each Oort cloud comet leisurely orbits the Sun at this vast distance until it interacts gravitationally with another member of the cloud. Such interactions, which cause orbital perturbations, may also involve nearby stars and the clouds of comets that probably surround them. Perturbations occurring within the Oort cloud can send a comet toward the inner solar system, where it will either pass through undisturbed in a long-period orbit or else be further perturbed into a short-period orbit (Figure 26.16). Although this model is generally accepted, Martin Duncan, Thomas Quinn, and Scott Tremaine have calculated that short-period comets are derived from a "cloud" just beyond the orbit of Neptune. These researchers believe that this cloud is disk-shaped, not spherical, and that a total mass of comets equivalent to 2% that of Earth's mass could supply the observed quantity of short-period comets. Unfortunately, comets at this distance, as well as those in the Oort cloud, are invisible to terrestrial observers, preventing direct verification of the existence of either comet swarm.

One of the more interesting theories about the Oort cloud involves periodic disturbances within it. It has recently been proposed that periodic events such as the passage of the Sun through the galactic plane as it orbits the center of the Milky Way or the close approach of a faint, as-yet-undiscovered solar companion star could cause severe disruptions of the Oort cloud. Such disruptions would send an inordinately large number of comets toward the Sun in short periods of time. The increased bombardment of the Earth at such times has been suggested as a mechanism for producing disruptions of the environment that led to the mass extinction of various terrestrial organisms, such as the dinosaurs. The fact that the fossil record seems to indicate a long-term periodicity of extinction events may be evidence for a corresponding periodicity in the major disruptions of the Oort cloud.

One problem about the Oort cloud that theorists face is explaining how the comets in it came to be located there in the first place. It is considered unlikely that they formed where they are presently located, because current theories of solar system formation indicate that there would have been insufficient material present to form comets at such an extreme distance from the Sun. Most astronomers currently believe that the comets formed in the cold outer solar system about where Uranus and Neptune are presently located. After they were formed, the comets

were subsequently ejected from this area by gravitational interactions with Uranus and Neptune, forming the Oort cloud. (Some of the comets may even have been given enough energy that they completely escaped from the solar system.) It is possible to predict, using mathematical models, the paths taken by comets after such gravitational encounters. These predictions indicate that comets would have been placed into orbits with all possible inclinations, not just those limited to the ecliptic plane. The comet population proposed by Duncan, Quinn, and Tremaine may be some of the original comets that were not ejected from their region of formation.

Since their formation, comets in the Oort cloud have been subjected to numerous collisions with each other. They have also collided with gas and dust as the Sun orbits the center of the galaxy. Alan Stern has estimated that a typical Oort cloud comet has lost 8 centimeters (3 inches) of material from its surface in this way since the formation of the solar system. The temperature in the Oort cloud is estimated to be only about 10 K (−440°F).

■ Figure 26.17 Atmospheric "holes" that have been proposed as resulting from the impact of small comets. This image was obtained by the Dynamics Explorer 1 satellite in 1981 using ultraviolet light. The inset at left shows an enlargement of one of the "holes." (Courtesy of Louis Frank.)

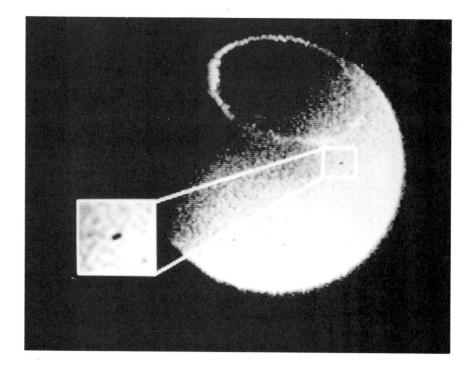

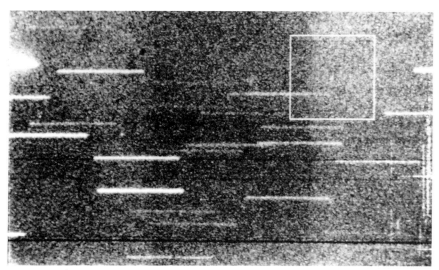

■ **Figure 26.18**   Photograph that possibly shows a "mini-comet" outside of the Earth's atmosphere. The horizontal lines are the trailed images of stars, whereas the box at upper right encloses a point-like image, probably produced by a "mini-comet." (Photograph courtesy of Clayne Yeates. From Yeates, Clayne. "Initial Findings From a Telescopic Search for Small Comets Near Earth." In *Planetary and Space Science*, vol. 37, p. 1185. Copyright 1989 by Pergamon Press, Oxford.)

■ Prospects for Future Knowledge

It is obvious from the discussion of the Oort cloud that many questions about comets remain unanswered. One of the most controversial new theories about comets was proposed in 1986 by Louis Frank, who suggested that Earth and the other terrestrial planets are being constantly bombarded by small (90,000-kilogram, or 100-ton) comets that vaporize in the upper atmosphere. Frank based his theory on the fact that ultraviolet images obtained from space of the Earth's upper atmosphere show "holes," which he suggested were produced by water vapor released by the comets (Figure 26.17). Shortly after Frank proposed his theory, Clayne Yeates attempted to photograph some of the "mini-comets" before they entered the atmosphere. Yeates obtained images of objects that appear to be the objects Frank predicted (Figure 26.18), but many researchers are unconvinced of the theory's validity, and the entire topic remains quite controversial.

The best prospect for improved understanding of comets is undoubtedly additional space probes to the vicinity of comets. The United States plans to launch the Comet Rendezvous Asteroid Flyby (CRAF) mission in 1995. After a flyby past asteroid Hamburga, CRAF will rendezvous with short-period Comet Kopff in 2000 and go into orbit around its nucleus. The mission will continue for two to five years, during which time CRAF will drop a penetrator probe to study the comet's nuclear material directly.

■ Chapter Summary

Comets, long considered mysterious, evil omens, were shown by Halley to be members of the solar system that orbit the Sun in unusual orbits. Although numerous comets have been discovered, many new ones are found annually, and "comet hunting" is enjoyed by many amateur astronomers. Several spacecraft have explored comets Halley and Giacobini–Zinner.

Comets are fairly small, icy objects. A comet's nucleus consists of ices and some rocky material. In the inner solar system, some of the ice vaporizes, producing a gaseous coma. The Sun's radiation pressure and the solar wind push some coma material outward, producing a comet's tail.

Comet orbits are generally eccentric and highly inclined. Long-period comets have periods over two centuries long; short-period comets have periods less than this. Short-period comets are believed to have once been long-period comets whose orbits were changed by gravitational perturbations.

Although they cannot be seen, there is indirect evidence that a vast swarm of comets, the Oort cloud, is located far beyond Pluto. This cloud is the source of the comets that visit the inner solar system. Current theories suggest that comets formed in the cold, outer solar system near Uranus and Neptune, then were ejected into the Oort cloud by gravitational interactions with these gas giants.

■ Chapter Vocabulary

clathrates	head	polymer	Tycho
coma	hydrates	short-period comet	type I tail
dirty snowball model	long-period comet	Sun-grazing comets	type II tail
Halley, Edmond	nucleus	tail	Whipple, Fred
Halley's comet	Oort, Jan	Tunguska event	zodiacal light
halo	Oort cloud		

■ Review Questions

1. How did the ancients feel about comets?
2. Explain how Halley helped increase early understanding of comets.
3. How are most comets discovered today, and what group of people makes the most discoveries?
4. Describe the appearance of a comet in the sky.
5. How does a comet change during its orbit as it approaches the Sun?
6. List and describe the basic physical parts of a comet.
7. Of what is a comet composed? How are comet compositions studied?
8. What would be the result of the collision of a comet tail or nucleus with the Earth?
9. What is the zodiacal light, and how is it related to comets?
10. Compare the orbits of long- and short-period comets.
11. How is a long-period comet changed into a short-period comet?
12. Why is the Oort cloud of comets believed to exist?
13. Explain how the Oort cloud is believed to have formed.

■ For Further Reading

Brandt, John C. *Comets.* Freeman, San Francisco, 1981.

Brandt, John C. and Robert C. Chapman. *Introduction to Comets.* Cambridge University Press, Cambridge, England, 1982.

Brandt, John C. and Malcolm B. Nieder, Jr. "The Structure of Comet Tails." In *Scientific American,* January, 1986.

Calder, Nigel. *The Comet Is Coming.* Viking, New York, 1981.

Chapman, Robert C. and John C. Brandt. *The Comet Book.* Jones and Bartlett, Boston, MA, 1984.

Kronk, Gary W. *Comets: A Descriptive Catalog.* Enslow, Hillside, NJ, 1984.

Seargent, David. *Comets: Vagabonds of Space.* Doubleday, New York, 1982.

Whipple, Fred. "The Nature of Comets." In *Scientific American,* February, 1974.

Whipple, Fred. "The Spin of Comets." In *Scientific American,* March, 1980.

Asteroids

The asteroid Eros, revealed by its motion with respect to the background stars during a 30-minute exposure. This photograph was taken on January 23, 1975, when Eros was only 14 million miles from Earth. (Photograph by George East, courtesy of Dennis Milon.)

Asteroids, which are also called planetoids or minor planets, are a group of relatively small members of the solar system. Like comets, the number of asteroids is uncountably large, but these two groups differ in composition and in orbital properties. Asteroids are composed of rock and metal, and most of them probably lack the icy materials found in comets. Most asteroids have low-inclination, low-eccentricity orbits, and the majority of them orbit the Sun in a belt occupying the space between the orbits of Mars and Jupiter. There are complications, however, as some objects appear to be intermediate between the two groups. Those asteroids orbiting in the outer solar system may contain ice, but unless they are seen to have a coma at some time, they are not considered comets.

■ Discovery

As discussed in Chapter 3, a curious numerical relationship describing the spacing of planetary orbits was developed by Titius and Bode in the 1760s. When Uranus was discovered in 1781 and found to be at the location predicted by Bode's law, astronomers began to take the relationship very seriously. (Recall that Bode's law is not a law in the true sense of the word because it is not based on any understood physical phenomenon. This is underscored by the fact that Neptune and Pluto are not located where Bode's law would predict them to be.) It was considered certain that an undiscovered planet was located between Mars and Jupiter, because Bode's law predicted a planet with a semi-major axis of 2.8 astronomical units.

In 1800, a group of German astronomers began searching for the "missing" planet. Unlike the later search for Neptune, there was no specific aim point, but various observers were assigned specific areas of the ecliptic in which to search for the object. Despite the coordinated

■ Table 27.1   The First Ten Asteroids Discovered

Number	Name	Date	Discoverer
1	Ceres	1/1/1801	Piazzi
2	Pallas	3/28/1802	Olbers
3	Juno	9/1/1804	Harding
4	Vesta	3/29/1807	Olbers
5	Astraea	12/8/1845	Hencke
6	Hebe	7/1/1847	Hencke
7	Iris	8/13/1847	Hind
8	Flora	10/18/1847	Hind
9	Metis	4/26/1848	Graham
10	Hygiea	4/12/1849	de Gasparis

search effort, the discovery was made accidentally by **Giuseppi Piazzi** (1746 – 1826) on the first night of the 19th century, January 1, 1801. Piazzi made several observations of the object, but it became "lost" by mid-year. Karl F. Gauss was able to calculate an orbit from Piazzi's observations, and the new solar system member was seen again in December of 1801. (Recall from Chapter

■ Figure 27.1   Photograph showing two asteroids. As the telescope is tracked to follow the stars, whose images are point-like, the asteroids reveal themselves by moving with respect to the background stars, yielding a linear image. (Yerkes Observatory Photograph.)

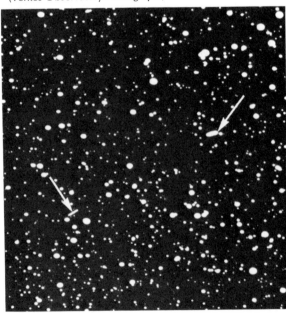

3 that Gauss developed the computational techniques for determining an orbit from three observations.) It was given the name **Ceres**, and its orbit, with a semi-major axis of 2.77 astronomical units, was considered appropriate for that of the missing planet.

When the German astronomers learned of Piazzi's discovery, some assumed Ceres was indeed the missing planet. However, others considered Ceres too small to be a true planet, so the German group continued the search. A second object, which was named Pallas, was found in 1802. Unfortunately, Pallas was even smaller than Ceres. A third object, Juno, was found in 1804; a fourth, Vesta, in 1807. (Vesta is the brightest asteroid and can sometimes be glimpsed with the unaided eye.) Because these objects have semi-major axes of 2.77, 2.67, and 2.36 astronomical units, respectively, it was soon obvious that one normal planet was not located between Mars and Jupiter; instead, several smaller objects occupied that region of space.

Although the fifth asteroid wasn't found until 1845, 300 were known by 1890. (The names and discovery dates of the first ten asteroids found are listed in Table 27.1.) As astronomical photography proliferated, the rate of discovery increased, as asteroids revealed themselves on time-exposure photographs as short streaks because of their motion relative to the background stars (Figure 27.1). Today, nearly 4000 asteroids have been observed well enough that their precise orbits can be calculated. Asteroids are named by their discoverers, and although mythological names were the rule at first, asteroids have been named for almost anything (and anyone) imaginable. In 1987, for example, seven recently discovered asteroids were named for the astronauts killed when the Space Shuttle *Challenger* was destroyed. In addition to its name, an asteroid is given a number designating its order of discovery (e.g., the 311th asteroid discovered was named Claudia, and its formal designation is 311 Claudia).

It is estimated that there are 100,000 asteroids within reach of current telescopes, but professional astronomers are too busy to spend time searching for new ones. Amateur astronomers generally try to find comets rather than asteroids because there is more prestige (and one's own name) attached to finding a comet. For this reason, most asteroids are discovered unintentionally when they appear on time-exposure photographs. Special attention has been given by some researchers, however, to discovering objects that might come very close to the Earth, and it has been proposed that serious steps be taken to provide advance warning of any asteroid on a collision course, so that preventative measures might be attempted.

■ Orbits

Mars and Jupiter are located 1.5 and 5.2 astronomical units from the Sun, respectively, but the asteroids are not ubiquitous throughout the region between them. The semi-major axes of most asteroid orbits lie between 2.1 and 3.3 astronomical units, in what is called the **main belt** (Figure 27.2). Unlike comets, most asteroids have fairly low eccentricity, low inclination orbits. Asteroid eccentricities range from 0.1 to 0.3, with an average value of 0.15. Inclinations typically range from near 0° to 30° and average 9.5°. There are no main-belt asteroids with retrograde orbits. (The orbital elements of some representative asteroids are listed in Table 27.2.)

Just as the particles in planetary ring systems have gaps as a result of orbital resonances with satellites, there are five regions of the asteroid belt with reduced asteroid populations caused by the powerful gravity of Jupiter. These regions are

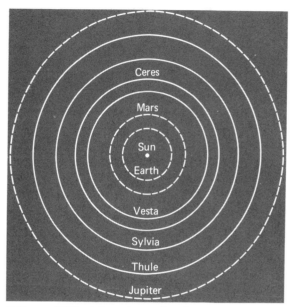

■ **Figure 27.2** Orbits of examples of main-belt asteroids.

called **Kirkwood gaps**, named for **Daniel Kirkwood** (1814–1895), who interpreted this phenomenon in 1866. Asteroids whose semi-major axes would be located at major resonance points, such as the 3:1 or 2:1 points, have been perturbed into different orbits, leaving relatively clear areas of the main belt. Jack Wisdom modeled orbital changes at the 3:1 resonance point and determined that a gap could be cleared there in only a few million years. Wisdom found that the perturbed asteroids were given high-eccentricity orbits, and many became Mars-crossing asteroids, to be described in the next section.

One unexpected effect of Jupiter's orbital resonances is that three of them cause the opposite effect, a concentration of asteroids. An example occurs at the 3:2 resonance 4 astronomical units from the Sun. The reason why some resonances produce concentrations instead of gaps is not yet understood. (A chart showing the distribution of asteroid orbits is shown in Figure 27.3.) The compositions of asteroids, which will be discussed later, vary according to their locations in the belt.

Although numerous asteroids orbit in the main belt, the space between individual objects is relatively large. This is demonstrated by the fact

■ Table 27.2   Orbital Elements of Selected Asteroids

Name	Semi-Major Axis (AU)	Period (yr)	Eccentricity	Inclination (°)
Main-Belt Asteroids				
Vesta	2.36	3.63	0.090	7.1
Juno	2.67	4.36	0.258	13.0
Ceres	2.77	4.60	0.078	10.6
Pallas	2.77	4.62	0.234	34.8
Sylvia	3.48	6.50	0.083	10.9
Thule	4.27	8.83	0.011	2.3
Planet-Crossing Asteroids				
Aten	0.97	0.95	0.183	18.9
Apollo	1.47	1.78	0.560	6.3
Amor	1.92	2.66	0.434	11.9
Trojan Asteroids				
Aneas	5.16	11.72	0.103	16.7
Hektor	5.17	11.76	0.023	18.2
Outer Solar System Asteroids				
Hidalgo	5.85	14.15	0.656	42.4
Chiron	13.64	50.39	0.380	6.9

Data from Kowal, Charles T. *Asteroids: Their Nature and Utilization.* Copyright 1988, John Wiley and Sons, Inc., New York.

■ Figure 27.3   The distribution of asteroids. Asteroid number is plotted against semi-major axis. Orbital resonance values are indicated. Note both Kirkwood gaps and resonances with increased concentrations. (From Abell, George, David Morrison, and Sidney Wolff. *Exploration of the Universe,* 5th ed. Saunders College Publishing, Philadelphia, 1987.)

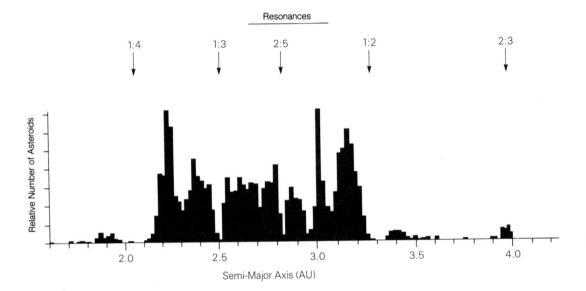

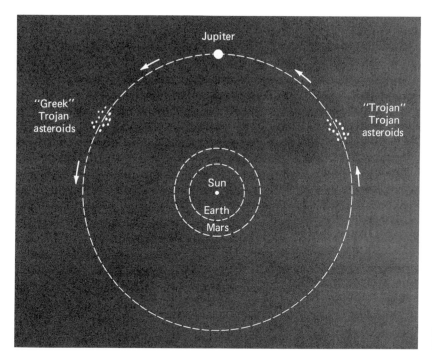

■ **Figure 27.4** The location of the Trojan asteroids.

that four space probes have traversed the belt without incident. Nevertheless, asteroidal collisions do occur, as evidenced by recovered meteorites that contain more than one type of material and bear evidence of high-pressure impact while still in space. (As we will learn in Chapter 28, meteorites are probably derived from fragments of asteroids that were placed into orbits that carried them toward collisions with the Earth.)

■ Asteroids with Unusual Orbits

Not all asteroids are main-belt objects, as several groups of asteroids are found in other regions of the solar system. One example is the group known as the **Trojan asteroids**, which are located at the two stable Lagrangian points of Jupiter's orbit (Figure 27.4). These objects normally orbit 60° from Jupiter but can wander away from that position by as much as 20°, primarily because of orbital perturbations caused by Saturn. Several dozen Trojans are currently known, but their total

number may be nearly 1000. Individual Trojan asteroids are named for characters from Homer's *Iliad,* which describes the legendary Trojan War between Greece and Troy. The leading group, which has more members and contains larger objects, has asteroids named for Greek heroes. Asteroids at the trailing Lagrangian point are named for Trojan heroes. Confusion results from the fact that this naming scheme was not devised until several Trojans in each group had been named for heroes of the "wrong" side, causing "spies" to be present in each group!

Several asteroids, the most notable of which are **Hidalgo** and **Chiron**, reside in the outer solar system beyond the orbit of Jupiter (Figure 27.5). Hidalgo's semi-major axis is larger than Jupiter's (5.85 versus 5.2 astronomical units), and its orbit is rather eccentric. Chiron has a large orbit, with a semi-major axis of 13.7 astronomical units, and stays entirely in the outer solar system, traveling from inside Saturn's orbit to the vicinity of Uranus' orbit. In recent years, it has been suggested that there may be belts of asteroids between the orbits of the gas giants as well as between the orbits of Mars and Jupiter. It would be

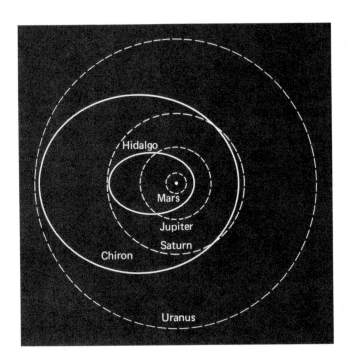

■ **Figure 27.5** Orbits of two outer solar system asteroids, Hidalgo and Chiron.

expected that few outer solar system asteroids would be visible from Earth because of their distance and faintness.

One problem that arises when studying asteroids in the outer solar system is distinguishing them from comet nuclei. Comets at these distances would not be expected to form comas, especially if considerable ice has been lost from their surfaces. The fact that Hidalgo, for example, has an eccentric, comet-like orbit supports the idea that it may not physically resemble a main-belt asteroid. Chiron revealed a tenuous coma in photographs taken in 1989, so it will probably be considered a comet rather than an asteroid. The diameter of Chiron's nucleus, 180 kilometers (110 miles), would make it the largest one yet known. Some researchers have expanded the concept of outer solar system asteroids by suggesting that Pluto and its satellite Charon should be considered asteroids instead of planet and satellite, but as mentioned in Chapter 25, the fact that Pluto has an atmosphere probably ensures that it will continue to be accorded planetary status.

In addition to these asteroids in the outer solar system, many are found closer to the Sun than in the main belt. Objects whose orbits cross those of Mars, Earth, or the other inner planets are of great interest because of the potential for future collisions. The numerous craters on inner solar system bodies are proof that many such collisions have occurred in the past. Fortunately, the majority of these **planet-crossing asteroids** are gone, having collided with planets long ago. Three major groups of planet-crossing asteroids are recognized, one that crosses only Mars' orbit and two that cross Earth's orbit. Each group is named for a specific asteroid that is a representative example of that group. Orbits of members of these groups are shown in Figure 27.6.

The Mars-crossing asteroids, called **Amor asteroids**, cross the orbit of Mars and have perihelia between 1.02 and 1.4 astronomical units. They do not cross Earth's orbit, but some of them have orbits eccentric enough that they also cross the orbit of Jupiter. Sixty-five Amor asteroids are currently known, and their total number may be in the thousands. Phobos and Deimos are possibly captured asteroids, and many of the Martian craters were caused by impacts of older members of the Amor group.

Members of the first group of Earth-crossing asteroids, called **Apollo asteroids**, spend most of their time outside of Earth's orbit (their semi-

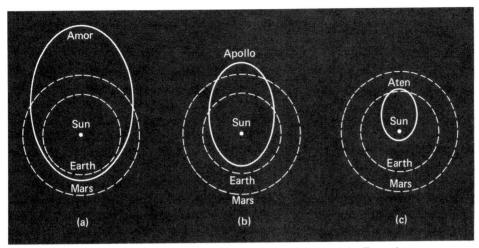

■ **Figure 27.6** Orbits of the Mars- and Earth-crossing asteroids Amor, Apollo, and Aten.

major axes are greater than 1 astronomical unit) but can come closer to the Sun than 1 astronomical unit at perihelion. One Apollo asteroid, Icarus, even has a perihelion distance that is smaller than Mercury's. Sixty-three Apollos are known. During the 1930s, three Apollo objects (Apollo, Adonis, and Hermes) came within 5 million kilometers (3 million miles) of Earth, but they were visible for such short times that their orbits were not accu-rately determined. Asteroid 1986JK (it has not yet been permanently named) also came this close in May of 1986, and its orbit was able to be deter-mined. On March 22, 1989, an asteroid desig-nated 1989FC passed within 690,000 kilometers (430,000 miles) of the Earth, the closest observed approach of any comet or asteroid. (Table 27.3 lists the closest observed approaches of asteroids and comets.) The asteroid 1989FC has a period of

■ **Table 27.3   Closest Observed Approaches of Asteroids and Comets**

Object	Type	Closest Approach	Date
1989FC	A	0.0046 AU	March 22, 1989
Hermes	A	0.005 AU	Oct 30, 1937
Hathor	A	0.008 AU	Oct 21, 1976
1988TA	A	0.009 AU	Sept 29, 1988
1491II	C	0.009 AU	Feb 20, 1491
Adonis	A	0.015 AU	Feb 7, 1936
Lexell	C	0.015 AU	July 1, 1770
Tempel–Tuttle	C	0.023 AU	Oct 26, 1366
1986JK	A	0.028 AU	May 29, 1986
1982DB	A	0.028 AU	Jan 23, 1982
Dionysius	A	0.031 AU	June 19, 1984
IRAS–Araki–Alcock	C	0.031 AU	May 11, 1983
Orpheus	A	0.032 AU	April 13, 1982
Aristaeus	A	0.032 AU	April 1, 1977
Halley	C	0.033 AU	April 10, 837

Abbreviations: A = asteroid, C = comet.
Modified from Sinnott, Roger W. "An Asteroid Whizzes Past Earth." In *Sky and Telescope*, July, 1989. Reproduced by permission of *Sky and Telescope* astronomy magazine, Cambridge, MA.

1.03 years and is believed to have a diameter somewhere between 200 and 400 meters (660 and 1320 feet).

Members of the second group of Earth-crossing asteroids, called **Aten asteroids**, have orbits with semi-major axes less than 1 astronomical unit. However, their orbits are sufficiently eccentric that they spend a brief time outside the Earth's orbit. Only nine Aten asteroids are known, and they are thought to comprise less than 10% of the total of Amor, Apollo, and Aten objects. Fortunately, estimates of the likelihood of collisions with members of all three groups of asteroids indicate that the risk to Earth is very small (although not nonexistent), about three collisions every million years. It is unclear why Amor, Apollo, and Aten asteroids still exist, because most would be expected to have already hit Mars, Earth, or Moon, given the age of the solar system. They are probably replenished by gravitational interactions or collisions within the asteroid belt, and some of them may be comets that have exhausted their ice supplies.

■ Physical Properties

Although no firsthand space exploration of asteroids has yet been completed, a great deal has been learned about them from Earth-based observations. Perhaps the best technique for studying asteroids is watching them occult stars. When such an occultation is predicted, observers gather in the area where it is likely to be visible. Accurate timings of how long it takes the asteroid to pass in front of the star can be used to determine the linear diameter of the asteroid, and by comparing timings from observers in different areas, the shape of the asteroid can be deduced. Most large asteroids are spherical, but many small ones have been found to be irregular in shape. For example, an occultation in 1975 showed that the asteroid Eros has dimensions of about 30 by 19 by 7 kilometers (19 by 12 by 4 miles). The most elongated asteroid, Geographos, is about six times longer than it is wide.

In addition to using occultation data, measurements of an asteroid's brightness and albedo can be used as a basis for estimating its diameter. Once an asteroid's albedo has been determined, one can calculate how large an object would be required to reflect the light necessary for its observed brightness. Most asteroids are very dark, with albedos under 5%, but others are much brighter. **Vesta**, which appears brighter in the sky than any other asteroid and can easily be seen with binoculars, has an albedo of 38%. (Vesta, like the planet Uranus, is sometimes bright enough to be faintly visible to the unaided eye.)

Measurements of asteroids indicate that only a few are relatively large, with only 29 having diameters greater than 200 kilometers (120 miles). The largest asteroid is Ceres, the first one to be discovered, and it is 932 kilometers (579 miles) across. Pallas, Vesta, and Hygiea are 528, 523, and 414 kilometers (328, 325, and 257 miles) across, respectively, but all of the other asteroids are smaller than 320 kilometers (200 miles) in diameter. (The ten largest asteroids and their diameters are listed in Table 27.4.) The majority of asteroids are no more than a few kilometers across, as are all of the known Earth-crossers.

The masses of asteroids are best calculated using values for their diameters and densities. Asteroid densities are estimated based upon compositions revealed by spectroscopy. Ceres, the largest asteroid, has a probable density of 2.7 grams per cubic centimeter and a mass of about

■ Table 27.4   The Ten Largest Asteroids

Number	Name	Diameter (km)	Diameter (mi)
1	Ceres	932	579
4	Vesta	528	328
2	Pallas	523	325
10	Hygiea	414	257
704	Interamnia	321	199
511	Davida	318	197
52	Europa	277	172
3	Juno	271	168
451	Patientia	267	166
31	Euphrosyne	257	159

Data from Cunningham, Clifford J. *Introduction to Asteroids.* Copyright © 1988, William-Bell, Inc., Richmond, VA.

1.2×10^{21} kilograms (2.6×10^{21} pounds), only 0.0002 that of Earth's. It is estimated that all of the other asteroids combined have a mass only 1.5 times that of Ceres. This means that all of the asteroids combined would not make a planet of any significant size. Of course, much asteroidal material that once existed has long since been destroyed by impacting into planets, and the original value for the total mass of asteroidal material is unknown.

The rotational rates of many asteroids have been determined by timing their periodic brightness variations. Most rates obtained in this manner range from a few hours to as much as one day. Eros, for example, has a rotational period of 5.3 hours. Indications are that larger asteroids tend to rotate faster than smaller ones. By watching the change in light fluctuations as an object moves with respect to the Earth, its obliquity can be determined as well. Most asteroid obliquities are high, undoubtedly because of the numerous collisions that they have experienced.

During the 1970s, some occultations seemed to indicate that a few asteroids had satellites. In recent years, this explanation has been discounted, and there is now speculation that these results were based upon observations of single objects that appeared double because they are actually peanut-shaped. The orbital evolution of asteroid–satellite pairs would be fairly rapid, and they would not be stable over long periods of time. Until additional evidence is gained, the existence of any true asteroid–satellite pairs must be considered doubtful.

We have been unable to obtain an image of the surface of any asteroid, so we are unable to describe their surface features. However, Phobos and Deimos, which may be captured asteroids, have rough, cratered surfaces and may be representative of other asteroids (see Figures 16.21 and 16.22). Radar reflectance studies of asteroids indicate that their surfaces are covered with regolith rather than bare rock. The surface composition of asteroids has been studied extensively in recent years by the use of visible- and infrared-light spectroscopy. Many common rocks and minerals exhibit characteristic spectral features in these regions, so we now have some information about the composition of various asteroids. The spectrum of Vesta, for example, resembles that of terrestrial basalt. Although not all asteroids can be categorized, a number of compositional groups have been established based upon spectroscopic studies. Most of these groups are given names based upon important minerals that appear to be components of asteroids in each group. For example, type E asteroids contain enstatite and type M asteroids contain metals. (Table 27.5 lists the basic properties of each group, and Figure 27.7 shows spectra typical for each group.) As indicated in Table 27.5, members of each group tend to be located in different areas of the main belt, and some groups, such as types D and Q, contain Trojan and planet-crossing asteroids, respectively. As we will see in Chapter 28, comparisons of asteroid and meteorite spectra often reveal similarities, indicating that many meteorites originate in the asteroid belt.

■ Table 27.5   Asteroid Spectroscopic Types

Type	Albedo	Composition and Properties	Location
E	0.33	Enstatite and other iron-free silicates	Inner edge of belt
S	0.09–0.24	Stony, made of various silicates	Inner to central belt
M	0.07–0.21	Metal (nickel–iron) plus silicates	Central belt
C	0.065	Carbonaceous materials and dark silicates	Outer belt
P	0.05	Metallic; similar to M but darker	Outer belt
D	0.05	Dark; contains clays, organic materials	Extreme outer belt, Trojans
Q	0.21	Similar to ordinary chondrites	Earth-crossing
A	0.12	Olivine with some pyroxene and metal	Rare; no trend
V	0.25	Basaltic	Vesta and a few others

Data from Cunningham, Clifford J. *Introduction to Asteroids.* Copyright © 1988, William-Bell, Inc., Richmond, VA.

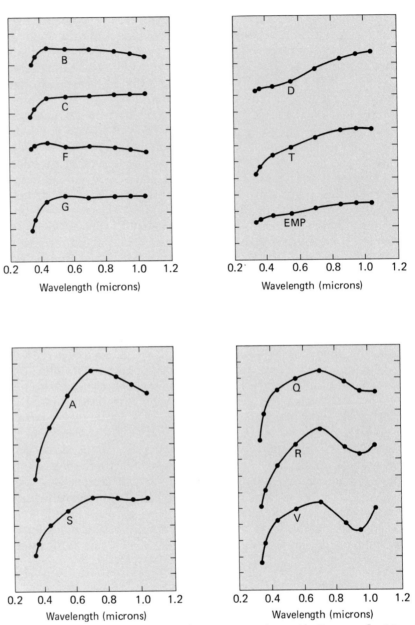

■ **Figure 27.7** Reflectance spectra of various types of asteroids. Percent reflectivity (vertical scale, offset for clarity) is plotted against wavelength in microns (1 micron = 10^{-6} meters). These spectra include those of the common asteroid types discussed in the text, as well as those of some less common types. See text for additional discussion. (Courtesy of David J. Tholen.)

■ Origin

Until fairly recently, asteroids were believed by some to be the remnants of a planet that had exploded. After World War II, a popular science fiction explanation for asteroids was that a planet had destroyed itself by nuclear war. Although the surviving mass of asteroids would not be enough to make a planet, much more material would have been present in the past. Nevertheless, modern theorists have shown that there would be no mechanism by which a fully formed planet could have been disrupted into the asteroids.

The modern theory of solar system formation, which we will discuss in Chapter 29, states that the planets formed from smaller solid objects joining together, a process known as **accretion.** In the asteroid belt, however, the bodies were prevented from accreting into a planet because of the disruptive influence of Jupiter's powerful gravity. As described earlier, many of these bodies were later removed from the belt by perturbations and collisions. Some of the asteroids removed in this way were placed into planet-crossing orbits and participated in the period of extreme bombardment that occurred early in the history of the solar system. (As mentioned earlier, some asteroids, especially those with high-eccentricity orbits, are probably comets that have lost their supply of volatile materials.) The fact that asteroid compositions in the main belt vary with distance from the Sun indicates that accreting material in the early solar system varied significantly over such a relatively short distance as the width of the asteroid belt.

Groups of asteroids with very similar orbits are known as **Hirayama families**, after **Kiyotsugu Hirayama,** who first described them in 1918. Hirayama identified five families, the Themis, Eos, Koronis, Maria, and Flora, each named for a representative asteroid member. At present, as many as 72 Hirayama families are recognized, and some of them have over 20 members. Nearly half of all asteroids currently known are thought to belong to one of the recognized families. The members of each family are considered to be the remnants of a single larger body that broke apart as a result of collisions within the asteroid belt. The existence of these asteroids of common origin indicates that many changes have occurred in the original bodies found in the asteroid belt over the billions of years since their initial formation, and that many of the present-day asteroids are much smaller than the objects that were once present in the main belt.

■ Exploration

Although the Pioneer and Voyager spacecraft passed through the asteroid belt, they did not study any asteroids. Future missions to the outer planets have been designed so that one or more asteroids will be studied during each trip. The Galileo mission to Jupiter, launched in October, 1989, will encounter the asteroids Gaspra and Ida, whereas Cassini, scheduled to be launched toward Saturn in 1996, will fly by Maja. The Comet Rendezvous Asteroid Flyby, scheduled for a 1995 launch toward Comet Kopff, will encounter at least one asteroid, Hamburga, while en route to the comet.

One reason why asteroid exploration is so exciting is that these huge rocky or metallic bodies may contain tremendous quantities of valuable materials. A single asteroid, even if only a few kilometers across, could contain billions of dollars worth of iron and nickel. Although "asteroid miners" have been a fixture in science fiction stories for many years, it is possible that rocket propulsion units may someday be attached to asteroids so that they can be brought close to the Earth for recovery of their resources. Obviously, such recovery of asteroid resources is many decades away.

■ Chapter Summary

Asteroids constitute a group of small rocky and metallic solar system members, many of which orbit between the orbits of Mars and Jupiter. Ceres, the first asteroid, was discovered in 1801 after astronomers realized that Bode's law suggested that a planet should exist between Mars and Jupiter. About 4000 asteroids are now known, but that is probably just a small fraction of the total that exist.

In addition to the main-belt asteroids, others exist at Lagrangian points of Jupiter's orbit, in orbits beyond the asteroid belt, and in orbits that cross those of the terrestrial planets. Some of the latter approach the Earth closely and could produce impacts in the future.

The largest asteroid has less than one-tenth of the Earth's diameter, and most are only a few kilometers across. Spectroscopy reveals most asteroids to be rocky or metallic, and several compositional groups have been defined. Current theories suggest that Jupiter prevented another terrestrial planet from accreting between it and Mars and was therefore responsible for the origin of the asteroids.

■ Chapter Vocabulary

accretion	Chiron	Kirkwood gaps	Trojan asteroids
Amor asteroids	Hidalgo	main-belt asteroids	Vesta
Apollo asteroids	Hirayama, Kiyotsugu	Piazzi, Giuseppi	
Aten asteroids	Hirayama families	planet-crossing	
Ceres	Kirkwood, Daniel	asteroids	

■ Review Questions

1. How do asteroids differ from comets?
2. Describe the events leading to the discovery of the first asteroids. How are most asteroids discovered today?
3. Describe the distribution of asteroids within the main belt. What is the cause of this distribution?
4. Describe the location of asteroids beyond the main asteroid belt.
5. List and describe the groups of asteroids that cross the orbits of the inner planets.
6. What information can be gained when watching an asteroid occult a star?
7. Summarize what has been learned about the physical properties of asteroids.
8. Explain the relationship between asteroid compositional groups and asteroid location.
9. Summarize the modern theory explaining the existence of the asteroids.
10. Explain how main-belt asteroids are emplaced into planet-crossing orbits.
11. Why is it likely that there were far more asteroids in existence in the past?
12. Why is the idea of firsthand space exploration of asteroids so exciting?

■ For Further Reading

Chapman, Clark. "The Nature of Asteroids." In *Scientific American*, January, 1975.

Cunningham, Clifford J. *Introduction to Asteroids*. Willmann–Bell, Richmond, VA, 1988.

Hartmann, William. "The Smaller Bodies of the Solar System." In *Scientific American*, September, 1975.

Kowal, Charles T. *Asteroids: Their Nature and Utilization.* Wiley, New York, 1988.

Sinnott, Roger W. "An Asteroid Whizzes Past the Earth." In *Sky and Telescope,* July, 1989.

Wetherill, George. "Apollo Objects." In *Scientific American,* March, 1979.

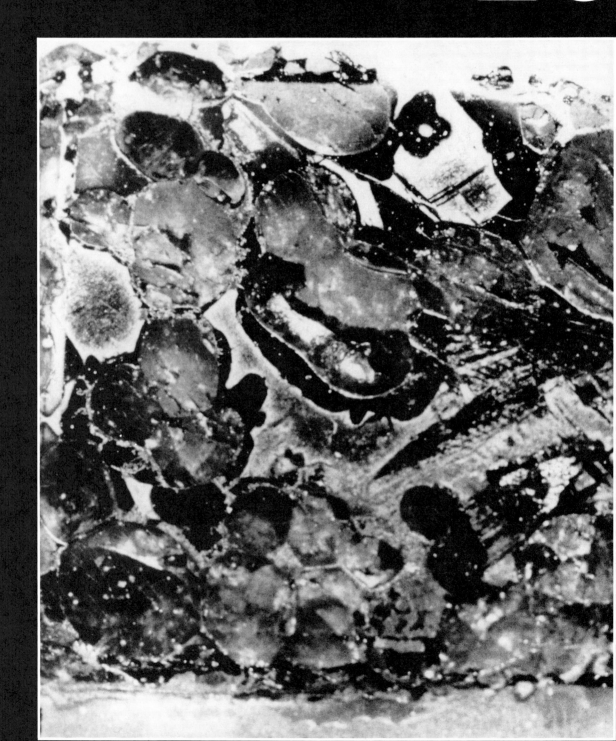

Meteorites

The polished surface of the Brenham pallasite, a stony–iron meteorite. (Photograph by the author.)

Two decades ago, billions of dollars were spent in obtaining the Apollo lunar samples and bringing them back to Earth, and their value to science has been enormous. However, these are not the only samples of extraterrestrial materials that we have. In fact, a great deal of material from space has been obtained at no cost at all! Meteorites that fall to Earth can be considered "free gifts," and their value to science has been no less than that of the costly lunar rocks. Most meteorites bear little resemblance to lunar samples and have added greatly to the variety of types of extraterrestrial materials available for study. The only disadvantage of meteorite samples is that, unlike lunar rocks, their exact source is not known. Even so, meteorites have added much to our knowledge of the solar system and its history. For example, about 1000 kilograms (two tons) of meteorite fragments landed in Allende, Mexico a few months before the first lunar landing in 1969. The **Allende meteorite** provided researchers with a large amount of what had previously been a very rare type of meteorite, and it became one of the best-studied meteorites in history (Figure 28.1). In fact, many researchers would argue that Allende taught us more about the early history of the solar system than all of the Apollo samples combined!

Because meteorites are obviously very important to science, the arrival of one on the Earth often marks the beginning of an extensive research program in an attempt to unlock its secrets. However, this arrival is only the final event in the long journey that the meteorite has taken from orbit in space to a blazing passage through the atmosphere to impact on our planet. In this chapter, we examine the phenomena associated with atmospheric entry and fall of meteorites, discuss how the sources of meteorites have been determined, and describe the types of meteorites.

(a)   (b)

■ **Figure 28.1** Two fragments of the Allende meteorite, which fell in Mexico in 1969. The sample at left has not been disturbed and shows the natural surface of the meteorite. The sample at right has been sliced in half and polished to show the meteorite's internal structure. Both samples are about 1 inch across. Allende is an example of the relatively rare type of meteorite known as carbonaceous chondrites. (Photographs by the author.)

■ Fall Phenomena

The story of a meteorite's fall begins far from Earth with a **meteoroid**, an object in space whose orbit will eventually cause it to enter the Earth's atmosphere. As will be described later in the chapter, most meteoroids are believed to be debris from comets and asteroids. (Since the advent of the space age, a considerable amount of "space junk" has also fallen to Earth from orbit, producing similar results.)

When a meteoroid enters the atmosphere, it is traveling many kilometers per second. As a result of this high speed, a great deal of heat is generated by friction, the surface of the meteoroid begins to melt, and the meteoroid is surrounded by a cloud of glowing gas. Someone watching from the Earth's surface observes a brief flash of light in the night sky, a phenomenon called a **meteor** (Figure 28.2). Although people often call meteors shooting stars or falling stars, this terminology is misleading because meteors have nothing at all to do with real stars. The light of the meteor, which results both from the glowing object itself and the trail of ionized gas pro-

duced by it, is typically generated at altitudes of 110 kilometers (70 miles). Most meteors are probably produced by very small meteoroids. In fact, it has been estimated that a 1-gram meteoroid is able to produce a meteor approximately as bright as the brightest stars.

Researchers have analyzed the spectra of meteor trails and have found that most of their light is produced by emission processes. Iron, calcium, magnesium, silicon, and other elements commonly found in silicate minerals are responsible for most of the emission features. Short-wave radio waves reflect from the ionized gas of meteor trails, allowing detection of meteors even during daylight hours, as well as measurements of the speed and altitude of the meteors. Some experimenters have transmitted communications signals over longer-than-normal ranges by bouncing them from meteor trails when there are many meteors entering the atmosphere.

On the average, an observer can see about 3 meteors every hour before midnight and 15 after midnight. The reason for the increase is that after midnight, the observer is on the leading side of the Earth as it moves in its orbit, and our planet

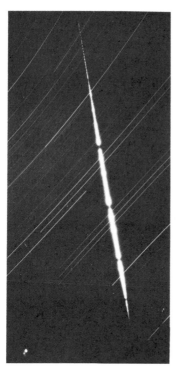

■ **Figure 28.2** The appearance of a meteor in the sky. The meteor streaked across the field of view of a camera as it was recording a time exposure of the stars, yielding star trails. The breaks in the meteor trail are timing marks used to determine the meteor's velocity. (Smithsonian Astrophysical Observatory Photograph.)

rection, the other person normally won't be able to look quickly enough to see it. However, unusually bright meteors, called **fireballs** or **bolides**, may leave trails that can linger as long as half an hour. Bright bolides, which probably result when large meteoroids fragment and expose large surface areas for gas generation, can sometimes even be seen in daylight! It has been estimated that the total number of meteors entering the Earth's atmosphere daily that would be bright enough to be seen with the unaided eye is about 25 million. Only a few of these would be bright enough to be considered bolides.

Most meteoroids that enter the atmosphere vaporize completely, and no trace of them remains. The exceptions are meteoroids that are either extremely small or large. Meteoroids smaller than grains of sand are able to survive by radiating away the heat generated while entering our atmosphere. Many of these small particles shower into the atmosphere every day but cannot normally be detected at the surface. In recent years, however, high-flying aircraft have sampled this "cosmic dust" high in the atmosphere for study and analysis (Figure 28.4). Meteoroids that are relatively large survive passage through the atmosphere because their inner portions remain cool as a result of insulation of the surrounding material, even though their outer layers melt and boil away. Therefore, these objects escape destruction and land on the Earth as **meteorites**. Meteorites range from extremely small objects to huge bodies that weigh many thousands of kilograms. Smaller meteorites are slowed during at-

overtakes more meteoroids than the number that overtake it before midnight (Figure 28.3). A meteor normally lasts only a second or so. This means that if you see a meteor and tell someone about it while he or she is looking in another di-

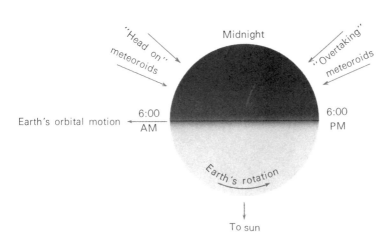

■ **Figure 28.3** Why more meteors are seen after midnight. Before midnight, an observer is on the trailing edge of Earth (*right*), and sees only those meteors caused by meteoroids that catch up with the Earth. After midnight, an observer is on the leading edge as the Earth plows into meteoroids (*left*). (From Abell, George, David Morrison, and Sidney Wolff. *Exploration of the Universe,* 5th ed. Saunders College Publishing, Philadelphia, 1987.)

■ **Figure 28.4** An extremely small meteorite, sampled by a high-flying aircraft. The particle is about 10^{-5} meters across. (NASA Photograph.)

mospheric passage to only a few hundred kilometers per hour and do very little damage when they hit. The largest meteorites are scarcely slowed during atmospheric passage and hit the ground with velocities of several kilometers per second, usually producing impact craters. When such crater-forming impacts occur, the meteorites themselves are generally destroyed by the energy released during impact. Fortunately, large impacts are rare, and many meteorites are small enough to be held in one's hand. The largest known meteorite is **Hoba** from Namibia (Southwest Africa), found in 1928. It weighs 55,000 kilograms (120,000 pounds) and, for obvious reasons, has not been moved from the point of discovery (Figure 28.5)! The reason why this huge object failed to form an impact crater is not yet known.

No one has been killed by an impacting meteorite, but Mrs. Hewlett Hodges of Sylacauga, Alabama was hit by a 3.7-kilogram (8.5-pound) meteorite that crashed through a house roof on November 30, 1954. She was badly bruised even though the meteorite was greatly slowed by the roof. There have been other instances of meteorites hitting houses, buildings, and even mailboxes along roads! In 1910, a dog was killed by a fragment of the Nakhla meteorite that fell in Egypt, the only meteorite-caused death in recorded history.

A meteorite that is seen to fall and is consequently recovered is called a **fall**. One that is discovered on the ground but whose arrival was not witnessed is called a **find**. Meteorites of both types are named for the locality in which they fell or were recovered. If reached quickly enough after their arrival, falls are often still warm as a result of their passage through the atmosphere. The material that melted on the object's surface cools and solidifies before landing to form a dark, glassy coating called a **fusion crust**. The interiors of large meteorites are often very cold, however, because the fiery but brief passage through the atmosphere may fail to warm the entire object.

Finds can usually be dated to learn how long they have been on the surface since they arrived from space. This is possible because **cosmic rays** (high-energy particles of uncertain origin that travel through space) hitting meteoroids in space cause radioactive isotopes to be formed in them. Once meteorites are on the Earth and protected from cosmic rays by its atmosphere, no new radioactive isotopes can form, and the isotopes formed in space decay. The longer the meteorite remains on the Earth, the smaller the proportion of radioactive isotopes present in it. Studies of meteorites show that some metallic ones have been recovered after being on the Earth as long as 100,000 years. Meteorites made of rocky material are less durable and last no more than a few thou-

■ **Figure 28.5** The Hoba iron meteorite, the largest meteorite ever found. (Smithsonian Institution Photograph, courtesy of Brian Mason.)

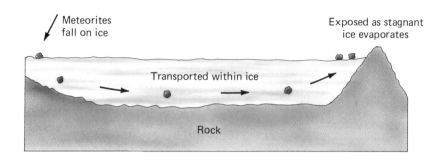

Meteorites
fall on ice

Exposed as stagnant
ice evaporates

Transported within ice

Rock

■ **Figure 28.6** Concentration of Antarctic meteorites. The meteorites fall on the ice (*left*) and flow within the glacial ice toward the right. When the glacier reaches a mountain (*right*), meteorites are exposed as the stagnant ice evaporates.

sand years. For this reason, and also because metallic meteorites look less like ordinary terrestrial rocks, most finds are of this type. However, most falls are of the stony variety.

The total amount of meteoroid material entering the atmosphere each day is estimated at 100,000 to 1,000,000 kilograms (100 to 1000 tons). Although this sounds like an enormous amount, the Earth is so massive that over its 4.6-billion-year history, this influx would have increased its weight by only one part in 10 million! Approximately 100 meteorite craters have been identified on the Earth's surface, but many more have, of course, been eroded away so completely that they are no longer recognizable (see Figure 10.14). Most of these craters have been identified in areas of the world with higher populations, as more geological exploration is normally done in these regions and there is less likely to be heavy forest cover, which easily masks such features.

Depending upon the strength of the material of which they are composed, many meteorites break into fragments before impacting. The Allende meteorite, mentioned earlier, is such an example. The combined weight of the Allende material was about 1000 kilograms (two tons), but it fell as hundreds of small fragments over an area of 150 square kilometers (60 square miles). In cases like this in which many fragments fall simultaneously, the individual fragments are not given separate names.

Since the early 1800s, when scientists began to realize that meteorites are indeed from beyond the Earth, over 2600 different meteorites have been recovered and identified. (Approximately 900 of these are falls, and 1700 are finds.) The total number of fragments found is in the tens of thousands, because one fall can consist of many different pieces. Most of these have been recovered in highly populated areas, especially those where people have been educated about the nature of meteorites and how to recognize them.

Since the early 1970s, a tremendous quantity of meteoritic material has been found in an unlikely place, the continent of Antarctica. Over 5000 fragments, believed to represent perhaps 50 to 500 actual meteorite arrivals, have been found in this frozen wasteland. Why is Antarctica such a treasure trove of meteorites? Meteorites fall on the glacial ice that covers most of the continent (Figure 28.6). As the glaciers move, they carry the meteorites along, until the glacial flow is stopped by mountain ranges. The stagnant ice is evaporated by the wind, and meteorites are exposed at the surface and are easily visible to searchers exploring these regions (Figure 28.7). Antarctic meteorite finds are significant, because the cold temperatures reduce weathering of the meteoritic minerals. Many examples of rare types have been found, as well as specimens of types that were previously unknown.

■ Sources of Meteorites

Although meteorites have fascinated mankind since ancient times, it has taken quite a long time for astronomers to realize where they originate. The story of meteorite origins begins in ancient times, as the fall of meteorites is mentioned in the Bible, as well as in ancient Chinese, Greek, and Roman writings. At that time, meteorites were considered an atmospheric phenomenon. Because hailstones fall as solid chunks of ice during thunderstorms and thunder-like noises often accompany a meteorite fall, it was considered rea-

(a)

(b)

■ **Figure 28.7** Recovery of Antarctic meteorites. (a) Researchers photograph a meteorite in place on the ice before retrieving it. (b) Photograph of a meteorite as found on the ice. The device above the specimen indicates its catalog number. (Photographs courtesy of William A. Cassidy, Department of Geology and Planetary Science, University of Pittsburgh.)

sonable that meteorites formed similarly from atmospheric dust.

Because of their mysterious origin, meteorites have been revered by many cultures. The Moslem holy city of Mecca contains a shrine, the Kaaba, in which a meteorite of unknown age is housed. A meteorite fell in the German town of Ensisheim on November 16, 1492, and much of the 127-kilogram (280-pound) mass is still preserved there in the town hall. Not all meteorites were given this treatment, as many metallic ones were used as sources of iron for swords, knives, and other implements. However, most of these objects were probably found rather than seen to fall.

Meteorites began to be less mysterious in the 1700s, when scientists started using chemical analysis to study meteorites and obtained unusual results. Until this time, most educated people scoffed at reports of stones from the sky, but their minds began to change as meteorites were found to contain concentrations of nickel much greater than those in any terrestrial rocks. In addition, the textures of meteorites were seen to be unlike those of any rocks found on Earth. A fall on L'Aigle, France in 1803, in which 3000 separate stones fell, was instrumental in convincing European scientists that meteorites were an actual phenomenon. By the end of the first decade of the 1800s, most people accepted the fact that meteorites indeed came from beyond the Earth. However, one important question remained: Where was the source of these objects that were landing on our planet?

The behavior of meteors in the sky gives one of the most basic clues about the origin of meteorites. Just as the rates of meteors vary during the night, there is annual variation as well. A number of **meteor showers** occur throughout the year, and during these times, as many as hundreds of meteors can be seen per hour (Figure 28.8). These showers occur when the Earth's orbit intercepts a high-density stream of meteoroids. Table 28.1 lists the best-known meteor showers.

During a shower, meteors appear to stream outward from a point in the sky called the **radiant**, which marks the point of intersection of the Earth's orbit and the orbit of the meteoroids (Figure 28.9). This phenomenon is similar to the appearance of snowflakes as seen through the windshield while driving through a snowstorm. The snowflakes appear to be coming straight toward the driver, but this is an illusion caused by the car's forward motion. The fact that the loca-

■ **Figure 28.8** Numerous meteors can be seen in this time-exposure photograph of the Leonid meteor shower in 1966. (Photograph by Dennis Milon.)

■ Table 28.1 Major Meteor Showers

Name	Date of Maximum*	Radiant Location (Constellation)	Associated Comet
Quadrantids	Jan 3	Draco	?
Lyrids	April 22	Lyra	Thatcher
Eta Aquarids	May 4	Aquarius	Halley
Delta Aquarids	July 28	Aquarius	?
Perseids	Aug 12	Perseus	Swift–Tuttle
Draconids	Oct 10	Draco	Giacobini–Zinner
Orionids	Oct 21	Orion	Halley
Taurids	Nov 2	Taurus	Encke
Leonids	Nov 17	Leo	Tempel–Tuttle
Andromedids	Nov 20	Andromeda	Biela
Geminids	Dec 14	Gemini	Phaeton†
Ursids	Dec 22	Ursa Minor	Tuttle

* Date may vary slightly from year to year. Meteors are usually seen a few days before and after.
† Asteroid that is probably an extinct comet.

tion of the radiant of a meteor shower does not change over the course of a few hours as the Earth rotates was another proof of the extraterrestrial origin of meteors.

In the 1860s, the source of shower meteoroids was found to be comet debris. In 1862, the orbit of Comet Swift–Tuttle was determined, and in 1866, Giovanni Schiaparelli noted that the Perseid meteor shower occurs when the Earth crosses the orbit of Comet Swift–Tuttle annually

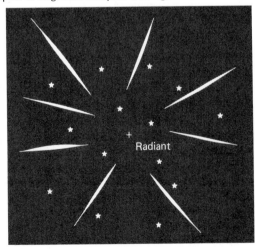

■ **Figure 28.9** Appearance of meteor shower radiant, which marks the point in the sky from which meteors appear to originate. Compare with Figure 28.8.

Radiant

in mid-August. Since then, cometary sources have been identified for most major meteor showers. As mentioned in Chapter 26, comets lose material during each orbit, and some of this matter eventually disperses and fills the orbital path of the comet. When the Earth reaches the point where its orbit intersects that of the comet, the shower occurs (Figure 28.10). Some showers display about the same number of meteors every year, indicating that for them, the material is uniformly distributed over the comet's orbit. Other showers are particularly rich at intervals corresponding to the orbital period of the associated comet, indicating that the debris is concentrated in a clump near the comet (Figure 28.11). An example is the Leonid shower, associated with Comet Temple–Tuttle, which is especially impressive every 33 years. In fact, the shower on November 13, 1833 was so impressive that many observers believed that the end of the world was occurring; in some places, over 100,000 meteors were visible during the course of that single night!

Although shower meteors are definitely related to comets, comets can be eliminated as the source of most meteorites. Falls of meteorites almost never occur during showers, which is understandable in view of the nature of cometary material. Most debris distributed along comet orbits is probably not durable enough to survive passage through the atmosphere. Nearly all me-

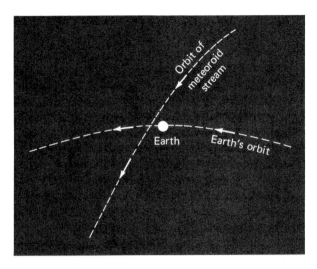

■ **Figure 28.10** Cause of meteor showers. The orbit of a meteoroid stream derived from comet debris intersects the orbit of the Earth. The relative motion of the meteoroids and the Earth causes meteors seen during the shower to stream outward from the radiant.

teorite falls are associated with meteors other than those that are seen during showers. These are called **sporadic meteors** because they cannot be predicted.

What is the origin of the sporadic meteors and the meteorites associated with them? At one time, the Moon and Mars were considered likely sources, as objects hitting these bodies could send fragments on collision courses with Earth. Alternately, explosive volcanic eruptions on these worlds might blast material into space. However, it is difficult to imagine how these processes could occur without destroying the fragments because of the energy involved; by the 1970s, few scientists considered that any meteorites were from either the Moon or Mars.

Surprisingly, certain meteorites found in Antarctica during the early 1980s were so similar in both composition and texture to lunar basalts that opinions have changed once again. There is now

■ **Figure 28.11** Distribution of meteoroids. (a) In this case, meteoroids are evenly spread along a comet's orbit. (b) In this case, they are concentrated near the comet. (From Abell, George, David Morrison, and Sidney Wolff. *Exploration of the Universe,* 5th ed. Saunders College Publishing, Philadelphia, 1987.)

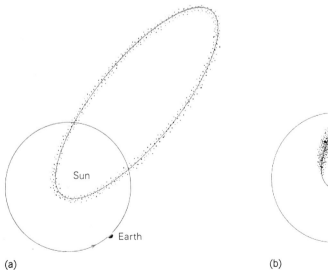

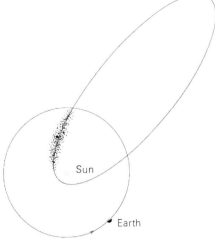

(a)   (b)

■ **Figure 28.12** An Antarctic meteorite that probably originated on the Moon. (NASA Photograph.)

little doubt that a few Antarctic meteorites have come from the Moon (Figure 28.12). Another group of meteorites is thought to be derived from Mars because its members are chemically similar to Martian materials analyzed by the Viking landers, but this hypothesis will not be confirmed until Martian samples are returned to Earth for study (Figure 28.13). Although it now seems that some meteorites have come from the Moon and Mars, they appear to be exceptional cases whose arrival cannot be completely explained.

The modern explanation for the source of the majority of meteorites is that they are derived from asteroid fragments. Orbital perturbations and collisions among asteroids send fragments of them into Earth-crossing orbits, as described in Chapter 27. Many meteorites bear evidence of shock from such collisions, as well as textures that appear to be derived from several distinct fragments that are now joined.

Two lines of experimental evidence link asteroids and meteorites. One is based upon the orbits of meteoroids before they entered the atmosphere. In the 1960s and 1970s, automatic meteor camera networks photographed numerous sporadic meteors. If a meteor can be photographed from three or more sites, the orbit of the meteoroid in space can be determined. Three meteorite falls that were recovered had their orbits determined in this way. In all three cases, the orbits had aphelia that were located in the asteroid belt (Figure 28.14).

The second line of evidence linking meteoroids with asteroids comes from spectroscopy, most of which has been done in the visible and infrared regions. The spectra of meteorites can

easily be obtained in the laboratory, and the spectra of asteroids can be determined by telescopic observations. Comparisons with spectra of terrestrial rocks and minerals aid in the understanding of the compositions of both meteorites and asteroids. Even more importantly, comparisons between meteorite and asteroid spectra often reveal similarities. These spectroscopic "matches" between meteorites and asteroids indicate that a specific meteorite may have originally been part of a specific asteroid or one similar to it. For example, the brightest asteroid, Vesta, is spectroscopically similar to the Kapoeta meteorite, and both of these objects are composed of the rock basalt. (Examples of spectroscopic matches are shown in Figure 28.15.) An asteroid from which a particular meteorite or group of meteorites appears to be derived is called that meteorite's **parent body.** It cannot be proved that a given meteorite actually came from a given asteroid, only that it came from one with similar composition.

In recent years, it has been shown that additional useful spectroscopic information is contained in the far-ultraviolet spectral region. Although meteorite spectra have been obtained in this region, the atmosphere is opaque at these wavelengths, and far-ultraviolet asteroid spectra will not be obtained until the Hubble Space Telescope measures them.

■ **Figure 28.13** An Antarctic meteorite that is considered likely to have originated on Mars. (NASA Photograph.)

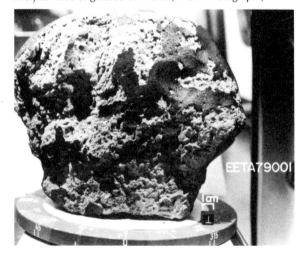

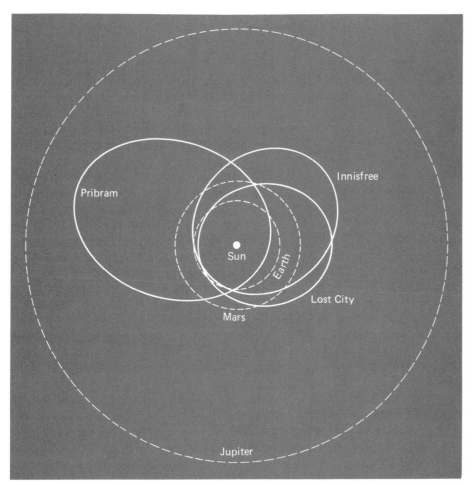

■ **Figure 28.14**   Orbits of three recovered meteorites. Note that their aphelia lie within the asteroid belt.

■ Types of Meteorites

Meteorites are classified into three main groups and numerous subgroups according to their compositions and textures (Table 28.2). Although meteorite composition varies widely, the main criterion for classification is whether a meteorite is primarily made of minerals (generally silicates) similar to those found in the Earth's crust and mantle, or nickel–iron metal, presumably similar to the material of the Earth's core. **Stony meteorites,** which constitute 94% of all falls, are made primarily of silicate minerals and therefore resemble terrestrial rocks. **Iron meteorites,** which constitute about 5% of all falls, are made almost en-

tirely of metal. Dense, durable, and unusual, they constitute almost half of all finds. **Stony–iron meteorites** contain both components in approximately equal amounts, and are rare, constituting only 1% of all falls. (Complete numbers of finds and falls of the three major meteorite types are found in Table 28.3.)

Although stony meteorites can contain a wide variety of both nonsilicate and silicate minerals, their most important components are silicates, especially olivines, pyroxenes, and feldspars. The mineralogy of meteoritic feldspars and especially pyroxenes is quite complicated, with a wide variety of names depending upon crystal structure and chemical composition. Unfortu-

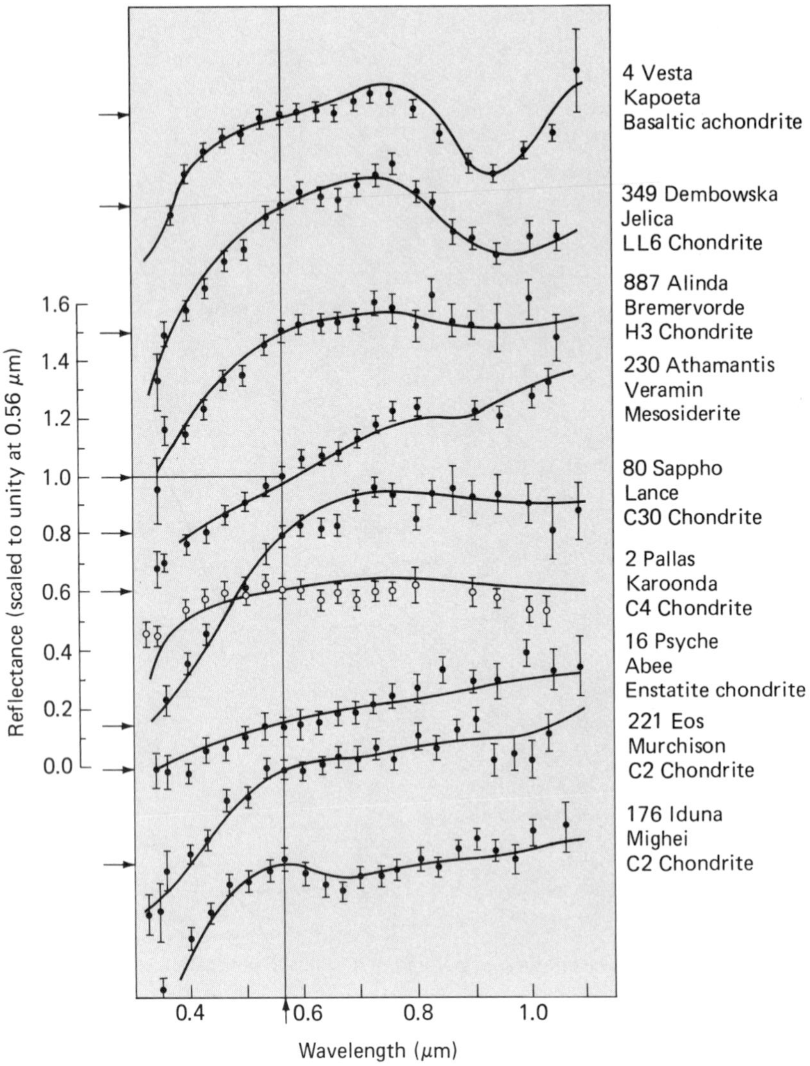

■ **Figure 28.15** Spectroscopic matches between meteorite spectra (*solid lines*) and asteroid spectra (*circles*). Percent reflectivity (vertical scale, offset for clarity) is plotted against wavelength in micrometers (10⁻⁶ meters). (From Chapman, C. R. "Asteroids as Meteorite Parent-Bodies: The Astronomical Perspective." In *Geochimica et Cosmochimica Acta*, vol. 40, p. 701. Copyright 1976 by Pergamon Press, Oxford.)

nately, understanding the classification of stony meteorites requires some familiarity with these names.

Stony meteorites are divided into two classes based upon whether they contain features known as **chondrules.** Chondrules are millimeter-sized spherical masses of silicates that are presumed to have formed when liquid droplets of material cooled very early in the history of the solar system (Figure 28.16). Stony meteorites with chondrules are called chondrites, whereas those without them are called achondrites. About 92% of stony meteorite falls are chondrites, and 8% are achondrites.

Members of the first class of stony meteorites, **chondrites,** are believed to have formed near the surfaces of asteroid parent bodies, where the lack of heat and pressure has caused relatively little change since the formation of the solar system. There are five chemical groups of chondrites, distinguishable because of minerals present and overall chemical composition. Each chemical group is subdivided into what are called petrologic types, classified according to the amount of heating and change the chondrite has experienced since formation.

Group C, or **carbonaceous, chondrites,** are so named because they include carbon-contain-

■ Table 28.2   Outline of Meteorite Classification*

I. Stony meteorites
  A. Chondrites
    1. Carbonaceous (group C)
    2. Ordinary
      a. High iron (group H)
      b. Low iron (group L)
      c. Low iron, low total metal (group LL)
    3. Enstatite (group E)
  B. Achondrites
    1. Calcium-poor
      a. Enstatite (aubrites)
      b. Hypersthene (diogenites)
      c. Olivine (chassignites)
      d. Olivine – pigeonite (ureilites)
    2. Calcium-rich
      a. Fassaite (angrites)
      b. Augite – olivine (nakhlites)
      c. Anorthite – pigeonite (eucrites)
      d. Maskelynite – pigeonite (shergottites)
      e. Anorthite – hypersthene (howardites)
II. Iron meteorites
  A. Hexahedrites
  B. Octahedrites
  C. Ataxites
III. Stony – iron meteorites
  A. Pallasites
  B. Mesosiderites
  C. Siderophyres
  D. Lodranites

* See text for complete discussion.

■ Table 28.3   Numbers of Meteorites of Various Types*

Type	Falls	Finds	Total
Iron	42	683	725
Stony – iron	10	63	73
Stony	853	960	1813
Chondrites	784	897	1681
Achondrites	69	63	132
All types	905	1706	2611

* Includes all meteorites known as of January, 1984, but omits most Antarctic meteorites found since 1977.
Data from Graham, A. L., A. W. R. Bevan, and R. Hutchison. *Catalogue of Meteorites,* 4th ed. Copyright 1985, University of Arizona Press, Tucson.

are collectively called **ordinary chondrites** because together they are the most abundant type of meteorite (Figure 28.17). Ordinary chondrites are composed mainly of the silicates olivine, pyroxene, and plagioclase, as well as other minerals, and they also contain some nickel – iron metal grains. Ordinary chondrites somewhat resemble the terrestrial rock known as peridotite, which is probably a major constituent of the Earth's mantle.

Enstatite (group E) **chondrites** contain mainly the low-iron, high-magnesium form of pyroxene called enstatite (Figure 28.18). The only

ing compounds, as well as silicate minerals, primarily those of the sheet-structure type (Figure 28.1). They are considered the most "primitive" meteorite type, which means that they are believed to have experienced very little change since their formation. Allende and some other carbonaceous chondrites contain grains of material believed to be some of the first solid material to have condensed from the hot gases of the solar nebula. Some carbonaceous chondrites have been found to contain amino acids, which are essential components of life on Earth, although not alive themselves. This indicates that progenitors of terrestrial life may have arrived from space, as will be discussed in Chapter 30.

Three groups of chondrites, group H (high iron content), group L (low iron content), and group LL (low iron and low total metal content),

■ **Figure 28.16** Chondrules. This photograph of a half-inch-wide polished section of an ordinary chondrite shows numerous circular chondrules. (NASA Photograph.)

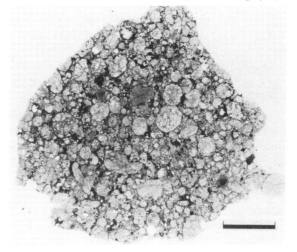

(a)

(b)

(c)

■ **Figure 28.17** Ordinary chondrites. (a) Nuevo Mercurio, a type H chondrite. (Photograph by the author.) (b) Bruderheim, a type L chondrite. (Photograph by the author.) (c) Yamato-792772, a type LL chondrite from Antarctica. (Japan Polar Research Association Photograph, courtesy of Tetsuya Torii.)

iron in these meteorites is present as grains of nickel-iron metal.

Members of the second class of stony meteorites, **achondrites,** are often called differentiated meteorites because they are believed to have been formed deep within the mantle regions of asteroid parent bodies. They are much different chemically from chondrites and appear to have experienced igneous and metamorphic processes due to the heat and pressure found inside their parent bodies. Many achondrites have experienced shock resulting from impacts occurring upon their parent bodies, and this shock manifests itself in the brecciation of achondrites and in the presence of high-pressure forms of minerals within them. There are nine groups of achondrites, each of which is given two names —

one based on the mineral content of the group, the other taken from the name of a representative example (Figure 28.19).

The first four groups of achondrites are calcium-poor and are often given that collective designation. The first is the enstatite achondrites, or aubrites. Aubrites are very similar to enstatite chondrites in that both are composed primarily of the mineral enstatite. Aubrites, however, lack chondrules and usually have a brecciated structure. Hypersthene achondrites, or diogenites, consist primarily of hypersthene, a form of pyroxene. All diogenites are breccias and are mainly fragments of pyroxene in a fine-grained matrix of the same material. Olivine achondrites, or chassignites, are rare achondrites composed almost entirely of olivine. The olivine–pigeonite achon-

■ **Figure 28.18** Yamato-691, an enstatite chondrite from Antarctica. (Japan Polar Research Association Photograph, courtesy of Tetsuya Torii.)

drites, or ureilites, are the only stony meteorites that contain diamonds and the only achondrites that contain appreciable amounts of nickel – iron metal. They are mostly olivine and pigeonite (a form of pyroxene) and are believed to be carbonaceous chondrites that have experienced shock. (The diamond in ureilites has been produced by shocking carbon.)

The remainder of the achondrite groups are calcium-rich. There is only one angrite, or fassaite achondrite, Angra dos Reis, which is composed mainly of fassaite, a high-calcium, high-titanium form of pyroxene. Augite – olivine achondrites, or nakhlites, are composed mainly of a variety of pyroxene known as augite and iron-rich olivine. The three final groups of achondrites are collectively called pyroxene – plagioclase achondrites. Eucrites, most of which are brecciated, are composed of the anorthite form of plagioclase and the pigeonite form of pyroxene, in approximately equal amounts. Eucrites are the most common achondrites. (The asteroid Vesta is considered a eucrite parent body.) Shergottites are similar to eucrites, except that their plagioclase has experienced shock and has been changed to a glassy, rather than crystalline, form. Howardites are brecciated, indicating that their parent bodies experienced considerable shock. These meteorites are composed of a form of plagioclase known as

anorthite and a form of pyroxene known as hypersthene. Shergottites, nakhlites, and chassignites, collectively called the SNC achondrites, are the meteorites that are believed to be of Martian origin.

Iron meteorites are composed primarily of nickel – iron metal, which can occur as two different minerals, kamacite (nickel-poor) and taenite (nickel-rich). Most iron meteorites contain interlocking crystals of the two minerals, which form the beautiful **Widmanstatten pattern**, which is found only in meteorites (Figure 28.20). The Widmanstatten pattern can be seen in an iron meteorite if it is cut, polished, and etched with an acid solution. The presence of the Widmanstatten pattern has revealed important facts about the origin of iron meteorites. Experiments have shown that the pattern forms only at pressures of less than 12,000 atmospheres, indicating that if iron meteorites formed in the cores of parent bodies, these bodies could have been no larger than 1600 kilometers (1000 miles) in diameter. However, it has been determined that the metal in most iron meteorites cooled at fairly rapid rates (1 to 100 K per million years), indicating that iron meteorites formed either in the interiors of small (20 to 400 kilometers, or 12 to 240 miles, in diameter) bodies or near the surfaces of large ones. The Widmanstatten pattern probably formed as the metal cooled between 1070 K and 770 K (1470°F and 930°F).

Iron meteorites are classified into three structural types based on the size and orientation of the metal crystals (Figure 28.21). Hexahedrites are entirely kamacite and have no visible structures other than linear striations, called **Neumann lines**, which can be seen on cut surfaces. Neumann lines, unlike the Widmanstatten pattern, do not result directly from mineral orientations but are produced by shock. Hexahedrites range from 5% to 6.5% nickel and are represented by about 50 examples. Octahedrites contain both types of nickel – iron and exhibit the Widmanstatten pattern, ranging from coarse in some to fine in others. The vast majority of iron meteorites are octahedrites, and their nickel content ranges from 6.5% to 13%. Ataxites are almost entirely pure taenite and lack structures visible to the unaided eye, although a Widmanstatten pattern can

(a)

(b)

(c)

(d)

■ **Figure 28.19** Achondrites found in Antarctica. (a) Allan Hills-78113, an enstatite achondrite (aubrite). (b) Yamato-791000, a hypersthene achondrite (diogenite). (c) Yamato-791839, a olivine – pigeonite achondrite (ureilite). (d) Yamato-791192, an anorthite – pigeonite achondrite (eucrite). (e) Yamato-790991, an anorthite – hypersthene achondrite (howardite). (Japan Polar Research Association Photographs, courtesy of Tetsuya Torii.)

(e)

be seen microscopically. Ataxites contain over 13% nickel, and only a few of them are known.

A newer classification system for iron meteorites has also been devised. It is based on the concentrations of trace elements such as germanium and gallium in the meteorites' metal. Slightly over a dozen groups of iron meteorites have been identified in this way. These groups have been given rather unexciting names such as IAB, IC, IIAB, IIC, IID, and so forth. Because it is believed that iron meteorites formed in the core regions of asteroid parent bodies or else in scattered pockets of metallic material inside one or more large asteroids, it is likely that each of the

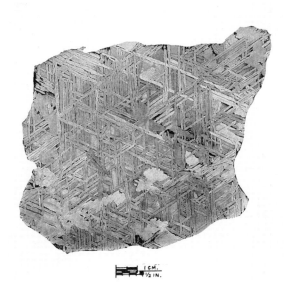

■ **Figure 28.20** The Widmanstatten pattern in a polished section of the Waingaromina octahedrite iron meteorite. (Smithsonian Institution Photograph, courtesy of Brian Mason.)

■ **Figure 28.21** Iron meteorites. (a) Cut section of the Bennett County hexahedrite. Note the Neumann lines. (Smithsonian Institution Photograph, courtesy of Brian Mason.) (b) Canyon Diablo, an octahedrite recovered near Meteor Crater in Arizona. (Photograph by the author.) (c) Yamato-791694, an Antarctic ataxite. (Japan Polar Research Association Photograph, courtesy of Tetsuya Torii.)

numerous chemical types of iron meteorites is derived either from a separate parent body or from a distinct area within an asteroid.

Stony–iron meteorites, which contain both silicates and metal, are usually explained as having been formed at the boundary between mantle and core regions of parent bodies. There are four different groups, distinguished according to the types of minerals present and their textural arrangement (Figure 28.22). Pallasites, which are composed of olivine grains embedded in metal, consist of approximately equal parts metal and olivine and are the most common type of stony–iron meteorites. Mesosiderites, which are mixtures of silicates and metal, lack a specific texture. Most mesosiderites are breccias, and their stony parts are similar in composition to howardite achondrites and to terrestrial basalts. There is only one lodranite, and it consists of granular olivine, granular pyroxene, and nickel–iron in approximately equal parts. There is also only one

(a)

(b)

(c)

(a)

(b)

■ **Figure 28.22** Stony–iron meteorites. (a) Salta, a pallasite. (Smithsonian Institution Photograph, courtesy of Brian Mason.) (b) Clover Springs, a mesosiderite. (Photograph by the author.)

siderophyre, and it is composed of a network of nickel–iron metal that encloses aggregates of pyroxene.

■ Conclusion

Although the last Apollo samples were returned to Earth in 1972, the discovery of the Antarctic treasure trove of meteorites has provided researchers with a windfall of new extraterrestrial material. So many new meteorites are being recovered in Antarctica that the relatively few scientists who specialize in studying meteorites will be kept busy studying them for many years! The fact that several meteorites of previously unknown types have been found indicates that our understanding of the interrelationships among the various meteorite types is far from complete. It is difficult to predict the future of meteorite research, because another unexpected fall as significant as Allende could provide an incredible wealth of new information. This unpredictability demonstrates that considerable luck is involved in meteorite research.

■ Chapter Summary

Meteorites are important sources of information for solar system researchers, as they provide "free" samples of extraterrestrial materials. They derive from meteoroids, which are fragments of debris in space in Earth-crossing orbits. When meteoroids enter the atmosphere, they burn and glow, producing a meteor. If a meteoroid is large enough, part of it may survive to reach the surface as a meteorite.

Meteor showers occur periodically and have been shown to be related to cometary debris. Most meteorites come from nonshower meteors, which have been proven to derive from asteroidal debris.

Meteorites can be classified into three groups. Stony meteorites are made primarily of silicates, whereas iron meteorites are made mainly of nickel–iron metal. Stony–iron meteorites contain both silicates and metal.

■ Chapter Vocabulary

achondrites
Allende meteorite
bolide
carbonaceous chondrite
chondrites
chondrules
cosmic rays

enstatite chondrite
fall
find
fireball
fusion crust
Hoba meteorite
iron meteorites

meteor
meteor shower
meteorite
meteoroid
Neumann lines
ordinary chondrite
parent body

radiant
sporadic meteor
stony meteorites
stony–iron meteorites
Widmanstatten pattern

■ Review Questions

1. How do meteorites and lunar rock samples compare in their value to science?
2. Distinguish among the terms meteoroid, meteor, and meteorite.
3. Describe the appearance of a meteor in the sky, and explain the cause of what is seen.
4. What is the difference between a meteorite fall and a meteorite find?
5. Summarize the events leading to the acceptance by science that meteorites were indeed objects of extraterrestrial origin.
6. Describe what is seen during a meteor shower.
7. Explain the evidence linking meteor showers to comets.
8. Explain why Antarctica has become an important resource for meteorite studies.
9. What lines of experimental evidence link asteroids and meteorites?
10. Describe the three major types of meteorites.
11. What are chondrules? Explain the classification of chondrites and achondrites.
12. Why are carbonaceous chondrites especially interesting to researchers?
13. What is the Widmanstatten pattern, and how does it form?

■ For Further Reading

Burke, John G. *Cosmic Debris: Meteorites in History*. University of California Press, Berkeley, 1986.

Dodd, Robert T. *Meteorites: A Petrologic–Chemical Synthesis*. Cambridge University Press, Cambridge, England, 1981.

Dodd, Robert T. *Thunderstones and Shooting Stars: The Meaning of Meteorites*. Harvard University Press, Cambridge, MA, 1986.

Hutchinson, Robert. *The Search for Our Beginning*. British Museum, London, 1983.

McSween, Harry Y. Jr. *Meteorites and Their Parent Planets*. Cambridge University Press, Cambridge, England, 1987.

Sears, Derek W. *The Nature and Origin of Meteorites*. Oxford University Press, New York, 1978.

Wagner, Jeffrey K. "An Overview of Meteorite History as Revealed by Isotopic Dating." In *The Astronomy Quarterly*, vol. 4, 1983, p. 171.

Wagner, Jeffrey K. "The Sources of Meteorites." In *Astronomy*, February, 1984.

Wasson, John T. *Meteorites: Classification and Properties*. Springer-Verlag, New York, 1974.

Wasson, John T. *Meteorites: Their Record of Early Solar-System History*. Freeman, New York, 1985.

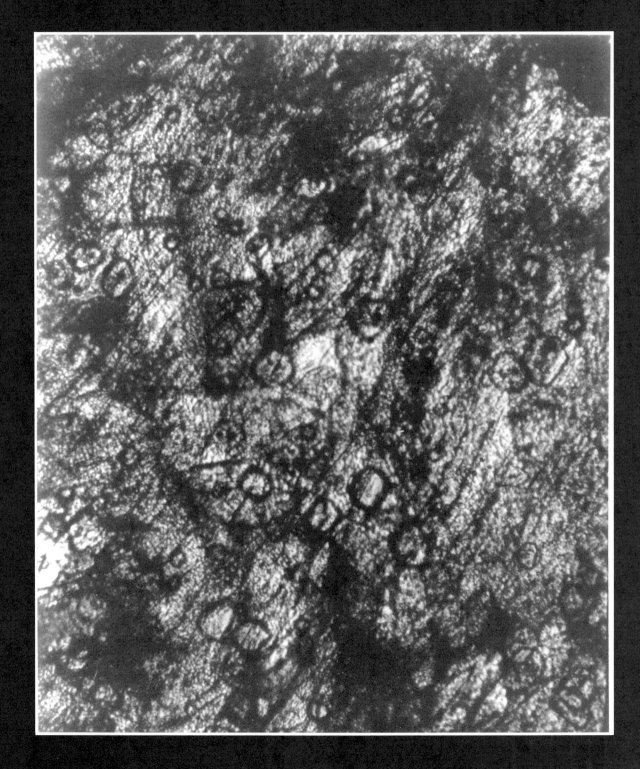

29

Cosmogony of the Solar System

CHAPTER OBJECTIVES

After studying this chapter, you should be able to

1. Understand how we can determine the age of the solar system.
2. List the important constraints on solar system formation.
3. Describe the solar nebula and what conditions were like as the solar system began to form.
4. Summarize the steps involved in the accretion of the planets.
5. Explain how we believe the Sun, comets, asteroids, meteoroids, and planetary satellites originated.
6. Understand why the Moon's origin is difficult to explain.

A polished section of the Allende carbonaceous chondrite, magnified 200 times, reveals grains of minerals that condensed from the solar nebula as the solar system was forming. (Photograph courtesy of William D. Partlow, III.)

As we near the end of our study of the solar system, it is hoped that we have gained an appreciation of its diversity and complexity. Centuries of research have given scientists a good understanding of the solar system's worlds and the phenomena occurring within them and on their surfaces. As a result of this understanding, researchers now have the ability to begin to answer questions of great interest and importance: How did the objects in the solar system originate? What processes shaped them and made them the way they are today? The great interest in understanding the origins of celestial objects has led to the development of a branch of astronomy called **cosmogony**, which deals with the origin and development of celestial objects. Some researchers who specialize in cosmogony study the origin of galaxies or the entire universe, but this chapter discusses current understanding of how our solar system formed.

Many theories have been proposed to explain how the solar system came to be. It is hard to test these theories, as we can't go back in time and watch the solar system being created. However, there are ways to tell whether our theories are right or wrong. One test is that any theory of solar system formation must account for the observed properties that solar system objects now display. For example, the presence of the ecliptic plane must be explained by our theories. A second way to test cosmogonical theories is to examine very old objects that have changed little since their formation, such as carbonaceous chondrites and some lunar rocks. Such objects contain "primitive" materials that can no longer be found on the Earth. A third source of information is the study of areas in space where other stars and solar systems may be forming today. We can assume that our solar system probably formed by the same processes as the ones that we currently witness in operation.

Although we cannot go back into the past and witness the formation of various solar system bodies, these objects carry within them informa-

423

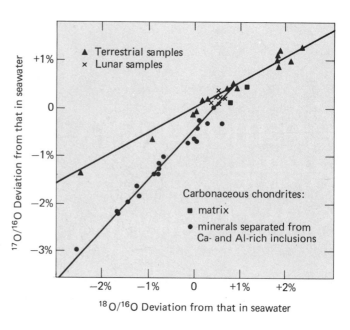

■ **Figure 29.1** Oxygen fractionation diagram. The ratios, compared to ocean water, plot along one line for the Earth and Moon, another for carbonaceous chondrites. (Graph originally appeared in a paper by R. N. Clayton et al. In *Science*, vol. 192, pp 485–8. Copyright 1973 by the American Association for the Advancement of Science. This version of the graph is taken from Wood, John A. "Meteorites." In Beatty, J. Kelly and Andrew Chaikin, editors. *The New Solar System*. Copyright 1990 by Sky Publishing Corporation. Reproduced with permission.)

tion that, if properly deciphered, can help us understand how they formed. One of the techniques used to decipher the past is radiometric age dating, which was discussed in Chapter 8. It is important because it can reveal the ages of objects. Another technique is the study of the elemental isotopes present in a body, which can reveal the relative locations in the solar system where objects formed. The fact that a single element can exist as several isotopes, even if radioactive decay is not involved, can provide evidence about solar system cosmogony. This is because the slightly different masses of isotopes of the same element cause them to behave somewhat differently in chemical reactions. As a result, different physical localities end up with different ratios of isotopes, a process called **fractionation**. For example, oxygen occurs as isotopes of weights 16, 17, and 18. A sample of material can be studied to determine the ratios of 17 to 16 and 18 to 16, and the values can be plotted on a graph (Figure 29.1). If such a plot is made for isotope ratios found in rocks from the Earth, Moon, and achondrites, the values are all found to lie on the same line, called a fractionation line. The values obtained from chondrites, however, do not lie along this line. This indicates that chondrites may have formed in a different area of the solar system

from the other objects, or else some other factor accounted for the difference.

■ When Was the Beginning?

One of the most basic questions about the origin of the solar system is when it actually began. In this section, we summarize how the currently accepted figure of about 4.6 billion years ago was determined.

Until the 1800s, nearly everyone, including scientists, believed that the Earth had been formed very recently, only about 6000 years ago. This figure was determined by counting the generations since the Creation of the Earth, as described in the Bible. However, as geology developed as a science in the early 1800s and geologists began to understand the Earth, this figure began to seem too small. Geologists saw that most current geological processes are slow and gradual, and they felt that these processes had operated in the same gradual way in the past. This idea, called the **Principle of Uniformity**, suggested that millions of years would have been required to shape the Earth's surface as we see it today. One application of uniformity was an attempt to use the salinity of the Earth's oceans to calculate the age of the Earth. Incoming river water contains low concentrations of dissolved salt, so minute amounts of salt are constantly being added to the oceans. Nineteenth-century calculations, which assumed that the oceans were salt-free when they first formed, indicated that this process could have built up ocean salinity to its present level in about 100 million years. Today we know that the reason salinity is not higher than it is, despite the Earth's great age, is that salt is constantly being removed from the oceans to form deposits of rock salt.

Meanwhile, evidence from other branches of science also seemed to indicate a very old Earth. In biology, the theory of evolution proposed that the development of the various forms of life on Earth was a gradual process requiring millions of years. Physicists also predicted an old Earth, based upon measurements of heat flow from the Earth's interior. They believed that the Earth had originally been entirely molten and calculated that 70 million years would have been required for it to cool to its present state. (They were not yet aware of internal radioactivity as a heat source.) Physicists also placed an upper limit on the Earth's age based upon the fact that gravitational contraction could power the Sun for only about 100 million years. Unfortunately, geologists and biologists began to feel that the Earth was even older than that.

As the 20th century began, new understanding of the atom solved the age problem. Nuclear fusion was learned to be the energy source for the Sun, and it was soon realized to be sufficient to enable the Sun to last for 10 billion years. Radiometric age dating was discovered, and as the century has progressed, older and older terrestrial rocks have been found. Currently, the record is held by some Canadian rocks, which are 3.96 billion years old. (Some Australian rocks contain mineral crystals that are 4.2 billion years old, but those minerals were freed by weathering from an undiscovered rock of that age and then incorporated into a new, younger rock.) Meteorites as old as 4.6 billion years and lunar samples of almost the same age have been found. These age determinations lead us to believe that all of the objects in the solar system are about the same age.

Another technique for estimating the Earth's age relies on isotopic abundance ratios. Currently, there is about 140 times more uranium-238 (which has a half-life of 4.5 billion years) than uranium-235 (which has a half-life of 1.3 billion years). However, it is calculated that these two isotopes are formed in approximately equal amounts in reactions inside stars and would have been present in nearly equal amounts in the material from which the Earth formed. Both isotopes have decayed in the time that they have been inside the Earth, and the differences in decay rates and the observed abundances can be used to calculate the age of the Earth, with the result again being about 4.6 billion years.

The modern scientific consensus is that the solar system formed about 4.6 billion years ago. Recall from Chapter 2 that this figure makes our solar system much younger than both the universe as a whole and the Milky Way galaxy. The age of the entire universe can be determined by examining one of its most basic characteristics, its expansion, and measuring the expansion rate. As described in Chapter 2, this expansion is the result of the Big Bang event, which probably occurred sometime between 15 and 20 billion years ago. After the Big Bang, individual galaxies, including the Milky Way, began to form from the hydrogen and helium produced during the Big Bang. The age of our galaxy can best be determined by studying the globular clusters of stars that surround it, and such determinations indicate that our galaxy is about 12 billion years old. This means that the Sun is a second- or third-generation star, having formed long after the first stars in our galaxy.

The fact that the Sun and the solar system are much younger than the Milky Way has an important implication regarding the origin of the solar system's constituent elements. This is because the Big Bang produced only the elements hydrogen, helium, and small amounts of lithium. Present theory indicates that heavier elements are produced only inside stars, both during their lives and as they explode as **supernovae**. Supernova explosions blast heavy elements into surrounding space, where they can eventually be incorporated into other stars and planets. By the time our second- or third-generation solar system formed, there were enough heavy elements present to form the terrestrial planets.

■ Constraints on Solar System Formation

Before we outline the sequence of events that modern researchers believe led to the formation of the solar system, we must examine the requirements that our theory must meet. In other words, what are some of the essential properties of the solar system that a successful theory must explain? Several properties involve orbital and rotational motion. The orbits of the planets are essentially coplanar, and the Sun's equator lies nearly in this same ecliptic plane. All planets revolve in the prograde direction, and all of the planets except Venus, Uranus, and Pluto have prograde rotation. (The Sun rotates in this direction as well.) The planets all have similar rotational periods, except where they have been slowed by tidal interactions, and most of the planets have low axial obliquities. Planet orbits are nearly circular, and the spacings of many of them obey Bode's law. Most of the angular momentum of the solar system is in the planets, not in the Sun. This is because the massive Sun rotates very slowly, whereas the planets, less massive but far away, move rapidly enough in their orbits that their angular momentum is greater. This is particularly true for the gas giants. (**Angular momentum** is a quantity for rotating or revolving objects that is somewhat analogous to momentum for objects moving in a straight line. The angular momentum, L, of an orbiting object is given by the equation

$$L = mav,$$

where m is the mass of the object, a is its semimajor axis, and v is its average orbital velocity. For a rotating object, angular momentum is given by

$$L = Cmr^2\omega,$$

where C is the object's moment of inertia coefficient, m is its mass, r is its radius, and ω is its rotational velocity in radians per second. Table 29.1 lists the angular momentum of various solar system objects.)

■ Table 29.1 Angular Momentum in the Solar System

Object	Angular Momentum (g cm²/sec)	Percentage of Total
Sun	1.6×10^{48}	0.5%
Mercury ⎫ Venus ⎬ Earth ⎪ Mars ⎭	4.95×10^{47}	0.2%
Jupiter	1.9×10^{50}	60.4%
Saturn	7.8×10^{49}	24.8%
Uranus	1.7×10^{49}	5.4%
Neptune	2.6×10^{49}	8.3%
Gas giant total	3.14×10^{50}	98.9%
Pluto	1.4×10^{48}	0.4%

Other important characteristics of the solar system involve the physical properties of its various objects. The planets differ in composition, with the controlling factor being distance from the Sun. The inner, terrestrial planets are rocky and metallic, the outer gas giants are primarily hydrogen and helium, and outermost Pluto is icy. The planets and their satellites resemble miniature solar systems, and most satellites orbit regularly in the equatorial plane of their planet. All of the solar system's solid bodies have experienced numerous impacts. Finally, the solar system appears to be surrounded at great distance by a vast cloud of comets.

This is certainly a long list of properties, and even more items could be added to it. However, the modern theory of solar system formation, which states that all of the bodies in the solar system formed as a by-product of the formation of the Sun itself, seems adequate to explain many of the solar system's features. Of course, some features of the solar system have not yet been adequately explained, but future work could answer many remaining questions.

■ Solar System Beginnings

Although several scientific theories have been proposed to explain the solar system's origin, only one has stood the test of time: that the entire solar system, including the Sun, formed at once.

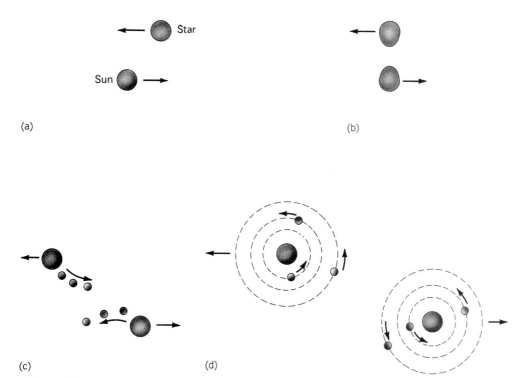

■ **Figure 29.2**   The near-miss theory of planetary formation. The close passage of the Sun and another star pulls material from each, which develops into planets around each star.

Although nearly all modern researchers accept this theory, a popular theory at one time was that the planets formed after the Sun did. This theory, which had many proponents, from Georges Louis de Buffon in 1745 to J. H. Jeans in 1917, suggested that the planets formed as a result of the passage of another star close to the Sun, with mutual gravitation drawing material outward from both stars. This material subsequently would have condensed to form planets around both stars (Figure 29.2). If true, this "near miss" theory would mean that the existence of the solar system was the chance result of an unlikely occurrence, but modern theorists have concluded that such a process would not be possible, either dynamically or chemically. The modern theory, which is the result of research by numerous 20th century scientists, is that instead of being a cosmic fluke, planetary formation seems to be a common by-product of the formation of stars.

In recent years, astronomers have learned much about the formation of stars. Although the actual formation of a single star takes too long to be witnessed completely from start to finish, we can observe stars in various stages of development and piece together the entire sequence of events. One of the most exciting discoveries is that evidence has been obtained that leads us to believe that planetary systems are currently being formed around some stars. We will describe the search for other planetary systems in the next chapter.

What has been learned about the formation of stars? Star formation appears to occur within clouds of gas and dust called **nebulae**. Most nebulae are about 99% hydrogen and helium gas and 1% dust, or grains of carbon and silicates. Some of the material in a nebula may be matter that has never before been incorporated into a star, whereas other nebular matter has been thrown into space from previously existing stars, either during their lifetimes or when they died by exploding as supernovae. The gas in many nebulae emits light because of excitation from nearby stars (Figure 29.3). A single nebula is normally the birthplace of many stars, which form a cluster or association. Some stars remain tightly bound to their siblings forever in a star cluster, whereas other groups of stars formed together eventually disperse. The Sun, for example, is no longer closely associated with other stars, and it is impossible to identify any stars that may have formed near to it in space and time.

Many young stars are seen in nebulae in the Milky Way today. These nebulae often contain dark areas, where nebular light is blocked by dense, opaque concentrations of dust (Figure 29.4). These dark sites are believed to be places where gravitational collapse of some of the nebula's material has occurred during the process of star formation. When a star forms in this way, it is initially invisible because of the remaining opaque material, but many such young stars have been detected and studied using infrared radiation, which can escape from these areas even though visible light cannot. Eventually, the leftover concentrated material disperses and the star becomes visible.

At least half of the stars in the Milky Way galaxy appear to be multiple stars with two, three, or even more components. Exactly why multiplicity appears to be favored when stars form is a mystery, but some theorists have proposed that the formational process leads to either a multiple star or else a single star with a planetary system. This is because the nebular material from which a star forms usually contains much more mass than is needed to form a single star. It is possible that planets could be found in a multiple-star system as well as in a single-star system. Recall from Chapter 17 that if Jupiter had been about 80 times more massive than it is, it would have provided a second star for our solar system.

Let us now trace the specific sequence of events that led to the formation of our solar system. The initial stage was the existence of the solar nebula, which was only a small portion of a larger nebula. We have no way of knowing at this point how large the total nebula was, or how many stars other than the Sun were formed from it. It is probable that most of the material in this nebula had been cycled through previously ex-

■ Figure 29.3 The Orion Nebula. This beautiful celestial object is a nebula in which star formation is currently occurring. (Photograph courtesy of the Observatories of the Carnegie Institution of Washington.)

■ **Figure 29.4**   Dark areas in the Rosette Nebula. These areas mark dense, dusty concentrations of nebular material where star formation may be beginning. (Palomar Observatory Photograph.)

isting stars, because it contained elements heavier than hydrogen and helium. Both the Sun and the planets contain large amounts of these heavy elements. As mentioned previously, these elements would not have been present immediately after the Big Bang.

The exact size and mass of the solar nebula are uncertain, but there had to be sufficient material present to account for all of the solar system's bodies, plus any volatile material that was driven off from the system. Estimates of the original mass suggest that the solar nebula had about twice the total mass of the present Sun. There are two lines of thought regarding temperatures within the nebula as formational events began. One suggestion is that the nebula was originally hot (about 1800 K) and cooled slowly. Another idea is that it was initially cold (about 300 K), then warmed after large bodies began to form within it, and finally cooled again.

The solar nebula existed for an undetermined period of time before the formation of the solar system began. All of the ingredients of the solar system had to be present before the formative process started. Evidence from radioactive

decay products found in some meteorites indicates that several short-lived radioisotopes (including iodine-129, whose half-life is 17 million years, and aluminum-26, whose half-life is 720,000 years) were incorporated into the meteorites when they formed. Obviously, these materials could not have been present in the nebula for billions of years prior to the solar system's formation. What could have produced them? Present theories state that one or two nearby supernovae injected quantities of newly formed, short-lived radioisotopes, as well as other materials, into the solar nebula just as the formational process began. These supernovae marked the deaths of stars that formed in the same overall nebula as the Sun but prior to it. These stars, however, were much more massive than the Sun, so they evolved quickly and died violently as supernovae. Another key result of the supernova (or supernovae) was that shock waves from the explosion(s), passing through the nebula, could have initiated the actual formational process.

How could a supernova have helped form the solar system? A nebular cloud is normally stable, with the gravitational attraction between

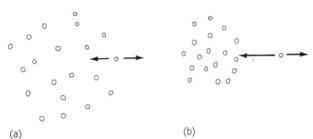

(a)   (b)

■ **Figure 29.5** Gravitational collapse. (a) The gravitational attraction among the particles in this nebula is balanced by their kinetic energy, so the nebula maintains its size. (b) The gravitational attraction (*long arrow*) among the particles in this nebula is greater than their kinetic energy (*short arrow*), so collapse occurs. Such collapse could be triggered by a supernova shock wave.

grains and molecules balanced by their kinetic energy of motion because of the cloud's temperature. According to the **virial theorem**, a gravitationally bound system is stable only if the magnitude of its total gravitational potential energy, P, equals twice that of its total kinetic energy, K. This condition for stability can be expressed by the equation

$$2K + P = 0.$$

(Recall from Chapter 3 that potential energy is negative.) An incoming shock wave would have caused the molecules to become closer together, enabling attractive forces to become predominant. At this point, a process called **gravitational collapse** would have begun (Figure 29.5). As a result of collapse, the solar nebula detached itself from the surrounding gas of the larger nebula and began to contract.

All nebular material tends to have rotational motion of some sort, and the law of conservation of angular momentum requires that as the solar nebula contracted, it would have begun to rotate faster. Forces produced by the increased rotational rate would have caused the collapsing nebula to assume a disk-like shape (Figure 29.6). In addition to conservation of angular momentum, the disk shape would have been caused in part by particle collisions, which average out the different directions of motion. (This is similar to why planetary ring systems form in a plane, as was described in Chapter 18.) As this disk formed, the

ecliptic plane and the prograde direction of motion predominant today in the solar system was established. In addition, the collapse caused pressure and temperature to increase inside the solar nebula. The most extreme conditions occurred in its center, where most of the material was concentrated and where the Sun would eventually form.

At this time, the nebula could best have been described as a hot gas. While the Sun formed from gas and has remained in that state, solid material was required for the development of many of the solar system's objects. It is believed that solid grains began to form from the nebular gas as it began to cool. The process whereby solid grains form directly from gas is called **condensation** and is exemplified by the formation of snowflakes inside clouds.

During the 1960s and 1970s, chemists determined the **condensation sequence** for various materials in the solar nebula (Table 29.2). For example, the first solids to condense were refractory materials such as titanium and aluminum, which are important components of some granular inclusions in chondritic meteorites. As the temperature fell, silicate minerals condensed, and at still lower temperatures, ices formed. Of course, at any given time, temperature was not uniform throughout the solar system; the inner solar system was always warmer than the outer, and ices probably never condensed there. This change in temperature as a function of distance from the center of the solar nebula is called the thermal gradient of the nebula. The basic compositional differences between objects in the inner and outer solar system are a direct result of the thermal gradient present at the time of the formation of these objects. This explains, for example, why there are no icy objects in the inner solar system.

■ Planetary Accretion

Although the various types of solar system bodies were probably formed simultaneously, for clarity we will discuss the origin of the various classes of solar system objects separately, beginning with the planets. The first step in the formation of

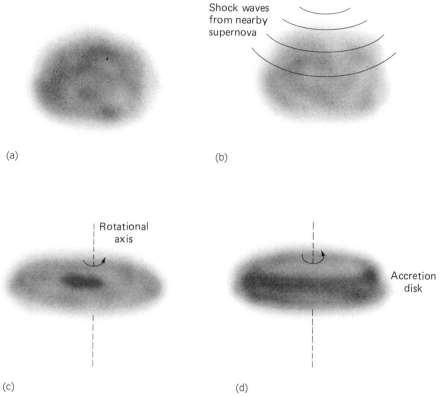

■ **Figure 29.6** Formation of a disk. A nebular region (a) is hit by a shock wave and begins to collapse (b). As it collapses, it begins to rotate faster (c) and assumes a disk-shaped form (d).

■ **Table 29.2**   Condensation Sequence in the Solar Nebula

Temperature (K)	Condensations and Reactions Occurring
1600	Condensation of refractory oxides such as CaO, Al_2O_3, TiO_2, and rare-earth oxides
1300	Condensation of metallic nickel–iron alloy
1200	Condensation of enstatite ($MgSiO_3$)
1200–490	Progressive oxidation of remaining metallic iron to FeO, which in turn reacts with enstatite to make olivine [$(Fe, Mg)_2SiO_4$]
1000	Reaction of sodium with Al_2O_3 and silicates to make feldspar and related minerals; condensation of potassium and other alkali metals
680	Reaction of H_2S with metallic iron to make troilite (FeS)
550	Combination of water vapor (H_2O) with calcium-bearing minerals to make tremolite
425	Combination of water vapor with olivine to make serpentine
175	Condensation of water ice
150	Reaction of ammonia gas (NH_3) with water ice to make the solid hydrate $NH_3 \cdot H_2O$
120	Partial reaction of methane gas (CH_4) with water ice to make the solid hydrate $CH_4 \cdot 7H_2O$
65	Condensation of argon and leftover methane gas into solid argon and methane
<25	Condensation of neon, hydrogen, and helium (temperature probably never fell this low, so this step did not occur in solar nebula)

Modified from Lewis, John S. "The Chemistry of the Solar System." In *Scientific American*, vol. 230 (3), 1974.

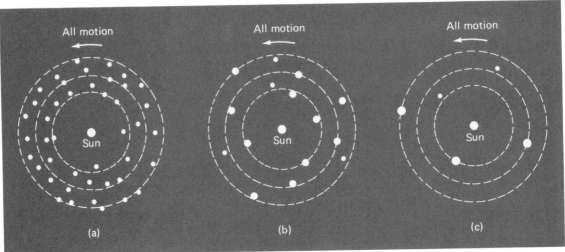

■ **Figure 29.7** The accretion process. (a) Numerous small solid bodies orbit the Sun. (b) The particles collide and form planetesimals. (c) Eventually, the planetesimals join to form the planets, with only a few smaller particles remaining.

planets was the condensation of the first solid grains from the solar nebula. Evidence gained from studying meteorites and from theoretical calculations leads us to believe that the terrestrial and icy planets, and the cores of the gas giant planets, were formed by a step-by-step growth process called **accretion**, the process occurring when smaller solid particles adhere to form larger objects. (As described in Chapter 17, it is believed that the gas giants grew to their present size by adding gas atoms and molecules individually to their accretion-formed cores by gravitational attraction, rather than by continued accretion of large, solid bodies.)

As the accretion process was beginning, the solar system was a far different place from what it is today. Other than the protostar Sun, or **protosun**, in the center, the solar system lacked large bodies. Orbiting the protosun in the ecliptic plane were countless small, newly condensed grains of material. Because nearly all of them moved in the prograde direction, grains often approached one another and joined together by **electrostatic attraction.** (The gravitational attraction between such minute grains is much less than their electrostatic attraction.) These larger grains subsequently attracted one another, and the process was repeated until particles had formed that were large enough that gravity began

to play a role in the accretion process. Even though the growing particles all orbited in the same direction, slight variations in inclination and eccentricity would have brought particles into contact with another. Eventually, rather large objects, called **planetesimals**, would have been formed as a result of gravitational accretion (Figure 29.7). Although it is not known with certainty just how large planetesimals were, the larger asteroids present today are probably similar to the objects that accreted to form the Earth and other terrestrial bodies.

The development of powerful computers has enabled researchers to simulate this accretion process. Thousands of "particles" can be studied to see exactly how their interactions lead to accretion. These calculations indicate that planetesimals as large as hundreds of kilometers across would have formed in the nebula only a few tens of thousands of years after the formational process began. Some of these accretionary model calculations give spacings and sizes of planets comparable to what is observed in the solar system. Although this cannot be considered proof of Bode's law, it indicates that there may be a physical basis for this relationship that is not yet fully understood.

As the planetesimals continued to accrete material, many became large enough that gravi-

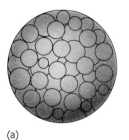
(a)

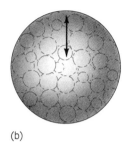
(b)

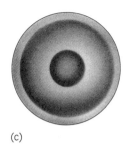
(c)

■ **Figure 29.8** Planetary differentiation. (a) A planet is initially composed of numerous accreted planetesimals. (b) As radioactive heat is generated, these planetesimals melt and their material separates as a result of density differences. (c) Eventually, a planet with several layers is produced.

tational forces began to give them a spherical shape. All of the planets (and most planetary satellites) have spherical shapes for this reason. As planets grew to their present size, internal heat was generated by the decay of the abundant radioactive elements present in the accreting material, as well as by gravitational potential energy released during the accretion process. As a result of this heating, the materials within the accreted objects became vertically mobile, and the process of differentiation occurred, with lighter materials floating to the surfaces and denser materials sinking to the cores of planets and larger satellites (Figures 29.8 and 29.9). Whether

■ **Figure 29.9** An everyday analogue of planetary differentiation: separation of a bottle of Italian salad dressing into layers. (a) Immediately after being shaken, the contents of the bottle are homogeneous, similar to a planetary interior just after accretion. (b) After several minutes, density differences cause the dressing to separate into layers. Differentiation of planetary interiors involved the same physical process. (Photographs by the author.)

(a)

(b)

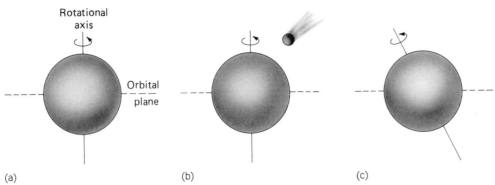

■ **Figure 29.10** Development of planetary obliquities. A planet, with its axis originally perpendicular to its orbital plane (a), is struck by a large body (b). This impact causes the planet's rotational axis to be oblique (c).

enough heat was generated for the terrestrial planets to have become entirely molten is uncertain, although most modern models suggest that they did not. Some researchers suggest that differentiation did not occur but that the planets were formed in a layered fashion. This explanation requires that the planets accreted different materials at different times — iron first, then silicates, and then volatiles — but the mechanism by which this could have occurred is unknown.

After the early episode of internal heating, the planets began to cool from their surfaces inward. Smaller objects, such as the Moon and Mars, appear to have cooled and solidified completely by the present time, unlike the Earth, which still retains a molten outer core. The internal heat of planets drove out any trapped volatile materials, which became the source of atmospheres and (in the case of Earth) oceans. Although most researchers believe that atmospheres formed in this way, there are some who believe that these materials were all accreted from outside after the solid planets had formed. No matter which theory is correct, planetary atmospheres have been changed considerably from their original state, as was discussed in Chapter 11.

As planets began to assume their present form, the remaining planetesimals that had not accreted earlier smashed into them during the era of intense bombardment. Because their crusts had solidified by this time, objects like the Moon

and Mercury were heavily scarred by this occurrence, and they still retain the marks formed over 3 billion years ago. More geologically active worlds, such as the Earth, were affected just as much during the heavy bombardment era, but they no longer bear the scars, because they have been erased by later geological activity.

It is possible that the events during this heavy bombardment era may have been responsible for the axial obliquities of the planets. It is likely that the newly accreted planets had obliquities very close to 0°. However, an impact by an extremely large body traveling in a highly inclined orbit could have disrupted the rotation of the planet that it struck (Figure 29.10). One difficulty with this theory is that it fails to explain why most planets with considerable obliquities have satellite systems that orbit in the inclined equatorial plane. (The Earth–Moon system is an exception.)

■ The Origin of Smaller Bodies

Because we have discussed the origin of planetary satellites, asteroids, meteoroids, and comets in previous chapters, we will limit our discussion here to a brief review of the important points already described. Although some planetary satellites are captured objects, the regularity of many satellite systems is considered evidence that they formed in place, in orbit around their planet. As each planet was accreting, some fragments would have approached in such paths that they would have gone into orbit around the planet rather than striking it and accreting. In this way, a secondary accretion disk would have developed around most planets, a process that was obviously most effective around the gas giants. Once in orbit around the planet, these particles would have collided, their inclinations and eccentricities would have cancelled, and they eventually would have accreted into satellites with nearly circular orbits in the planet's equatorial plane. Tidal effects during the years since their formation have produced an important characteristic of most satellites, namely, that most of them have perfect spin-orbit coupling with the planets around which they orbit. Another characteristic of most satellites is that they are much smaller than the planets around which they orbit, with the Earth–Moon and Pluto–Charon systems as notable exceptions. For this and other reasons, the origin of the Moon is not well understood; we will consider the special problem of its origin shortly.

Satellites, like planets, were intensely bombarded and heavily cratered. Unlike planets, however, they experienced two bombardment episodes. They first were struck by material still in orbit around the Sun that was leftover from the formation of the planets, and then they were hit by particles in orbit around the planet that were leftover from the accretion of satellites. Because of their generally smaller sizes and less internal heating, relatively few satellites have been resurfaced by internal geological heating.

Although most planetesimals participated in the formation of the planets and their satellites, some of them remained in their initial state and became the asteroids. Apparently, Jupiter had reached its present size before the planetesimals now in the asteroid belt could accrete into a single planet, and its extreme gravity then prevented such accretion from occurring. As mentioned in Chapter 28, evidence from meteorites indicates that many collisions have occurred within the asteroid belt since the formation of these objects. These collisions have sent asteroid fragments on their way to the Earth and other planets, forming meteoroids.

Although even the largest asteroids are much smaller than planets, enough heating occurred within some of them that metamorphism and differentiation occurred. Evidence of this heating is found in meteorites, and the melting observed to have occurred within differentiated meteorites was probably caused by the intense period of radioactive decay of short-lived isotopes such as aluminum-26 in these bodies. Evidence indicates that this era of melting occurred between 4.6 and 4.4 billion years ago and that the objects had cooled enough to begin trapping gaseous decay products by 4.5 to 3.0 billion years ago. Individual meteoroid fragments have formed throughout the history of the solar system since asteroid collisions have continued until the present.

As mentioned in Chapter 26, the origin of the Oort cloud of comets is not yet fully understood. Most theorists consider it unlikely that the comets formed at their present extreme distance from the Sun, because this region of the solar nebula would have been far too tenuous. Current theories indicate that comets formed closer to the Sun and were then transported to the Oort cloud. The large quantities of volatile materials present in comets indicate that they were formed in the cold outer solar system, probably in the vicinity of Uranus and Neptune, which themselves are rich in ices. Because comets are fairly small, they could be considered icy planetesimals that failed to be accreted into the outer planets or their satellites. After the outer planets formed, there were many opportunities for gravitational interactions between comets and the two outermost gas giants. Such interactions could have ejected numerous comets from the orbits in which they first formed into their present orbits far beyond the rest of the solar system. Calculations indicate that even if all comets had originally formed in the ecliptic plane, these interactions would have sent them off in all directions, explaining the fact that the Oort cloud is not limited to the ecliptic plane.

■ The Origin of the Sun

As mentioned earlier, the conditions at the center of the solar nebula were most extreme, and the high concentration of material there resulted in the Sun, the most massive object in the solar system. The Sun grew by gravitational collapse, rather than piece-by-piece accretion, and it is estimated that the collapse to the **protostar** stage required only about 100,000 years from the nebular state. As the protosun formed, its interior temperature rose and its rate of rotation increased. However, the modern Sun rotates very slowly, indicating that some process transferred most of its angular momentum to the planets. It is thought that this transfer may have been accomplished by a process called **magnetic braking**, which occurred because the hot solar nebula contained many ionized atoms. The rapidly rotating magnetic field of the protosun then began dragging the ionized portion of the disk along with it. This accelerated the material in the disk but also slowed the Sun's rotation as energy was transferred.

Continued gravitational collapse of the protosun caused its central temperature to rise dramatically. As the collapse occurred, the protosun warmed internally and radiated energy, although initially only at long wavelengths. When the Sun's core temperature reached about 15 million K, hydrogen fusion began, making the Sun a full-fledged star that began generating energy by nuclear fusion. Collapse did not continue after this point because the internal pressure generated by fusion helped to balance the gravitational forces promoting collapse.

It is estimated that fusion began in the Sun about 10 million years after the protostar stage began. The first few tens of millions of years of a star's life after fusion starts are marked by instability, after which the star settles down to an essentially constant light output. Newly formed stars that are still unstable can be observed today and are called **T Tauri stars** after a well-known example. During the T Tauri stage, there is an extremely strong stellar wind of gas flowing outward from the star's surface, and a T Tauri star can lose between 10^{-7} and 10^{-8} times the mass of the Sun each year. Many T Tauri stars observed today appear to be surrounded by dense gas and dust, which is either a remnant of the nebula from which the stars formed or else material ejected from them by their stellar winds. However, recent observations have indicated that numerous T Tauri stars lack this surrounding material. As many as 90% of T Tauri stars are thought to be "naked" in this way, and it has therefore been suggested that only 10% of low-mass stars like the Sun may form planetary systems. It is believed that the T Tauri phase of the young Sun was instrumental in dispersing the gaseous remnant of the original solar nebula, but the exact length of time that this would have required is not known. After leaving the T Tauri stage, the Sun became a stable star. Because a star of the Sun's mass is thought to have a lifetime of about 10 billion years, our star is probably halfway through its life.

■ The Origin of the Moon

One of the aspects of the Earth–Moon system that is most difficult to explain is why the Moon is so large in relation to the Earth. All other planet–satellite systems in the solar system, with the exception of Pluto and Charon, are marked by satellites whose diameters are only a few percent of those of the planet they orbit. The Moon, however, has a diameter over one-quarter that of the Earth's. How did such a "double planet" form? Prior to the Apollo explorations, three simple theories of lunar formation had been proposed. As is often the case in science, the Apollo results seemed to raise more questions than they answered, and the solution to the problem of the Moon's origin is still not known.

The first serious theory of lunar origin was proposed in 1898 by George Darwin (1845–1912), who stated that the Moon had broken away from the Earth shortly after both formed as a single unit. Early proponents of this **fission theory** believed that the Earth's huge Pacific Ocean basin marked the site where the Moon had torn from the Earth, but we know today that the Pacific Ocean basin is not a permanent feature but simply reflects the current situation of global plate tectonics. A substantial objection to the fission theory is the extreme difficulty in explaining how such a split would have been possible dynamically. Another objection is that lunar surface rocks are sufficiently different from those at the Earth's surface that a common origin seems unlikely. The fission theory does, however, have one point in its favor: if the split occurred after the Earth had differentiated, a mantle source for the Moon could explain its lower density; however, if the Moon is as old as the Earth, there probably would not have been sufficient time for the Earth to separate into layers prior to the split. The many problems of the fission theory make it unlikely that the entire Moon broke away from the Earth.

Another theory proposes that the Moon formed elsewhere in the solar system, only to end up in an Earth-crossing orbit, from which it was captured by our planet. One line of evidence supports this **capture theory**: the Moon's orbit lies approximately in the ecliptic plane rather than in Earth's equatorial plane, which is the orientation of most satellite orbits. However, there are many lines of evidence against capture. One is that the probability of capture without impact or gravitational disruption is very low. Another is that the Moon's orbit is prograde and nearly circular, which would be unlikely if capture had occurred. Finally, the variation within the solar nebula would have led to a bigger difference in composition between the two bodies had they been formed significantly far apart. This is not observed to be the case, as the Earth and Moon have very similar ratios of oxygen isotopes. For these reasons, it is considered unlikely that the entire Moon was captured after it had formed.

A third model of lunar formation is that the Earth and Moon formed simultaneously at about the same places where they are today. One problem with this **mutual formation theory** is explaining why the accreting material was so nearly evenly divided between the two bodies rather than primarily building up the Earth. Another problem is that a mutual formation would make it difficult to account for the observed chemical and isotopic differences between the two bodies. Specifically, the Moon is depleted in volatile materials and siderophile elements (elements usually found in association with iron) compared to the Earth.

As we have seen, each of these three simple theories has shortcomings. The modern, complex theory of lunar formation can best be described as a combination of some of the aspects of the simple ones. The Moon may have begun when a portion of the Earth was broken off and thrown into orbit by a large impact. This object then grew into the Moon by accreting particles already in orbit around the Earth. Alternately, the initial Moon may have been a smaller body that was gravitationally captured. Even if it started out in an unusual orbit, the impacts associated with the accretion process would have altered its orbit to the present form. A "giant impact" scenario, recently proposed by William K. Hartmann and Alastair G. W. Cameron, hypothesizes that the impact of a large, planetesimal-sized body onto the Earth sprayed a huge amount of small fragments of material into Earth orbit. They suggest that this material would have subsequently accreted into the Moon.

No matter how the Moon originally formed, we do know that it has been slowly moving away from the Earth because of tidal effects. Another result of this tidal interaction is that the Earth's rotation has slowed through time. As described in Chapter 15, the era of intense bombardment on the Moon occurred about 4 billion years ago, and the formation of the maria as the large impact basins were filled occurred between about 3.9 and 3.1 billion years ago.

■ Conclusion

Although many questions remain about the origin of the solar system, prospects for additional insight are good. Most of our new knowledge will be gained by exploration of additional solar system bodies and analysis of material returned from them. Other knowledge will be added by theoreticians modeling conditions in the solar nebula. Most exciting of all is the fact that disks of material that are apparently in the process of accreting into planets have been discovered around several other stars. Careful study of these accretion disks may give us additional information about what conditions in our own solar system were like at the time of its origin.

■ Chapter Summary

Cosmogony, the branch of astronomy that studies origins, has developed a modern theory of solar system formation that states that the solar system's various bodies formed from a large nebula about 4.6 billion years ago. The solar system's current properties provide specific constraints on its original formation.

Current evidence suggests that a nearby supernova triggered the solar nebula's collapse. The Sun formed at the center of the nebula, whereas solid grains condensed from the hot gas in the surrounding areas. Those solid grains eventually accreted, forming the planets. Accretion of particles orbiting the planets yielded satellites. Comets formed in the outer solar system and were ejected into the Oort cloud, whereas asteroids are planetesimals that failed to accrete. Although several theories have been proposed to explain the origin of the Earth's Moon, none has been universally accepted.

■ Chapter Vocabulary

accretion
angular momentum
capture theory
condensation
condensation sequence

cosmogony
electrostatic attraction
fission theory
fractionation
gravitational collapse
magnetic braking

mutual formation theory
nebula
planetesimal
Principle of Uniformity

protostar
protosun
supernova
T Tauri star
virial theorem

■ Review Questions

1. What is cosmogony?
2. In what ways can theories of the solar system's origin be tested?
3. Explain how fractionation data reveal information about the solar system's origin.
4. What are the currently accepted values for the ages of the Earth, the Milky Way galaxy, and the entire universe? How have these figures been determined?
5. What basic properties of the solar system must be explained by any theory of its origin?
6. Why is it now considered unlikely that the solar system was produced by a "freak accident?"
7. Describe what has been learned about the formation of stars.
8. Describe the conditions in the nebula from which the solar system formed.
9. Why is it considered likely that a nearby supernova explosion had something to do with the solar system's formation?
10. Explain how the early solar nebula formed into a disk shape.
11. Describe the processes that led to the formation of the planets.
12. Describe the process of differentiation and how it led to planetary atmospheres.
13. How did planetary satellites form?
14. Summarize the formation of the solar system's minor bodies (comets, asteroids, and meteoroids).
15. Summarize the formation of the Sun.
16. Summarize the simple theories of Moon formation, and explain why a complex formation model is more reasonable.

■ For Further Reading

Cameron, Alastair G.W. "The Origin and Evolution of the Solar System." In *Scientific American,* September, 1975.

Hoyle, Fred. *The Cosmogony of the Solar System.* Enslow, Short Hills, NJ, 1978.

Lewis, John. "Putting it All Together." In Beatty, J. Kelly and Andrew Chaikin, *The New Solar System,* Sky Publishing, Cambridge, MA, 1990.

Schramm, David N. and Robert N. Clayton. "Did a Supernova Trigger the Formation of the Solar System?" In *Scientific American,* October, 1978.

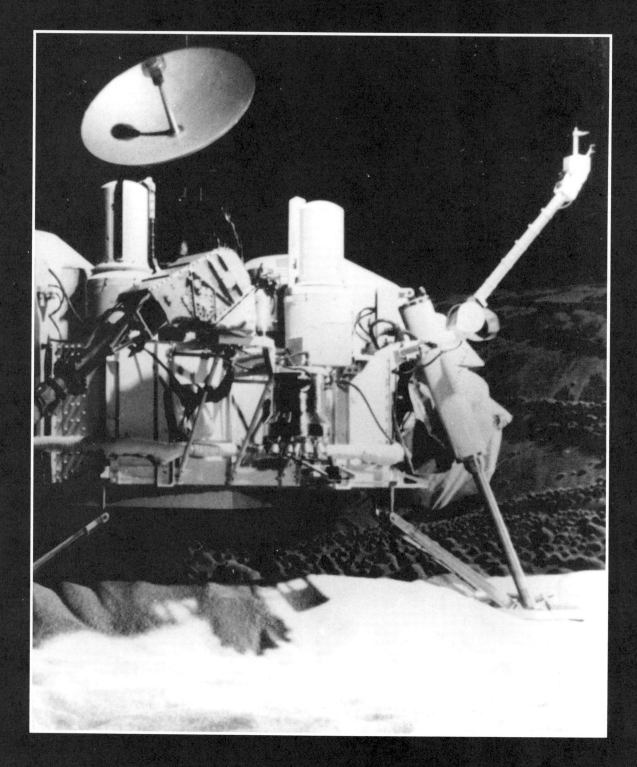

30

The Search for Other Planetary Systems and Extraterrestrial Life

CHAPTER OBJECTIVES

After studying this chapter, you should be able to

1. Understand how scientists define "life."
2. List the theories that have been proposed to explain the origin of terrestrial life.
3. Summarize the characteristics of life and the unique traits of humans.
4. Explain why other life in the solar system is considered unlikely.
5. Describe the evidence indicating that other stars may have planets.
6. Understand the concept of a star's habitable zone.
7. Describe why detection of life in other planetary systems would be difficult.

The first "biologist" to visit another planet, Viking. This test model is identical to the two spacecraft that landed on Mars in 1976 and investigated soil samples for evidence of metabolism. (NASA/JPL Photograph.)

A s we look at the myriad stars visible in the sky on a dark, moonless night, we cannot help but wonder how many of them have planetary systems like our own solar system. An even more intriguing question is whether Earth alone of all of the universe's bodies is the home of living organisms. Although we have not yet identified any other planets around other stars, there is increasing evidence that they do exist. However, the question "Is there life elsewhere than on Earth?" cannot yet be answered with any degree of certainty. Obviously, the answer is either yes or no, and whichever proves to be correct, the result will certainly revolutionize the way we think about ourselves and the universe around us. (If we modify the question to read "Is there intelligent life elsewhere than on Earth?" it becomes even more intriguing, but the answer is still unknown.)

■ The Nature of Life

Before we describe the search for life beyond Earth, it is necessary that we have a good understanding of what life on Earth is like. Evidence from fossils indicates that terrestrial life has existed for at least 3.5 billion years. Exactly what is life? Although philosophers may debate the meaning of life, scientists might define life itself as a series of chemical reactions occurring within a living being known as an **organism**. Organisms take in gases and water from the environment around them. Animals also take in food, but green plants produce their own food by the process of **photosynthesis**. The materials taken into the organism interact in a series of chemical reactions called **metabolism**, from which the organism obtains energy and nourishment. The solid, liquid, and gaseous waste products resulting from metabolism are subsequently expelled from the organism. The ultimate result of metabolism is growth of the organism and reproduction.

Terrestrial organisms are primarily composed of the elements carbon, hydrogen, oxygen, and nitrogen. Terrestrial life is often called "**carbon-based**," and it is no accident that this is so, because carbon has a very active and complex chemistry. Liquid water is necessary for all life on Earth, and all terrestrial organisms contain water. In discussing extraterrestrial life, we often use the term "**life as we know it**" to describe life with the same basic properties as terrestrial life. Although it may be argued that there is possibly life "not as we know it" based on chemistries involving other elements, we obviously have no idea at present how such organisms would function. Furthermore, we might not even be able to identify organisms based upon completely different chemistries as being alive, even if we found them!

■ The Origin and History of Terrestrial Life

Perhaps the biggest mystery in science is how terrestrial life formed. Many ideas regarding the origin of life on our planet have been proposed. Some people believe that life originated as the result of the actions of a supernatural being, a process usually called **special creation.** Of course, if life originated in this way, it would not be a natural phenomenon and therefore would be outside of the laws of science. Until the 1880s, many people believed that living things were even now being formed from nonliving material. The evidence for this "**spontaneous creation**" was the observation that rotten meat soon became infested with flies and maggots, and this spontaneous creation was considered to be a residue of the original process of special creation. Louis Pasteur (1822 – 1895) proved in the 1880s that spontaneous creation does not occur. He did this by isolating decaying materials from the air, which prevented the infestation of flies and maggots.

Some suggestions have been made that factors beyond our planet may have been responsible for the origin of terrestrial life. One of them is the idea called **panspermia**, which states that spores and other minute organisms travel through interstellar space propelled by the radiation pressure of stars. These organisms then propagate when they reach planets with suitable conditions. Of course, panspermia does not address the question of the ultimate formation of life, which necessarily would have occurred sometime on some planet.

Even if living beings or their dormant spores don't travel through space, there is evidence that the formation of life on Earth could have been aided by materials from space. As shown in Table 30.1, many complex organic molecules have been found in nebulae from which stars like the Sun form. Presumably, these molecules were present in the solar nebula and could have been incorporated into the Earth and other planets. Even more importantly, **amino acids** have been found in some carbonaceous chondrite meteorites. As described in Chapter 28, amino acids are not alive, but they are important components of terrestrial organisms. Their presence in meteorites shows that some of the building blocks of life might have been present originally in the solar nebula and incorporated into the Earth as it formed or brought to Earth after its formation.

Unless some sort of supernatural event was involved or panspermia brought organisms to Earth, the ultimate origin of terrestrial life must have been the biochemical development of living material from nonliving components. Even though Pasteur proved that spontaneous creation does not occur, most researchers today believe that life on Earth developed spontaneously from nonliving material. How might this have occurred? The early Earth had organic compounds present, perhaps derived from the solar nebula. Given the proper conditions, more and more complex molecular structures could have been synthesized naturally as these organic compounds reacted with one another. At some point, these molecular systems could have gained the ability to replicate their structure, and life would have begun at that point, because the most basic definition of a living organism is that it can reproduce itself. Until recently, most researchers believed that the first life on Earth formed in the ocean, because terrestrial organisms have a high water content and all of the earliest fossils are found in marine rocks. However, a new suggestion is that the beginning of life may have oc-

■ Table 30.1   Examples of Molecules that Have Been Found in Nebulae

Complexity	Inorganic		Organic	
Diatomic	Hydrogen	(H_2)	Methylidyne radical	(CH)
	Deuterized hydrogen	(HD)	Methylidyne ion	(CH^+)
	Hydroxyl radical	(OH)	Cyanogen radical	(CN)
	Silicon monoxide	(SiO)	Carbon monoxide	(CO)
	Silicon monosulfide	(SiS)	Carbon monoxide ion	(CO^+)
	Nitrogen monosulfide	(NS)	Carbon monosulfide	(CS)
	Sulfur monoxide	(SO)	Carbon	(C_2)
	Nitric oxide	(NO)		
Triatomic	Water	(H_2O)	Ethynyl radical	(CCH)
	Heavy water	(HDO)	Hydrogen cyanide	(HCN)
	Imidyl ion	(N_2H^+)	Hydrogen isocyanide	(HNC)
	Hydrogen sulfide	(H_2S)	Deuterium cyanide	(DCN)
	Sulfur dioxide	(SO_2)	Deuterium isocyanide	(DNC)
	Nitroxyl	(HNO)	Formyl radical	(HCO)
4-atomic	Ammonia	(NH_3)	Formaldehyde	(H_2CO)
			Hydrocyanic acid	(HNCO)
			Thioformaldehyde	(H_2CS)
			Acetylene	(HC_2H)
5-atomic			Methane	(CH_4)
			Cyanamide	(H_2NCN)
			Formic acid	(HCOOH)
			Cyanoacetylene	(HC_3N)
			Ketene	(H_2C_2O)
6-atomic			Methyl alcohol	(CH_3OH)
			Methyl cyanide	(CH_3CN)
			Formamide	($HCONH_2$)
7-atomic			Methylamine	(CH_3NH_2)
			Cyanodiacetylene	(HC_5N)
8-atomic			Methyl formate	($HCOOCH_3$)
			Methyl cyanoacetylene	(CH_3C_3N)
9-atomic			Ethyl alcohol	(CH_3CH_2OH)
			Cyanotriacetylene	(HC_7N)
11-atomic			Cyanotetraacetylene	(HC_9N)
13-atomic			Cyanopentaacetylene	($HC_{11}N$)

Adapted from Zeilik, Michael and Elske v. P. Smith. *Introductory Astronomy and Astrophysics,* 2nd ed. Saunders College Publishing, Philadelphia, 1987.

curred in moist clays. It is considered possible by some researchers that the atomic structure of clay minerals could have served as a catalyst for the development of life.

An attempt to duplicate the conditions of the early Earth at the time the first life appeared was conducted by Stanley Miller and Harold Urey (1893 – 1981) in the early 1950s. They produced an artificial early atmosphere, with no oxygen and some methane and ammonia, and placed it in a chamber along with water. (We know that the Earth's early atmosphere lacked oxygen because iron contained in the oldest rocks is not completely oxidized.) Everything had been sterilized, and energy simulating early sunlight was introduced by spark discharges within the system (Figure 30.1). The result of the **Miller – Urey experiment** was that, after several days, amino acids were produced in the water. As mentioned earlier, although amino acids are not alive, they are

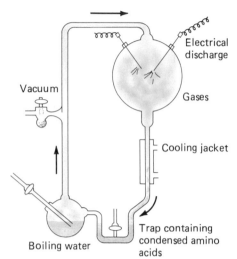

■ **Figure 30.1** The Miller–Urey experiment. A sterile chamber contained water and gases similar to those in the Earth's early atmosphere. A spark generated ultraviolet light. (From Levin, Harold L. *The Earth Through Time,* 3rd ed. Saunders College Publishing, Philadelphia, 1988.)

important components of living organisms. The implication of the Miller–Urey experiment is that given the right conditions, an important step toward forming terrestrial life could have occurred naturally. Since this experiment, more advanced organic structures have been synthesized starting with amino acids, but no one has ever created life in the laboratory from nonliving material. For this reason, the exact steps by which life began will possibly remain a mystery.

The basic unit of life on Earth is an entity called the **cell.** The simplest organisms on Earth, including those that were probably the first to appear, are composed of only a single cell, but the term simple here is misleading, because a single cell typically contains 1 trillion atoms! Single-celled organisms reproduce asexually by splitting in half to form two new organisms. Remarkable microscopic fossils over 3 billion years old actually preserve single-celled organisms in the process of dividing!

All but the simplest cells on Earth contain a **nucleus,** which stores the complete genetic information about the organism in which it is found. Genetic information, which provides each organism its specific characteristics and controls growth and development, is contained in deoxyribonucleic acid **(DNA),** which consists of long molecules shaped like a helix or twisted ladder. The DNA molecule splits down the middle and replicates itself during cell division, so that each new cell contains DNA that is a duplicate of the original (Figure 30.2). (Sometimes a mistake, called a **mutation,** occurs during the replication process.)

The majority of terrestrial organisms are multicelled. For example, there are about 100 trillion cells in the human body. This huge number of cells has led to a complex system of specialization, whereby various types of cells in the body perform different tasks. However, each cell must still take in and give off materials, just as the entire organism does, and each cell contains a nucleus

■ **Figure 30.2** Replication of DNA. A strand of DNA *(left)* splits down the center and replicates its structure, resulting in a duplicate of the original *(right).* The letters A, C, G, and T represent various components of the DNA molecule. (From Solomon, Eldra P. and P. William Davis. *Human Anatomy and Physiology.* Saunders College Publishing, Philadelphia, 1983.)

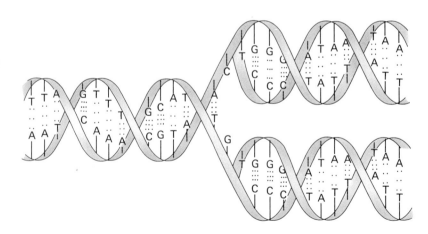

and divides. The fossil record shows us that multicelled organisms appeared only about 1 billion years ago. Since then, their complexity and diversity have increased with time. Most multicelled organisms reproduce sexually, with two parents required, and an offspring is a duplicate of neither parent but receives genetic material from both.

Biologists believe that currently there may be over 5 million different forms of life on Earth. In addition, there are many more that once lived but are now extinct. How can we explain the vast diversity of living organisms? In 1859, Charles Darwin (1809–1882) proposed that a process now called **evolution** causes gradual changes in organisms over many generations, requiring long periods of time, and that such changes can lead to the development of entirely new forms of life. Mutations during reproduction can sometimes cause new characteristics to appear in offspring. In each new generation, there are more young produced than can survive to reach maturity, and Darwin proposed that the best of these offspring survive and pass on their new, favorable traits to succeeding generations. Although evolution was once believed to be a constant, ongoing process, a newer proposal is that evolutionary change is sporadic, occurring fairly rapidly in response to an environmental crisis of some sort. For example, when the ice age began a few million years ago, some types of mammals developed thick fur and a larger size to enable them to retain heat better.

The evidence from fossils indicates that the diversity of terrestrial life began to increase rapidly about 570 million years ago, leading to the development of the most advanced animals, the vertebrates. Although there were too many milestones in the history of life to enumerate them all here, it is appropriate to discuss human beings, who have been on Earth for only a short time, geologically speaking. Humans are mammals, which are vertebrates characterized by having body hair, by being born alive rather than from eggs, and by having young that are nourished by their mother's milk. We belong to the primate group of mammals, whose specific adaptations are good vision with color sensitivity and depth perception, as well as grasping hands and feet.

Humans have several specific adaptations that are not found in other primates. We possess bipedalism (the ability to walk upright), which frees our hands, which are therefore able to use tools. Humans also have large, complex brains that enable us to control our bodies well enough to walk bipedally and to utter speech.

As humans, we call ourselves "intelligent" and often speak of finding intelligent life elsewhere in the universe. Exactly what is **intelligence?** Evidence of intelligence includes self-awareness, inquiry about the surrounding world, and the use of abstract concepts. Although no one can tell for certain, it is highly unlikely that a single-celled organism wonders about where it came from and ponders things like love and hate. Although other organisms besides humans communicate and even use tools of some sort, no other creatures have developed these abilities to the sophisticated level that humans have.

Several human developments have set us completely apart from other animals. One of these is **civilization**, an organized structure of society that allows our knowledge and ideas to be shared and transmitted from one generation to the next. Marks of civilization, such as cities and trade, have existed for thousands of years. **Technology**, the use of tools, has been part of our heritage since the first person picked up a rock and used it in some fashion, but advanced technology has existed only since the industrial revolution of the 1700s. Electronic mass communications began with AM radio broadcasts during the 1920s. One result of these and later broadcasts is that we are now publicizing our existence to anyone within 70 light years of Earth who has the ability and interest to listen! Although the foregoing are certainly brilliant accomplishments, our century has also seen the development of nuclear weapons, advanced rockets, and computers. These three advances together have given us the ability to destroy civilization and possibly even all terrestrial life. As we discuss communication with other intelligent civilizations, a very real question is, "How long will our civilization last?" We have possessed the ability to communicate with other civilizations for only a few decades. Will we survive until contact is made?

■ Other Life in the Solar System

As mentioned earlier, our search for life elsewhere is limited to "life as we know it," because we might not recognize life based on an alternate chemistry even if we saw it. This also means that our search must be limited to areas that are somewhat Earth-like. However, life on Earth is found virtually everywhere, from the hottest, driest deserts, to the coldest mountain tops, to the bottom of the deepest oceans. A planet would not have to be an exact duplicate of Earth in order to support life. However, we are limiting our discussion of extraterrestrial life to planets and planetary satellites that possess atmospheres. It is hard to imagine any other environment, such as a star or the void of space itself, where life could exist.

Most of the objects in the solar system can be eliminated as possible sites for life. All airless bodies are undoubtedly lifeless, at least on their surfaces, and many atmospheres are of compositions that would not support life. Venus, for example, is much too hot for any terrestrial life. We discussed the search for life on Mars in Chapter 16 and noted that although the Viking landers found evidence that could be interpreted as indicating life, this activity could also be explained as the result of natural, although somewhat unusual, chemical reactions. Many terrestrial organisms survive when placed into environmental chambers simulating Martian surface conditions, but this, of course, is no proof that life exists on Mars. The final word on Martian life will not be heard until either humans visit the red planet or material is brought back by an automated probe for study.

Beyond Mars, the cold conditions of the outer solar system make the prospects for life even poorer. The gas giants are unlikely to have life, but there has been some speculation that life may exist on some of their satellites. It has been proposed that Jupiter's moon Europa has an ocean of liquid water under its icy outer crust and that some form of life might exist in this ocean. Although Europa's surface is very cold, it is possible that life in Europa's watery depths, if it exists, obtains heat from the satellite's interior. Such an occurrence is not without precedent, as some organisms in Earth's deep oceans live at the mid-oceanic ridges, obtaining heat and nourishment from the material coming out from the ocean floor at these points — the only terrestrial organisms independent of the Sun! It is considered likely that the clouds of Titan may contain some organic material, but Titan's atmosphere is so cold that life there is probably impossible.

The preceding discussion shows that although no one yet knows the answer with certainty, the overall odds are probably very high against finding life elsewhere in our solar system. Therefore, we must look beyond it.

■ The Search for Planets of Other Stars

Because life is probably limited to the surfaces of planets and their satellites, in order to find life beyond our solar system, we must determine whether other stars have planets. Unfortunately, this question is not easy to answer. If planets existed around another star, they would be so faint and so close to the star that even the most powerful telescope could not detect them. Even though direct observations of extrasolar planets may always be impossible, there is considerable indirect evidence that such planets exist. Unfortunately, this indirect evidence often works best at detecting planets much larger than Jupiter. This leads to an important question: Just where do we draw the line between planet and star? The Sun is 10 times larger in diameter and 1000 times more massive than Jupiter, and intermediate objects are often called "substars" or "brown dwarf" stars. Some claims of planet discoveries have been disputed primarily because the object found is considered too large to be a true planet. It is likely that a planet no larger than Jupiter will have to be detected in order to convince everyone that extrasolar planets indeed exist.

One line of evidence that some stars have planets is that material is present around stars and seems to be in the process of forming planetary systems. Although T Tauri stars have long been

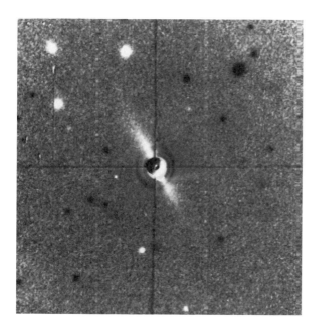

■ **Figure 30.3** A photograph of the accretion disk around the star Beta Pictoris. In order to obtain this image, most of the light from the star itself was blocked. (NASA/JPL Photograph.)

known to be enveloped in dust clouds, in 1983 the first observations were made of actual **accretion disks.** These were detected first by the Infrared Astronomy Satellite around many stars, including the bright star Vega. Later in 1983, an accretion disk was photographed from Earth around the star Beta Pictoris, which is 50 light years distant (Figure 30.3). The disk is viewed approximately edge-on and appears to be 900 to 1150 astronomical units in radius. Spectroscopic evidence indicates that it contains dark, dust-sized, carbon-rich particles, and it is unknown how advanced the Beta Pictoris system is in the accretion process, because individual large particles cannot be detected. Because an accretion disk is believed to be an essential step in the formation of planets, it is likely that planetary systems are commonplace around other stars.

Another line of evidence for planets comes from studies of the proper motions of other stars. Although we normally think of stars as maintaining fixed positions with respect to one another, close examination over many years reveals gradual motion, particularly for nearby stars. This movement, called **proper motion,** occurs because both the Sun and other stars are moving as they orbit the center of the Milky Way. The proper motion of most stars is so small (generally only a few seconds of arc per century) that their positions do not change noticeably in periods comparable to a human lifetime. Nonetheless, proper motions can be measured for many stars by observing them over long periods of time and carefully determining their positions. When the proper motions of some stars are plotted, their paths are seen to be wave-like rather than perfectly straight. The interpretation of this phenomenon is that such stars have unseen companions.

When a star is orbited by a companion of substantial mass, the **barycenter,** or combined center of gravity of the star and the orbiting companion, moves in a straight line. However, because the companion is too faint to be visible, only the star itself is seen as it oscillates back and forth around the barycenter (Figure 30.4). Most proper motion "wobbles" already detected are the result of faint stellar companions, but some could be caused by planets. Although several research studies have claimed to find planets somewhat bigger than Jupiter from proper motion determinations, the results have been questioned, and the entire field of research is very controversial. The experimental difficulties of conducting these measurements from Earth-based tele-

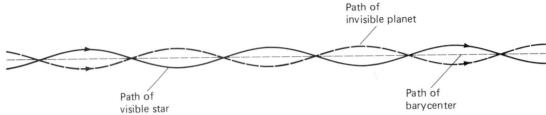

Path of
invisible planet

Path of
visible star

Path of
barycenter

■ **Figure 30.4** Irregular stellar proper motion. The *wavy solid line* indicates the path of the visible star. The path of the planet, which cannot be seen, is also indicated. Note that the barycenter moves in a straight line.

scopes are numerous, but the Hubble Space Telescope is expected to increase greatly the accuracy of these measurements. An appreciation of the skill required for making proper motion observations may be gained by noting that the "wobble" of the Sun's proper motion seen by an observer 30 light years away would be only 0.0005 seconds of arc!

In addition to proper motion, stars typically possess radial velocities toward or away from the Earth. Radial velocities are measured using the Doppler effect, and the presence of an unseen companion would cause periodic variability of this velocity, just as with proper motion. In 1988, researchers noted a star with an 84-day periodicity in radial velocity, which they believe is caused by a planet perhaps ten times more massive than Jupiter.

As mentioned in Chapter 29, most of the angular momentum of the solar system resides in the planets rather than in the Sun. Many stars otherwise similar to the Sun rotate very rapidly, and many researchers assume that they lack planets, because the process of planetary formation seems to transfer angular momentum away from the star. Sun-like stars with correspondingly slow rotation are thought to have formed planetary systems and to have lost angular momentum in so doing.

Even the statistics of stars favor the presence of other planetary systems. The fact that nearly half of all stars appear to be single stars presents

■ **Figure 30.5** The habitable zone. This hypothetical star has eight planets, but only two of them lie within the habitable zone *(shaded area)* where liquid water can exist.

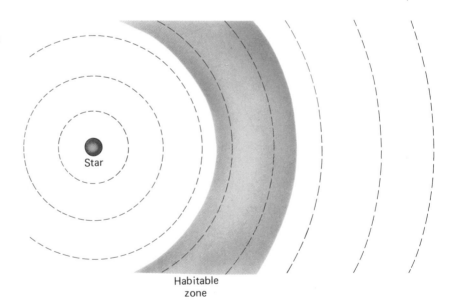

Star

Habitable
zone

an interesting question: What happened to the nebular material remaining after the formation of such stars? Perhaps the formation of planets must occur in order to use up the "leftovers" in cases in which a second star does not develop.

Although none of the preceding lines of evidence alone is proof that other planets exist, they collectively suggest that other planetary systems are likely. However, even the direct detection of planets around other stars would not prove that life exists on them. There are certainly many requirements to be met in order for life to be possible on a planet. To understand them, we must define the concept of a star's **habitable zone**. The habitable zone of a star is the area around it in which a planet would be heated to a temperature that allows liquid water to exist (Figure 30.5). The Sun's habitable zone encompasses Venus (which would be cooler if not for the greenhouse effect), Earth, and Mars. What about other stars?

In order for a star to have a satisfactory habitable zone, several requirements must be met. First, the star must have a mass between about one-third and 1.5 times that of the Sun. A star that is too low in mass would probably not emit enough energy to produce a satisfactory habitable zone. A star that is too massive would emit too much harmful ultraviolet light and would also have a lifetime so short that life would probably lack sufficient time to develop. A second requirement is that a life-supporting star could not be appreciably variable, because variability would change the location of the habitable zone. Finally, the star would probably have to be single, because a multiple star system would not be likely to have a zone of habitability in which a planet could orbit. It is estimated that only 6% of the stars in our galaxy meet these requirements. Although this is a small percentage, over 10 billion stars in the Milky Way would meet these requirements! Of course, whether any of these other than the Sun actually have planets remains unknown.

A planet must also meet certain requirements in order to be capable of supporting life. Its orbit must be within the habitable zone and of low enough eccentricity to avoid significant variation in temperature. The planet must also have sufficient mass to hold a substantial atmosphere. We can't estimate the number of planets that might fill these requirements, but of the nine planets in our solar system, two do — Earth and Venus. (Recall that Venus would be much cooler were it not for its extreme greenhouse effect.) However, some researchers believe that Venus does not actually lie in the Sun's habitable zone because it developed a runaway greenhouse effect.

◼ Is There Life in Other Planetary Systems?

From the foregoing discussion, it is certainly realistic to believe that there are planets orbiting other stars, and that many of these planets could support life. How can we tell if life is actually present in other planetary systems? There are only two ways to find extraterrestrial life on extrasolar planets: detection of the life by going to the planets, or communication with any intelligent life present on them. Let us examine each case separately.

Although travel between stars often requires only days or weeks in science fiction, the current state of science and engineering makes such rapid travel impossible. The distance between stars is so great that a round trip to the nearest star, even at the speed of light, would require almost a decade. Unfortunately, our current technology allows us to travel at only a small fraction of that speed. This means that for human travel to the stars to occur using present technology, the travelers would need to be frozen somehow and revived at journey's end (another science fiction idea that is far from current reality) or else a multigenerational journey would have to be undertaken. At present, neither alternative seems likely to be attempted soon.

Is it reasonable to assume that future breakthroughs could allow travel at velocities greater than the speed of light? Unfortunately, hope of reaching or exceeding the speed of light is minimal, because Einstein's theory of relativity states that this velocity cannot be attained. This "ulti-

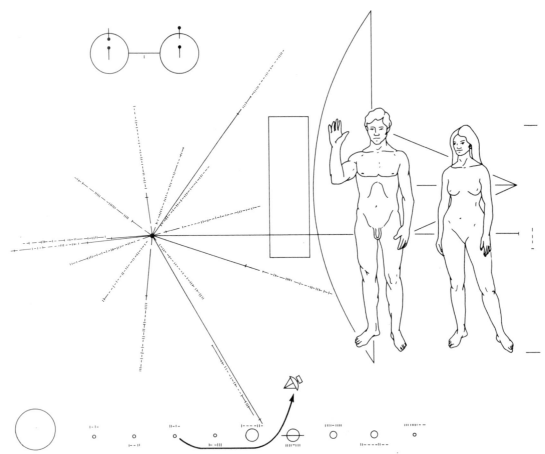

■ **Figure 30.6** The plaque that was carried by the Pioneer 10 and 11 spacecraft, the first to leave the solar system. The plaque showed human beings against the silhouette of the craft for scale, a map of the solar system with the craft's course, the hydrogen molecule, and a map of the Sun's location in the Milky Way. (NASA received some criticism for portraying the humans unclothed!) (NASA.)

mate speed limit" may mean that rapid travel between stars will be forever out of the question. The only such travel is likely to be conducted by relatively slowmoving unmanned craft, the first of which were the Pioneer and Voyager spacecraft, which were decorated with plaques illustrating humans and the solar system (Figure 30.6). (In addition, each Voyager craft contained a phonograph record of "the sounds of Earth" and instructions for constructing a device with which to play it!) Unfortunately, these craft will cease transmitting long before reaching the vicinity of the nearest stars along their courses thousands of

centuries from now, and the chance of anyone finding these craft in the vastness of the galaxy is certainly remote.

Some people argue that sightings of unidentified flying objects (UFOs) and claims by some people that they have seen alien beings and their vessels are proof that other civilizations have mastered interstellar travel. What are we to make of these claims? Some of the reported incidents are outright lies and frauds. Most of the others are probably honest mistakes made by people who have seen phenomena in the sky that they cannot understand and have jumped to conclusions about their interpretation. It is likely that a civilization advanced enough to be able to travel to Earth would either have no qualms about publicizing the arrival of their vessels, or else would have ways to prevent anyone from detecting their presence if they desired anonymity.

If travel to other stars can be ruled out, the only other exploratory technique is communication, which limits us to contact with intelligent life from civilizations that have advanced to a level of technology comparable to our 20th century or beyond. We have been sending messages outward unintentionally since broadcasting began, but there has been no effort to send messages toward specific destinations at frequencies that would be universally used for radio astronomy studies. In the past few decades, a number of researchers have listened for messages, either by scanning the entire sky or pointing radio telescopes at nearby Sun-like stars. However, nothing resembling a message from an intelligent civilization has been detected. These efforts continue, and advanced technology now allows a single radio telescope to scan many frequencies simultaneously, greatly increasing the chances of detecting an incoming signal. Unfortunately, these attempts at communication would fail to reveal life that was not intelligent, intelligent life that lacked our communications technology, or intelligent life with communications abilities who have no desire to contact other civilizations. Even if contact were made with another civilization, there would still be problems, such as language differences and the fact that radio signals are limited to the speed of light, eliminating rapid conversations.

Could we be alone? An argument can be made for suggesting that we may be the only civilization in the galaxy. Recall that the Milky Way was at least 5 billion years old at the time the solar system formed. This means that technological

civilizations could have preceded ours by billions of years. A planet's surface is a limited area and can support only a finite population. The number of people on Earth is growing incredibly fast, and we will eventually run out of room. Unless we desire to stop population growth at that point, the only alternative will be the colonization of other worlds. Because the other planets in our solar system are not extremely favorable, we may be faced with the necessity of developing colonization ships to carry people on long journeys toward other stars to live on any habitable planets that might be found. In 5 billion years, the population of Earth, growing at its present rate, would be sufficient to fill every habitable planet in the Milky Way! Although we ourselves may not survive long enough to do this, any forerunner civilization that preceded us by several billion years could have filled the galaxy in this way long before intelligent life appeared on Earth. It has been argued by some astronomers that the fact that Earth was not colonized in this way can be considered evidence that no such civilizations have developed.

■ Conclusion

It is fitting that we close this book with the unanswered question of whether we are alone. Although we have learned many things about the solar system and the larger universe, it is well to remember that there are many things that we still do not understand. (One advantage of the incompleteness of our knowledge is that if we already knew everything, most scientists would find themselves out of work!) When we look beyond the solar system and speculate about such things as other civilizations, we realize that we have barely scratched the surface in our study of the universe.

■ Chapter Summary

Although we often wonder whether other stars have planets and if they might contain life, neither question can yet be answered with certainty.

Terrestrial life, which is carbon-based, has existed for at least 3.5 billion years, but its origin has not yet been adequately explained. Most scientists believe that life may have developed spontaneously from nonliving material. Humans have only been on the Earth a short time, but our intelligence has enabled us to develop our modern technological civilization.

It is unlikely that life exists elsewhere in our solar system. However, there is considerable indirect evidence that other stars may have planets in orbit around them. How many of these other planets might support life, however, is unknown.

Even if we assume that planets around other stars contain life, verification of that fact would be difficult. Human travel to other stars is, and may remain, impossible, and the finite speed of light would hamper communication with extraterrestrial intelligences.

■ Chapter Vocabulary

accretion disk	DNA	metabolism	panspermia
amino acids	evolution	Miller–Urey	photosynthesis
barycenter	habitable zone	experiment	proper motion
carbon-based life	intelligence	mutation	special creation
cell	life	nucleus	spontaneous creation
civilization	life as we know it	organism	technology

■ Review Questions

1. What is life? Describe the major characteristics of life on Earth.
2. Explain the theories of special creation, spontaneous creation, and panspermia.
3. Explain how life might have originated from biochemical development.
4. What was the Miller–Urey experiment, and what were its results?
5. Describe a cell, and explain the difference between single-celled and multicelled organisms.
6. Describe the theory of evolution.
7. Describe the characteristics of human beings, and explain what civilization is.
8. Why is it unlikely that there is other life within the solar system?
9. How is a search made for other planetary systems? What evidence is there that other planetary systems may exist?
10. What is a habitable zone? Describe the conditions necessary for a star to have a habitable zone.
11. Describe the conditions necessary for a planet to be able to support life.
12. Describe difficulties that would be encountered in attempting to communicate with, or travel to, an inhabited planet around a distant star.
13. Why is it sometimes argued that we are probably the only intelligent civilization in the Milky Way?

■ For Further Reading

Anderson, Poul. *Is There Life on Other Worlds?* Collier Books, New York, 1963.

Goldsmith, Donald. *The Quest for Extraterrestrial Life: A Book of Readings.* University Science Books, Mill Valley, CA, 1980.

Goldsmith, Donald and Tobias Owen. *The Search for Life in the Universe.* Benjamin/Cummings, Menlo Park, CA, 1980.

Hartman, H. et al, editors. *Search for the Universal Ancestors.* NASA, Washington, DC, 1985.

Horowitz, Norman H. *To Utopia and Back: The Search for Life in the Solar System.* Freeman, New York, 1986.

Ovenden, Michael W. *Life in the Universe: A Scientific Discussion.* Doubleday, Garden City, NY, 1962.

Rood, R. and John S. Trefil. *Are We Alone?* Scribner's, New York, 1981.

Sagan, Carl and Frank Drake. "The Search for Extraterrestrial Intelligence." In *Scientific American,* May, 1975.

Appendixes

■ Appendix 1
Scientific Notation and Astronomical Measurements

Scientific Notation

Because of the nature of the objects that astronomers study, they often use extremely large numbers to express quantities or dimensions. At the same time, they also discuss very small distances, such as the wavelength of light. Scientists use a shorthand way of expressing such numbers, called **scientific notation.** Let us examine how it works.

First, we need to familiarize ourselves with large and small numbers (Table A1.1). The key is the number ten. If we multiply ten times ten (we call this ten squared, written 10^2, where the superscript 2 is called an **exponent**), we obtain 100. Ten times itself three times (ten cubed, written 10^3) yields 1000, and so forth. (Note that the number of zeros in the long form of these numbers is the same as the exponent.) Negative exponents are used to give decimal fractions: 10^{-1} is one-tenth (0.1), 10^{-2} is one one-hundredth (0.01), and so on. (Note that with these numbers, the exponent is the number of places to the right of the decimal point that the numeral 1 is located.)

To express a number in scientific notation, ten with the proper exponent is multiplied by a number between one and ten, yielding the correct final result. For example, the number of seconds in a year is about 31,500,000. This is equal to 3.15 multiplied by 10,000,000, or 10^7, so the scientific notation form is 3.15×10^7. Writing numbers in this fashion is much simpler than expressing them using numerous zeros, and a quick glance at the exponent allows one to rapidly appreciate how large or small a quantity is.

In measuring very large quantities, scientists are often unable to determine precise values, but they can usually determine them to within a factor of ten. A value that may not be precise but is accurate within a factor of ten either way is called an order-of-magnitude approximation. For example, the number of stars in our galaxy is about 200 billion (2×10^{11}). Although the precise figure may be closer to 100 billion or 400 billion, it is very unlikely that it is as low as 20 billion or as high as 2 trillion. Therefore, we call 10^{11} an order-of-magnitude approximation of the number of stars in the galaxy.

■ TABLE A1.1   Large and Small Numbers

Number	Long Form	Exponent Form	Metric Prefix	Prefix Abbreviation
Trillion	1,000,000,000,000	10^{12}	tera-	T
Billion	1,000,000,000	10^{9}	giga-	G
Million	1,000,000	10^{6}	mega-	M
Thousand	1,000	10^{3}	kilo-	k
Hundred	100	10^{2}	hecto-	h
Ten	10	10^{1}	deca-	da
Tenth	0.1	10^{-1}	deci-	d
Hundredth	0.01	10^{-2}	centi-	c
Thousandth	0.001	10^{-3}	milli-	m
Millionth	0.000001	10^{-6}	micro-	μ
Billionth	0.000000001	10^{-9}	nano-	n
Trillionth	0.000000000001	10^{-12}	pico-	p

Astronomical Measurements

As mentioned in Chapter 1, we will make many quantitative measurements of solar system objects, generally using metric units. The purpose of this section is to familiarize you with the physical quantities described in this book, as well as with the metric system.

The **metric system** has only one basic unit for each quantity such as length and mass, and one of the system's features is that each basic unit can be modified by attaching a prefix (see Table A1.1). For example, a kilometer is 1000 meters and a milligram is 0.001 gram.

Dimensional length is an important measurement used to express the distances of objects, their diameters, and sizes of their surface features. The basic unit of length is the **meter** (39.37 inches), and the kilometer (0.62 miles) is commonly used in solar system measurements. The nanometer (10^{-9} meters) is commonly used in measuring extremely small distances, such as the wavelength of light, but the **Angstrom** (10^{-10} meters) is also used for this purpose. On a larger scale, we often use the **astronomical unit** as a standard unit within the solar system. One astronomical unit is defined as the average distance between the Earth and the Sun, 149.6 million kilometers (92.9 million miles). Another large unit is the **light year**, the distance light travels in one year, about 9.5 trillion kilometers (6 trillion miles).

Closely related to length are **area** and **volume**. Area measures the extent of a surface, and volume measures the amount of space contained within an object. The basic unit of area is a square whose sides are of unit length. Examples include square kilometers and square miles. The basic unit of volume is a cube whose edges are of unit length. Examples include cubic centimeters and cubic miles.

Mass is a measure of how much material an object contains. Mass is not the same as **weight**, which is a measure of the downward force exerted by an object as the result of the acceleration of gravity. Although we can generally ignore this distinction on the Earth, there are situations in which it is important; for example, an astronaut in orbit is weightless but has the same mass as on Earth. The basic metric unit of mass is the **gram,** which is 0.035 ounces. There are 453.6 grams in a pound, and 1 kilogram is 2.2 pounds. (In this book, we will use pounds as mass units, unless otherwise specified.) The basic metric unit of weight or force is the **dyne,** which is the force necessary to give a 1-gram mass a constant increase in speed (acceleration) of 1 centimeter per second every second. A 1-gram mass at the Earth's surface has a weight of 980 dynes. The **newton** is the force necessary to give a 1-kilogram mass a constant acceleration of 1 meter per second every second. A 1-kilogram mass at the Earth's surface has a weight of 9.8 newtons.

Pressure describes the amount of force applied to an area and is measured in units of force (weight) per unit area. One metric pressure unit is the dyne per square centimeter. One million dynes per square centimeter is equal to 1 **bar.** Another pressure unit is pounds per square inch. The Earth's atmosphere, for example, exerts a force of 14.7 pounds per square inch on everything at the Earth's surface. (This pressure, called 1 **atmosphere,** is also used as a unit of pressure.) Coincidentally, the Earth's atmospheric pressure is almost exactly 1 bar. (The precise relationship is 1 atmosphere = 1.013 bar.) Pressure is useful in describing planetary atmospheres, and pressures inside solar system objects are also frequently discussed.

Density is a quantity describing the mass per unit volume of an object. It is very useful in solar system studies because it allows us to compare the compositions of various objects. The standard unit used for density measurements is grams per cubic centimeter. The metric system was devised so that 1 cubic centimeter of water has a mass of 1 gram, so water has a density of 1 gram per cubic centimeter. By comparison, the average density of the Earth is 5.5 grams per cubic centimeter, whereas that of air at the Earth's surface is only 0.0013 grams per cubic centimeter. (The term **specific gravity** is sometimes used instead of the term density. Specific gravity is the ratio between the density of an object and the density of water; therefore, it is numerically the same as density, but lacks units.) When discussing the density of a planet, the term **uncompressed density** is sometimes used. The tremendous pressure within a planet compresses its internal material so much that its interior is much denser than it would be normally. Uncompressed density is the density that a planet's interior would have if it were not under high pressure. The uncompressed average density of the Earth would be only 4.0 grams per cubic centimeter.

Time is an important quantity in solar system studies and is most commonly used to measure periods of rotation or revolution and the ages of solar system objects and their features. The basic unit of time, the **second,** is used in all measurement systems, as are familiar units such as days, months, and years.

Velocity is a quantity used to measure rate of motion, but it is not synonymous with speed. Speed is simply a measurement of how fast an object is moving, but velocity also takes into account the direction of motion. A car may maintain a constant speed while negotiating a curve, but its velocity will change. Velocity and speed are measured in units of distance divided by time, such as kilometers per hour or miles per second. A special speed merits our attention: the speed of light, 300,000 kilometers per second (186,000 miles per second). This quantity is given the symbol "c" and, according to Einstein's theory of relativity, cannot be attained by any moving object with mass. As we shall learn, c plays a role in several important calculations.

Energy is a measure of the ability to do work. (**Work,** in turn, is the product of the amount of force exerted and the distance that an object is moved by that force.) In our everyday lives, there are many familiar forms of energy, such as sound, light, and heat, which can change objects and perform work. Two possibly less familiar forms of energy are of particular importance in solar system studies, especially when considering orbits. One is **kinetic energy,** the energy of motion. Any object in orbit around the Sun has kinetic energy as a result of that motion. Another form of energy is **potential energy,** the energy of position. If an object is lifted above Earth's surface, its potential energy increases. If the object were dropped, its potential energy would become kinetic energy of motion as it fell. Solar system objects also have potential energy determined by their orbital position "above" the Sun. Astronomers normally use an energy unit known as the erg, which is the kinetic energy of a 2-gram mass moving at a speed of 1 centimeter per second. The **erg** is a small quantity and can be compared to the energy of a collision with a slow-flying mosquito! The explosion of a ton of dynamite releases about 4×10^{16} ergs of energy.

The **momentum** of an object moving in a straight line is the product of its mass and velocity. **Angular momentum** is a quantity for rotating or revolving objects that is somewhat analogous to momentum for objects moving in a straight line. The angular momentum, L, of an orbiting object is given by the equation

$$L = m\,a\,v,$$

where m is the mass of the object, a is its semimajor axis, and v is its average orbital velocity. For a rotating object, angular momentum is given by

$$L = C\,m\,r^2\,\omega,$$

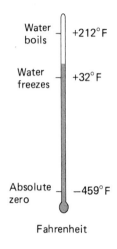

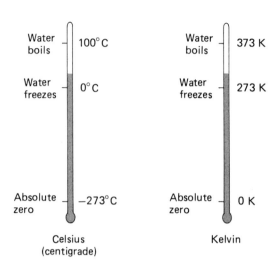

■ Figure A1.1 Comparison of temperature scales. The values of the boiling point of water, freezing point of water, and absolute zero are given for each scale.

where C is the object's moment of inertia coefficient, m is its mass, r is its radius, and ω is its rotational velocity in radians per second. Angular momentum is normally expressed in units of grams–centimeters squared per second.

   Temperature is a measurement related to the speed of atomic motion. All atoms and molecules are constantly moving, and as temperature increases, they move faster. Our nervous system is able to judge the temperature of the air around us because the warmer it is, the more energetic are the air molecules hitting us. At a low temperature, called **absolute zero**, molecular motion reaches a minimum value. There are several systems of temperature measurement (Figure A1.1). The **Fahrenheit** scale has the boiling point of water at 212°F, the freezing point of water at 32°F, and absolute zero at −459°F. The **centigrade**, or **Celsius**, scale has the boiling point at 100°C, the freezing point at 0°C, and absolute zero at −273°C. The **Kelvin** scale, which is most often used in astronomy, has the boiling point at 373 K, the freezing point at 273 K, and absolute zero at 0 K. (Note that the degree symbol is not used here.) The three temperature scales may be converted using the following equations:

$$F = \frac{9}{5} C + 32$$

$$F = \frac{9}{5} (K - 273) + 32$$

$$C = \frac{5}{9} (F - 32)$$

$$C = K - 273$$

$$K = \frac{5}{9} (F - 32) + 273$$

$$K = C + 273$$

The term **heat** is applied to the thermal energy contained in a system. For example, a glass of water and a large lake may be at the same temperature, but the lake will have a higher heat content because more water molecules are present.

   Angular measure describes the apparent diameter of an object as it is viewed. It is easy to measure but doesn't reveal true dimensions unless the distance to the object is also known. The basic unit is the **degree** (°), of which there are 360 in a circle. The Sun and Moon each have an angular diameter of about half a degree. A smaller unit is the **minute** ('), of which there are 60 in a degree. The smallest unit is the **second** (''), of which there are 60 in a minute.

   Angular measure forms the basis for specifying the locations of points on the surfaces of spherical (or nearly spherical) objects. The line halfway between the poles of rotation of an object is called its **equator** and forms the starting point for measuring **latitude**, the north–south component of location. Latitude is 0° at the equator and reaches 90° north latitude at the north pole and 90° south latitude at the south pole. Measurement of location in the east–west sense is called **longitude**. Unfortunately, there is no natural starting point for longitude measurements corresponding to the equator for latitude.

Therefore, an arbitrary **prime meridian,** or line connecting the north and south poles, must be selected for every planet or other object. On Earth, this is the Greenwich meridian, which passes through London, England, which was selected for political reasons. Longitude is measured from the prime meridian westward to 180° west longitude and eastward to 180° east longitude. Minutes and seconds are also used to describe latitude and longitude with greater accuracy.

Review Questions

1. Convert the following numbers to scientific notation: 12,600, 1,340,000, 15.5, 201,000,000.

2. Convert the following to ordinary numbers: 1.45×10^6, 6.2×10^9, 1.11×10^{10}, 3.3×10^3.

3. Explain the difference between mass and weight.

4. Explain the difference between speed and velocity.

5. Match the quantity with the unit:

Energy	Bar
Density	Gram
Pressure	Angstrom
Mass	Erg
Length	Grams per cubic centimeter

Appendix 2
Planetary Data

■ Table A2.1 Properties of the Planets

	Mercury	Venus	Earth
Astronomical symbol	☿	♀	⊕
Semi-major axis (AU)	0.387	0.723	1.000
Semi-major axis (10^6 km)	57.91	108.21	149.60
Semi-major axis (10^6 mi)	35.98	67.24	92.96
Eccentricity	0.206	0.007	0.017
Perihelion distance (AU)	0.307	0.718	0.983
Perihelion distance (10^6 km)	46.00	107.48	147.10
Perihelion distance (10^6 mi)	28.59	66.78	91.41
Aphelion distance (AU)	0.467	0.728	1.017
Aphelion distance (10^6 km)	69.82	108.94	152.10
Aphelion distance (10^6 mi)	43.38	67.70	94.51
Orbital period (yr)	0.241	0.615	1.000
Orbital period (days)	87.971	224.709	365.256
Average daily motion (deg)	4.0923	1.6021	0.9856
Average orbital velocity (km/sec)	47.88	35.03	29.79
Average orbital velocity (mi/sec)	29.74	21.76	18.51
Inclination of orbit (deg)	7.004	3.394	—
Longitude of ascending node (deg)	48.094	76.500	—
Argument of perihelion (deg)	29.050	54.790	—
Longitude of perihelion (deg)	77.144	131.290	102.596
Equatorial diameter (km)	4878	12104	12756.28
Equatorial diameter (mi)	3031	7521	7926.75
Equatorial diameter (Earth = 1)	0.382	0.949	1
Polar diameter (km)	4878	12104	12713.51
Polar diameter (mi)	3031	7521	7900.18
Oblateness	0	0	0.0033528
Rotational period (days)	58.6462	−243.01	0.9972697
Rotational period (hr)	1407.5	−5832.2	23.934472
Obliquity of axis (deg)	0.0	2.7	23.45

Mars	Jupiter	Saturn	Uranus	Neptune	Pluto
♂	♃	♄	♅	♆	♇
1.524	5.203	9.555	19.218	30.110	39.442
227.94	778.29	1429.37	2874.99	4504.33	5900.52
141.64	483.63	888.21	1786.52	2798.99	3666.53
0.093	0.049	0.056	0.046	0.009	0.248
1.381	4.950	9.023	18.328	29.839	29.660
206.65	740.57	1349.88	2741.82	4463.78	4437.14
128.41	460.19	838.81	1703.76	2773.79	2757.19
1.666	5.455	10.086	20.108	30.381	49.223
249.23	816.01	1508.87	3008.17	4544.88	7363.76
154.87	507.07	937.61	1869.28	2824.19	4575.77
1.881	11.862	29.458	84.012	164.796	247.711
686.990	4332.754	10759.61	30686.1	60193	90478
0.5240	0.0831	0.0336	0.01175	0.00599	0.003979
24.13	13.06	9.66	6.81	5.44	4.74
14.99	8.12	6.00	4.23	3.38	2.95
1.850	1.304	2.489	0.773	1.772	17.150
49.403	100.252	113.489	73.877	131.561	109.941
286.288	273.757	339.177	98.859	276.307	113.031
335.691	14.010	92.665	172.736	47.867	222.972
6786.8	142796	120000	50800	48600	2245
4217.3	88733	74568	31567	30200	1395
0.532	11.194	9.407	3.982	3.810	0.176
6434.80	133542	107085	49276	47341	2245
3998.59	82983	66543	30620	29148	1395
0.051865	0.0648088	0.1076209	0.030	0.0259	0
1.0259568	0.41354	0.4375	−0.718	0.669	−6.3867
24.622962	9.92496	10.5	−17.24	16.05	−153.28
25.19	3.12	26.73	82.14	29.56	62 (?)

■ Table A2.1   (continued)

	Mercury	Venus	Earth
Surface area (Earth = 1)	0.147	0.902	1
Volume (Earth = 1)	0.056	0.857	1
Mass (10^{24} kg)	0.33022	4.8690	5.9742
Mass (10^{24} lb)	0.72800	10.734	13.171
Mass (Earth = 1)	0.055	0.815	1
Mass (Sun = 1)	1/6023600	1/408523.5	1/328900.5
Surface gravity (Earth = 1)	0.370	0.876	1
Escape velocity (km/sec)	4.2	10.4	11.2
Escape velocity (mi/sec)	2.6	6.4	6.9
Density (g/cm³)	5.43	5.24	5.515
Uncompressed density (g/cm³)	5.31	3.95	4.03
Moment-of-inertia coefficent	0.4	0.34	0.331
Magnetic moment (Gauss/cm³)	2.4×10^{22}	$<4 \times 10^{21}$	8×10^{25}
Surface field strength (Gauss)	0.002	<0.00002	0.31
Magnetic axis obliquity (deg)	2.3	?	11.9
Sun's apparent size (arcmin)	83	44	32
Solar constant (Earth = 1)	6.68	1.91	1
Geometric albedo (%)	0.106	0.65	0.367
Surface temperature (K)	440	730	288
Surface temperature (°F)	333	855	59
Number of satellites	0	0	1
Closest to Earth (AU)	.516	.255	—
Farthest from Earth (AU)	1.484	1.745	—
Minimum angular diameter (arcsec)	4.5	9.6	—
Maximum angular diameter (arcsec)	13.0	65.4	—
Average magnitude (opposition)	−1.80	−4.30	—
Synodic period (days)	115.98	583.46	—

Mars	Jupiter	Saturn	Uranus	Neptune	Pluto
0.274	120.201	82.40	15.58	14.297	0.031
0.143	1316.24	745.4	61.46	54.05	0.0055
0.64191	1898.8	568.50	86.625	102.78	0.015
1.4152	4186.1	1253.3	190.97	226.59	0.033
0.107	317.8	95.16	14.50	17.20	0.0025
1/3098710	1/1047.4	1/3498.5	1/22869	1/19314	1/131000000
0.381	2.637	1.151	1.059	1.426	0.03 (?)
5.0	59.6	35.6	21.3	23.8	1.2 (?)
3.1	37.0	22.1	13.3	14.8	0.7 (?)
3.94	1.33	0.70	1.30	1.76	1.84 – 2.14
3.71	—	—	—	—	?
0.365	0.26	0.22	0.20	0.27	?
2.5×10^{22}	1.6×10^{30}	4.7×10^{28}	4×10^{27}	?	?
0.0006	4.28	0.21	0.25	?	?
?	9.6	0.0	60	50	?
21	6.1	3.3	1.7	1.1	0.8
0.43	0.037	0.011	2.7×10^{-3}	1.1×10^{-3}	6.4×10^{-4}
0.15	0.52	0.47	0.51	0.41	0.3
218	110	80	50	59	30
−67	−260	−315	−370	−353	−405
2	16	17	15	8	1
.364	3.933	8.006	17.311	28.822	28.643
2.683	6.472	11.103	21.125	31.398	50.24
3.5	30.4	14.9	3.3	2.1	0.1
25.7	50.0	20.7	4.0	2.3	0.1
−2.01	−2.70	0.67	5.52	7.84	15.12
779.85	398.88	378.09	369.66	367.49	366.74

Sources of Data:
U. S. Naval Observatory. *The Astronomical Almanac, 1990.* Government Printing Office, Washington, DC, 1989.
Duffett-Smith, Peter. *Practical Astronomy with Your Calculator,* 2nd ed. Cambridge University Press, Cambridge, England, 1981.
Pollack, James B. "Atmospheres of the Terrestrial Planets," in Beatty, J. Kelly and Andrew Chaikin, editors. *The New Solar System.* Sky Publishing Corporation, Cambridge, MA, 1990. Reproduced with permission.
Kaula, William M. "The Interiors of the Terrestrial Planets: Their Structure and Evolution," in Kivelson, Margaret G., editor. *The Solar System, Observations and Interpretations.* Copyright 1986. Table 4–1 on page 124 used as a source by permission of Prentice-Hall, Inc., Englewood Cliffs, NJ.

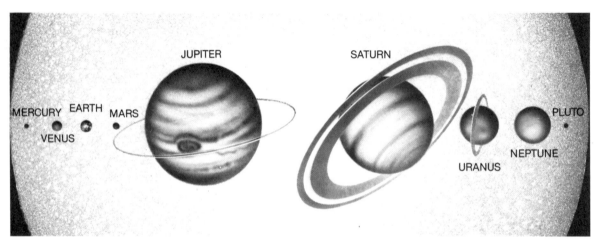

■ **Figure** A2.1   The relative sizes of the planets, as well as the Sun, part of which is shown in the background. (From Pasachoff, Jay M. *Contemporary Astronomy,* 4th ed. Saunders College Publishing, Philadelphia, 1989.)

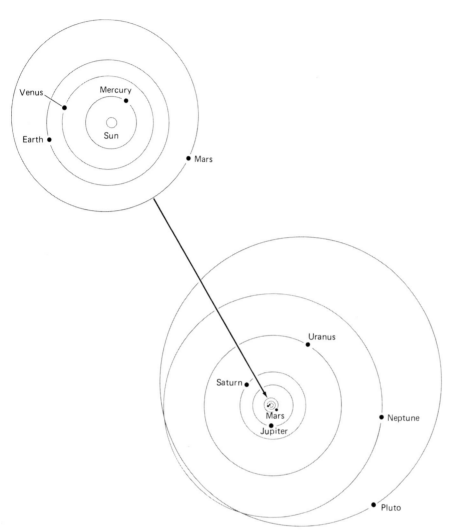

■ **Figure A2.2**   The relative sizes of the orbits of the planets. (From Abell, George, David Morrison, and Sidney Wolff. *Exploration of the Universe*, 5th ed. Saunders College Publishing, Philadelphia, 1987.)

Appendix 3
Planetary Satellite Data

■ Table A3.1   Properties of Planetary Satellites

	Semi-Major Axis			Orbital Period (days)	Eccentricity	Inclination (°)
	R_p	10^3 mi	10^3 km			
Earth						
Moon	60.3	238.9	384.4	27.3217	0.05490	*
Mars						
Phobos	2.76	5.827	9.378	0.319	0.015	1.02
Deimos	6.91	14.577	23.459	1.263	0.00052	1.82
Jupiter						
Metis	1.7922	79.51	127.96	0.2948	<0.004	~0
Adrastea	1.8065	80.15	128.98	0.2983	~0	~0
Amalthea	2.539	112.66	181.3	0.4981	0.003	0.40
Thebe	3.108	137.89	221.90	0.6745	0.015	0.8
Io	5.905	261.98	421.6	1.769	0.0041	0.040
Europa	9.397	416.90	670.9	3.551	0.0101	0.470
Ganymede	14.99	665	1070	7.155	0.0006	0.195
Callisto	26.37	1170	1883	16.689	0.007	0.281
Leda	155.4	6894	11094	238.72	0.148	27
Himalia	160.8	7134	11480	250.57	0.158	28
Lysithea	164.2	7283	11720	259.22	0.107	29
Elara	164.4	7293	11737	259.65	0.207	28
Anake	296.9	13174	21200	−631	0.169	147
Carme	316.5	14044	22600	−692	0.207	163
Pasiphae	329.1	14603	23500	−735	0.378	148
Sinope	331.9	14727	23700	−758	0.275	153
Saturn						
Atlas	2.281	85.53	137.64	0.602	~0	~0
Prometheus	2.310	86.59	139.35	0.613	0.0024	0.0
Pandora	2.349	88.05	141.70	0.629	0.0042	0.0
Epimetheus	2.510	94.09	151.422	0.694	0.009	0.34
Janus	2.511	94.12	151.472	0.695	0.007	0.14
Mimas	3.075	115.28	185.52	0.942	0.0202	1.53
Enceladus	3.945	147.91	238.02	1.370	0.0045	0.02
Tethys	4.884	183.10	294.66	1.888	0.0000	1.09
Telesto	4.884	183.10	294.66	1.888	~0	~0
Calypso	4.884	183.10	294.66	1.888	~0	~0
Dione	6.256	234.52	377.40	2.737	0.0022	0.02
Helene	6.256	234.52	377.40	2.737	0.005	0.2
Rhea	8.736	327.50	527.04	4.518	0.0010	0.35
Titan	20.25	759.26	1221.85	15.945	0.0292	0.33
Hyperion	24.55	920.36	1481.1	21.277	0.1042	0.43
Iapetus	59.03	2213.0	3561.3	79.331	0.0283	7.52
Phoebe	214.7	8048.4	12952	−550.48	0.163	175.3

■ Table A3.1   *(continued)*

	Semi-Major Axis			Orbital Period (days)	Eccentricity	Inclination (°)
	R_p	10^3 mi	10^3 km			
Uranus						
Cordelia	1.90	30.91	49.75	0.336	<0.001	0
Ophelia	2.05	33.41	53.77	0.377	0.01	0
Bianca	2.26	36.76	59.16	0.435	<0.001	0
Cressida	2.36	38.38	61.77	0.465	<0.001	0
Desdemona	2.39	38.39	62.65	0.476	<0.001	0
Juliet	2.47	40.16	64.63	0.494	<0.001	0
Portia	2.52	41.07	66.10	0.515	<0.001	0
Rosalind	2.67	43.45	69.93	0.560	<0.001	0
Belinda	2.87	46.76	75.25	0.624	<0.001	0
Puck	3.28	53.44	86.00	0.764	<0.001	0
Miranda	4.95	80.66	129.8	1.413	0.0027	4.22
Ariel	7.30	118.8	191.2	2.520	0.0034	0.31
Umbriel	10.15	165.3	266.0	4.144	0.0050	0.36
Titania	16.64	270.8	435.8	8.706	0.0022	0.14
Oberon	22.24	362.0	582.6	13.463	0.0008	0.10
Neptune						
1989N6	1.98	30.0	48.2	0.296	~ 0	4.5
1989N5	2.06	31.1	50.5	0.313	~ 0	<1
1989N3	2.16	32.6	52.5	0.333	~ 0	<1
1989N4	2.55	38.5	62.0	0.396	~ 0	<1
1989N2	3.03	45.7	73.6	0.554	~ 0	<1
1989N1	4.84	73.1	117.6	1.121	~ 0	<1
Triton	14.0	220.2	354.3	−5.877	<0.0005	159.0
Nereid	219	3426	5513.4	360.16	0.7483	27.6
Pluto						
Charon	12.7	11.87	19.1	6.387	0	94.3

* Varies from 18.28 to 28.58.

■ Table A3.1   (continued)

	R (days)	Mass (10²⁰lb)	Mass (10²⁰kg)	Mass (P=1)†	Diameter (mi)	Diameter (km)
Earth						
Moon	S	1620.2	734.9	0.0123	2160	3476
Mars						
Phobos	S	2.78	1.26	1.5E−8	16 × 13 × 12	26 × 21 × 19
Deimos	S	3.97	1.8	3E−9	9 × 7.5 × 7	15 × 12 × 11
Jupiter						
Metis	—	.00209	.00095	53−11	? × 25 × 25	? × 40 × 40
Adrastea	—	.00042	.00019	1E−11	16 × 13 × 9	25 × 20 × 15
Amalthea	S	0.159	.0722	3.8E−9	168 × 102 × 93	270 × 164 × 150
Thebe	—	0.017	.0076	4E−10	? × 68 × 56	? × 110 × 90
Io	S	1971	894	4.68E−5	2256	3630
Europa	S	1058	480	2.52E−5	1950	3138
Ganymede	S	3267.9	1482.3	7.80E−5	3270	5262
Callisto	S	2373.5	1076.6	5.66E−5	2983	4800
Leda	—	.00013	.000057	3E−12	10	16
Himalia	0.4	.209	.095	5E−9	112	180
Lysithea	—	.0017	.00076	4E−11	25	40
Elara	0.5	.017	.0076	4E−10	50	80
Anake	—	.00084	.00038	2E−11	19	30
Carme	—	.0021	.00095	5E−11	27	44
Pasiphae	—	.0042	.0019	1E−10	43	70
Sinope	—	.0017	.00076	4E−11	25	40
Saturn						
Atlas	—	—	—	—	24 × ? × 17	38 × ? × 28
Prometheus	—	—	—	—	87 × 62 × 46	140 × 100 × 74
Pandora	—	—	—	—	68 × 53 × 41	110 × 86 × 66
Epimetheus	S	—	—	—	87 × 73 × 62	140 × 118 × 110
Janus	S	—	—	—	137 × 118 × 99	220 × 190 × 160
Mimas	S	0.84	0.38	8.0E−8	245	394
Enceladus	S	1.76	0.8	1.3E−7	312	502
Tethys	S	16.8	7.6	1.3E−6	651	1048
Telesto	—	—	—	—	? × 15 × 14	? × 24 × 22
Calypso	—	—	—	—	19 × 16 × 10	30 × 26 × 16
Dione	S	23.1	10.5	1.85E−6	695	1118
Helene	—	—	—	—	22 × ? × 19	36 × ? × 30
Rhea	S	54.9	24.9	4.4E−6	949	1528
Titan	—	2966.7	1345.7	2.238E−4	3200	5150
Hyperion	—	0.37	0.17	3E−8	217 × 149 × 124	350 × 240 × 200
Iapetus	S	41.4	18.8	7E−10	892	1436
Phoebe	0.4	—	—	—	143 × 137 × 130	230 × 220 × 210

■ Table A3.1   *(continued)*

	R (days)	Mass (10^{20}lb)	Mass (10^{20}kg)	Mass (P=1)†	Diameter (mi)	Diameter (km)
Uranus						
Cordelia	—	—	—	—	30	50
Ophelia	—	—	—	—	30	50
Bianca	—	—	—	—	30	50
Cressida	—	—	—	—	37	60
Desdemonda	—	—	—	—	37	60
Juliet	—	—	—	—	50	80
Portia	—	—	—	—	50	80
Rosalind	—	—	—	—	37	60
Belinda	—	—	—	—	37	60
Puck	—	—	—	—	106	170
Miranda	S	1.56	0.71	2E−6	301	484
Ariel	S	31.7	14.4	1.8E−5	721	1160
Umbriel	S	26.0	11.8	1.2E−5	739	1190
Titania	S	75.6	34.3	6.8E−5	994	1600
Oberon	S	63.3	28.7	6.9E−5	963	1550
Neptune						
1989N6	—	—	—	—	34	54
1989N5	—	—	—	—	50	80
1989N3	—	—	—	—	110	180
1989N4	—	—	—	—	95	150
1989N2	—	—	—	—	120	190
1989N1	—	—	—	—	250	400
Triton	S	2866	1300	1.3E−3	1680	2705
Nereid	—	0.45	0.204	2E−7	210	340
Pluto						
Charon	—	41.34	18.75	0.125(?)	695	1119

† To save space, fractional mass is given in the form aE-b, which is short for a$\times 10^{-b}$.

■ Table A3.1 *(continued)*

	Density (g/cm³)	Geometric Albedo (%)	AS (° ′ ′′)	M	Discovery Date	Discovered by
Earth						
Moon	3.34	0.12	—	−12.74	—	—
Mars						
Phobos	2.2	0.06	25	11.3	1877	Hall
Deimos	1.7	0.06	1 02	12.40	1877	Hall
Jupiter						
Metis	—	‡	42	17.5	1979	V1/2
Adrastea	—	‡	42	19.1	1979	V1/2
Amalthea	—	0.05	59	14.1	1892	Barnard
Thebe	—	‡	1 13	15.6	1979	V1/2
Io	3.57	0.61	2 18	5.02	1610	Galileo
Europa	2.97	0.64	3 40	5.29	1610	Galileo
Ganymede	1.94	0.42	5 51	4.61	1610	Galileo
Callisto	1.86	0.20	10 18	5.65	1610	Galileo
Leda	—	—	1 00 39	20.2	1974	Kowal
Himalia	—	0.03	1 02 46	14.84	1904	Perrine
Lysithea	—	—	1 04 04	18.4	1938	Nicholson
Elara	—	0.03	1 04 10	16.77	1904	Perrine
Anake	—	—	1 55 52	18.9	1951	Nicholson
Carme	—	—	2 03 31	18.0	1938	Nicholson
Pasiphae	—	—	2 08 26	17.03	1908	Melotte
Sinope	—	—	2 09 31	18.3	1914	Nicholson
Saturn						
Atlas	—	0.5	22	—	1980	V1
Prometheus	—	0.5	23	—	1980	V1
Pandora	—	0.5	23	—	1980	V1
Epimetheus	—	0.5	24	—	1979	P11
Janus	—	0.5	24	—	1966	Dollfus
Mimas	1.17	0.77	30	12.9	1789	Herschel
Enceladus	1.24	1.00	38	11.7	1789	Herschel
Tethys	1.26	0.80	48	10.2	1684	Cassini
Telesto	—	0.6	48	—	1980	Reitsema
Calypso	—	0.9	48	—	1980	Pascu
Dione	1.44	0.55	1 01	10.4	1684	Cassini
Helene	—	0.6	1 01	—	1980	Lecacheux
Rhea	1.33	0.65	1 25	9.7	1672	Cassini
Titan	1.88	0.2	3 17	8.28	1655	Huygens
Hyperion	—	0.25	3 59	14.19	1848	Bond
Iapetus	1.21	§	9 35	11.1	1671	Cassini
Phoebe	—	0.06	34 51	16.45	1898	Pickering

■ Table A3.1   *(continued)*

	Density (g/cm³)	Geometric Albedo (%)	AS (° ′ ′ ′)	M	Discovery Date	Discovered by
Uranus						
Cordelia	—	0.05	4	—	1986	V2
Ophelia	—	0.05	4	—	1986	V2
Bianca	—	0.05	4	—	1986	V2
Cressida	—	0.04	5	—	1986	V2
Desdemona	—	0.04	5	—	1986	V2
Juliet	—	0.06	5	—	1986	V2
Portia	—	0.09	5	—	1986	V2
Rosalind	—	0.05	5	—	1986	V2
Belinda	—	0.05	6	—	1986	V2
Puck	—	0.06	7	—	1985	V2
Miranda	1.26	0.22	10	16.5	1948	Kuiper
Ariel	1.65	0.38	14	14.4	1851	Lassell
Umbriel	1.44	0.16	20	15.3	1851	Lassell
Titania	1.59	0.23	33	13.98	1787	Herschel
Oberon	1.50	0.20	44	14.23	1787	Herschel
Neptune						
1989N6	—	0.06	2	—	1989	V2
1989N5	—	0.06	2	—	1989	V2
1989N3	—	0.06	2	—	1989	V2
1989N4	—	0.05	3	—	1989	V2
1989N2	—	0.06	3	—	1989	V2
1989N1	—	0.06	6	—	1989	V2
Triton	5 (?)	0.6 – 0.9	17	13.69	1846	Lassell
Nereid	—	0.14	4 21	18.7	1949	Kuiper
Pluto						
Charon	0.8 (?)	—	< 1	16.8	1978	Christy

‡ Value is uncertain, between 0.05 and 0.10.
§ Bright side value is 0.50; dark side value is 0.04.

Abbreviations: AS = angular separation observed between planet and satellite at opposition, M = magnitude at opposition, P = Planet, P11 = Pioneer 11, R = sidereal rotational period, R_p = planetary radii, S = synchronous, V1 = Voyager 1, V2 = Voyager 2.

Sources of Data:

U. S. Naval Observatory. *The Astronomical Almanac, 1990.* Government Printing Office, Washington, DC, 1989.

Beatty, J. Kelly. "Getting To Know Neptune." In *Sky and Telescope,* February 1990, vol. 79, p. 146. Reproduced by permission from *Sky and Telescope* astronomy magazine, Cambridge, MA.

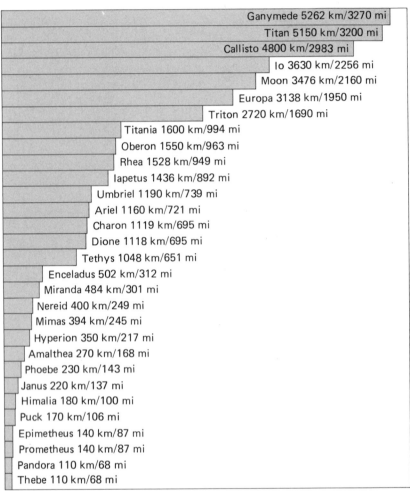

Ganymede 5262 km/3270 mi
Titan 5150 km/3200 mi
Callisto 4800 km/2983 mi
Io 3630 km/2256 mi
Moon 3476 km/2160 mi
Europa 3138 km/1950 mi
Triton 2720 km/1690 mi
Titania 1600 km/994 mi
Oberon 1550 km/963 mi
Rhea 1528 km/949 mi
Iapetus 1436 km/892 mi
Umbriel 1190 km/739 mi
Ariel 1160 km/721 mi
Charon 1119 km/695 mi
Dione 1118 km/695 mi
Tethys 1048 km/651 mi
Enceladus 502 km/312 mi
Miranda 484 km/301 mi
Nereid 400 km/249 mi
Mimas 394 km/245 mi
Hyperion 350 km/217 mi
Amalthea 270 km/168 mi
Phoebe 230 km/143 mi
Janus 220 km/137 mi
Himalia 180 km/100 mi
Puck 170 km/106 mi
Epimetheus 140 km/87 mi
Prometheus 140 km/87 mi
Pandora 110 km/68 mi
Thebe 110 km/68 mi

■ **Figure A3.1** Relative diameters of planetary satellites larger than 100 kilometers in diameter. (For irregularly shaped objects, the longest dimension is given.)

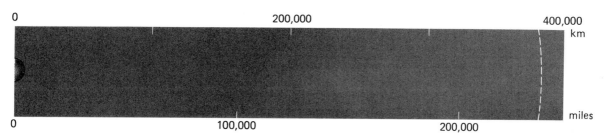

(a) Earth + orbit of Moon

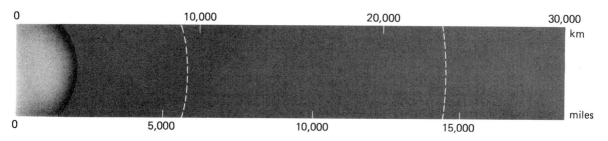

(b) Mars + orbits of Phobos, Deimos

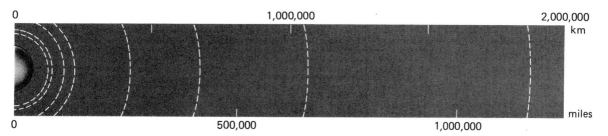

(c) Jupiter + orbits of Metis, Adrastea, Amalthea, Thebe, Io, Europa, Ganymede, Callisto

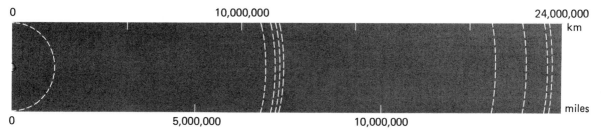

(d) Jupiter + orbits of Callisto, Leda, Himalia, Lysithea, Elara, Anake, Carme, Pasiphae, Sinope

■ **Figure A3.2**   Orbital spacings of planetary satellite systems. Each map has a scale in kilometers above it and a scale in miles below it. The planet is shaded and is to scale, but satellite diameters are not indicated. Satellites are listed below each map in order of increasing semi-major axis, and their orbits are shown in this order from left to right.

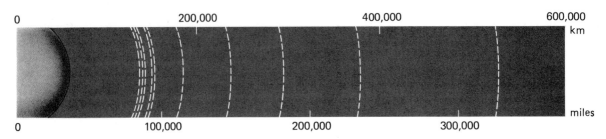

(e) Saturn + orbits of Atlas, Prometheus, Pandora, Epimetheus, Janus, Mimas, Enceladus, Tethys, Dione, Rhea

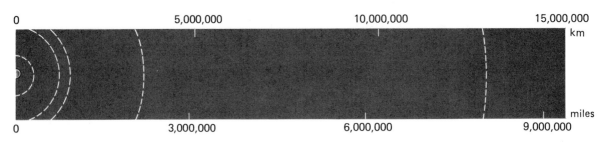

(f) Saturn + orbits of Rhea, Titan, Hyperion, Iapetus, Phoebe

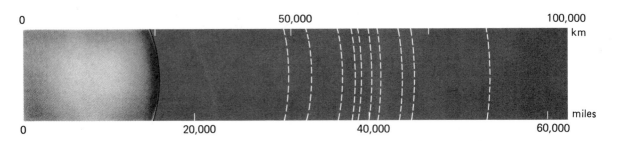

(g) Uranus + orbits of Cordelia, Ophelia, Bianca, Cressida, Desdemona, Juliet, Portia, Rosalinda, Belinda, Puck

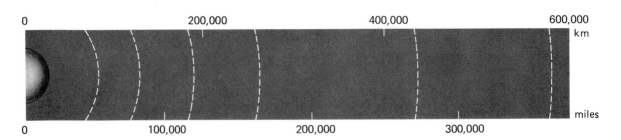

(h) Uranus + orbits of Puck, Miranda, Ariel, Umbriel, Titania, Oberon

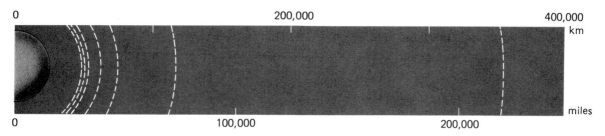

(i) Neptune + orbits of 1989N6, 1989N5, 1989N3, 1989N4, 1989N2, 1989N1, Triton

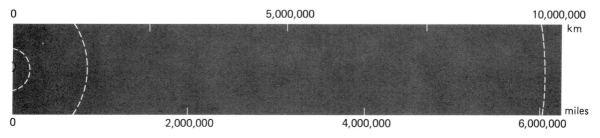

(j) Neptune + orbits of Triton, Nereid (perigee distance), Nereid (apogee distance)

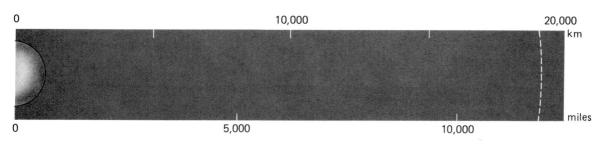

(k) Pluto + orbit of Charon

■ Appendix 4
Minerals Mentioned in This Book

I. Nonsilicates
   Sulfur   S
   Graphite   C
   Diamond   C
   Kamacite $Fe_{0.93-0.96}Ni_{0.07-0.04}$
   Taenite $Fe_{<0.8}Ni_{>0.2}$
   Calcite $CaCO_3$
   Halite (salt)   NaCl

II. Silicates
   A. Separate SiO_4 tetrahedron structure
      Olivine $(Mg,Fe)_2SiO_4$
      Fayalite (an olivine) Fe_2SiO_4
      Forsterite (an olivine) Mg_2SiO_4
   B. Single chains of SiO_4 tetrahedra structure
      Pyroxene = Collective name for the following:
      Augite $(Ca,Na)(Mg,Fe,Al,Ti)(Si,Al)_2O_6$
      Enstatite $MgSiO_3$
      Fassaite $Ca(Mg,Fe,Al)(Si,Al)_2O_6$
      Hypersthene $(Mg,Fe)_2Si_2O_6$
      Pigeonite $(Mg,Fe,Ca)(Mg,Fe)Si_2O_6$
   C. Sheets of SiO_4 tetrahedra structure
      Montmorillonite $(Na,Ca)_{0.33}(Al,Mg)_2Si_4O_{10}(OH)_2 \cdot nH_2O$
      Nontronite $Na_{0.33}Fe_2(Si,Al)_4O_{10}(OH)_2 \cdot nH_2O$
      Saponite $(Ca/2,Na)_{0.33}(Mg,Fe)_3(Si,Al)_4O_{10}(OH)_2 \cdot 4H_2O$
   D. Three-dimensional framework of SiO_4 tetrahedra structure
      Quartz SiO_2
      Leucite $KAlSi_2O_6$
      Scapolite $(Na,Ca,K)_4Al_3(Al,Si)_3Si_6O_{24}(Cl,F,OH,CO_3,SO_4)$
      Feldspar = Collective name for the following:
      Orthoclase $KAlSi_3O_8$
      Plagioclase $(Na,Ca)Al(Al,Si)Si_2O_8$
      Albite (a plagioclase) $NaAlSi_3O_8$
      Anorthite (a plagioclase) $CaAl_2Si_2O_8$

Glossary

A ring — The outermost ring of Saturn visible from Earth using a small telescope.

Aberration of starlight — The bending of starlight caused by Earth's orbital motion. The discovery of this phenomenon by Bradley in the 18th century was an observational proof that the Earth orbits the Sun.

Absolute zero — The temperature at which atomic motion reaches its minimum value. It is 0 on the Kelvin scale, $-273°$ on the Celsius scale, and $-459°$ on the Fahrenheit scale.

Absorption spectrum — A spectrum containing dark lines formed when gases absorb light passing through them.

Acceleration — The rate of change of velocity.

Accretion — The process of planetary formation during which the planets developed from small particles that combined to form larger bodies (planetesimals), which then combined to form the planets.

Accretion disk — A disk of material, in orbit about a star, in which particles are in the process of accreting into planets.

Achondrite — A stony meteorite that lacks chondrules. Achondrites appear to have experienced heating and metamorphism during their history.

Acidic magma — Magma that is rich in quartz, light in color, and very viscous. It forms composite cone volcanoes on Earth but appears to be uncommon or absent on other solar system bodies.

Active volcano — A volcano that is in the process of erupting.

Adams, John — One of the two astronomers who independently deduced the existence of Neptune and calculated where it would be located.

Airy, George — The British Astronomer Royal who was unsuccessful in his attempt to locate Neptune in the location predicted by Adams.

Alba Patera — A huge, gently sloping Martian volcano that is the largest known volcanic feature in the solar system.

Albedo — The percentage of light incident upon an object that is reflected by it.

Algae — Primitive forms of single-celled and multi-celled organisms. They are the oldest terrestrial organisms preserved as fossils.

Allende meteorite — A large carbonaceous chondrite meteorite that fell as numerous fragments in Mexico in 1969.

Alluvial fan — Fan-shaped body of sediment deposited at the base of a steep slope by running water.

Alpine glacier — A type of glacier whose flow is restricted to a limited area, usually a mountain valley. Such glaciers tend to increase the relief of the land affected by them.

Amazonian era — The most recent era of Martian geological history, from 1.8 billion years ago until the present.

Amino acids — Organic compounds that are important components of terrestrial organisms.

Ammonia — Gas of composition NH_3 that is a component of gas giant atmospheres. It also commonly occurs as an ice in the outer solar system.

Ammonium hydrosulfide — Compound of composition NH_4SH that constitutes the middle (white and brown) cloud layer in gas giant atmospheres.

Amor asteroids — Asteroids whose orbits intersect the orbit of the planet Mars.

Amplitude — The height of a wave. The greater the amplitude of a light wave, the greater the intensity of the light.

Angstrom — A unit of length corresponding to 10^{-10} meter.

Angular measure — The measurement of the apparent size of an object. It depends on the object's physical dimensions and its distance.

Angular momentum — A physical quantity possessed by an orbiting or rotating object. It is analogous to linear momentum for an object moving in a straight path.

Anorthosite — A rock, composed mainly of the feldspar anorthite, that is a major component of the crust of the Moon.

Antonindi, E. M. — Telescopic observer of Mars who felt that Martian canals did not actually exist.

Aperture — The diameter of the light-gathering lens or mirror of a telescope.

Aphelion — The point in a planet's orbit that is farthest from the Sun.

Aphrodite Terra — One of the highland continental regions of Venus.

Apollo — American space program of the 1960s and 1970s that accomplished the goal of a successful manned lunar landing.

Apollo asteroids — Asteroids whose orbits intersect Earth's orbit while having a semi-major axis greater than 1 astronomical unit.

Area — Measurement of the two-dimensional extent of a surface.

Argument of perihelion — The position of a planet's perihelion point, measured as an angle from the ascending node.

Ariel — One of Uranus' five major satellites.

Aristarchus — One of the few ancient Greek astronomers who supported the heliocentric concept.

Aristotle — An ancient Greek astronomer who was one of the main proponents of the geocentric concept.

Arsia Mons — A shield volcano in Mars' Tharsis volcanic province.

Ascraeus Mons — A shield volcano in Mars' Tharsis volcanic province.

Ashen lights — Mysterious lights observed on the night side of Venus, probably as a result of lightning or auroras.

Asteroid belt — The area of the solar system between Mars and Jupiter in which most asteroids occur.

Asteroids — Rocky and metallic objects that orbit the Sun, mostly in a belt located between the orbits of Mars and Jupiter.

Asthenosphere — The lower region of the Earth's mantle. It is semi-liquid, and convection within it is responsible for plate tectonic motion.

Astronomical unit — The standard "yardstick" for solar system measurements, defined as the average distance between the Earth and Sun.

Astronomy — The science that studies the universe beyond the Earth.

Astrophysics — The combination of astronomy and physics in which attempts are made to understand the physics of celestial objects.

Atalanta Planitia — The most extensive lowland area on Venus.

Aten asteroids — Asteroids whose orbits intersect Earth's orbit while having a semi-major axis smaller than 1 astronomical unit.

Atmosphere — (1) A unit of pressure equal to the standard atmospheric pressure at Earth's surface. (2) A layer of gases surrounding a planet.

Atom — The basic unit of matter, composed of a central nucleus surrounded by a cloud of electrons.

Atomic number — The number of protons in the nucleus of an atom. The atomic number determines which element a particular atom is.

Aurora — A glow in a planet's upper atmosphere caused by interactions between incoming charged particles and the planet's ionosphere.

Average radius — The starting point for making elevation measurements on a planet lacking an ocean or atmosphere. It is defined as the average distance from the planet's center of all points on its surface.

B ring — The most conspicuous of the rings of Saturn visible from Earth through a small telescope.

Backward scattering — The process of light scattering that sends light back toward the general direction of the source of illumination.

Bar — A unit of pressure equal to 1 million dynes per square centimeter.

Barycenter — The common center of gravity of a system containing one object in orbit about another.

Basalt — A common igneous rock formed when dark, basic magma erupts and cools at a planetary surface. It is characterized by its dark color and fine-grained texture.

Basic magma — Magma that is low in quartz content, dark in color, and of low viscosity.

Basins — The term applied to lunar impact craters larger than about 220 kilometers (135 miles) in diameter.

Batholith — A large, irregularly shaped deposit of intrusive igneous rock.

Belts — Dark-colored atmospheric bands that encircle gas giant planets. Those of Jupiter are especially prominent.

Bessel, Friedrich — Nineteenth-century astronomer whose most significant accomplishment was making the first successful measurement of stellar parallax.

Beta Regio — One of the highland continental regions of Venus.

Biermann, Ludwig — Astronomer who first proposed the existence of the solar wind based upon observations of the behavior of comet tails.

Big Bang — (1) The theory stating that the universe originated with an explosive event 15 to 20 billion years ago. (2) The explosive event that began the universe.

Billion — One thousand million, written 1,000,000,000 or 10^9.

Binary star — A star that appears as one star to the unaided eye but is actually two stars in orbit about their common center of gravity.

Blink comparator — A device that compares two photographs taken at different times by rapidly "blinking" back and forth between images. Pluto was discovered in this way, because the images revealed its motion.

Bode, Johann — The astronomer who popularized the relation now called by his name, and who also proposed the name "Uranus" for the seventh planet.

Bode's law — A numerical sequence that gives the spacing for the planets out to Uranus and the main-belt asteroids. It is not a "law" in the strict sense of the word.

Bolide — An extremely bright meteor.

Bonding — The combination of atoms to form a molecule.

Bradley, James — Eighteenth-century astronomer who discovered aberration of starlight, an effect that proves that the Earth orbits the Sun.

Breccia — A rock that consists of fragments of rocks that have been shattered and welded together. Many lunar rocks have been brecciated by impacts.

Brecciation — The fragmentation of rocks and subsequent welding together of the fragments.

Buffon, Georges Louis de — One of the early proponents of the theory that the planets were formed as a result of a chance gravitational encounter between the Sun and another star.

C ring — The faintest and innermost ring of Saturn normally visible from Earth. It is also called the Crepe ring.

Caldera — Large opening at the summit of a shield volcano from which the magma emanates.

Callisto — The outermost of the four Galilean satellites of Jupiter.

Caloris basin — A huge impact basin on Mercury. It is 1300 kilometers (800 miles) across.

Canals — Linear markings on Mars, first seen in the 19th century, that were believed by some to be the work of intelligent Martian life. However, they proved to be nonexistent optical illusions.

Capture theory — A theory of lunar origin suggesting that the Moon was formed elsewhere in the solar system and subsequently captured by the Earth.

Carbonaceous chondrite — A group of stony meteorites that are considered primitive, or relatively unchanged since their formation. They are so named because they contain carbon compounds.

Carbonaceous material — Material containing carbon or its compounds.

Carbon-based life — A term sometimes applied to terrestrial life, which is based upon the complex chemistry of the element carbon.

Carbon dioxide — Gas of composition CO_2. It is the primary component of the atmospheres of Venus and Mars and is found in small quantities in the atmosphere of Earth.

Cassini, Giovanni D. — Seventeenth-century astronomer who discovered the gap in Saturn's rings that is now known by his name.

Cassini division — The most conspicuous gap in Saturn's rings. It separates the A and B rings and is easily visible from Earth in relatively small telescopes.

Celestial mechanics — The study of orbital motion.

Celestial sphere — An imaginary sphere surrounding the Earth, often used for representing the night sky.

Cell — The basic unit of life. Some terrestrial organisms are single-celled, but most contain many cells.

Celsius scale — A temperature scale with absolute zero at $-273°$, the freezing point of water at $0°$, and the boiling point of water at $100°$.

Cenozoic era — The portion of terrestrial geological time between about 65 million years ago and the present.

Centigrade scale — An older name for the Celsius scale, derived from the fact that there are 100 degrees between freezing and boiling.

Central peak — Mountainous peak located in the center of an impact crater.

Centripetal force — The term coined by Newton to describe a "center-seeking" force. Gravity is the centripetal force that keeps planets in their orbits.

Ceres — The first asteroid to be discovered, and the one with the largest diameter.

Challis, James — Nineteenth-century astronomer who assisted George Airy in searching for Neptune. He and Airy found Neptune only after it had already been discovered by Galle.

Charon — The satellite of Pluto.

Chiron — An asteroid orbiting in the outer solar system.

Chondrites — Stony meteorites containing chondrules.

Chondrules — Small, millimeter-sized spheres found in some stony meteorites. They are believed to be objects formed when molten material solidified in the early solar nebula.

Christy, James — Astronomer who discovered Charon in 1978.

Chromosphere — The layer of the Sun's atmosphere immediately above the photosphere. It emits pink light by the emission process but is too faint to be visible except during a total solar eclipse.

Chryse Planitia — Martian region that was the landing site of the Viking 1 spacecraft in 1976.

Cinder cone — Type of volcano that is composed largely of pyroclastic material. Cinder cones are generally small and erode rapidly.

Circle — Conic section with an eccentricity of 0. An object in a circular orbit is gravitationally bound.

Circular, coplanar, collisionless orbits — A description of the orbits of ring particles. These characteristics are the result of orbital interactions early in the history of a ring system.

Civilization — An organized system of society found among humans and a possible attribute of intelligent life on other planets, if any such life exists.

Clastic — A type of sedimentary rock composed of pieces of pre-existing rocks that have been cemented or compacted together.

Clathrates — Compounds in which water is trapped in voids within the molecular structure.

Clay minerals — A group of silicate minerals that are considered probable constituents of the Martian regolith.

Cleavage plane — A flat surface along which a mineral breaks. The cleavage of a mineral results from a plane of weakness in its crystal structure.

Climate — The long-term average weather conditions of a planet or region of a planet.

Cloud — Generally opaque region within a planetary atmosphere that contains liquid droplets or solid crystals.

Cloud top — The point in a gas giant atmosphere at which pressure is 100 millibars. By definition, this is the point from which elevation is measured on gas giants, because they lack solid surfaces.

Colluvium — Material that has moved downslope during an episode of mass wasting.

Coma — The cloud of gases surrounding the solid nucleus of a comet.

Comet — An object, consisting largely of frozen gases and some rock, orbiting the Sun, generally in an eccentric and highly inclined orbit.

Composite cone — The most spectacular type of volcano, typically high and steep-sided. Such volcanoes are formed by the release of acidic magma and seem to be limited to the Earth.

Compound — A molecule that contains two or more different chemical elements.

Compound telescope — A telescope that contains both a lens and mirror. The most common example is the Schmidt–Cassegrain telescope.

Concordant age — A radiometric age value determined for a rock that is supported by consistent values from more than one mineral in it.

Condensation — The formation of a liquid droplet or solid particle from a gas.

Condensation sequence — A theoretically derived listing of the order in which solid substances condensed from the solar nebula during the formation of the solar system.

Conduction — A method of heat transport involving atom-to-atom contact, generally through a solid.

Conic sections — The collective term given to circles, ellipses, parabolas, and hyperbolas. The name derives from the fact that each figure can be obtained by "slicing" through a cone. All orbital paths follow one of these sections.

Conjunction — The event that occurs when a planet is near the Sun in the sky and therefore cannot be seen from the Earth.

Constellation — A star pattern in the sky. The sky is divided into 88 constellations, many of which originated in ancient times.

Contact metamorphism — A form of metamorphism that occurs when rock is changed by contact with a hot body of magma.

Continent — Area of the Earth's surface that extends above sea level. The term is also applied to highland areas on other planets, such as Venus, which rise above the terrain surrounding them.

Continental crust — The portion of the Earth's crust comprising the continents. It is composed primarily of granite but is covered in most places by a thin veneer of sedimentary rock.

Continental glacier — A thick glacier that covers a large land area and lowers the relief of the land over which it travels.

Continental shelf — The portion of a continent extending beyond the shoreline and covered by relatively shallow ocean water.

Continuous spectrum — A spectrum containing all wavelengths of light.

Convection — A method of heat transport accomplished by circulation within a gas or liquid.

Convective zone — A region within the Sun in which thermal energy is transported by convection. It consists of the outer 20% of the Sun's interior.

Co-orbital satellites — Two satellites of Saturn (Janus and Epimetheus) that occupy virtually the same orbit and periodically exchange places with one another.

Copernican era — The era of time encompassing the past 1 billion years of lunar history.

Copernicus, Nicholas — Renaissance astronomer who became the first modern proponent of the heliocentric theory.

Core — The central region of a star, planet, or other celestial object.

Coriolis effect — Forces experienced on a rotating body that cause objects moving on it to deviate from a straight-line trajectory.

Corona — The hot, tenuous, outer layer of the Sun's atmosphere. It emits light but is so faint that it cannot be seen with the unaided eye except during a total solar eclipse.

Coronae — Oval-shaped bands of ridges and grooves found on the surface of Uranus' satellite Miranda.

Coronal holes — Areas of the Sun's corona where its temperature and density are lower than normal. These sites appear to be the sources of unusually strong flows of the solar wind.

Cosmic rays — Mysterious high-energy particles that originate beyond the solar system.

Cosmogony — The branch of astronomy that studies the origin of celestial objects, from the solar system to the universe as a whole.

Covalent bonding — Type of bonding in which atoms share electrons.

Crater counting — The process of estimating the age of a planetary surface by determining the density of craters on it.

Cratered plains — Name applied to a type of surface terrain found on Saturn's satellite Dione.

Cratered terrain — Name applied to a type of surface terrain found on Saturn's satellite Enceladus.

Creep — A form of mass wasting consisting of the gradual downhill sliding of regolith.

Crust — The outermost layer of a solid planet.

Crystal — A solid object marked by flat surfaces in a regular arrangement, resulting from the regular internal arrangement of the atoms in the solid.

Crystal lattice — The orderly arrangement of atoms in a mineral.

D ring — The innermost ring of Saturn. It is invisible from Earth and was discovered because it partially obscured light from the planet below.

D' ring — The term applied to rings outside Saturn's A ring when they are observed from Earth.

Dark, cratered terrain — Surface unit of Jupiter's satellite Ganymede that appears to be its original crust.

Darwin, Charles — Nineteenth-century naturalist best known for developing the theory of evolution.

Darwin, George — Researcher who proposed that the Moon originated by breaking free from the Earth.

Daughter isotope — An isotope produced by radioactive decay of a "parent" isotope.

Davis, Raymond — Modern researcher who supervised the first attempt to measure solar neutrinos.

Decametric radiation — Radio radiation of approximately 10-meter wavelength. Jupiter's radiation belts produce this type of radiation.

Decimetric radiation — Radio radiation of approximately 0.1-meter wavelength. Jupiter's radiation belts produce this type of radiation.

Deferent — The primary portion of an object's orbit around the Sun in the geocentric model. The deferent supposedly carried the smaller epicycle.

Deflation basin — A low area produced by the removal of regolith by wind.

Deformation — A change in the shape, location, or orientation of a body of rock. Folds and faults are deformations.

Degree — The basic unit of angular measurement, equal to 1/360 of a circle.

Deimos — The outer and smaller of Mars' two satellites.

Delta — A deposit of sediment formed when a river reaches standing water such as a lake or the ocean.

Density — The measurement of the mass per unit volume of an object. The density of water is 1 gram per cubic centimeter.

Depressed lowlands — The areas of Venus' surface lying below the planet's average radius.

Diana Chasma — The lowest point on the surface of Venus. It appears to be a fault-related feature, possibly evidence of tectonic activity.

Differential gravitational force — Force caused by the unequal attraction of gravity on different portions of an object. Such forces are responsible for tides.

Differential rotation — Rotation of a nonsolid body in which different latitude regions of the body rotate at different rates.

Differentiation — The separation of a planet into distinct internal layers because of density differences.

Diffraction grating — A sheet of glass or plastic with closely spaced inscribed lines that cause light hitting them to be dispersed into a spectrum.

Dike — A body of intrusive igneous rock that cuts across pre-existing rock layers.

Dinosaur — Collective term given to two groups of mostly large, land-dwelling reptiles that lived during the Mesozoic era.

Dione — One of the large satellites of Saturn.

Dipole field — The term applied to a normal magnetic field because of the presence of two magnetic poles.

Dirty snowball model — The theory suggesting that a comet is composed mostly of ices with some rocky material also present.

Discontinuity — Boundary within a solid planet between two layers with different properties. Physical conditions change rapidly over the boundary.

DNA — Acronym for deoxyribonucleic acid, the organic compound that stores the genetic information in all terrestrial organisms.

Doppler, Christian — Nineteenth-century physicist who first described the effect that now bears his name.

Doppler effect — The change in apparent wavelength of a light or sound source caused by relative motion between it and the observer.

Dormant volcano — A volcano that is not presently active but has the potential for renewed activity in the future.

Dry ice — Ice of carbon dioxide.

Dune — A deposit of wind-blown sand.

Dynamo theory — Theory that attempts to explain the origin of the Earth's magnetic field as the result of circulation within its liquid outer core.

Dyne — Basic metric unit of force, the weight of a 1-gram mass.

E ring — The outermost ring of Saturn. It is wide and tenuous, was discovered by spacecraft, and is possibly associated with the satellite Enceladus.

Earthquake — Vibrational waves, caused by fault motion, that travel through the Earth or other planets.

Earthshine — The faint illumination of the crescent Moon caused by light reflected from the nearly full Earth.

Eccentricity — A quantity describing the shape of an orbit. A circle has an eccentricity of 0, an ellipse between 0 and 1, a parabola of 1, and a hyperbola greater than 1.

Eclipse — An event occurring when the shadow of a solar system object touches or engulfs another object.

Ecliptic plane — The plane of the solar system, defined by the plane of the Earth's orbit. All planetary orbits lie close to the ecliptic.

Einstein, Albert — Physicist active in the first half of the 20th century. He is best known for the development of the theory of relativity.

Ejecta — Material splashed outward from a crater during the impact event that formed it.

Electromagnetic radiation — General term for light, including both visible light and forms invisible to the human eye.

Electromagnetic spectrum — Term applied to all forms of light, including both visible light and the forms that our eyes cannot detect.

Electron — Negatively charged particle that can be thought of as orbiting an atomic nucleus.

Electrostatic attraction — Force of attraction, caused by static electricity, that allows particles to join together even when their gravitation is too small to cause attraction.

Element — Specific variety of atom characterized by a particular atomic number, or number of protons in the nucleus.

Elliot, James — Astronomer who led the research team that discovered Uranus' rings in 1977.

Ellipse — Conic section with an eccentricity between 0 and 1. Objects in elliptical orbits are gravitationally bound, and all planetary orbits are ellipses.

Elliptical galaxy — Galaxy whose overall form is approximately spherical or elliptical.

Elongation — The angle between a planet and the Sun. Inferior planets are best seen when at maximum elongation. Superior planets are best seen when their elongation is 180°.

Elysium — One of the three Martian volcanic provinces.

Emission spectrum — Spectrum containing bright lines of light at particular wavelengths, rather than light of all wavelengths. Such spectra commonly are produced by hot gases.

Enceladus — One of the large satellites of Saturn and the solar system body with the highest albedo.

Energy — The ability to do work.

Enstatite chondrites — Stony meteorites that contain chondrules and are composed primarily of iron-free pyroxene, or enstatite.

Epicycle — Small, secondary orbit proposed by supporters of the geocentric concept to explain retrograde motion.

Epsilon ring — The most conspicuous ring of Uranus.

Equator — The line midway between the poles of a planet.

Equilibrium — A condition of balance, typically between opposing forces, such as gravity and internal pressure within a star.

Eratosthenian era — The era of lunar history occurring between 3.2 and 1.0 billion years ago.

Erg — The basic metric unit of energy that is normally used in astronomy.

Erosion — The wearing down and removal of rock and soil by surface agents of geological change.

Escape velocity — The velocity required for an object to escape completely the gravity of another object.

Europa — The smallest of the four Galilean satellites of Jupiter.

Evaporation — Change of state from liquid to gas.

Evolution — Theory proposing that the form and function of adult living organisms change gradually over many generations.

Exosphere — The outer edge of the thermosphere in the Earth's atmosphere.

External process — Process that attacks a planet's surface from the outside and normally results in the lowering of the surface.

Extinct volcano — Volcano that is no longer capable of erupting.

Extrusion — Deposit of igneous rock formed at the surface of a planet.

Extrusive — Igneous rock formed at the surface of the Earth and characterized by having fine-grained texture.

F ring — Narrow ring of Saturn lying just outside its A ring.

Faculae — Small, bright emission areas above the Sun's photosphere.

Fahrenheit scale — A temperature scale with absolute zero at −459°, the freezing point of water at 32°, and the boiling point of water at 212°.

Fall — A meteorite that is seen to fall and is subsequently recovered.

Fault — Rock deformation consisting of a break in rock along which motion has occurred. Such motion is responsible for causing earthquakes.

Find — Meteorite that is not seen to fall but is recognized on the ground some time after its arrival.

Fireball — An extremely bright meteor.

First quarter — Phase of the Moon midway between new moon and full moon when half of the side of the Moon facing us is illuminated.

Fission theory — Theory suggesting that the Moon broke away from the Earth early in our planet's history.

Flare — Eruptive events on the Sun generally associated with sunspots.

Flyby mission — Space mission to another solar system object which studies the object during a brief flight past it.

Focus — (1) One of the two points on which an ellipse is based. The Sun is located at one focus point of each planet's orbit. (2) The point at which fault motion occurs during an earthquake.

Fold — Rock deformation consisting of bending of rock layers.

Foliation — Parallel alignment of mineral grains in a metamorphic rock.

Formation — A distinct unit of rock that can be studied and mapped and is usually given a characteristic name.

Forward scattering — The process of light scattering that sends light onward in the same general direction as before.

Fossil — Remnant of ancient life or trace of its activity that has been preserved in rock since prehistoric times.

Foucault, J. B. L. — Nineteenth-century scientist best known for his demonstration, using the pendulum now named for him, that the Earth rotates.

Foucault pendulum — Pendulum whose motion can be used to demonstrate that the Earth rotates on its axis.

Fractionation — The variation in ratios between isotopes of an element in different physical locations.

Fractionation line — A straight line on a graph of isotope fractionation ratios. Materials lying on the same line normally are believed to have originated in the same physical region.

Fraunhofer, Joseph — Researcher who first noted dark lines in the Sun's spectrum during the early 19th century.

Fraunhofer lines — Dark lines in the Sun's spectrum that are caused by absorption.

Frequency — The number of waves passing a point every second.

Fretted channels — Wide, steep-walled Martian channels that have smooth, flat floors.

Full moon — Lunar phase occurring when the Moon is opposite the Sun in the sky and all of the side facing us is illuminated.

Fusion crust — Dark, glassy coating on a meteorite's surface caused by melting and cooling of material as it entered the Earth's atmosphere.

G ring — Faint, outer ring of Saturn discovered by spacecraft. Only the E ring is farther from the planet.

Galaxy — Huge aggregate of stars, typically numbering hundreds of billions, that are held together by their mutual gravitation.

Galilean satellites — The four large satellites of Jupiter which were discovered by Galileo in 1610. Their names are Io, Europa, Ganymede, and Callisto.

Galileo — (1) Astronomer who made the first telescopic observations in the early 1600s. (2) Spacecraft that will orbit Jupiter and send a probe into Jupiter's atmosphere.

Galileo Regio — The largest area of dark, cratered terrain on Jupiter's satellite Ganymede.

Galle, Johann — German astronomer who discovered Neptune in 1846 at the location predicted by Leverrier.

Gamma rays — Extremely short wavelength, high energy form of electromagnetic radiation.

Ganymede — The third Galilean satellite of Jupiter and the largest planetary satellite in the solar system.

Gas-exchange experiment — One of the experiments aboard the Viking Mars landers that was designed to search for Martian life by detecting evidence of metabolism.

Gas giant planets — The planets Jupiter, Saturn, Uranus, and Neptune, which are very large and composed predominantly of gaseous materials such as hydrogen and helium.

Gauss, Karl F. — Mathematician who developed techniques for determining orbital elements based on three observations of an object's position.

Geocentric concept — One of the theories developed by the ancient Greeks to explain the motions of solar system objects. It proposed that the Earth was in the center of the universe and was orbited by all other objects.

Geologic map — A map showing the distribution of rocks of various types and ages.

Geology — The science that studies the Earth, including the materials of which it is composed, its surface features, its internal features, and its history.

Geosynchronous orbit — Orbit in which an orbiting object has a velocity equal to the rotational speed of the object it orbits. An object in such an orbit appears stationary over a point on a planet's surface.

Geosynchronous satellite — A satellite that orbits the Earth once every 24 hours and therefore appears fixed in the sky.

Gibbous — Phase of the Moon when it is between quarter and full phase.

Giordano Bruno — A relatively young-appearing lunar crater whose formation may have been witnessed in 1178. It is named for an Italian astronomer who lived during the Renaissance.

Glacier — A large body of ice, derived from snow, that moves over a land surface.

Globular star cluster — A cluster containing stars arranged in a spherical distribution.

Goddard, Robert — Researcher who successfully launched the first liquid-fueled rocket during the 1920s.

Gossamer ring — Faint component of Jupiter's ring system. It extends outward beyond the main ring.

Graben — A low area or valley produced by the downward movement of a block of rock by faulting.

Gram — The basic metric unit of mass.

Grand tour — Space mission during which a spacecraft flies past more than one planet, gaining a gravitational boost as it encounters each.

Granulation — Mottled appearance of the Sun's photosphere resulting from the convection cells that bring thermal energy to the surface.

Gravitational collapse — The process whereby an extended cloud of gas or dust shrinks in size because of the gravitational attraction of the atoms or grains within it.

Gravitational contraction — The process that would cause the continued shrinking of a star such as the Sun if it were not balanced by the star's internal pressure.

Gravity — The force, first described by Newton, that causes attraction between any objects with mass.

Gravity anomaly — An area of a planet's surface where variation in subsurface rock type or structure causes the force of gravity to be slightly higher or lower than the planet-wide average.

Great Red Spot — Red atmospheric storm on Jupiter that has been observed for several centuries.

Great Dark Spot — Atmospheric storm on Neptune. It was discovered by Voyager 2 in 1989.

Greatest elongation — The point at which an inferior planet is oriented so that it has the maximum angular separation from the Sun and is therefore easiest to observe.

Greenhouse effect — The trapping of heat by carbon dioxide or water vapor in a planetary atmosphere. As a result of this effect, Venus has temperatures much higher than would otherwise be expected.

Ground water — Water contained beneath the Earth's surface within openings in rock and soil.

Guardian satellites — Satellites that orbit just inside or outside the edge of a planetary ring. Their gravity causes the edges of the ring to be sharp and distinct.

Guinan, Edward — Astronomer whose occultation studies of Neptune first gave evidence that that planet has a ring system.

Habitable zone — The area of space around a star where a planet would have surface temperature conditions that would allow liquid water to exist.

Hadley circulation — Movement of air in a planetary atmosphere characterized by rising air at the equator, sinking air at the poles, and air moving poleward at altitude and toward the equator at the surface.

Half-life — The time required for half of a given quantity of a radioactive substance to undergo decay to the product material.

Hall, Asaph — The American astronomer who discovered Mars' satellites Phobos and Deimos.

Halley, Edmond — Astronomer who first suggested that comets were regular solar system members whose orbits obeyed Kepler's and Newton's laws. He predicted the 1758 return of the comet that was subsequently named for him.

Halley's comet — Comet that was realized by Halley to be periodic. He predicted its 1758 return, and the comet was named in his honor.

Halo — (1) The portion of Jupiter's ring system located closer to the planet than the main ring and extending above and below the main ring plane. (2) Huge cloud of gas atoms surrounding the head of a comet.

Haze — Atmospheric phenomenon similar to a cloud, except that haze droplets tend to be smaller than cloud droplets.

Head — The portion of a comet containing its nucleus and coma.

Heat — A measure of the amount of thermal energy present in a system.

Heat flow — The movement of heat from within a planet toward its surface. Measurements of heat flow give important information about conditions within a planet.

Heavily cratered terrain — A type of landscape found on Mercury and on Saturn's satellite Dione.

Heliocentric concept — Model of the solar system stating that the Sun is in its center. This model first was proposed by a few ancient Greeks, but it was not generally accepted until the time of Copernicus, Kepler, and Newton.

Heliopause — The outer boundary of the heliosphere.

Heliosphere — The vast region of space in which the solar wind is found.

Hellas — (1) Martian impact basin that is the solar system's largest. (2) The oldest of Mars' three volcanic provinces.

Herschel — Huge crater on Saturn's satellite Mimas.

Herschel, William — Astronomer who is well known for his observational skills and especially for discovering Uranus in 1781.

Hertz — Unit used in measuring wave frequency. One Hertz is one wave per second.

Hesperian era — The middle era of Martian geological history, from 3.5 until 1.8 billion years ago.

Hidalgo — An asteroid orbiting in the outer solar system.

Highlands — (1) Elevated continental regions on Venus. (2) Old, rough, light-colored regions on the Moon believed to be the original lunar crust.

Hilly and lineated terrain — Type of terrain on Mercury located on the opposite side of the planet from the Caloris basin and thought to have been produced by forces generated from the Caloris impact.

Hilly, cratered terrain — Type of terrain found on Saturn's satellite Tethys.

Hipparchus — Ancient Greek astronomer who discovered precession and developed the system of stellar magnitude.

Hirayama, Kiyotsugu — Astronomer who suggested that numerous asteroids may be remnants of an old asteroid that broke apart and formed them.

Hirayama family — Group of asteroids believed to have once been part of a single asteroid.

Hoba meteorite — The largest meteorite that has been found. It is an iron meteorite that fell in Africa.

Hooke, Robert — The first observer to see Jupiter's Great Red Spot, in 1664.

Horrocks, Jeremiah — The first person to observe a transit of Venus.

Hubble, Edwin — Modern astronomer who discovered the expansion of the universe.

Hubble's law — Relationship that states that the farther away from us a galaxy is, the faster it is receding from us.

Huygens, Christiaan — The first person to explain adequately the phenomenon of Saturn's rings.

Hydrates — Compounds that contain water as part of their molecular structure.

Hydrocarbons — Compounds of hydrogen and carbon. Most of them are dark in color, and planetary surfaces containing them usually have low albedos.

Hyperbola — Conic section with an eccentricity greater than 1. Objects in hyperbolic orbits have velocities greater than the escape velocity and are therefore not gravitationally bound to the central object around which they orbit.

Hyperion — Satellite of Saturn that is very irregular in shape.

Hypothesis — Proposed solution to a scientific problem that must be verified by experiment, observation, or calculation.

Iapetus — The next-to-outermost satellite of Saturn.

Ice — Term applied to any frozen gas.

Ice age — Period of time during which the areal extent of continental glaciers is much greater than normal.

Icy planets — Planets (or satellites) that are composed largely of ice, but usually have a rocky interior. Pluto is an example.

Igneous rock — Rock that forms from the cooling of hot, molten rock called magma.

Image processing — The process of enhancing a photographic image returned by a spacecraft in order to reveal the maximum amount of information.

Imbrian era — The era of lunar history occurring between 3.85 and 3.2 billion years ago.

Impact crater — Circular depression formed by the explosive impact of a meteorite.

Impact melt rocks — Igneous rocks produced from the cooling of rock melted by the heat released during an impact event.

Inclination — The angle between the plane of an object's orbit and the plane of the ecliptic.

Inertia — The property of an object that causes it to remain at rest if at rest or remain in motion if in motion.

Inferior conjunction — The event that occurs when Mercury or Venus passes between the Earth and the Sun.

Inferior planet — Planet whose orbit is between the orbit of the Earth and the Sun.

Infrared light — Form of electromagnetic radiation whose wavelength is slightly longer than red visible light.

Inner core — The region of the Earth that is hot, dense, solid, and composed of nickel–iron metal.

Intelligence — The characteristic of a living organism that is aware of its own existence and capable of abstract thought.

Intercrater plains — Terrain type on Mercury that is smooth and gently rolling.

Internal process — Geological process occurring within a planet. Such processes tend to build up the surface of the planet.

Interplanetary medium — The gas and particles that pervade the space between the planets.

Interstellar medium — The gas and dust that are found in the space between stars. Stars form from concentrations of the interstellar medium.

Intrusion — Body of igneous rock that has formed underground.

Intrusive — Igneous rock that has formed underground and is characterized by having a coarse-grained texture.

Inversion — Characteristic of an atmospheric layer in which temperature rises with increasing altitude.

Io — The innermost of the four Galilean satellites of Jupiter. It is the only place other than the Earth and Triton where volcanoes are known to be currently active.

Ion — Atom that has either gained or lost electrons and is therefore electrically charged.

Ionic bonding — Type of bonding in which atoms join together because of electrical attraction.

Ionosphere — Region of an atmosphere in which ionized atoms are present.

Iron meteorite — Meteorite that is composed largely of nickel–iron metal.

Irregular galaxy — Galaxy that lacks a clear structure such as that found in an elliptical or spiral galaxy.

Ishtar Terra — One of the highland continental regions of Venus.

Isostasy — (1) Vertical motion of crustal material due to density differences. (2) The theory that planetary crusts are in vertical balance because of such motion.

Isotopes — Varieties of a specific element that have different numbers of neutrons and thus different atomic masses.

Ithaca Chasma — A huge, branching valley system on Saturn's satellite Tethys.

Jansky, Karl — Researcher who developed the first radio telescope.

Jeans, J. H. — One of the last proponents of the theory that the planets were formed as a result of a chance gravitational encounter between the Sun and another star.

Joint — Crack in rock.

Jovian planets — Another term for the gas giant planets.

Karst topography — Type of landscape characterized by containing numerous caves and sinkholes. It occurs in moist areas that have limestone rocks.

Keeler, James — Astronomer who used the Doppler effect to measure the orbital velocity of particles in Saturn's rings. He also discovered a gap in Saturn's A ring, which is now called the Keeler division.

Keeler division — The largest division in Saturn's A ring.

Kelvin scale — A temperature scale with absolute zero at 0 K, the freezing point of water at 273 K, and the boiling point of water at 373 K.

Kelvin–Helmholtz contraction — Mechanism of stellar energy generation produced by gradual shrinking of the star's size. This was believed to be the source of the Sun's energy prior to the discovery of nuclear fusion.

Kepler, Johannes — Renaissance astronomer who first proposed that planetary orbits are elliptical and who developed three great laws of planetary motion.

Kepler's first law — The orbits of the planets are ellipses, with the Sun occupying one of the focus positions.

Kepler's second law — Planets move fastest when closest to the Sun and slowest when farthest from the Sun and sweep out equal areas in equal times.

Kepler's third law — There is a proportional relationship between the square of a planet's orbital period and the cube of its semi-major axis. The sum of the masses of the Sun and the planet also is part of the relationship.

Kinetic energy — The energy that an object possesses as a result of its motion.

Kirkwood, Daniel — Nineteenth-century astronomer who explained that the gaps in the spacing of main-belt asteroids are the result of orbital resonances with Jupiter.

Kirkwood gaps — Areas within the asteroid belt where there are few asteroids because of orbital resonances with Jupiter.

Kuiper Airborne Observatory — A specially modified aircraft from which astronomical observations are made. It is useful for making observations of celestial events that must be observed over oceans or above clouds.

Labeled release experiment — One of the experiments aboard the Viking Mars landers that was designed to search for Martian life by detecting evidence of metabolism.

Lagrange, J. L. — Mathematician who calculated the positions associated with an orbiting object at which other objects could be located.

Lagrangian points — Positions with respect to an orbiting object where another object can get a "free ride." Two stable Lagrangian points exist, 60° preceding and 60° behind the primary orbiting object.

Lagrangian satellite — Satellite that occupies a stable Lagrangian point, such as in the orbits of Saturn's satellites Tethys and Dione.

Landslide — Form of mass wasting characterized by the rapid downhill sliding of rock or soil.

Lapse rate — The rate of change of temperature as a function of elevation in an atmosphere.

Last quarter — Phase of the Moon midway between full moon and new moon when half of the side of the Moon facing us is illuminated.

Latitude — System of measurement of location north and south of an object's equator.

Lava flow — A type of volcanic activity produced when low-viscosity magma floods a large area of the surface, forming a level surface.

Law — Scientific theory that has been tested, verified, and believed to have universal application. Many laws are stated mathematically.

Layered terrain — Terrain type found in Martian polar regions, consisting of light and dark bands of material deposited by the polar caps.

Leading hemisphere — The hemisphere of a 1 : 1 spin-orbit coupled object that is on the leading side as the object orbits.

Leap year — 366-day year used every fourth year to compensate for the fact that the length of a terrestrial year is about 365 $\frac{1}{4}$ days.

Least-energy orbit — An elliptical orbit used by a spacecraft traveling from one planet to another. It is designed to use the minimum amount of energy necessary to reach the destination.

Leverrier, Urbain — One of the two astronomers who independently deduced the existence of Neptune and calculated where it would be located.

Libration — The variation of the Moon's orientation as seen from Earth. As a result of the combination of several motions of the Moon and Earth, we can see 59% of the Moon's total surface despite its perfect 1 : 1 spin-orbit coupling.

Life — The system of complex biochemical reactions that occur in any living organism.

Life as we know it — The term applied to life similar to terrestrial life, that is, life composed primarily of the elements carbon, hydrogen, oxygen, and nitrogen and that requires water to live.

Light — Radiant energy.

Light, grooved terrain — The most abundant terrain type on Jupiter's satellite Ganymede.

Light year — The distance that light travels in one year, approximately 9.5 trillion kilometers or 6 trillion miles.

Limb — The edge of a spherical celestial object as it is observed from the Earth or from space.

Limb darkening — The phenomenon causing the edge of the Sun's photosphere to appear darker than its center.

Line of nodes — The line connecting the two points where an object's orbit intersects the ecliptic.

Linear ridges — Straight ridges on Mercury that are probably fault scarps.

Linear rilles — Straight features on the Moon that are probably fault scarps.

Linear troughs — Long surface features on Mercury that are probably grabens.

Liquid-fuel rocket — A rocket whose engine burns liquid fuel and oxidizer. Their advantage over solid-fuel rockets is that their engines can be throttled as well as stopped and restarted.

Lithosphere — The term applied to the material in Earth's crustal plates, consisting of the crust and upper mantle.

Lobate scarps — Rounded features on Mercury that are probably the result of reverse faulting.

Local Group — The cluster of galaxies of which the Milky Way is a member.

Lomonosov, Mikhail V. — Russian astronomer who noted Venus' hazy appearance during its 1761 transit of the Sun and concluded that it had a substantial atmosphere.

Longitude — System of measurement of location east and west of an object's prime meridian.

Longitude of ascending node — The angle between the vernal equinox and the point in an object's orbit where it crosses the plane of the ecliptic from south to north.

Long-period comet — A comet whose orbital period is longer than two centuries.

Low Earth orbit — The type of orbit into which most Earth satellites are placed, approximately 160 kilometers (100 miles) above the surface.

Lowell, Percival — American astronomer best known for starting the observational program that led to the discovery of Pluto. He is also remembered for his fanciful theories of intelligent Martian life.

Luna — (1) Another name for the Earth's Moon. (2) A Soviet lunar exploration program during the 1960s and 1970s.

Lunar dome — Dome-shaped feature on the Moon that is considered to be the result of volcanic activity.

Lunar Orbiter — American space program during the 1960s that sent five spacecraft into orbit around the Moon to photograph its surface.

Magellanic Clouds — The two irregular galaxies that are satellites of the Milky Way.

Magma — Molten rock generated in a planetary crust or mantle.

Magnetic braking — The slowing of the Sun's rotational rate as it lost magnetic energy while its field swept through the solar nebula.

Magnetic field — The invisible pattern of forces generated by a magnet.

Magnetometer — A device that measures the strength and orientation of a magnetic field.

Magnetopause — The outer boundary of a magnetosphere.

Magnetosheath — The transition region between a magnetopause and the solar wind.

Magnetosphere — The region around the Earth or other planet in which its magnetic field has more influence than the Sun's.

Magnitude — A number that measures the brightness of a star, planet, or other celestial object.

Main-belt asteroids — Asteroids that orbit the Sun in a zone between the orbits of Mars and Jupiter.

Main ring — The major component of Jupiter's ring system.

Major axis — The longest line segment that can be drawn through an ellipse, intersecting both focus points.

Mantle — The region of a planet's interior located between its crust and core.

Mare — Planetary surface feature consisting of a large impact basin that has been flooded with basaltic magma.

Mariner — American program of planetary exploration spacecraft. Successful Mariner craft were 2, 5, and 10 to Venus; 4, 6, 7, and 9 to Mars; and 10 to Mercury.

Mascon — Acronym for "mass concentration," an area under some lunar maria where complete isostatic adjustment has failed to occur.

Mass — The measure of the quantity of material contained within an object.

Mass wasting — The downhill movement of soil or rock due to gravity.

Maunder, E. W. — Researcher who made a study of the long-term pattern of sunspot variation and discovered that very few sunspots had occurred between 1645 and 1715.

Maunder minimum — The period of time between 1645 and 1715 when very few sunspots were seen and the sunspot cycle apparently was not occurring.

Maxwell, James C. — Nineteenth-century physicist, one of whose many discoveries was that a solid ring orbiting Saturn would be dynamically unstable.

Maxwell Montes — Mountain on Venus that is that planet's highest surface feature and is possibly volcanic in origin.

Mesosphere — The region of the Earth's atmosphere located between the stratosphere and the thermosphere.

Mesozoic era — The portion of terrestrial geological time between about 245 and 65 million years ago.

Metabolism — The reactions occurring within a living organism that consume materials taken into the body, emit wastes, and produce energy for growth and reproduction.

Metallic hydrogen — Hydrogen in which molecules are not present but the nuclei are surrounded by common electrons that make the material highly conductive.

Metamorphic rock — Rock that has experienced change, primarily due to heat and pressure.

Metamorphism — The change in texture or mineralogy of a rock brought about mainly by heat and pressure.

Meteor — Flash of light visible in the sky as a meteoroid enters the atmosphere and burns.

Meteor shower — Period of time when an unusually large number of meteors can be seen. These events occur at times during the year when the Earth intersects the debris-filled orbit of a comet.

Meteorite — Fragment of a meteoroid that survives passage through the atmosphere and lands on the Earth.

Meteoroid — Fragment of comet or asteroid debris orbiting the Sun in such a path that it could eventually hit the Earth.

Meter — The basic metric unit of length, equal to about 39.37 inches.

Methane — Gas of composition CH_4 that is present in many planetary and satellite atmospheres and, as an ice, in comets, satellites, and Pluto.

Metric system — Modern system of measurement used in virtually all scientific applications.

Microwaves — Region of electromagnetic radiation intermediate in wavelength between infrared and radio.

Milky Way — The spiral galaxy in which the solar system is located.

Miller, Stanley — Researcher noted for conducting experiments simulating conditions under which the first terrestrial life may have formed.

Miller–Urey experiment — Experiment that simulated early terrestrial conditions. During the course of the experiment, amino acids were synthesized from the atmospheric materials present.

Million — One thousand thousand, written 1,000,000 or 10^6.

Mimas — The innermost of Saturn's large satellites.

Mineral — A solid, naturally occurring chemical compound that has a crystalline structure and a definite chemical formula.

Minor axis — The shortest line segment that can be drawn through an ellipse. It is perpendicular to the major axis.

Minor planets — Another term for asteroids.

Minute — Unit of angular measurement equal to 1/60 degree.

Miranda — The innermost of Uranus' large satellites.

Molecular hydrogen — Hydrogen consisting of pairs of atoms comprising the H_2 molecule.

Molecule — Combination of two or more atoms.

Moment of inertia — A measure of the inertia of a rotating object. It is controlled by the distribution of mass within it.

Moment-of-inertia coefficient — Number that describes the distribution of mass within a rotating object.

Mottled terrain — Type of landform found on Jupiter's satellite Europa.

Mountain material — Blocks of rock reaching high elevation on Jupiter's satellite Io.

Multiringed basin — Large impact basin that contains several concentric rings produced by the faulting caused by energy released during the impact event.

Mutation — A mistake in the replication of DNA during the reproductive process in an organism.

Mutual formation theory — The theory suggesting that the Moon formed at its present location, in orbit around the Earth.

Nebula — A large, relatively dense cloud of gas and dust in space.

Nectarian era — The era of lunar history occurring between 3.9 and 3.85 billion years ago.

Nereid — The outermost of Neptune's satellites. It had the largest eccentricity of any planetary satellite.

Neumann lines — Linear striations visible on cut surfaces of hexahedrite iron meteorites.

Neutrinos — Massless subatomic particles which move at the speed of light. They are produced during nuclear fusion reactions.

Neutron — Chargeless subatomic particle found in atomic nuclei.

New moon — Lunar phase occurring when the Moon is between the Sun and the Earth and is therefore invisible.

Newton — A basic metric unit of weight.

Newton, Isaac — British scientist known for the development of laws of motion, the law of gravitation, and calculus.

Newton's laws of motion — Three laws derived by Newton that explain inertia, the relationship between force and acceleration, and the principle of action and reaction.

Noachian era — The era of Martian history lasting from the planet's origin until 3.5 billion years ago.

Node — A point at which the orbit of a planet intersects the plane of the ecliptic.

Nonclastic — A type of sedimentary rock derived from dissolved minerals rather than rock fragments.

Nuclear fission — A reaction involving an atomic nucleus in which the nucleus splits into two or more simpler nuclei.

Nuclear fusion — The combination of two or more atomic nuclei into a more complex nucleus.

Nucleosynthesis — The process of forming the nuclei of various elements. Nucleosynthesis occurred at the time of the Big Bang and presently occurs inside stars.

Nucleus — (1) The central region of an atom, where most of its mass is contained. (2) The central portion of a comet, composed of ice and some rocky material. (3) The central region of a cell, where genetic material is contained.

Nutation — Periodic variation in the Earth's precession, caused by irregularities in the Moon's orbit.

Oberon — The outermost satellite of Uranus.

Objective — The primary lens or mirror of an astronomical telescope.

Oblateness — The degree of polar flattening of a planet or satellite. It is obtained by dividing the difference between a planet's polar and equatorial radii by the equatorial radius.

Obliquity — The angle between a planet's rotational axis and the direction perpendicular to the plane of its orbit.

Occultation — An event that occurs when an object with a larger apparent diameter blocks the view of an object with a smaller apparent diameter. This occurs when the Moon blocks the view of a star or planet and when a planet blocks the view of a star.

Ocean basin — A low portion of the Earth's surface having a surface elevation averaging 2 to 3 miles below sea level and filled with ocean water.

Ocean current — Pattern of large-scale circulation of ocean water caused by the Earth's global wind circulation.

Oceanic crust — The portion of the Earth's crust underlying the ocean basins. It is composed of basalt and is thinner and denser than the continental crust.

Ockham's (or Occam's) Razor — A statement recommending that if two or more competing scientific theories attempt to explain a phenomenon, the one that is simplest and makes the fewest assumptions should be preferred.

Odysseus — Huge crater found on Saturn's satellite Tethys.

Ohmic heating — Mechanism proposed for explaining the heating of small solar system bodies. Internal heat was generated as conducting bodies passed through the solar magnetic field.

Olympus Mons — Large, extinct shield volcano on Mars that is the largest shield volcano in the solar system.

Oort, Jan — Astronomer who deduced the existence of a vast swarm of comets surrounding the solar system.

Oort cloud — Huge cloud of comets that surrounds the solar system, located approximately 50,000 astronomical units from the Sun.

Open star cluster — Cluster of stars that typically contains hundreds or thousands of members and is found in the plane of the Milky Way.

Opposition — An event that occurs when a superior planet is located 180° from the Sun in the sky.

Orbital elements — A set of seven quantities that describe a planetary orbit and allow the planet's position at any time to be calculated.

Orbital resonances — Relationships between two orbits that occur when the period of one orbit is a simple numerical fraction of the other. Resonances with massive objects often cause smaller objects to change orbits.

Orbital velocity — The velocity of an object in orbit around another body.

Ordinary chondrites — The group of stony meteorites that are the most common meteorites to fall.

Organism — Any object that is alive.

Orographic clouds — Clouds that form from condensation caused by the upward flow of the atmosphere over high landforms.

Outer core — The region of the Earth's interior that is hot, molten, and composed of nickel – iron metal.

Outflow channels — Large Martian stream-related features that appear to have involved the release of large quantities of water at some time in the past.

Outgassing — The release of gas from a planetary interior due to internal heating.

Ozone layer—Region in the Earth's atmosphere that contains ozone (O_3), which serves to block much of the Sun's harmful ultraviolet radiation.

Paleomagnetism—The study of the Earth's past magnetic field using remanent magnetism in rocks. It is one of the best lines of evidence supporting the theory of plate tectonics.

Paleozoic era—The portion of terrestrial geological time between about 570 and 245 million years ago.

Palimpsest—Crater that can be identified from above but lacks topographical relief, which has been removed by isostatic rebound.

Pangaea—The name given to the "supercontinent" that existed when all of the Earth's continents were linked together as a result of plate tectonic activity.

Panspermia—The theory proposing that terrestrial life originated from dormant spores that drift throughout interstellar space.

Parabola—Conic section with an eccentricity of 1. Objects in parabolic orbits have velocities equal to the escape velocity and are therefore not gravitationally bound to the central object around which they orbit.

Parallax—The shift in apparent position of an object caused by differences in viewing location.

Parent body—The asteroid from which a specific meteorite has been derived.

Parent isotope—Radioactive isotope from which a daughter isotope is produced by decay.

Pasteur, Louis—French scientist who proved that life is not produced by spontaneous creation.

Patera—Form of volcano apparently unique to Mars. Paterae are similar to shield volcanoes, except that they are much larger and have very low relief.

Pavonis Mons—A shield volcano in Mars' Tharsis volcanic province.

Pendulum—Physical device consisting of a weight suspended from a rope or cable. Once it is set swinging back and forth, it continues to oscillate in the same plane.

Penumbra—(1) The lighter region of a sunspot. (2) A region of a two-part shadow in which the light is not completely blocked.

Perihelion—The point in a planet's orbit that is closest to the Sun.

Perturbations—Changes in a planet's orbit caused by the gravitational influence of another planet.

Phases—Changes in an object's apparent shape caused by variation in solar illumination due to changing positions.

Phobos—The inner and larger of Mars' two satellites.

Phoebe—The outermost satellite of Saturn.

Photodissociation—The separation of an atmospheric gaseous molecule into its component atoms as a result of incident solar energy.

Photon—The basic component of light, consisting of a bundle or packet of waves.

Photosphere—The region of the Sun that emits the light visible when we look at it.

Photosynthesis—The method whereby green plants produce their own food. It requires sunlight to operate.

Piazzi, Giuseppi—The astronomer who discovered the first asteroid, Ceres, in 1801.

Pioneer—American program of interplanetary exploration missions. Pioneer spacecraft sent to planetary targets were Pioneer 10, the first spacecraft to explore Jupiter; Pioneer 11, which also explored Jupiter and became the first spacecraft to explore Saturn; and Pioneer Venus, which included two spacecraft, an orbiter and an atmospheric entry probe.

Plains terrain—Flat, lightly cratered terrain on Saturn's satellite Tethys.

Plains units—The most widespread topographic type on Jupiter's satellite Io.

Planet—One of the nine large objects that orbit the Sun.

Planetary geology—The science of studying the planets; a combination of astronomy and geology.

Planetary satellite—Solar system object that orbits a planet.

Planet-crossing asteroids—Asteroids whose orbits intersect the orbit of a planet.

Planetesimal—Intermediate-sized object formed during the planetary accretion process that later joined with other objects its size to form the planets.

Planetoids—Another term for asteroids.

Planetology—The science of studying the planets; a combination of astronomy and geology.

Plasma—Hot gas in which all atoms are ionized.

Plate tectonics—Theory suggesting that the Earth's large-scale surface features are produced by the gradual horizontal movement of large sections of its crust.

Polar caps—Thin frost layers of water ice and carbon dioxide ice at the poles of Mars.

Polar holes—Regions near Venus' poles where the cloud tops occur at lower elevations than normal.

Polar hood—Haze that forms over the Martian poles as that planet's polar caps are forming.

Polar units—Layered, ice-related deposits found near Mars' poles.

Polaris—The star that is currently the Earth's pole star.

Polarization—The alignment of light waves so that the vibration direction in all of them is parallel.

Pole star—The star toward which the axis of the Earth or any other planet is pointed. Because of this orientation, the star remains fixed in the planet's sky.

Polymer—Chain of linked molecules.

Posteclipse brightening—Phenomenon observed on Jupiter's satellite Io. When it emerges from Jupiter's shadow, Io is initially much brighter than normal, apparently as a result of changes in the structure of sulfur as it cools, which in turn change its albedo.

Potential energy—The energy that an object possesses as a result of its position.

Precambrian era—The portion of terrestrial geological time between the formation of the Earth and about 570 million years ago.

Precession—The gradual "wobble" of the axis direction of the Earth or other planet, brought about by tidal forces produced by another object.

Precipitation—The formation of liquid droplets or solid crystals in a gas or of solid crystals in a liquid.

Pre-Nectarian era—The era of lunar history lasting from its formation until 3.85 billion years ago.

Pressure—The measurement of the force applied to a unit surface area.

Primary—The main light-gathering mirror of a reflector telescope.

Primates—The group of mammals of which humans are a member.

Principle of uniformity—One of the key concepts of geology, stating that geological change is brought about by gradual processes acting over long periods of time and that past geological events presumably occurred in the same manner as those today.

Prograde—Orbital or rotational motion that occurs in the counterclockwise direction as seen from north of the ecliptic plane.

Prominence—A loop of gas that is suspended above the surface of the Sun, presumably by magnetic forces.

Proper motion—The gradual movement of a star across the sky, caused both by its and the Sun's orbital motion around the center of the galaxy.

Proton—A positively charged subatomic particle contained in atomic nuclei.

Proton–proton chain—The specific form of hydrogen fusion that occurs inside the Sun and other stars similar to it.

Protostar—A star that is emitting some energy due to the heat generated by gravitational contraction, but is not yet undergoing nuclear fusion.

Protosun—The Sun while in the protostar stage.

Ptolemy—Greek astronomer who wrote a summary of all important Greek astronomical research but sided with those favoring the geocentric orbital concept.

Puck—The largest of the ten new satellites of Uranus discovered by Voyager 2 in 1985 and 1986.

Pyroclastic material—Material composed of ash and rock fragments that is spewn forth from certain types of volcanic eruptions.

Pyrolitic release experiment—One of the experiments aboard the Viking Mars landers that was designed to search for Martian life by detecting evidence of metabolism.

Quadrature—The orientation of a superior planet when it is located 90° from the Sun in the sky.

Radar—A form of radio observation in which a signal is emitted and the reflected signal is received and studied.

Radial spokes—Radial markings in Saturn's rings that appear to be fine dust suspended above the ring plane by electromagnetic forces.

Radial velocity—The speed at which an object is moving either toward or away from an observer.

Radiant—The point in the sky from which meteors appear to radiate outward during a meteor shower.

Radiation—A method of heat transport in which the energy is carried by photons.

Radiation pressure—The pressure that is generated by photons during the process of radiation.

Radiative zone—The portion of the Sun between about 20% and 80% of the way from the center to the surface in which solar energy is carried outward by radiation.

Radio—Form of electromagnetic radiation with the longest wavelengths.

Radioactive—An element or isotope whose nuclei spontaneously change into another type of nucleus is said to be radioactive.

Radioactive decay—The process whereby a radioactive nucleus spontaneously changes into a completely different nucleus, releasing energy in the process.

Radiogenic—A material that has been produced by radioactive decay.

Radioisotope—An isotope that is radioactive.

Radiometric age dating—The technique that determines the age of a rock by measuring the amounts of radioisotopes and decay products in it.

Ranger—American space program of the 1960s that crashed several spacecraft onto the Moon after transmitting photographs of its surface.

Rays—Streaks of light-colored ejecta radiating outward from a recently formed crater.

Recovery—The first observation of a periodic comet during a particular approach to the Sun.

Red shift—A shift in the wavelength of a photon to a longer value as a result of relative motion of the source and observer away from one another.

Reflector—A telescope that gathers light by means of a mirror.

Refractor—A telescope that gathers light by means of a lens.

Regional metamorphism—Form of metamorphism in which rocks over a large area are affected, generally as a result of an episode of mountain building.

Regolith—General term for soil covering the rocks on a planet's surface.

Regression of nodes—The precession of the line of nodes of the Moon's orbit. A complete circuit around the sky takes 18.6 years.

Relative dating—Arranging geological events into order without knowing the actual times at which they occurred. This is the only type of dating that can be conducted for planets whose rocks have not yet been age dated.

Remanent magnetism—Magnetism that a rock acquires at the time of its formation.

Remote sensing—Viewing or studying a celestial object from afar, as opposed to landing upon it for firsthand observation.

Resolution—The ability of a telescope to discern fine detail on a celestial object. It is normally expressed using angular measure.

Retrograde—Orbital or rotational motion that is clockwise when viewed from north of the ecliptic.

Retrograde loop—Pattern of motion made by a planet in the sky as the Earth overtakes it in its orbit (or is overtaken by it).

Rhea—One of the large satellites of Saturn.

Rhea Mons—One of the two shield volcanoes that constitute Venus' Beta Regio highland continent.

Ridged plains—One of the four major landform types on Saturn's satellite Enceladus.

Ring system—Belt of fine particles orbiting a planet in its equatorial plane. All four of the solar system's gas giant planets have ring systems.

Ringlets—The thousand of fine bands that compose Saturn's ring system.

Roche, Edouard—Researcher who investigated the dynamics of tidal forces on orbiting objects.

Roche limit—The closest point that an orbiting body can approach the object around which it orbits without being broken apart by tidal forces.

Rock—Solid substance consisting of a mixture of minerals. Rocks are the primary components of the crusts of terrestrial planets.

Rock-forming mineral—Mineral that is an important constituent of rocks.

Rocket—Propulsion system that is based upon Newton's third law of motion.

Rockfall—Form of mass wasting in which individual rocks fall down a slope.

Roemer, Olaus—Scientist who first proved in 1675 that the speed of light is finite, not infinite.

Rotation—The spinning of a celestial object around an axis.

Runoff channels—Stream-related features on Mars whose increase in width downstream causes them to resemble those on Earth.

Satellite—An object that orbits another celestial object. The term is most commonly applied to objects orbiting planets.

Saturation—Situation that results when either a gas or a liquid contains the maximum amount of evaporated or dissolved material that it can contain. As a result, condensation or precipitation occurs.

Scarp—A steep wall caused by vertical fault motion.

Schiaparelli, Giovanni—Observer who proposed 1:1 spin-orbit coupling for Mercury and Venus, noted linear markings or canals on Mars, and established the relationship between comets and meteor showers.

Schroeter, Johann H.—Astronomer who first noted Venus' crescent phase cusp extension, one of the proofs that the planet has an atmosphere.

Schwabe, Heinrich—Nineteenth-century astronomer who discovered the Sun's 11-year activity cycle.

Science—The observation and explanation of natural phenomena.

Scientific method—The step-by-step process of investigation used in scientific inquiry.

Scientific notation—Shorthand notation used to express both large and small numbers.

Seasons—Periodic variation in weather caused by the obliquity of a planet's axis.

Secchi, Angelo—The first observer to see linear markings on the surface of Mars.

Second—(1) The basic unit of time measurement. (2) Unit of angular measurement equal to 1/60 minute.

Secondary crater—Smaller impact crater associated with a larger one and formed by a large chunk of ejecta from the main crater.

Sediment — Body of rock fragments that have been moved from their place of origin and deposited.

Sedimentary rock — Rock composed of material derived from pre-existing rocks.

Seeing — Term used in observational astronomy to describe the steadiness of the image in the telescope.

Seismic wave — Vibrational wave caused by an earthquake.

Seismograph — Instrument that detects and measures seismic waves.

Semi-major axis — The average distance between an orbiting object and the object around which it orbits. It is equal to half the major axis of the elliptical orbit.

Shapley, Harlow — Astronomer who determined the approximate size of the Milky Way galaxy and the location of the solar system within it.

Shatter cone — Cone-shaped structure found in rocks near the site of a large meteorite impact. The presence of shatter cones indicates that an impact occurred, even if the crater formed was subsequently eroded.

Shield volcano — Large volcano with gently sloping sides. This type of volcano is produced by basic magma.

Short-period comet — A comet whose orbital period is less than two centuries.

Sidereal day — The length of time required for the Earth (or other planet) to rotate, measured with respect to the fixed stars. It is also called the sidereal rotational period.

Sidereal period — The length of time required for an object to complete an orbit, measured with respect to the fixed stars. The sidereal period of the Earth is called the sidereal year.

Siderophile elements — Elements that tend to be found in association with iron.

Silicate — A mineral that contains both silicon and oxygen.

Silicate tetrahedron — Structural unit, consisting of one silicon atom and four oxygen atoms, that provides the framework for all silicate minerals.

Sill — Igneous intrusion in which the magma deposit is parallel to the boundaries of pre-existing rock layers.

Sinkhole — Depression found in rock, usually caused either by collapse of a cave roof or solution of rock from the surface down.

Sinuous rille — Curved feature on the Moon that appears to be the result of lava flowing over the lunar surface.

Slingshot effect — Term applied to the use of a planet's gravity to boost a passing spacecraft to a higher velocity.

Slump — Form of mass wasting consisting of material sliding downhill over a curved surface of rupture.

Smooth plains — Term applied to a terrain type found both on Mercury and Saturn's satellites Enceladus and Dione.

Solar constant — The amount of solar energy received at a point just above the Earth's atmosphere.

Solar day — The length of time required for the Earth (or another object) to complete a rotation, measured with respect to the Sun. It is the length of time between successive appearances of the Sun at the highest point in the sky.

Solar mass — Unit of mass measurement with the Sun assigned a value of 1.

Solar system — Region of the Milky Way galaxy consisting of the Sun and all of the objects in orbit around it.

Solar wind — The continuous outward stream of gas from the Sun's outer atmosphere.

Solid-fuel rocket — A rocket whose engine burns a paste-like mixture of fuel and oxidizer.

Sounding rocket — Rocket that is launched vertically and then returns its payload to the Earth.

Special creation — The idea that terrestrial life was created by some form of supernatural event.

Specific gravity — A term that is essentially synonymous with density.

Spectroscopy — The analysis of the spectrum of light received from an object. One of its primary purposes is to determine the composition of the object whose light is studied.

Spectrum — The breakdown of light into its component parts or wavelength regions, as occurs when sunlight is dispersed into the rainbow.

Speed — The rate of motion of an object.

Speed of light — The speed at which light travels in a vacuum. It is believed to be the fastest attainable speed.

Spicule — Region extending upward from the Sun's chromosphere. Spicules are believed to be associated with the phenomenon of supergranulation.

Spin-orbit coupling — The phenomenon that occurs when the orbital period and rotational period of an object are either identical or related to each other by a simple ratio such as 1:2 or 2:3.

Spiral galaxy — A galaxy, such as the Milky Way, in which most of the stars are distributed in a pinwheel fashion in a generally flat disk.

Spontaneous creation — The once-held idea that living material can be generated from nonliving material.

Sporadic meteor — Meteor not associated with any meteor shower.

Star—Large, hot, spherical mass of hydrogen and helium gas that generates energy by nuclear fusion.

Stickney—Relatively large crater found on Mars' satellite Phobos.

Stony meteorite—Meteorite composed largely of silicate minerals.

Stony–iron meteorite—Meteorite composed of both silicate minerals and nickel–iron metal.

Straight Wall—Impressive linear feature on the Moon that is probably a large fault scarp.

Strata—Layers of rocks.

Stratigraphic column—The sequence of rock layers found in a given area.

Stratosphere—The second of Earth's four atmospheric layers. It is in inversion and contains the ozone layer.

Subduction—The disappearance of oceanic crust as it sinks back down into the Earth's mantle during plate tectonic activity.

Substar—An object, such as Jupiter, that emits more energy than it receives but initially lacked the mass required to become a star, in which fusion occurs.

Sulfuric acid—Extremely corrosive acid that is the primary component of Venus' clouds.

Sun-grazing comets—Long-period comets that approach very close to the Sun. Some have been seen to fall into it.

Sunspot—Dark region of the solar photosphere that is 1500 K cooler than its surroundings. Sunspots are believed to be produced by variations in the Sun's magnetic field.

Sunspot cycle—The 11-year cycle during which both the number and location of sunspots vary.

Super-rotating atmosphere—Atmospheric layer that rotates faster than the planet underneath it.

Supergranulation—Large-scale effect of solar convection that manifests itself in the solar chromosphere.

Superior conjunction—The event that occurs when Mercury or Venus is aligned on the opposite side of the Sun from the Earth.

Superior planet—Planet whose orbit is further from the Sun than the Earth's.

Supernova—A massive star that explodes at the end of its life, becoming extremely bright and spewing newly created heavy elements into the interstellar medium.

Superoxides—Unusual chemical compounds that are considered likely components of Martian surface materials because they would have reacted in a life-mimicking way, as did the reactions observed during the Martian life-detection experiments.

Surface field strength—The strength of a magnetic field at the surface of a planet or other object.

Surveyor—American space program during the 1960s that soft-landed spacecraft on the Moon.

Synodic period—The time required between successive oppositions or conjunctions of a planet.

T Tauri star—Stage in the early life of a star during which an especially strong stellar wind clears away any material remaining from the nebula from which it formed.

Tail—The part of a comet consisting of a stream of gas directed away from the Sun because of radiation pressure and the solar wind.

Technology—The use of tools and instruments, one of the characteristics of intelligence and civilization.

Tektite—A fragment of glass formed when a meteorite impact splashes material into the air. The molten material cools and solidifies before striking the surface.

Telescope—Instrument used for astronomical observation that gathers light and increases resolution.

Temperature—The measurement of the rate of atomic or molecular motion in a substance.

Terminator—The line on a planetary surface marking the sites where sunrise or sunset is occurring.

Terrestrial planets—Planets that are relatively small, composed largely of rock and metal, possess a solid surface and possibly an atmosphere, and have few satellites and no ring systems.

Tethys—One of the major satellites of Saturn.

Tharsis—The youngest of Mars' three volcanic provinces.

Theia Mons—One of the two shield volcanoes that constitute Venus' Beta Regio continental highland.

Theory—Proposed solution to a scientific problem, generally of broader application than a hypothesis.

Thermosphere—The outermost of the Earth's four major atmospheric layers. It is in inversion.

Tholus—Type of small, steep volcano found on Mars.

Thrust—The amount of force generated by a rocket engine.

Tidal forces—Differential forces exerted by one object upon another due to the variation in gravitational attraction with distance.

Tides—The periodic rise and fall of Earth's ocean water level due to tidal forces from the Moon and Sun.

Time of perihelion passage — Orbital element that measures the time at which an object reaches its closest approach to the Sun.

Titan — The largest satellite of Saturn and one of only two satellites in the solar system known to have atmospheres.

Titania — One of the major satellites of Uranus.

Titius, Johann — Astronomer who first developed the numerical relationship predicting the spacing of the planets. It was popularized by Bode and is now called Bode's law.

Tombaugh, Clyde — The astronomer who discovered the planet Pluto in 1930.

Topographic province — An area of a planetary surface marked by characteristic crustal type and landforms.

Torus — A doughnut-shaped figure. The term is applied to a cloud of sodium gas around the planet Jupiter that has this shape.

Total magnetic moment — The total strength of the magnetic field generated within a solar system object.

Trailing hemisphere — The hemisphere of a 1:1 spin-orbit coupled object that is on the trailing side as the object orbits.

Transfer orbit — An orbit used to carry a spacecraft from one planet to another.

Transient phenomena — Mysterious, short-lived phenomena on the Moon believed to result from the emanation of gas because of lunar volcanic activity.

Transit — Event that occurs when an inferior planet perfectly aligns with the Sun at inferior conjunction and can be seen moving across the Sun's disk.

Trans-Neptunian planet — The term applied to the planet beyond Neptune while it was being sought.

Transparency — Term used in observational astronomy to describe how well the atmosphere transmits light.

Triangulation — The determination of distance by observing a distant object from two separate points.

Trigonometric parallax — Shift in the location of stars caused by the Earth's orbital motion around the Sun. It is a small shift and cannot be seen without a telescope.

Trillion — One million million, written 1,000,000,000,000 or 10^{12}.

Triton — The largest satellite of Neptune and one of only two satellites in the solar system known to have atmospheres.

Trojan asteroids — Asteroids located at the Lagrangian points of Jupiter's orbit.

Tropical year — The time between the beginnings of a terrestrial season in successive years.

Troposphere — The lowermost layer of the Earth's atmosphere. It is the layer in which all weather phenomena occur.

Tuff — Rock composed of pyroclastic material that has been welded together.

Tunguska event — Mysterious explosive event that occurred in Siberia in 1908. It is now considered likely that it resulted from the entry of a comet's nucleus into the Earth's atmosphere.

Tycho — Renaissance astronomer who developed a solar system model that was a curious mix of the geocentric and heliocentric concepts. The last great naked-eye observer, Tycho startled the world with the discovery of a new star, now known to have been a supernova. He also proved that comets are more distant than the Moon. A relatively young lunar crater is named for him.

Type I tail — Comet tail that contains ionized gases and is bluish in color.

Type II tail — Comet tail that contains dust particles and is yellowish in color.

Ultraviolet light — Form of electromagnetic radiation that is just shorter in wavelength and higher in energy than visible light.

Umbra — (1) The dark, central portion of a sunspot. (2) The dark portion of a two-part shadow in which all incoming light is blocked.

Umbriel — One of the five major satellites of Uranus.

Uncompressed density — The average density which a planet or other object would have if pressure did not compress its interior.

Upland rolling plains — The most abundant landform on the planet Venus.

Urey, Harold — Chemist whose numerous contributions to science included participation in the Miller–Urey experiment, which attempted to reproduce the conditions under which terrestrial life may have developed.

Utopia Planitia — Martian region that was the landing site of Viking 2 in 1976.

Valhalla — Multiringed basin that is the largest surface feature on Jupiter's satellite Callisto.

Valles Marineris — Huge system of rift valleys on Mars.

Van Allen, James A. — Researcher who discovered the radiation belts around the Earth that were named for him.

Van Allen belts — The set of two belts of trapped subatomic particles that circle the Earth in its equatorial plane.

Velocity — A quantity that measures the rate of movement and the direction of movement of an object.

Vent material — Surface unit of Jupiter's satellite Io consisting of volcanic products.

Ventifact — A rock that has been shaped by the sandblasting effect of wind-blown materials.

Vernal equinox — The time of the year when the Sun crosses the ecliptic plane from south to north, marking the beginning of spring. Also, the Sun's direction in space at that instant.

Vertebrate — Animal possessing an internal skeleton, including a backbone.

Vertical relief — The difference between the lowest and highest points in an area or an entire planet.

Vertical structure — The change in temperature as a function of altitude in an atmosphere and the associated layering that such changes mark.

Vesta — The brightest asteroid and the one with the highest albedo.

Viking — American space program that successfully soft-landed two spacecraft on Mars in 1976.

Virial theorem — Theorem that describes the behavior of a nebula or other gas cloud. It states that for a bound (stable) system, the long-term average of the kinetic energy is one-half of the potential energy.

Visible light — The form of electromagnetic radiation that can be seen by human eyes.

Volatiles — Term used in planetary science to describe materials with relatively low melting and boiling points.

Volcano — Landform that is produced by extrusive igneous activity.

Volume — Measurement of the space contained within an object.

Voyager 1 — Spacecraft that visited Jupiter in 1979 and Saturn in 1980.

Voyager 2 — Spacecraft that visited Jupiter in 1979, Saturn in 1981, Uranus in 1986, and Neptune in 1989.

Waning — Term applied to lunar phases during which the illuminated portion of the Moon is decreasing in size.

Wave — Cyclical oscillation, as occurs in light, earthquakes, bodies of water, and in many other natural phenomena.

Wave clouds — Martian clouds that form behind obstacles to wind flow.

Wavelength — The distance between successive crests or troughs of a wave.

Waxing — Term applied to lunar phases during which the illuminated portion of the Moon is increasing in size.

Weather — Temperature, pressure, and other atmospheric conditions present in an area at a given time.

Weathering — The breaking down of rocks into soil particles.

Weight — The measurement of the force an object feels as a result of gravity.

Whipple, Fred — Modern researcher who first proposed that comets are composed of ices containing some rocky material.

Widmanstatten pattern — Pattern found in many iron meteorites. It consists of small interlocking crystals of kamacite and taenite.

Wien, Wilhelm — Nineteenth-century physicist who determined the relationship between the temperature of a light-emitting object and the wavelength at which it emits the maximum amount of light.

Wien's law — The relationship between the temperature of a light-emitting object and the wavelength at which it emits the maximum amount of light.

Wind — The horizontal movement of atmospheric gases.

Wind streaks — Light or dark markings behind Martian surface features that are believed to be caused by the deposition of wind-blown material.

X-rays — Forms of electromagnetic radiation that is intermediate between ultraviolet and gamma rays.

Yardangs — Streamlined hills of rock sculpted by wind erosion.

Zeeman effect — The splitting of spectroscopic lines as a result of magnetic fields.

Zodiac — The band of 12 constellations through which the ecliptic plane passes.

Zodiacal light — Faint light visible near the horizon after sunset or before sunrise. It is produced by sunlight scattered by comet debris lying in the ecliptic plane.

Zonal jets — Winds in a gas giant atmosphere, roughly parallel to the equator, that are responsible for producing atmospheric markings.

Zones — Light-colored atmospheric bands that encircle gas giant planets.

Index

Astronomical Constants

Pi	$= 3.14159$
Gravitational Constant	$G = 6.670 \times 10^{-11}$ N m²/kg²
Speed of Light	$c = 2.998 \times 10^5$ km/s
Planck's Constant	$h = 6.626 \times 10^{-27}$ erg s
Wien's Law Constant	$= 2.898 \times 10^6$ nm K
Boltzmann's Constant	$k = 1.381 \times 10^{-16}$ erg/K
Astronomical Unit	$AU = 1.496 \times 10^8$ km
Light Year	$LY = 9.46 \times 10^{12}$ km
Mass of Earth	$M_E = 5.977 \times 10^{24}$ kg
Radius of Earth (Equatorial)	$R_E = 6378$ km
Mass of Sun	$M_S = 1.989 \times 10^{30}$ kg
Radius of Sun	$R_S = 695{,}990$ km
Luminosity of Sun	$L_S = 3.827 \times 10^{33}$ ergs/s